APPLIED BIOTECHNOLOGY FOR SITE REMEDIATION

Edited by

Robert E. Hinchee
Battelle, Columbus, Ohio

Daniel B. Anderson and F. Blaine Metting, Jr.
Battelle, Richland, Washington

Gregory D. Sayles
USEPA Risk Reduction Engineering Laboratory, Cincinnati, Ohio

Library of Congress Cataloging-in-Publication Data

Catalog record is available from the Library of Congress

This book represents information obtained from authentic and highly regarded sources. Reprinted material is quoted with permission, and sources are indicated. A wide variety of references are listed. Every reasonable effort has been made to give reliable data and information, but the author and the publisher cannot assume responsibility for the validity of all materials or for the consequences of their use.

Direct all inquiries to CRC Press, Inc., 2000 Corporate Blvd., N.W., Boca Raton, Florida 33431.

Lewis Publishers is an imprint of CRC Press

International Standard Book Number 0-87371-982-4
Printed in the United States of America 2 3 4 5 6 7 8 9 0
Printed on acid-free paper

CONTENTS

Articles

Technical Notes

FOREWORD

Bioremediation as a whole remains an emerging and rapidly changing field. Since the first symposium in 1991, significant advances have been made. In 1991, only one paper was devoted to biofilters, but in 1993 biofiltration was of significant interest to generate an entire session. Natural attenuation, also little discussed in 1991, was of great interest in 1993. Increased interest in surfactant-enhanced biodegradation, bioventing, air sparging, and metals bioremediation also was apparent. Interest in chlorinated solvent bioremediation remained steady but strong, and a trend toward field application has developed. The ratio of laboratory to field studies has clearly moved toward the field. This ratio is an indication of a technology that is beginning to mature, but by no means indicates that bioremediation is a mature technology. Clearly, many knowledge gaps and needs exist. Well-developed relationships between laboratory bench-scale testing and field practice frequently are lacking. Although much encouraging laboratory research work has been done with chlorinated and many other recalcitrant organics, field practice of bioremediation is still largely directed at petroleum hydrocarbon-contaminated sites. The number of well-documented field demonstrations of bioremediation is increasing; however, before bioremediation can mature into a readily accepted and widely understood technology area, many more field demonstrations will be necessary.

This book and its companion volumes, *Bioremediation of Chlorinated and Polycyclic Aromatic Hydrocarbon Compounds* and *Hydrocarbon Bioremediation*, represent the bulk of the papers arising from the Second International Symposium on In Situ and On-Site Bioreclamation held in San Diego, California, in April 1993. Two other books, *Air Sparging* and *Emerging Technology for Bioremediation of Metals*, also contain selected papers on these topics.

The symposium was attended by more than 1,100 people. More than 300 presentations were made, and all presenting authors were asked to submit manuscripts. Following a peer review process, 190 papers are being published. The editors believe that these volumes represent the most complete, up-to-date works describing both the state of the art and the practice of bioremediation.

The symposium was sponsored by Battelle Memorial Institute with support from a wide variety of other organizations. The cosponsors and supporters were:

Bruce Bauman, *American Petroleum Institute*
Christian Bocard, *Institut Français du Pétrole*
Rob Booth, *Environment Canada, Wastewater Technology Centre*
D. B. Chan, *U.S. Naval Civil Engineering Laboratory*
Soon H. Cho, *Ajou University, Korea*
Kate Devine, *Biotreatment News*
Volker Franzius, *Umweltbundesamt, Germany*
Giancarlo Gabetto, *Castalia, Italy*
O. Kanzaki, *Mitsubishi Corporation, Japan*

Dottie LaFerney, *Stevens Publishing Corporation*
Massimo Martinelli, *ENEA, Italy*
Mr. Minoru Nishimura, *The Japan Research Institute, Ltd.*
Chongrak Polprasert, *Asian Institute of Technology, Thailand*
Lewis Semprini, *Oregon State University*
John Skinner, *U.S. Environmental Protection Agency*
Esther Soczo, *National Institute of Public Health and Environmental Protection, The Netherlands*

In addition, numerous individuals assisted as session chairs, presented invited papers, and helped to ensure diverse representation and quality. Those individuals were:

Bruce Alleman, *Battelle Columbus*
Christian Bocard, *Institut Français du Pétrole*
Rob Booth, *Environment Canada, Wastewater Technology Center*
Fred Brockman, *Battelle Pacific Northwest Laboratories*
Tom Brouns, *Battelle Pacific Northwest Laboratories*
Soon Cho, *Ajou University, Korea*
M. Yavuz Corapcioglu, *Texas A&M University*
Jim Fredrickson, *Battelle Pacific Northwest Laboratories*
Giancarlo Gabetto, *Area Commerciale Castalia, Italy*
Terry Hazen, *Westinghouse Savannah River Laboratory*
Ron Hoeppel, *U.S. Naval Civil Engineering Laboratory*
Yacov Kanfi, *Israel Ministry of Agriculture*
Richard Lamar, *U.S. Department of Agriculture*
Andrea Leeson, *Battelle Columbus*
Carol Litchfield, *Keystone Environmental Resources, Inc.*
Perry McCarty, *Stanford University*
Jeff Means, *Battelle Columbus*
Blaine Metting, *Battelle Pacific Northwest Laboratories*
Ross Miller, *U.S. Air Force*
Minoru Nishimura, *Japan Research Institute*
Robert F. Olfenbuttel, *Battelle Columbus*
Say Kee Ong, *Polytechnic University, New York*
Augusto Porta, *Battelle Europe*
Roger Prince, *Exxon Research and Engineering Co.*
Parmely "Hap" Pritchard, *U.S. Environmental Protection Agency*
Jim Reisinger, *Integrated Science & Technology*
Greg Sayles, *U.S. Environmental Protection Agency*
Lewis Semprini, *Oregon State University*
Ron Sims, *Utah State University*
Marina Skumanich, *Battelle Seattle*
Jim Spain, *U.S. Air Force*
Herb Ward, *Rice University*
Peter Werner, *University of Karlsruhe, Germany*
John Wilson, *U.S. Environmental Protection Agency*
Jim Wolfram, *Montana State University*

The papers in this book have been through a peer review process, and the assistance of the peer reviewers is recognized. This typically thankless job is essential to technical publication. The following people peer-reviewed papers for the publication resulting from the symposium:

Jens Aamand, *Water Quality Institute*
Nelly M. Abboud, *University of Connecticut*
Daniel A. Abramowicz, *GE Corporate R&D Center*
Dan W. Acton, *Beak Consultants Ltd., Canada*
William Adams, *Monsanto Co., U4E*
Peter Adriaens, *University of Michigan*
C. Marjorie Aelion, *University of South Carolina*
Robert C. Ahlert, *Rutgers University*
David Ahlfeld, *University of Connecticut*
Hans-Jorgen Albrechtsen, *Technical University of Denmark*
Bruce Alleman, *Battelle Columbus*
Richelle M. Allen-King, *University of Waterloo*
Sabine E. Apitz, *NCCOSC RDTE DIV 521*
John M. Armstrong, *The Traverse Group*
Boris N. Aronstein, *Institute of Gas Technology*
Mick Arthur, *Battelle Columbus*
Erik Arvin, *Technical University of Denmark*
Steven D. Aust, *Utah State University*
Serge Baghdikian
M. Talaat Balba, *TreaTek - CRA Co.*
D. Ballerini, *Institut Français du Pétrole*
N. Bannister, *University of Kent, England*
Jeffrey R. Barbaro, *University of Waterloo*
James F. Barker, *University of Waterloo*
Morton A. Barlaz, *North Carolina State University*
Denise M. Barnes, *Ecosystems Engineering*
Edward R. Bates, *U.S. Environmental Protection Agency*
Tad Beard, *Battelle Columbus*
Cathe Bech, *SINTEF Applied Chemistry, Norway*
Pamela E. Bell, *Hydrosystems, Inc.*
Judith Bender, *Clark Atlanta University*
James D. Berg, *Aquateam Norwegian Water Technology Centre A/S*
Christopher J. Berry, *Westinghouse Savannah River Company*
Sanjoy K. Bhattacharya, *Tulane University*
Jeffery F. Billings, *Billings & Associates*
James N. P. Black, *Stanford University*
Joan Blake, *U.S. Environmental Protection Agency*
Robert Blanchette, *University of Minnesota*
Bert E. Bledsoe, *U.S. Environmental Protection Agency*
Christian Bocard, *Institut Français du Pétrole*
Gary Boettcher, *Geraghty & Miller, Inc.*
David R. Boone, *Oregon Graduate Center*
James Borthen, *ECOVA Corporation*
Edward J. Bouwer, *Johns Hopkins University*
John P. Bowman, *University of Tennessee*
Joan F. Braddock, *University of Alaska*

A. Braun-Lullemann, *Forstbotanisches Institut der Universität Göttingen, Germany*
Susan E. Brauning, *Battelle Columbus*
Alec W. Breen, *U.S. Environmental Protection Agency*
James A. Brierley, *Newmont Metallurgical Services*
Fred Brockman, *Battelle Pacific Northwest Laboratories*
Kim Broholm, *Technical University of Denmark*
Thomas M. Brouns, *Battelle Pacific Northwest Laboratories*
Edward Brown, *University of Northern Iowa*
Guner Brox, *EIMCO Process Equipment*
Gaylen R. Brubaker, *Remediation Technologies, Inc.*
Wil P. de Bruin, *Wageningen Agricultural University, The Netherlands*
Robert S. Burlage, *Oak Ridge National Laboratory*
David Burris, *Tyndall Air Force Base*
Timothy E. Buscheck, *Chevron Research and Technology Company*
Larry W. Canter, *University of Oklahoma*
Jason A. Caplan, *ESE Biosciences*
Peter J. Chapman, *U.S. Environmental Protection Agency*
Abe Chen, *Battelle Columbus*
G. O. Chieruzzi, *Keystone Environmental Resources*
Soon Hung Cho, *Ajou University, Korea*
Patricia J. S. Colberg, *University of Wyoming*
Edward Coleman, *MK Environmental*
Ronald L. Crawford, *University of Idaho*
Steven L. Crawford, *DPRA Inc.*
Craig Criddle, *Michigan State University*
Jon Croonenberghs, *Coors Brewing Company*
Scott Cunningham, *DuPont Central Research and Development*
Mohamed F. Dahab, *University of Nebraska*
Lois Davis, *Sybron Chemicals, Inc.*
Wendy J. Davis-Hoover, *U.S. Environmental Protection Agency*
Peter Day, *Rutgers University*
Sue Markland Day, *University of Tennessee*
Mary F. DeFlaun, *Envirogen*
Richard A. DeMaio, *Mycotech Corporation*
Dave DePaoli, *Oak Ridge National Laboratory*
Allen Deur, *Polytechnic University*
Kate Devine, *Biotreatment News*
L. Diels, *Vlaamse, Instelling voor Technologisch Onderzoek, Belgium*
Greg Douglas, *Battelle Ocean Sciences*
Douglas C. Downey, *Engineering-Science, Inc.*
David Drahos, *SPB Technologies*
Murali M. Dronamraju, *Tulane University*
Jean Ducreux, *Institut Français du Pétrole*
James Duffy, *Occidental Chemical Corporation*
Ryan Dupont, *Utah State University*
Geraint Edmunds
Elizabeth A. Edwards, *Beak Consultants Ltd., Canada*
Richard Egg, *Texas A&M University*
David L. Elmendorf, *University of Central Oklahoma*
Mark Emptage, *DuPont Company*
Burt D. Ensley, *Envirogen*
Michael V. Enzien, *Westinghouse Savannah River Site*

David C. Erickson, *Harding Lawson Associates*
Richard A. Esposito, *Southern Company Services*
J. van Eyk, *Delft Geotechnics, The Netherlands*
Brandon J. Fagan, *Continental Recovery Systems, Inc.*
Liv-Guri Faksness, *SINTEF Applied Chemistry, Norway*
John Ferguson, *University of Washington*
J. A. Field, *Wageningen Agricultural University, The Netherlands*
Pedro Fierro, *Geraghty & Miller, Inc.*
Stephanie Fiorenza, *Amoco Corporation*
Paul E. Flathman, *OHM Remediation Services Corporation*
John Flyvbjerg, *Water Quality Institute, Denmark*
Cresson D. Fraley, *Stanford University*
W. T. Frankenberger, *University of California, Riverside*
James Fredrickson, *Battelle Pacific Northwest Laboratories*
David L. Freedman, *University of Illinois*
Ian V. Fry, *Lawrence Berkeley Laboratory*
Clyde W. Fulton, *CH2M HILL*
Kathryn Garrison, *Geraghty & Miller, Inc.*
Edwin Gelderich, *U.S. Environmental Protection Agency*
Richard M. Gersberg, *San Diego State University*
John Glaser, *U.S. Environmental Protection Agency*
Fred Goetz, *Mankato State University*
C. D. Goldsmith, *EnvironTech Mid-Atlantic*
James M. Gossett, *Cornell University*
Peter Grathwohl, *University of Teubingen, Germany*
Charles W. Greer, *Biotechnology Research Institute, Canada*
Christian Grøn, *Technical University of Denmark*
D.R.J. Grootjen, *DSM Research BV, The Netherlands*
Matthew J. Grossman, *Exxon Research & Engineering*
Ipin Guo, *Alberta Environmental Centre, Canada*
Haim Gvirtzman, *The Hebrew University of Jerusalem*
Paul Hadley, *California Environmental Protection Agency*
John R. Haines, *U.S. Environmental Protection Agency*
Kenneth Hammel, *U.S. Department of Agriculture*
Mark R. Harkness, *GE Corporate R&D Center*
Joop Harmsen, *The Winand Staring Center for Integrated Land, Soil and Water Research, The Netherlands*
Zachary Haston, *Stanford University*
Gary R. Hater, *Chemical Waste Management, Inc.*
Tony Hawke, *Groundwater Technology Canada Ltd.*
Caryl Heintz, *Texas Tech University*
Barbara B. Hemmingsen, *San Diego State University*
Stephen E. Herbes, *Oak Ridge National Laboratory*
Gorm Heron, *Technical University of Denmark*
Ronald J. Hicks, *Groundwater Technology, Inc.*
Franz K. Hiebert, *Alpha Environmental, Inc.*
E. L. Hockman, *Amoco Corporation*
Robert E. Hoffmann, *SiteRisk Inc.*
Desma Hogg, *Woodward-Clyde Consultants*
Brian S. Hooker, *Tri-State University*

Kevin Hosler, *Wastewater Technology Centre, Canada*
M. Akhter Hossain, *Atlantic Environmental*
Perry Hubbard, *Integrated Science & Technology, Inc.*
Michael H. Huesemann, *Shell Development Company*
Scott G. Huling, *U.S. Environmental Protection Agency*
Jasna Hundal, *CH2M HILL*
Peter J. Hutchinson, *The Hutchinson Group, Ltd.*
Aloys Huttermann, *Forstbotanisches Institut der Universität Göttingen, Germany*
Mary Pat Huxley, *U.S. Naval Civil Engineering Laboratory*
Charles E. Imel, *Ecosystems Engineering*
Danny R. Jackson, *Radian Corporation*
Peter Jaffe, *Princeton University*
Trevor James, *Woodward Clyde Ltd., New Zealand*
D. B. Janssen, *University of Groningen, The Netherlands*
Minoo Javanmardian, *Amoco Oil Company*
Ursula Jenal-Wanner, *Western Regional Hazardous Substance Research Center, Stanford University*
Bjorn K. Jensen, *Water Quality Institute, Denmark*
Douglas E. Jerger, *OHM Remediation Service Corporation*
Randall M. Jeter, *Texas Tech University*
Richard L. Johnson, *Alberta Environmental Centre, Canada*
George Johnson, *Stillwater, Inc.*
C. D. Johnston, *CSIRO, Australia*
Donald L. Johnstone, *Washington State University*
E. Fraser Johnstone, *Exxon Company*
K. C. Jones, *Lancaster University*
Warren L. Jones, *Montana State University*
E. de Jong, *Wageningen Agricultural University, The Netherlands*
Linda de Jong, *University of Washington*
Miryan Kadkhodayan, *University of Cincinnati*
Don Kampbell, *U.S. Environmental Protection Agency*
Yacov Kanfi, *Israel Ministry of Agriculture*
Chih-Ming Kao, *North Carolina State University*
Leslie Karr, *U.S. Naval Civil Engineering Laboratory*
Keith Kaufman, *RESNA Industries, Inc.*
S. Keuning, *Bioclear Environmental Biotechnology, The Netherlands*
T. Kent Kirk, *U.S. Department of Agriculture*
Michael D. Klein, *EG&G Rocky Flats, Inc.*
Calvin A. Kodres, *U.S. Naval Civil Engineering Laboratory*
Simeon J. Komisar, *University of Washington*
Raj Krishnamoorthy, *Keystone Environmental Resources, Inc.*
M. Kuge, *Industrial and Fine Chemicals Division, Environmental Technology*
Debi Kuo, *University of Tennessee*
Bruce E. LaBelle, *California Environmental Protection Agency*
William F. Lane, *Remediation Technologies, Inc.*
Margaret Lang, *Stanford University*
Robert LaPoe, *U.S. Air Force*
Barnard Lawes, *DuPont Company*
Maureen E. Leavitt, *IT Corporation*
Clifford Lee, *DuPont Environmental Remediation Services*
Kun Mo Lee, *Ajou University, Korea*

Michael D. Lee, *DuPont Environmental Remediation Services*
Richard F. Lee, *Skidaway Institute of Oceanography*
Paul LeFevre, *Coors Brewing Company*
Robert Legrand, *Radian Corporation*
Terrance Leighton, *University of California*
Sarah K. Leihr, *North Carolina State University*
M. Tony Lieberman, *ESE Biosciences*
Carol D. Litchfield, *Chester Environmental*
Kenneth H. Lombard, *Bechtel Savannah River, Inc.*
Sharon C. Long, *University of North Carolina*
Charles R. Lovell, *University of South Carolina*
Ja-Kael Luey, *Battelle Pacific Northwest Laboratories*
J. S. Luo, *Center for Environmental Biotechnology*
Stuart Luttrell, *Battelle Pacific Northwest Laboratories*
John Lyngkilde, *Technical University of Denmark*
Ian D. MacFarland, *EA Engineering, Science, and Technology, Inc.*
Joan Macy, *University of California*
Andzej Majcherczyk, *Forstbotanisches Institut der Universität Göttingen, Germany*
David Major, *Beak Consultants Ltd., Canada*
Pryodarshi Majumdar, *Tulane University*
Leo Manzer, *E. I. DuPont, De Nemours & Co., Inc.*
Nigel V. Mark-Brown, *Woodward-Clyde International, New Zealand*
Donn Marrin, *InterPhase Environmental, Inc.*
Dean A. Martens, *University of California, Riverside*
Michael M. Martinson, *Delta Environmental Consultants, Inc.*
Perry L. McCarty, *Stanford University*
Gloria McCleary, *EA Engineering*
Linda McConnell, *Logistics Management Institute*
Mike McFarland, *Utah Water Research Laboratory, Utah State University*
Ilona McGhee, *University of Kent, England*
David H. McNabb, *Alberta Environmental Centre, Canada*
Sally A. Meyer, *Georgia State University*
Kathy Meyer-Schulte, *Computer Science Corporation*
Robert Miller, *Oklahoma State University*
Ali Mohagheghi, *Solar Energy Research Institute*
Peter Molton, *Battelle Pacific Northwest Laboratories*
Ralph E. Moon, *Geraghty & Miller, Inc.*
Jim Morgan, *The MITRE Corporation*
Frederic A. Morris, *Battelle Seattle Research Centers*
Pamela J. Morris, *University of Florida*
Klaus Müller, *Battelle Europe*
Julie Muolyta, *Stanford University*
H. S. Muralidhara, *Cargill, Inc.*
Reynold Murray, *Clark Atlanta University*
Karl W. Nehring, *Battelle Columbus*
Christopher H. Nelson, *Groundwater Technology, Inc.*
Per H. Nielsen, *Technical University of Denmark*
Dev Niyogi, *Battelle Marine Research Laboratory*

Robert Norris, *Eckenfelder, Inc.*
John T. Novak, *Virginia Tech*
Evan Nyer, *Geraghty & Miller*
Joseph E. Odencrantz, *Lavine-Fricke Consulting Engineers*
Laurra P. Olmsted, *Brown & Root Civil, England*
Brian O'Neill, *Dearborn Chemical Company, Ltd., Canada*
Richard Ornstein, *Battelle Pacific Northwest Laboratories*
David Ostendorf, *University of Massachusetts*
Donna Palmer, *Battelle Columbus*
Anthony V. Palumbo, *Oak Ridge National Laboratory*
Sorab Panday, *HydroGeologic Inc.*
Joel W. Parker, *The Traverse Group*
John H. Patterson, *Continental Recovery Systems*
Richard E. Perkins, *DuPont Environmental Biotechnology Program*
James N. Peterson, *Washington State University*
Erik Petrovskis, *University of Michigan*
Brent Peyton, *Battelle Pacific Northwest Laboratories*
Frederic K. Pfaender, *University of North Carolina at Chapel Hill*
S. M. Pfiffner, *University of Tennessee*
George Philippidis, *National Renewable Energy Laboratory*
Peter Phillips, *Clark Atlanta University*
C.G.J.M. Pijls, *TAUW Infra Consult B.V., The Netherlands*
Keith R. Piontek, *CH2M HILL*
Michael Piotrowski, *Biotransformations, Inc.*
Augusto Porta, *Battelle Europe*
Roger C. Prince, *Exxon Research and Engineering Co.*
Parmely "Hap" Pritchard, *U.S. Environmental Protection Agency*
Jaakko A. Puhakka, *University of Washington*
Santo Ragusa, *CSIRO Division of Water Resources, Australia*
Ken Rainwater, *Texas Tech University*
Svein Ramstad, *SINTEF Applied Chemistry, Norway*
Grete Rasmussen, *University of Washington*
Mark E. Reeves, *Oak Ridge National Laboratory*
Roger D. Reeves, *Massey University, New Zealand*
H. James Reisinger, *Integrated Science & Technology, Inc.*
Charles M. Reynolds, *U.S. Army Cold Regions Research and Engineering Laboratory*
Hanadi S. Rifai, *Rice University*
Derek Ross, *ERM Inc.*
J. J. Salvo, *GE Corporate, R&D Center*
Réjean Samson, *Biotechnology Research Institute, Canada*
Erwan Saouter, *Center for Environmental Diagnostics and Bioremediation*
Bruce Sass, *Battelle Columbus*
Eric K. Schmitt, *ESE Biosciences, Inc.*
Gosse Schraa, *Wageningen Agricultural University, The Netherlands*
Alan G. Seech, *Dearborn Chemical Co., Ltd., Canada*
Robert L. Segar, *University of Texas at Austin*
Douglas Selby, *Las Vegas Valley Water District*
Patrick Sferra, *U.S. Environmental Protection Agency*
Daniel R. Shelton, *U.S. Department of Agriculture*
Tatsuo Shimomura, *Ebara Research Co. Ltd., Japan*
Mark Silva, *American Proteins, Inc.*

Thomas J. Simpkin, *CH2M HILL*
Judith L. Sims, *Utah State University*
Rodney S. Skeen, *Battelle Pacific Northwest Laboratories*
George J. Skladany, *Envirogen*
Marina Skumanich, *Battelle Seattle Research Centers*
Lawrence Smith, *Battelle Columbus*
Gregory Smith, *ENSR Consulting and Engineering*
Darwin Sorenson, *Utah State University*
Jim Spain, *Tyndall Air Force Base*
Gerald E. Speitel, *University of Texas at Austin*
D. Springael, *Vlaamse, Instelling voor Technologisch Onderzoek, Belgium*
Thomas B. Stauffer, *Tyndall Air Force Base*
Robert J. Steffan, *Envirogen*
H. David Stensel, *University of Washington*
Jan Stepek, *EA Engineering Science and Technology*
David Stevens, *Utah Water Research Laboratory, Utah State University*
Gerald W. Strandberg, *Oak Ridge National Laboratory*
Janet Strong-Gunderson, *Oak Ridge National Laboratory*
John B. Sutherland, *U.S. Food & Drug Administration*
C. Michael Swindoll, *DuPont Environmental Remediation Services*
Robert D. Taylor, *The MITRE Corporation*
Alison Thomas, *U.S. Air Force*
Francis T. Tran, *Diocese Loire-Atlantique, Seminaire Des Carmes*
Mike D. Travis, *RZA-AGRA Engineering and Environmental Services*
Sarah C. Tremaine, *Hydrosystems, Inc.*
Jack T. Trevors, *University of Guelph, Canada*
Marleen A. Troy, *OHM Remediation Services Corporation*
Mark Trudell, *Alberta Research Council, Canada*
Michael J. Truex, *Battelle Pacific Northwest Laboratories*
Samuel L. Unger, *Groundwater Technology, Inc.*
J. P. Vandecasteele, *Institut Français du Pétrole*
Ranga Velagaleti, *Battelle Columbus*
Albert D. Venosa, *U.S. Environmental Protection Agency*
Stephen J. Vesper, *University of Cincinnati*
Bruce Vigon, *Battelle Columbus*
John S. Waid, *La Trobe University, Australia*
Terry Walden, *BP Research*
Mary E. Watwood, *Idaho State University*
Lenly Joseph Weathers, *University of Iowa*
Marty Werner, *Washington State University*
Mark Westray, *Remediation Technologies Inc.*
David C. White, *University of Tennessee*
Patricia J. White, *Battelle Marine Research Laboratory*
Jeffrey Wiegand, *Alton Geoscience*
J. W. Wigger, *Amoco Corporation*
Peter Wilderer, *Technische Universität München, Germany*
Barbara H. Wilson, *Dynamac Corporation*
John T. Wilson, *U.S. Environmental Protection Agency*
Roger M. Woeller, *Water & Earth Science Associates, Ltd., Canada*
Arthur Wong, *Coastal Remediation*

Jack Q. Word, *Battelle Marine Research Laboratory*
Darla Workman, *Battelle Pacific Northwest Laboratories*
Brian A. Wrenn, *University of Cincinnati*
Lin Wu, *University of California*
Robert Wyza, *Battelle Columbus*
Andreas Zeddel, *Forstbotanisches Institut der Universität Göttingen, Germany*
Gerben Zylstra, *Rutgers University*

The editors wish to recognize some of the key contributors who have put forth significant effort in assembling this book. Lynn Copley-Graves served as the text editor, reviewing every paper for readability and consistency. She also directed the layout of the book and production of the camera-ready copy. Loretta Bahn worked many long hours converting and processing files, and laying out the pages. Karl Nehring oversaw coordination of the book publication with the symposium, and worked with the publisher to make everything happen. Gina Melaragno coordinated manuscript receipts and communications with the authors and peer reviewers.

None of the sponsoring or cosponsoring organizations or peer reviewers conducted a final review of the book or any part of it, or in any way endorsed this book.

Rob Hinchee
June 1993

IN SITU BIOREMEDIATION IN EUROPE

A. Porta, J. K. Young, and P. M. Molton

ABSTRACT

Site remediation activity in Europe is increasing, even if not at the forced pace of the United States. Although there is a better understanding of the benefits of bioremediation than of other approaches, especially about in situ bioremediation of contaminated soils, relatively few projects have been carried out full scale either in Europe or in the United States. Some engineering companies and large industrial companies in Europe are investigating bioremediation and biotreatment technologies, in some cases to solve their internal waste problems. Technologies related to the application of microorganisms to the soil, release of nutrients into the soil, and enhancement of microbial decontamination are being tested through various additives such as surfactants, ion exchange resins, limestone, or dolomite. New equipment has been developed for crushing and mixing or injecting and sparging the microorganisms, as have new reactor technologies (e.g., rotating aerator reactors, biometal sludge reactors, and special mobile containers for simultaneous storage, transportation, and biodegradation of contaminated soil). Some work also has been done with immobilized enzymes to support and restore enzymatic activities related to partial or total xenobiotic decontamination. Finally, some major programs funded by public and private institutions confirm that increasing numbers of firms have a working interest in bioremediation.

INTRODUCTION

This paper contains a discussion of the status of bioremediation efforts in Europe, including the status of regulations, market size, and innovative approaches, and how these approaches might be applied in the United States. Remediation activity in Europe is growing. Progress has been made in applying microorganisms to the soil and enhancing microbial decontamination through various additives such as surfactants, ion exchange resins, and limestone or dolomite. New equipment is available from industry for crushing or mixing the soil and injecting and sparging the microorganisms. New reactor technologies in Europe include rotating aerator reactors, biometal sludge reactors, and special mobile containers for storage, transportation, and biodegradation of contaminated soil. Some work has been done on using immobilized enzymes to support and restore enzymatic activities with regard to xenobiotic decontamination. Some major programs are now being funded either publicly or privately, but the lack of a unified regulatory framework in Europe for bioremediation activities is a serious hindrance to progress in this area.

REGULATIONS

In Europe, there is no standard methodology for classifying a contaminated site. Three regions — The Netherlands, Denmark, and Germany — have high levels of public awareness that influence sound environmental legislation. These regions and the United Kingdom have spent considerable time and money identifying hazardous waste sites.

In The Netherlands, sites are classified by soil quality guidelines. The Dutch "ABC" list addresses heavy metals, organics, and pesticides in soils, groundwater, surface water, and drinking water. In 1991, the list was updated to address soil pollutants by group. Various länders in Germany published different soil evaluation procedures in the 1980s. Due to lack of regulatory standards for soil in many other European countries, soil pollution is not officially recognized until the existence of contamination is noted in the underlying groundwater. In many remaining European countries, well-established standards will be needed to select and prioritize contaminated soil sites and remediation activities.

MARKET SIZE

Market size information is extremely tentative, because initial investigations on the degree of pollution in soils have not been completed in many European countries. The Netherlands, Germany, and Denmark have established a complete list of polluted sites. Finland, Italy, France, Norway, and Sweden have prepared only a preliminary list of known or suspected sites. The United Kingdom has prepared such a list, but has decided not to publish it, fearing adverse effects on property values.

Estimates of the total number of sites are difficult to obtain because of the different criteria used for classification, insufficient knowledge of the extent and depth of pollution, and a perceived lack of urgency to clean up individual sites. The figures reported by different sources vary considerably, and estimation of the number of actual sites is likely to be low in several countries due to the limited knowledge about the situation.

The estimated total and yearly expenditures on remediation by country are shown in Table 1. In Germany, the largest market in Europe, the estimate ranges from $10 to $239 billion (U.S). The Netherlands is the second largest market; markets in France, Italy, and some of the Scandinavian countries are still marginal due to limited emphasis on cleanup.

TABLE 1. Current and expected expenditures for bioremediation by country (Europe).

Countries	Present yearly expenditures for remediation (US $)	Estimated total cost for remediation (US $)
Denmark	100 million	~3.8 billion
France	35-70 million	10.5-12.5 billion
Germany	3-6 billion	10-230 billion
Great Britain	60 million	30 billion
Italy	15-18 million	3 billion
Sweden	40 million	
Other Nordic Countries	20 million	
Switzerland		1.5 billion
The Netherlands	~260 million	2.7 billion

Factors that will encourage market growth in Europe are public opinion, better knowledge of the state of soil pollution and what other countries are doing about it, introducing concise regulations, protecting drinking water supplies from groundwater, recognizing the necessity for preserving the integrity of limited soil resources, and developing cheap remediation technologies and in situ tools for screening analysis.

On the other side, delays in market growth will result from difficulty in identifying who is responsible for the contamination and who will assume the costs. Treatment costs are still much too high, which exacerbates the problem, especially for old sites. Finding the necessary financial resources will require that consortia, industrial associations, sectorial institutions, and private companies use financial measures such as self-taxation, mixed participation, joint ventures, and guaranteed mutual funds.

BIOREMEDIATION APPROACHES

As various pollution problems are addressed in Europe, the scope and diversity of in situ bioremediation technology continue to grow. The large projected expenditures for soil remediation in Germany, The Netherlands, and Denmark make it likely that progress will continue in bioremediation research and development. In situ bioremediation is attractive because it costs less to clean up large areas of polluted land.

Table 2 provides an overview of European organizations active in bioremediation and their major technologies. This is not a complete list. Many other countries have performed in situ bioremediation actions and have been involved in full-scale demonstrations. In situ technologies being developed in Europe include biotreatment with air-stripping and various microbial treatments.

TABLE 2. European organizations active in bioremediation and their major technologies.

Company or University	Location	State of Development	Soil Treatment	Ground-water Treatment	In Situ Treatment	Onsite and/or Offsite Treatment
Germany						
Gertec GmbH	Essen					
Trautmann GmbH	Essen					
HP Biotechnologie GmbH	Witten					
Trischler GmbH	Darmstadt	Demonstration projects	X		X	X
Caro Biotechnik	Aachen					
Hochtief AG	Essen	Laboratory & industrial scale	X		X	X
Messer Griesheim GmbH	Krefeld	Test field completed	X		X	
Xenex Gesellschaft zur biotechnischen Schadstoffsanierung	Iserlohn		X		X	
TGU Technologieberatung Grundwasser und Umwelt GmbH	Koblenz	Field tests	X	X	X	
Santec GmbH	Berlin	Pilot test	X		X	
BCE	Koblenz					

TABLE 2. (continued)

Company or University	Location	State of Development	Soil Treatment	Ground-water Treatment	In Situ Treatment	Onsite and/or Offsite Treatment
EBI	Karlsruhe					
Argus Umweltbiotechnologie GmbH	Berlin	Industrial-scale demonstration	X		X	X
IBL	Heidelberg					
Umweltschutz Nord	Ganderkesee	Full-scale demonstration projects	X	X	X	X
Fraunhofer Institut für Grenzflächen und Bioverfahrenstechnik (FhIGB)	Stuttgart	Pilot tests	X			
Gesselschaft für Boden und Grundwassersanierung mbH	Kirchheim/ Tech					
GBF Gesellschaft für Biotechnologische Forschung mbH	Braunschweig		X		X	
Degussa AG	Hanau	Research at laboratory	X		X	
Institut für Technische Chemie	Munich					
GSF Forschungszentrum für Umwelt und Gesundheit Institut für Bodenökologie	Neuherberg					

TABLE 2. (continued)

Company or University	Location	State of Development	Soil Treatment	Ground-water Treatment	In Situ Treatment	Onsite and/or Offsite Treatment
Engler Bunte Institut						
University of Karlsruhe	Karlsruhe					
Linde AG	Hollriegelstreuth	Research and field demonstration	X	X	X	X
Institute für Gewässerschutz	Kiel					
Hamburg Botanical Institute	Hamburg	Research	X			
Hamburger Wasserwerke Consulaqua	Hamburg	Full-scale demonstration	X			
Westfäüsche Wilhelms Universitat	Münster	Pilot tests	X			
Institute für Mikrobiologie	Braunschweig					
Este GmbH	Hamburg	Full-scale demonstration (Shell Bioreg)	X	X		X
Herbst Umwelttechnik		Pilot scale		X		X
Institut für Molekularbiologie und Analytik GmbH	Zeppelinheim	Field demonstration completed	X		X	

TABLE 2. (continued)

Company or University	Location	State of Development	Soil Treatment	Ground-water Treatment	In Situ Treatment	Onsite and/or Offsite Treatment
Groth U.CO.	Pinneberg	Full-scale demonstration	X	X		X
Senator Projekt Service GmbH	Düsseldorf	Demonstration full-scale project (GDS process)	X	X		
Kloeckner Oecotec GmbH	Duisburg	Full-scale demonstration	X	X	X	X
LFU, Labor für Umweltanalytik GmbH	Berlin	Full-scale demonstration	X		X	
Philipp Holzmann AG	Düsseldorf	Pilot and field study (Shell Bioreg)	X	X	X	X
Rethmann Städtereinigung GmbH	Selm	Pilot project	X			X
Anakat, Institut für Biotechnologie	Berlin	Full-scale demonstration	X		X	X
Bauer Spezialtiefbau GmbH	Schrobenhausen	Full-scale demonstration	X			X
Biodetox Gesellschaft zur biologischen Schadstoffentsorgung GmbH	Ahnsen/b. Bückegurg	Industrial-scale	X	X	X	X

TABLE 2. (continued)

Company or University	Location	State of Development	Soil Treatment	Ground-water Treatment	In Situ Treatment	Onsite and/or Offsite Treatment
Bonnenberg & Drescher	Aldenhoven	Pilot plant	X		X	
Deutsche Shell	Hamburg	Pilot projects (Shell Bioreg)	X			X
Biolipsia GmbH	Markkleeberg					
CBA GmbH - Chemie, Biotechnologie Analytic	Sonneberg					
COMCO MARTECH Deutschland GmbH	Halle					
Fichtner GmbH	Dresden					
Ign.-Büro Grünzel GmbH	Dessau					
In Situ GmbH Gessellsch. für Boden und Grundwassersanierung	Sonneberg/ Thür					
Ökotec GmbH	Belzig					
Santec GmbH - Ing.-Büro für Sanierungs-technologien	Ketzin					

TABLE 2. (continued)

Company or University	Location	State of Development	Soil Treatment	Ground-water Treatment	In Situ Treatment	Onsite and/or Offsite Treatment
France						
Elf Atochem	Orléans					
Institut Français du Pétrole	Rueil Malmaison					
A.T.E.	Meyzieu					
BRGM	Orléans					
Geoclean	Dardilly (Lyon)					
IBS France	St. Michel sur Orge					
Burgeap	Paris					
Enviromax	Les Ulis					
Pollution Service	Lyon					
Serpol	Vénissieux					
Solentanche	Nanterre					

TABLE 2. (continued)

Company or University	Location	State of Development	Soil Treatment	Ground-water Treatment	In Situ Treatment	Onsite and/or Offsite Treatment
The Netherlands						
TAUW Infra Consult BV	Deventer	Full-scale demonstration	X	X	X	
Rijksinstituut voor Volsgezondheid en Milieuhygiene (RIVM)	Bilthoven	Cleanup on demonstration scale	X	X	X	
TNO Environment and Energy	Apeldorn	Cleanup on demonstration scale	X	X	X	
Ecolyse	Groningen	Small-scale demonstration	X	X	X	
Ballast Nedam Milieutechniek BV	Lekkerkerk					
Paques BV	Ab Balk	Full-scale installation		X		X
Delft Geotechnics	Delft	Experimental field project completed	X	X	X	
Ecotechniek BV	Utrecht	Research	X			X

TABLE 2. (continued)

Company or University	Location	State of Development	Soil Treatment	Ground-water Treatment	In Situ Treatment	Onsite and/or Offsite Treatment
Heidemij Reststoffendiensten BV Afdeling Milieutechniek	Waalwijk	Developed technology	X			X
Heijmans Milieutechniek BV	Rosmalen		X			X
HWZ-Milieu	Gouda	Research	X		X	
Mourik Groot-Ammers BV	Groot-Ammers	Full-scale demonstration	X	X	X	X
De Ruiter Milieutechnologie	Halfweg and Zwanenburg	Full-scale demonstration	X		X	X
Witteveen + Bos-Consulting Engineers	Deventer	Production scale trials	X			X
Scandinavia						
Aquateam Norwegian Water Technology Center	Oslo, Norway	Full-scale project	X			X
Terrateam A/S	Oslo, Norway	Full-scale project	X			X
Senter for Industriforskning	Norway					

TABLE 2. (continued)

Company or University	Location	State of Development	Soil Treatment	Ground-water Treatment	In Situ Treatment	Onsite and/or Offsite Treatment
Danish Geotechnical Institute	Denmark					
Bioteknisk Jordens KK Miljöteknik	Kalundborg, Denmark		X		X	X
Alko	Finland					
Neste Oil	Finland					
Skanska and Consultants	Sweden					
Banverket	Sweden					
VBB-VIAK	Sweden					
ANOX	Sweden	Pilot project	X			
FUNGINOVA	Sweden					
AGA	Sweden					
Neste Oxo	Sweden					
Abitec Ab	Sweden					
Other Regions						
Department of Biotechnology Institute of Microbiology and Virology	Kiev, Ukraine					

TABLE 2. (continued)

Company or University	Location	State of Development	Soil Treatment	Ground-water Treatment	In Situ Treatment	Onsite and/or Offsite Treatment
Prague Institute of Technical Chemistry, Institute of Microbiology and Biochemistry	Prague, Czech Republic					
GS Geological Services	Podebrady, Czech Republic					
AREA	Prague, Czech Republic					
Agricultural University	Gödöllö, Hungary					
Pyrus Environmental Services, Ltd.	Hungary					
Proterra Umwelttechnik GmbH	Vienna, Austria					
Universität für Bodenkultur	Vienna, Austria					
VITO Boeretang 200	Mol, Belgium					
ENEL (Italian National Center Electricity Board)	Brindisi, Italy	Research	X			
ISMES	Bergamo, Italy	Research	X			

TABLE 2. (continued)

Company or University	Location	State of Development	Soil Treatment	Ground-water Treatment	In Situ Treatment	Onsite and/or Offsite Treatment
Castalia	Genoa, Italy					
Eniricerche	Rome, Italy	Research	X			
Groundwater Technology Int., Ltd.	Epsom, U.K.	Industrial-scale remediation	X	X	X	
Land Restoration Systems	Slough, U.K.	Experimental installation	X	X	X	X
Biotal	Cardiff, Wales	Development of microbial products	X	X		
DDH Désinfection/Dépollution/Hygiène	Saxon, Switzerland	Pilot	X		X	
Ebiox AG	Sursee, Switzerland	Full-scale projects	X	X	X	X
Optima Kosmetik	Prilly, Switzerland					
MBT Umwelttechnik AG	Zurich, Switzerland		X	X		

Germany

Germany has spent more time and money than any other country identifying environmental problems, and thus has the largest number of companies working on bioremediation, at 22 remediation centers (Table 3). A number of these companies are located in the former East Germany. The list of contaminated sites and needed remedial actions has increased dramatically by the German reunification. Risk sites include vehicle workshops, airports, traffic and parking areas, waste dumps, fuel storage and transfer points, and munitions sites.

According the Bundesministerium für Forschung und Technologie (BMFT, the Ministry for Research and Technology), in Germany, 28 bioremediation techniques have been applied there. BMFT has sponsored the 16 projects summarized in Table 4 with a total funding of 20 million DM ($12.5 million U.S.). The German Research Association also has conducted projects in enzymatic dehalogenation of contaminants using *Pseudomonas*, *Streptomyces*, and thermophilic microorganisms, and in biodegradation for "dioxin-like" substances.

TABLE 3. Planned or existing remediation centers for offsite remediation in Germany.

	Status		Treatment		
Location	Planned	Realized or In Use	Thermal	Physico-chemical	Biological
Hamburg-Veddel		X		X	
Hamburg-Billbrook	X			X	X
Hamburg-Elmsbüttel		X		X	
Hamburg-Peute		X		X	
Itzehoe		X		X	
Ganderkesee		X			X
Bremen		X			X
Ahnsen		X			X
Hildesheim	X		X	X	X
Northeim-Göttingen		X			X
Berlin-Gronau		X			X
Berlin-Tiergarten			X		
Grosskreuz				X	X
Münster	X				X
Hattingen	X			X	
Bochum	X				X
Duisburg	X		X		
Dresden	X			X	X
Gröben (bei Meissen)	X			X	X
Schwarze Pumpe	X		X	X	X
Neunkirchen			X	X	X
Frankfurt			X	X	X

TABLE 4. BMFT-sponsored bioremediation technologies.

Sponsored Institutions	Time	Cost, DM
Stadt Hamburg	1985-1988	804,450
IWL Köln	1986-1987	58,200
Probiotec	1986-1988	331,785
Inst. für Umweltanalytik und Biotechnologie	1986-1988	438,470
TU Hamburg	1986-1992	1,064,174
TU Braunschweig	1987-1989	1,933,000
TU Göttingen	1987-1991	1,030,505
Ruhrkohle Öl und Gas GmbH	1988-1993	817,373
Biodetox	1988-1991	647,000
DMT	1988-1993	4,811,958
HDI	1988-1992	1,253,103
Land Hessen	1989-1992	2,494,042
Uni Karlsruhe	1990-1993	2,945,210
Tu Braunschweig	1990-1992	959,920
Bauer-Spezial-Tiefbau	1990-1993	20,845,804

A number of companies conduct polycyclic aromatic hydrocarbon (PAH) decontamination using microbes. De Ruiter Milieutechnologie, Halfweg, conducted a demonstration project involving aliphatic or aromatic hydrocarbons to study the influence of pH, nutrient addition (potassium, nitrate, and others), and inoculation of adapted microorganisms. The German bioremediation firm, Argus Umweltbiotechnologie GmbH, uses infiltration of air and addition of nutrients to degrade hydrocarbons in situ. The Chemisches Laboratorium E. Wessling-Altenberge blows ozone through contaminated soil to degrade PAHs.

Some innovative technologies are being developed at the Fraunhofer Institute. Researchers at the Department of Chemical Microbiology of the Fraunhofer Institute of Interface Technology and Biotechnology have focused on the microbial and engineering aspects of bioremoving of xenobiotic compounds from wastewaters and exhausted air. In particular, they have demonstrated that PAH biodegradation can be achieved in airlift bioreactors and accelerated using water-soluble solvents as lipophilic mediators to facilitate mass transfer. The biological process in airlift reactors is carried out in an organic-aqueous mixed phase.

Wilhelm Universität of Muenster and the Technical University of Munich studied the application of specially developed, immobilized microorganisms to xenobiotically degrade soil contamination. These immobilized microorganisms have better resistance to soil microflora, because they are affixed to a microporous support that provides a habitat promoting reproduction of microbial cells yet allowing release of cells from the support.

Work is under way in Germany to introduce nutrients into the soil using explosive cartridges. Soil-mixing machines expedite mixing the soil with ion exchange resins, dolomite or limestone (to adjust pH), and nutrients. Microorganisms and enzymes are immobilized on wood chips, granular clay, anthracite, and synthetic polymers to assist their establishment in the soil matrix. The use of earthworms to biotransform pesticides is being examined by the Institut für Bodenokologie.

The Tardecon process significantly raises the rates of decontamination by mixing activated sludge with soil contaminated with mineral oil and polycyclic hydrocarbons. The State of Baden-Württenberg is conducting a development program to evaluate new remedial techniques. An abandoned dump near Heidelberg has been selected for demonstrating in situ decontamination of the soil column using steel pipes inserted horizontally into the ground by vibration. The so-called "old site" program in the former East Germany has been set up, and more than 20 projects have been initiated using soil-venting and bioventing. With the large number of problem areas in Eastern Germany, risk assessments are under way to identify remedial measures to block pathways, lower the toxic content, and control exposure risks.

The Netherlands

The Netherlands and Denmark are leaders in establishing nationwide programs for decontaminating thousands of sites. A number of well-established companies are located in the Netherlands, and a significant number of sites have been cleaned up since 1982. Soil pollution is an environmental problem of the highest priority because of the limited land area and proximity to sea level. In situ bioreclamation is one of several methods available for treating oily wastes and PAH in sediments. Delft University of Technology has demonstrated venting-assisted evaporation of contaminants. A petroleum-contaminated site at Asten was used to evaluate the feasibility of in situ bioremediation and showed good prospects for remediation of the petroleum spill if hydrogen peroxide was added as a chemical alternative to oxygen.

A biological method for water treatment is available that uses controlled biological oxidation in sulfide reactors. A full-scale biological treatment facility that uses the Thiopaq process started in mid-1992 at Budelco BV (a zinc manufacturer). The process treats the groundwater, highly polluted with sulfate and heavy metals, underneath the property. Sulfur compounds are reduced to hydrogen sulfide using anaerobic sulfate-reducing bacteria, and heavy metals are precipitated as metal sulfides. The remaining sulfide is oxidized to elemental sulfur using aerobic sulfide-oxidizing bacteria, and elemental sulfur is then separated from the water.

TAUW Infra Consult B.V. has developed Biopur®, an innovative bioreactor for simultaneous cleanup of groundwater and soil vapor contaminated with xenobiotic compounds. Biopur® is a fixed-film bioreactor filled with polyurethane as carrier material for the biomass.

Scandinavia

In Denmark, a company (Bioteknisk Jordens) treated 130,000 tons of soil by biological methods. In Finland, Alko specialized in the biological removal of chlorophenols from soil. They have piloted the method on more than nine sites. Sweden and Norway have conducted projects on abandoned wood-treating and cokework sites.

Other Regions

Other European countries working to address contamination issues include Italy, France, the Spanish province of Catalonia, Switzerland, and the United Kingdom. These countries have made efforts to identify contaminated sites (the United Kingdom reportedly has 50,000 to 100,000 contaminated waste sites) but have not yet defined nationwide decontamination measures, selected technical approaches, or planned large decontamination projects. Meanwhile, Spain, Portugal, Greece, and Ireland are just beginning to assess contamination problems and sites to be remediated.

An interesting development in France is the use of algal cultures in aqueous solutions to stabilize cesium and strontium in the soil. These cultures are used primarily for shallow surface contamination, but adaptations may be possible to extend the technology to groundwater and subsurface contamination. Experimental programs are being conducted in collaboration with the former USSR.

The French DVM (Decontaminating Vegetal Network) process is a biomechanical method for removing soil contamination using plants that create a dense root network that traps the contaminated soil particles. Removing the turf then removes the contaminated soil. Biosurfactant-producing microorganisms have been used to increase the removal of contaminants using soil washing.

In Eastern Europe, several Czech companies offer reasonably advanced bioremediation services. A microbial mixed population is being studied by the University of Prague to treat surface contamination in an abandoned site polluted with petroleum hydrocarbons.

COMPARISON WITH U.S. BIOREMEDIATION TECHNOLOGY

The status of U.S. bioremediation technologies is reviewed here for comparison purposes. Routine applications of in situ bioremediation in the United States are limited mostly to small-scale treatment of surface and near-surface contaminated soils and groundwater. Contaminants are degraded with native microorganisms and topical application of nutrients.

Research programs are under way to increase the capabilities of bioremediation to deep, extensive, subsurface contamination due to chlorinated hydrocarbons and complex mixed wastes, including soils and groundwater. The U.S. Environmental Protection Agency (EPA) is focusing on the waste types at 1,200 National Priorities List sites, including organic solvents, wood-preserving chemicals, halogenated aromatic hydrocarbons, pesticides, and munitions waste. Technology development funded by the U.S. Department of Energy (DOE) treats

TABLE 5. Selected in situ bioremediation projects in the United States.

Name/Location	Technology	Contaminant	Status
Allied Chemical & Ironton Coke, Pennsylvania	Bioremediation of lagoon sediments	PAHs	Predesign completed in winter 1993
French Ltd, Texas	In situ lagoon	VOCs, PAHs	In design
Fairfield Coal and Gas, Iowa	In situ sludge; injection of H_2O_2 and other nutrients	BTEX, naphthalene	Field-scale pilot test completed in January 1994
Libby Groundwater, Montana	Injection of H_2O_2 and potassium tripolyphosphate	Benzene, PCP, and creosote	Operational
Kelly AFB, Texas	Injection of H_2O_2 and addition of ammonium and phosphate salts	TCE	System operational for 9 months
Savannah River Site, South Carolina	Horizontal air injection and extraction wells	TCE	Testing began in July 1990
Cabot Carbon/Koppers, Florida	Nutrient addition in groundwater and soils above and below	PCP, *bis*(2-ethylhexyl)phthalate, DNT, dimethylphenol, PAH	Design will be completed in September 1994

volatile organic compounds in both arid and nonarid soils. The DOE plans to demonstrate bioremediation technology in actual field conditions.

Injecting air into the vadose zone or aquifers (at depths below the water table) is becoming a practical alternative for subsurface soil and groundwater treatment in the United States. Cometabolites, nitrate, and inocula may also be injected (in conjunction with oxygen for aerobic processes) to stimulate degradation of chlorinated organics. Horizontal wells transport gas-phase nutrients through tight soils at sufficiently low flowrates to prevent transport of volatile organics to the surface. When soils are so tightly bound that movement of oxygen and nutrients is severely restricted (as is the case in saturated zones), hydrofracturing is used to modify the soil to create transport passages. While horizontal wells for nutrient delivery are being tested at DOE sites such as Savannah River and Hanford, further engineering will be required before they can be considered reliable in situ treatment technologies. Bioremediation of contaminated sediments and sludges is in the early stages of development, and much research will be required to design viable field-scale processes.

Field tests currently being conducted by EPA include fungal treatment of pentachlorophenol (PCP), bioventing of contaminated vadose soils, and bioremediation of an aquifer contaminated with solvent. Field demonstration data will be made available through EPA's Alternative Treatment Technologies Information Center (ATTIC) database for many of the tests being conducted. Treatability studies and testing protocol are currently being developed by the EPA so that the efficacy of various bioremediation strategies can be evaluated in advance. Table 5 presents some selected U.S. in situ bioremediation projects.

Recommendations

The American Academy of Microbiology (AAM) has concluded that enough knowledge is now available for field trials of bioremediation technology for organic compounds. Research is needed for the following classes of environmental pollutants: metals, metalloids, radionuclides, and complex polycyclic hydrocarbons. In all of these areas, Europe offers promising technologies.

REMEDIATION OF AQUEOUS-PHASED XENOBIOTIC CONTAMINATION BY FRESHWATER BIVALVES

P. J. Hutchinson

ABSTRACT

The Asian clam, *Corbicula fluminea*, as determined by controlled bioreactor experiments, displays the ability to absorb certain monosubstituted phenylic compounds from aqueous media and possibly satisfy some energy requirements from catabolization of these organic compounds. Axenozoic absorption is a generalized feeding strategy that employs carrier-mediated endothelial transport mechanisms for the uptake of dissolved organic molecules. Xenocatabolically aggressive bivalves mineralize the organic contaminant through the cytochrome *P*-450 mono-oxygenase system. These bivalves can be used to remediate low levels of anthropogenic chemical contamination of groundwater supplies. Bivalves show chemospecific effects to xenobiotic exposure that vary with the time of exposure, the concentration of contamination, and the innate tolerance of the species.

INTRODUCTION

The occurrence of xenobiotic organic and inorganic compounds within the living tissue of aquatic invertebrates supports the contention that mechanisms for biological absorption of organic compounds exist. The Mussel Watch Program uses the manifestation of these mechanisms to monitor the occurrence of pollution within the biosphere (Goldberg 1986). However, researchers have also discovered that absorbed compounds can be catabolized by the organism and used as a metabolic energy source (Livingstone 1985, Sindermann 1985, Wright 1982).

Pütter (1909) initiated a controversy by suggesting that marine organisms could actually survive on dissolved organic matter (DOM). Since Pütter's time, sampling techniques have improved, radiolabeling has been perfected, and epithelial carrier-mediated mechanisms have been discovered. Additionally, the oceans are an enormous reservoir of DOM (all organic carbon >1,000 daltons), and this reservoir of reduced carbon exists in quantities sufficient to sustain life (Benner et al. 1992).

Bivalve gills are ideal structures for the uptake of organic and inorganic solutes because of the extensive surface area, cell monolayer, and highly developed vascularization (Owen 1974). Gupta (1977) observed that the gills are ideally suited for the uptake of ions through the enormous surface area due to large numbers of filaments and to the continuous renewal of water accomplished by the terminal cilia. Reid (1980) found that, for the gutless protobranch, *Solemya* sp., an obligate feeder of dissolved organic compounds, internal organs consist mostly of gill tissue by weight. Indeed, at least 75% of the bivalves listed in Table 1 have been discovered to absorb organic compounds through the gills. Further, Anderson et al. (1992) observed that absorption of small molecules (< 1 kilodalton) through the epithelium is affected by caveolae acting with glycosylphatidylinositol-anchored membrane proteins to take up a variety of small molecules in a process termed potocytosis.

Recently, numerous researchers have observed the uptake and catabolism of organic compounds through enzymatic processes (Table 1). Absorbed DOM can undergo intracellular enzymatic catalyzed metabolism (Stewart 1979). Riley et al. (1981) observed what appeared to be metabolites of naphthalene in the marine bivalve *Ostrea edulis*. Anderson (1978) demonstrated the generation of PCB, 3-methylcholanthrene, and benzo[a]pyrene metabolites catalyzed by the enzymatic actions of aryl hydrocarbon hydroxylase (AHH). Livingstone (1985) contends that mechanisms exist for the detoxication and/or elimination of xenobiotic contaminants from marine bivalves and divided the metabolism of xenobiotic compounds into two reactions; biotransformation and conjugation through monooxygenase-catalyzed reactions.

Axenic Organisms

Axenic bivalves can absorb and possibly metabolize organic compounds without the aid of a bacterial intermediary (Table 1). Research into the organic-compound absorption phenomenon has followed three paradigms: in vitro absorption, in vivo uptake, and enzymatic activity.

Researchers have observed in vitro uptake of DOM by certain bivalve tissues, notably the gills, hepatopancreas, and digestive gland (Table 1). These experiments include radiotagging and chemical-specific tests to detect the loss of the organic compounds from solution. The uptake of dissolved free amino acids by the gills has been carefully documented for *Mya arenaria*, *Mytilus californianus*, and *M. edulis*.

In vivo studies encompass elegant models and sampling strategies, normally involving quantified uptake by analyzing the whole tissue or the gills. The work presented herein and other in vivo studies establish real-time loss of the xenobiotic contaminant but do not determine if catabolization can supply nutritional requirements. Moore et al. (1987) and Anderson (1978) document the increase in enzymatic activity upon exposure to a xenobiotic organic compound and identified metabolites of enzyme-catalyzed reactions. Enzyme-catalyzed metabolism has been documented to occur in the gill, digestive gland, and hepatopancreas (Table 1).

TABLE 1. Table of bivalves by family determined by experimental analysis to absorb or metabolize organic compounds. Key to abbreviations: Blood (BL); digestive gland (DG); dissolved free amino acids (DFAA); dissolved organic matter (DOM); freshwater (F); gills (G); hepatopancreas (HP); mantle (MA); marine (M); catabolic process (P); pinocytosis (PI); uptake (U); whole tissue (WT).

Organism	Tissue	Envir.	Chemical	Process	Reference
ARCIDAE					
Anadara granosa (Linné)	G	M	naphthalene, acetone	U	Patel & Eapen 1989b
	G	M	naphthalene, acetone	U	Patel & Eapen 1989a
CARDIIDAE					
Cardium (*Cerastoderma*) *edule* (Linné)	G	M	benzo[a]pyrene	P	Livingstone 1985
	G	M	L-alanine;L-lycine	U	Bamford & McCrea 1975
	DG,WT,HP	M	crude oil	P	Moore et al. 1987
CORBICULIDAE					
Corbicula fluminea (Müller)	WT	F	Organic chemical & sewage effluent	P	Cantelmo-Cristini et al. 1985
C. manilensis (Philippi)	WT	F	DFAA	P	Gainey 1978
	WT	F	organochlorine pesticides	P?	Hartley & Johnston 1983
	WT	F	DDT	P	Leard et al. 1980
Polymesoda caroliniana (Bosc)	WT	F	DFAA	P	Gainey 1978
DREISSENIDAE					
Dreissena polymorpha (Pallas)	G	F	algal toxins; DDT	P	Birger & Malarevskaja 1977
	G	F	algal toxins	P	Birger et al. 1978
D. bugensis	G	F	algal toxins	P	Birger et al. 1978
ISOGNOMONI					
Isognomon alatus	MA	M	thorium dioxide	PI	Nakahara & Bevelander 1967
MACTRIDAE					
Rangia cuneata (Gray)	G	M	glycine	U	Anderson & Bedford 1973
MYIDAE					
Mya arenaria (Linné)	G	M	DFAA	U	Dupaul & Webb 1971
	WT	M	crude oil	P	Gilfillan et al. 1985
	G	M	alanine	U	Stewart & Bamford 1975
	G	M	DFAA	U	Stewart & Bamford 1976
	G	M	DFAA	U	Stewart 1978
	G	M	L- and D-alanine	U	Stewart 1979
	HP	M	benzo[a]pyrene	P	Stegeman 1981

TABLE 1. (Continued).

Organism	Tissue	Envir.	Chemical	Process	Reference
MYTILIDAE					
Geukensia (Modiolus) demissa	G	M	DFAA	U	Crowe et al. 1977
	G	M	DFAA	U	Wright & Stephens 1978
	G	M	DFAA	U	Wright 1979
	G	M	DFAA	P	Wright 1982
G. modiolus (Linné)	G	M	DFAA	P	McCrea 1976
Mytilus californianus (Conrad)	G	M	aldrin	P	Krieger et al. 1979
	G	M	DFAA	U	Swinehart et al. 1980
	G	M	cycloleucine	P	Wright et al. 1975
	G	M	DFAA	U	Wright & Stephens 1977
	G	M	DFAA	U	Wright & Stephens 1978
	G	M	DFAA	U	Wright 1979
	G	M	DFAA	U	Wright et al. 1980
	G	M	DFAA	P	Wright 1982
M. edulis (Gray)	G	M	L-phenylalanine	U	Bamford & Campbell 1976
	BL,DG,G	M	aldrin,anthracene	P	Bayne et al. 1979
	WT	M	2- & 3-ringed aromatic hydrocarbons	P	Bayne 1989
	G	M	1-naphthol	P	Ernst 1979
	WT	M	crude oil	P	Gilfillan et al. 1985
	G	M	diesel	P	Livingstone et al. 1985
	G	M	DFAA	U	Manahan et al. 1982
	DG,WT,HP	M	crude oil	P	Moore et al. 1987
	G,MA	M	DFAA,glucose	P	Péquignat 1973
	G	M	DFAA	U	Seibers & Winkler 1984
	HP,DG	M	benzo(a)pyrene	P	Stegeman 1981
	G	M	DFAA	U	Wright & Stephens 1978
M. galloprovincialis	DP,G	M	crude oil	P	Gilewicz et al. 1984
OSTREIDAE					
Crassostrea gigas	G	M	DFFA	P	McCrea 1976
	G	M	D-glucose;D-galactose	P	Bamford & Gingles 1974
Crassostrea virginica (Gmelin)	DG	M	benzo(a)pyrene, 3-methylcholanthrene, PCBs	P	Anderson 1978
	G	M	free fatty acids	U	Bunde & Fried 1978
	G	M	carbohydrates	U	Collier et al. 1953
	G	M	DFAA	P	McCrea 1976
	G	M	No. 2 fuel oil	P	Stegeman & Teal 1973
	DG	M	benzo(a)pyrene	P	Stegeman 1981
Ostrea edulis (Linné)	G	M	naphthalene	U	Riley et al. 1981
	G	M	DFAA	U	Rice et al. 1980

TABLE 1. (Continued).

Organism	Tissue	Envir.	Chemical	Process	Reference
PECTINIDAE					
Chlamys opercularis (Linné)	G	M	DFAA	P	McCrea 1976
PTERIIDAE					
Pinctada radiata	MA	M	thorium dioxide	PI	Nakahara & Bevelander 1967
SOLEMYIDAE					
Solemya borealis (Totten)	G	M	DOM	P	Reid & Bernard 1980
	G	M	DOM	P	Reid 1980
S. sp.	G	M	DOM	P	Reid & Bernard 1980
	G	M	DOM	P	Reid 1980
TEREDINIDAE					
Bankia gouldi (Bartsch)	G,MA	M	L- and D-alanine	U	Stewart & Dean 1980
Tridacna elongata	MA	M	Leucine	U	Fankboner 1971
UNIONIDAE					
Amblema costata (Rafinesque)	WT	F	DDT	P	Leard et al. 1980
Anodonta piscinalis	WT	F	organochlorines	P	Herve 1991
Anodonta sp.	G	F	aldrin	P	Kahn et al. 1972
Elliptio crassidens (Lamarck)	WT	F	DDT	P	Leard et al. 1980
Lampsilis anadontoides	WT	F	DDT	P	Leard et al. 1980
L. claibornensis	WT	F	DDT	P	Leard et al. 1980
Megalonaias gigantea (Barnes)	WT	F	DDT	P	Leard et al. 1980
Plectomerus dombeyanus (Valenciennes)	WT	F	DDT	P	Leard et al. 1980
VENERIDAE					
Macrocallista maculata	MA	M	carmine	PI	Bevelander & Nakahara 1966

Wolf (1982) identified two types of microsomally generated enzymes involved in intracellular oxidation of xenobiotic organic compounds: dehydrogenases and oxygenases. Dehydrogenase enzymes are the most common and catalyze the oxidation of organic compounds by the reduction of oxygen to water. Oxygenases, the cytochrome *P*-450-dependent monooxygenases, catalyze the incorporation of molecular oxygen into the organic compound (Stegeman 1981).

The cytochrome *P*-450-dependent monooxygenase system catalyzes the oxidation of the organic compound to an oxide (Sindermann 1985). The cytochrome *P*-450 system consists of a heterogeneous population of enzymes that display different chemical specificities (Wolf 1982). For example, cytochrome *P*-450-mediated oxidation of an aromatic compound produces an arene oxide. The arene oxide isomerizes to a phenol or dihydrodiol followed by nonenzymatic oxidation to quinone (Jakoby et al. 1982).

The oxidation of the organic compound continues until complete mineralization is achieved. Processes to complete mineralization include additional dehydrogenase and oxygenase enzymes (Stegeman 1981). Endocytotic catabolism of amino acids in bivalves is poorly understood (Bishop et al. 1983); however, nearly 30 papers document the uptake of dissolved free amino acids (Table 1). Many of these studies and others suggest that dissolved free amino acids act as a nutritional source (Table 1; de Zwaan & Wijsman 1976, Somero & Bowlus 1983). If catabolism of amino acids supplies nutritional requirements, then other organic molecules can be affected by enzymes of the cytochrome *P*-450 system in a capacity to also supply energy and nutritional requirements.

Jones & Jacobs (1992) suggest that clams are facultative in their feeding modes. Bivalves, demonstrating facultative feeding modes, can be an effective tool in remediation of low levels of anthropogenic groundwater contamination. In previous studies, a salt (potassium bromide), phenol, and toluene were attenuated from aqueous solutions in static and flowthrough reactors, allegedly through the filter-feeding actions of the Asian freshwater bivalve, *Corbicula fluminea* (Müller) (Hutchinson 1992, Hutchinson et al. 1993a & b). The present study documents, through mass balance, the loss of aqueous organic solutes from a sealed flowthrough system mediated by *C. fluminea* and evaluates the results in the context of groundwater remediation.

MATERIALS AND METHODS

Flowthrough System

A flowthrough system was designed to deliver the specific concentrations of xenobiotic solutes in a water-tight, aerobically-sealed container (i.e., no head-space) through peristaltic pumps (Figure 1). Each reactor consists of a 9-cm-i.d. by 34-cm-tall plexiglass housing (volume = 2.1 L) with three tiers vertically staged at 10-cm higher levels. All tubing and fittings are Tygon™ or glass and all equipment was cleaned or replaced between experiments through standard EPA praxis (Todd et al. 1976).

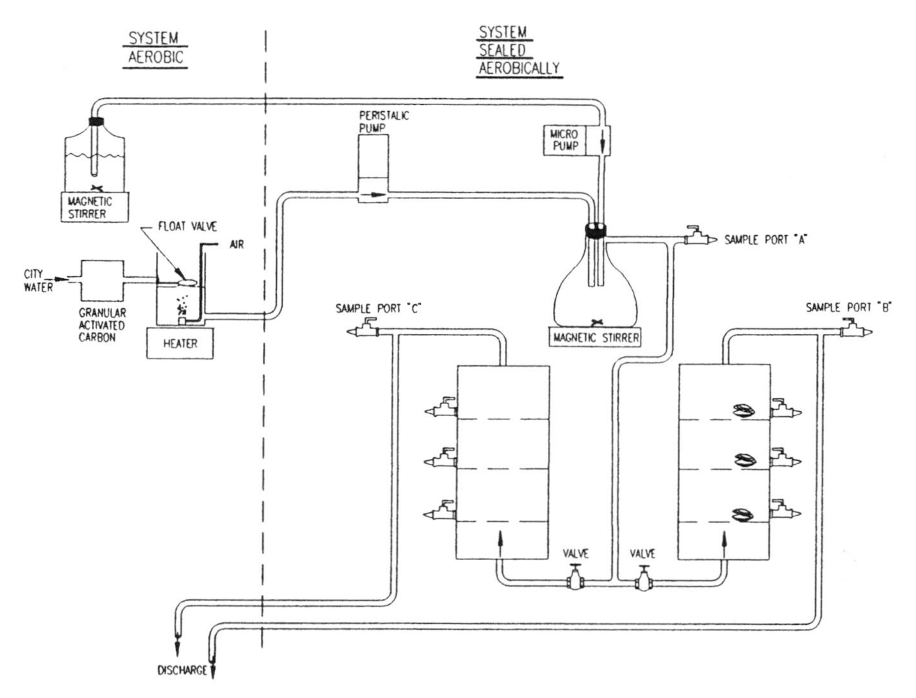

FIGURE 1. Flowthrough system designed to control evaporative flux of volatile organic compounds. Three-tier bioreactor with one bivalve per tier is offset by the three-tiered control reactor.

Peristaltic pumps (Cole-Parmer Instruments model 7553-30) supplied predetermined constant flow to the reactors and produced a turnover rate depending upon experiment of between 8 to 70 times per day at flowrates of 0.7 to 6.1 L/hr, respectively. The minimum flowrate of oxygen-saturated water necessary for unstressed respiration of three 33-mm-long clams is estimated at 0.5 L/hr (8 mL/min) (Hutchinson 1992).

Flowrates were determined for the concentrated solution and for the feedwater to produce the desired concentration of xenobiotic contamination for the experiment. Flowrates were verified via timed flow measurements several times throughout the experiment. Water samples were analyzed by commercial laboratories by purge-and-trap gas chromatography (Hewlett-Packard 5890A with FID: 1 µg/L sensitivity; EPA Method 5030A and 8020, see Franson 1990). Duplicate and blank samples were analyzed pursuant to U.S. Environmental Protection Agency protocols (U.S. EPA 1979).

Unpreserved samples were drawn periodically via the petcock ports by the pressure supplied by the peristaltic pump (no back- or gravity-fill) into prelabeled 40-mL volatile-organic-analysis vials (All-Pak, Inc., Pittsburgh, PA). The vials were maintained at 3°C until chain-of-custody delivery to a commercial laboratory.

Bivalves

The freshwater Asian clam, *Corbicula fluminea* (Müller), filters water at rates varying from 500 mL/hr to 900 mL/hr (Buttner & Heidinger 1981, Haines 1979). Therefore, three clams during filter feeding can turnover reactor water at least once per hour under no-flow conditions.

Solutes

Two phenylic compounds were used to test the efficacy of bivalves for the removal of aqueous-phase contamination, toluene ($C_6H_5CH_3$) and bromobenzene (C_6H_5Br). Toluene was used because (1) it is not nocent at low concentrations, (2) it is very reactive microbiologically (Worsey & Williams 1975), and (3) it is a light, nonaqueous-phase liquid (LNAPL). Bromobenzene was used because (1) it is not known to be biologically reactive, (2) the catabolic pathways have been reported for mammals (Mathews 1982), and (3) it is a dense, nonaqueous-phase liquid (DNAPL).

A series of short-term experiments were conducted and consisted of exposing three clams staged in bioreactors to bromobenzene (BBZ) at 50 mg/L and at 50 μg/L and to toluene (TLN) at 5 mg/L and at 50 mg/L.

RESULTS

Bromobenzene

In the two BBZ experiments, concentrations of BBZ were attenuated 75% and 56% more effectively from the bioreactor than from the control reactor (Table 2).

TABLE 2. Percentage of solute reduction from inlet to outlet during observed clam filtration periods.

Experiment	Ave. Diel Loss in Bioreactor	Ave. Diel Loss in Control	$\Delta\sigma$	Observed Filtration by Clams
BBZ-50 mg/L	35.1%	8.8%	74.9%	C
BBZ-50 μg/L	19.9%	8.8%	55.8%	C
TLN-50 mg/L	56.8%	47.8%	15.8%	I
TLN-5 mg/L	NA	NA	12.7%	N

NOTE: C = clams observed continuously pumping water.
I = clams observed pumping water intermittently.
N = clams never observed to be open.
NA = not available.
a = loss between influent and effluent at 72-hour event only.

The bioreactors experienced an average diel decline in solute concentration of 20% to 35%, whereas the control reactor registered only a 9% decline in concentration from influent to effluent. However, due to low numbers and variability of the samples no statistically significant difference can be assigned to the apparent difference between the concentration of BBZ in the influent and the effluent of the bioreactor.

The first experiment introduced approximately 50 mg/L of BBZ to the reactor for a 72-hour period and the bioreactor displayed a 75% reduction in BBZ concentrations compared to the control reactor. The influent water sample at the 24-hour mark had a markedly lower concentration of BBZ than the subsequent diel sampling events, probably due to initial solute-solvent immiscibility. The reduction in solute concentration for each diel sampling event ranged from 26.4% to 47.4%. No mortality was observed during the 72-hour test; however, at the termination of the experiment, the clams appeared torpid and did not respond vigorously to tactile stimulation.

The second experiment exposed another set of three clams to 50 μg/L of BBZ for 6 days (142 hours). Additionally, the clams were exposed to 50 mg/L of the bacteriostat, chloramphenicol, for 24 hours on the fourth day (hour 68 to 92) to determine if contaminant attenuation is bacterially mediated. The average diel loss from influent to effluent sampling ports ranged from 13% to 33%. The bioreactor was 56% more efficient at removing the solute than was the control reactor. The average diel loss of solute before antibiotic treatment was 16.5%, and the average loss after treatment was 24.9%; consequently, the antibiotic had no apparent dilatory effect upon the ability of the clams to absorb the BBZ and may have increased their ability to absorb the solute. The clams appeared healthy and responsive at the termination of the experiment.

Toluene

Two experiments consisted of exposing clams to 50 mg/L and 5 mg/L of TLN for 48 hours and 72 hours, respectively (Table 2). The experiment exposing the clams to 50 mg/L of TLN indicated a 16% more effective loss of toluene in the bioreactor than in the control reactor (Table 1). The first diel sample for the TLN-50 mg/L experiment indicated that the bioreactor was 28.6% more effective at removing the solute than the control reactor. In this experiment, the effluent of the biotower was lower than the effluent of the control reactor at the 75% confidence level. The second diel sampling event displayed a 17% greater removal rate for the bioreactor in comparison with the samples from the control reactor. Although there were no mortalities, clams were torpid and did not respond to tactile stimulation at the conclusion of the experiment.

The experiment exposing the clams to 5 mg/L of TLN only showed a net reduction in concentration of solute from the bioreactor compared to the control reactor for the 72-hour event. During this experiment the clams did not open, burrow, or display any of the activities demonstrated by clams in the other experiments, except prior to the introduction of the contamination and the last day of the experiment. However, samples of the effluent from the bioreactor after 72 hours indicated that there was a 12.7% decrease in concentration compared to the control reactor for that same period. The clams did not display any detrimental effects from the exposure at the conclusion of the experiment.

DISCUSSION

This series of experiments shows that clams can remove certain monosubstituted phenylic compounds from freshwater. Through the use of an aerobically sealed and mass-balanced reactor, the loss of the contaminant from the aqueous medium was observed consistently within the reactor containing actively filtering bivalves. These short-term experiments did not display as dramatic a reduction in solute levels as had been observed in earlier work where the solute (TLN) was completely removed from the solvent (Hutchinson et al. 1993a); however, the loss of solute from the bioreactor in comparison to the control reactors is significant.

The relatively short term of these experiments, the sparse data, the low numbers of bivalves (and their predisposition to feed at differing times) and the sample collection methodology precluded the calculation of statistical significance; however, the 50 mg/L TLN experiment displayed a significant difference between the effluent of the bioreactor and the control reactor at the 75% confidence level. Additionally, direct comparisons between the concentrations in the effluent of the bioreactor in comparison to the effluent of the control reactors, indicate that there is an apparent greater loss of contaminant from the bioreactor than the control reactor. This loss is generally attributed to the actions of the bivalves.

CONCLUSIONS

Axenotrophy represents a generalized, facultative condition of Bivalvia where holozoic feeding is complemented or replaced by absorption. The ability to absorb dissolved organic and inorganic compounds may be ubiquitous to Bivalvia; indeed, the experiments presented herein, regarding the freshwater bivalve, *C. fluminea,* indicate that this organism can absorb at least two monosubstituted phenyl compounds with limited nocent physical effects at low concentrations and limited exposure.

The ability to satisfy or to supplement nutritional requirements through the uptake of organic compounds affords clams an unique ability to survive under a wide variety of conditions; 10% of the extant families of mollusks have been documented to absorb dissolved organic compounds. Smaller, saturated compounds are more readily converted to water-soluble molecules and oxidized by the mixed function oxidase system than by the polycyclic aromatic hydrocarbons (PAH), which tend to be partitioned to the lipid tissue (Wolf 1982). PAH are more lipophilic and hydrophobic, and the cost for catabolization is high.

Gilewicz et al. (1984) examined marine bivalves exposed to Arabian light crude, a heterogeneous liquid composed of light to heavy paraffins and PAH, and found no light paraffins (<C_{15}; representing 35% of the original paraffin concentration) and recorded a loss of 35% of the aromatic hydrocarbons (primarily phenylic compounds). The authors attribute the loss to sampling; however, these molecules, known to be readily biometabolized (Armstrong et al. 1991, Nelson & Montgomery 1988, Worsey & Williams 1975), may have been mineralized by the bivalve.

Although nocent compounds stress organisms; limited exposure may not be lethal. Increased concentrations can be lethal but if maintained at or below transient levels for limited periods can be tolerated (Figure 2). Under long-term, moderate exposure, the organism may survive due to its ability to catabolize the organic compound (i.e., remove the contaminant from the system). If the contaminant is catabolized and lost from the system, the organism can remain exposed to the contaminant in a transient phase without immediate mortality. However, if the expenditure of energy for catabolization is too high the organism will ultimately slip into the final stage (i.e., death).

Unfortunately, during the course of metabolic transformations, reactive electrophilic intermediates may form that are more toxic, mutagenic, or carcinogenic than the parent compound (Livingstone 1985). For example, benzo[a]pyrene may be tolerated at certain concentrations but one of its metabolites, 7,8 diol-9,10-epoxide, is very carcinogenic at very reduced rates and is more deadly than the parent compound (Anderson 1978). Chlorinated aliphatic hydrocarbons are readily catabolized by chemospecific microbes and produce the more toxic and carcinogenic compound, vinyl chloride (Barcelona et al. 1990). The generation of a more toxic daughter compound or potentiation from the two could shift the organism from the transient phase to death.

Bivalves that display axenic properties may be useful in the remediation of groundwater contaminated with low levels of xenobiotic organic compounds. Pump-and-treat systems that employ bacteria to remove xenobiotic contamination

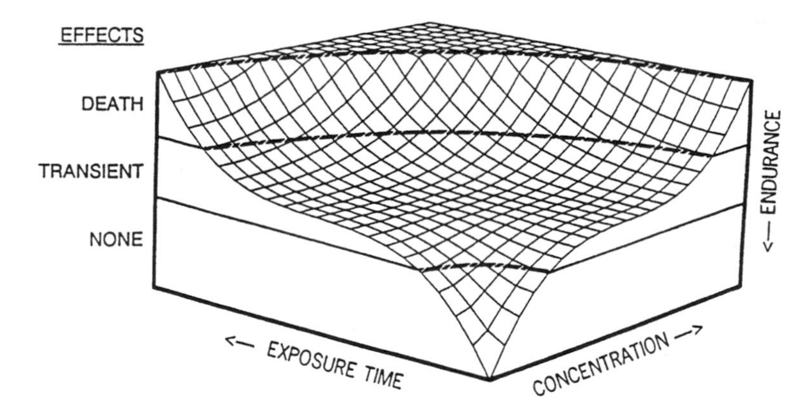

FIGURE 2. Three-dimensional plane of xenobiotic endurance, correlating time of chemical exposure, concentration, and endurance to physical effects on the organism.

eventually become ineffective as the concentration of xenobiotic contaminant in the abstracted groundwater decreases. Consequently, alternative strategies must be employed and usually at greater expense. Bivalves with the capacity to catabolize specific organic compounds may be effective in the remediation of these compounds. Research continues to quantify the clam's ability to catabolized organic compounds and to determine the potential as a contaminated-groundwater treatment strategy.

REFERENCES

Anderson, J. W., and W. B. Bedford. 1973. "The physiological response of the estuarine clam, *Rangia cuneata* (Gray), to salinity. II. Uptake of glycine." *Biol. Bull. 144*(2): 229-247.

Anderson, R. G. W., B. A. Kamen, K. G. Rothberg, and S. W. Lacey. 1992. "Potocytosis: Sequestration and transport of small molecules by caveolae." *Science 255*: 410-411.

Anderson, R. S. 1978. *Benzo[a]pyrene metabolism in the American oyster Crassostrea virginica.* U.S. Environmental Protection Agency Ecol. Res. Ser. Monogr. EPA-600/3-78-009. 18 pp.

Armstrong, A. Q., R. E. Hodson, H. M. Hwang, and D. L. Lewis. 1991. "Environmental factors affecting toluene degradation in groundwater at a hazardous waste site." *Journal of Environmental Engineering 10*: 147-158.

Bamford, D. R., and E. Campbell. 1976. "The effect of environmental factors on the absorption of L-phenylalanine by the gill of *Mytilus edulis*." *Comp. Biochem. Physiol. 53A*: 295-299.

Bamford, D. R., and R. Gingles. 1974. "Absorption of sugars in the gill of the Japanese oyster, *Crassostrea gigas*." *Comp. Biochem. Physiol. 49A*: 637-646.

Bamford, D. R., and R. McCrea. 1975. "Active absorption of neutral and basic amino acids by the gill of the common cockle, *Cerastoderma edule*." *Comp. Biochem. Physiol. 50A*: 811-817.

Barcelona, M., A. Wehrmann, J. F. Keely, and W. A. Pettyjohn. 1990. *Contamination of Ground Water: Prevention, Assessment, Restoration.* Noyes Data Corporation, Park Ridge, NJ.

Bayne, B. L. 1989. "Measuring the biological effects of pollution: The Mussel Watch approach." *Water Sci. Tech. 21*: 1089-1100.

Bayne, B. L., M. N. Moore, J. Widdows, D. R. Livingstone, and P. Salkeld. 1979. "Measurement of the responses of individuals to environmental stress and pollution: Studies with bivalve molluscs." *Phil. Trans. R. Soc. Lond. B 286*: 563-581.

Benner, R., J. D. Pakulski, M. McCarthy, J. I. Hedges, and P. G. Hatcher. 1992. "Bulk chemical characteristics of dissolved organic matter in the ocean." *Science 255*: 1561-1564.

Bevelander, G., and H. Nakahara. 1966. "Correlation of lysomal activity and ingestion by mantle epithelium." *Biol. Bull. 131*: 76-82.

Birger, T. I., A. Y. Malarevskaja, O. M. Arsan, V. D. Solomatina, and Y. M. Gupalo. 1978. "Physiological aspects of adaptions of mollusks to abiotic and biotic factors due to blue-green algae." *Malacol. Rev. 11*: 100-102.

Birger, T. I., and A. Y. Malarevskaja. 1977. "Toxicology and radioecology of water: Some biochemical mechanisms involved in the resistance of invertebrates to toxic substances." *Hydrobio. J. 13(6)*: 58-61.

Bishop, S. H., L. L. Ellis, and J. M. Burcham. 1983. "Amino acid metabolism in molluscs." In P. W. Hochachka (Ed.), *The Mollusca, Volume 1: Metabolic Biochemistry and Molecular Biomechanics*, pp. 243-327. Academic Press, Inc., New York, NY.

Bunde, T. E., and M. Fried. 1978. "The uptake of dissolved free fatty acids from the seawater by a marine filter feeder, *Crassostrea virginica*." *Comp. Biochem. Physiol. 60A*: 139-144.

Buttner, J. K., and R. C. Heidinger. 1981. "Rate of filtration in the asiatic clam, *Corbicula fluminea*." *Trans. Ill. State Acad. Sci. XX*: 13-17.

Cantelmo-Cristini, A., F. E. Hospod, and R. J. Lazell. 1985. "An in situ study on the adenylate energy charge of *Corbicula fluminea* in a freshwater system." In F. J. Vernberg, F. P. Thurberg, A. Calabrese and W. B. Vernberg (Eds.), *Marine Pollution & Physiology: Recent Advances*, pp. 83-106. Univ. of South Carolina Press, SC.

Collier, A., S. M. Ray, A. W. Magnitzky, and J. O. Bell. 1953. "Effect of dissolved organic substances on oysters." *Fishery Bulletin 54*: 167-185.

Crowe, J. H., K. A. Dickson, J. L. Otto, R. D. Colón, and K. K. Farley. 1977. "Uptake of amino acids by the mussel *Modiolus modiolus*." *J. Exp. Zool. 202*: 323-332.

de Zwaan, A., and T. C. M. Wijsman. 1976. "Anaerobic metabolism in Bivalvia (Mollusca): Characteristics of anaerobic metabolism." *Comp. Biochem. Physiol. 54B*: 313-324.

Dupaul, W. D., and K. L. Webb. 1971. "Free amino acid accumulation in isolated gill tissue of *Mya arenaria*." *Arch. 1 Internat. Physiol. Biochem. 79*: 327-336.

Ernst, W. 1979. "Metabolic transformation of (1-^{14}C) naphthol in bioconcentration studies with the common mussel *Mytilus edulis*." *Veröff. Inst. Meeresforsch. Bremerhaven. 17*: 233-240.

Fankboner, P. V. 1971. "Intercellular digestion of symbiontic zooxanthellae by host amoebocytes in giant clams (Bivalvia: Tridacnidae), with a note on the nutritional role of the hypertrophied siphonal epidermis." *Biol. Bull. 141*: 222-234.

Gainey, L. F., Jr. 1978. "The response of the corbiculidae (Mollusca: Bivalvia) to osmotic stress: The cellular response." *Physiol. Zool. 51*: 79-91.

Gilewicz, M., J. R. Guillaume, D. Carles, M. Leveau, and J. C. Bertrand. 1984. "The effects of petroleum hydrocarbons on the cytochrome P_{450} content of the mollusc bivalve *Mytilus galloprovincialis*." *Mar. Biol. 80*: 155-159.

Gilfillan, E. S., D. S. Page, D. Vallas, L. Gonzalez, E. Pendergast, J. C. Foster, and S. A. Hanson. 1985. "Relationship between glucose-6-phosphate dehydrogenase and aspartate amino transferase activities, scope for growth and body burden of Ag, Cd, Cu, Cr, Pb and Zn in populations of *Mytilus edulis* from a polluted estuary." In F. J. Vernberg, F. P. Thurberg, A. Calabrese, and W. B. Vernberg (Eds.), *Marine Pollution & Physiology: Recent Advances*, pp. 107-124. University of South Carolina Press, SC.

Goldberg, E. D. 1986. "The mussel watch concept." *Envir. Monit. Assess. 7*: 91-103.

Gupta, A. S. 1977. "Observations on the gill of *Viviparus bengalensis* in relation to calcium uptake and storage." *Acta Zool.* (Stockh.) *58*: 129-133.

Haines, K. C. 1979. "The use of *Corbicula* as a clarifying agent in experimental tertiary sewage treatment process on St. Croix, US. Virgin Island." In J. C. Britton, J. S. Mattice, C. E. Murphy, and L. W. Newland (Eds.), *Proceedings of the First International Corbicula Symposium*, pp. 165-175. American Malacological Bulletin, Special Edition No. 1. Texas Christian University, Fort Worth, TX.

Hartley, D. M., and J. B. Johnston. 1983. "Use of the freshwater clam *Corbicula manilensis* as a monitor for organochlorine pesticides." *Bull. Environ. Contam. Toxicol. 31*: 33-40.

Herve, S. 1991. "Mussel incubation method for monitoring organochlorine compounds in freshwater recipients of pulp and paper industry." Ph.D. Dissertation, University of Jyväskylä, Jyväskylä, Finland.

Hutchinson, P. J. 1992. "Detection and remediation of aqueous-phased xenobiotic contamination mediated by *Corbicula fluminea*." Ph. D. Dissertation, University of Pittsburgh, Pittsburgh, PA.

Hutchinson, P. J., H. B. Rollins, and R. Prezant. 1993a. "Detection of xenophobic response in the periostracum of *Corbicula fluminea* through laser-induced mass spectrometry." *Arch. Environ. Contam. Toxicol. 24*: 258-267.

Hutchinson, P. J., H. B. Rollins, R. Prezant, J. Sharkey, Y. Kim, and D. Hercules. 1993b. "A freshwater bioprobe: Periostracum of the Asian clam, *Corbicula fluminea* (Müller) combined with laser microprobe mass spectrometer." *Environ. Poll. 78*: 95-100.

Jakoby, W. B., J. R. Bend, and J. Caldwell. 1982. "Introduction." In W. B. Jakoby, J. R. Bend, and J. Caldwell (Eds.), *Metabolic Basis of Detoxication: Metabolism of Functional Groups*, pp. 1-3. Academic Press, New York, NY.

Jones, D. S., and D. K. Jacobs. 1992. "Photosymbiosis in *Clinocardium nuttalli*: Implications for tests of photosymbiosis in fossil molluscs." *Palaios 7*: 86-95.

Kahn, M. A. Q., A. Kamal, R. J. Wolin, and J. Runnels. 1972. "In vivo and in vitro epoxidation of aldrin by aquatic food chain organisms." *Bull. Environ. Contam. Toxicol. 8*: 219-228.

Krieger, R. I., S. J. Gee, L. O. Lim, J. H. Ross, A. Wilson, C. Alpers, and S. R. Wellings. 1979. "Disposition of toxic substances in mussels (*Mytilus califorianus* [sic]): Preliminary metabolic and histologic studies." In M. A. Q. Khan, J. J. Lech, and J. J. Menn (Eds.), *Pesticide and Xenobiotic Metabolism in Aquatic Organisms*, pp. 259-277. Amer. Chem. Soc. Symp. Ser., *99*, Washington, DC.

Leard, R. L., B. J. Grantham, and G. F. Pessoney. 1980. "Use of selected freshwater bivalves for monitoring organochlorine pesticide residues in major Mississippi stream systems, 1972-73." *Pest. Monitor. J. 14(2)*: 47-52.

Livingstone, D. R. 1985. "Responses of the detoxication/toxication enzyme systems of molluscs to organic pollutants and xenobiotics." *Marine Poll. Bull. 16(4)*: 158-164.

Livingstone, D. R., M. N. Moore, D. M. Lowe, C. Nasci, and S. V. Farrar. 1985. "Responses of the cytochrome *P*-450 monooxygenase system to diesel oil in the common mussel, *Mytilus edulis* and the periwinkle, *Littorina littorea* L." *Aquat. Toxicol. 7*: 79-91.

Manahan, D. T., G. C. Wright, G. C. Stephens, and M. A. Rice. 1982. "Transport of dissolved amino acids by the mussel, *Mytilus edulis*: Demonstration of net uptake from sea water by HPLC analysis." *Science 215*: 1253-1255.

Mathews, H. B. 1982. "Aryl Halides." In W. B. Jakoby, J. R. Bend, and J. Caldwell (Eds.), *Metabolic Basis of Detoxication: Metabolism of Functional Groups*, pp. 51-68. Academic Press, New York, NY.

McCrea, S. R. 1976. "Comparative studies of amino acid absorption in bivalve gill in relation to environmental factors." Ph. D. Dissertation, The Queen's University of Belfast, Belfast, Ireland.

Moore, M. N., D. R. Livingstone, J. Widdows, D. M. Lowe, and R. K. Pipe. 1987. "Molecular, cellular and physiological effects of oil-derived hydrocarbons on molluscs and their use in impact assessment." *Phil. Trans. R. Soc. Lond. B316*: 603-623.

Nakahara, H., and G. Bevelander. 1967. "Ingestion of particulate matter by the outer surface of the mollusc mantle." *J. Morph. 122(2)*: 139-146.

Nelson, M. J. K., and S. O. Montgomery. 1988. "Trichloroethylene metabolism by microorganisms that degrade aromatic compounds." *Appl. Environ. Microb. 54(2)*: 604-606.

Owen, G. 1974. "Feeding and digestion in the Bivalvia." In O. Lowenstein (Ed.), *Advances in Comparative Physiology and Biochemistry*, pp. 1-35. Academic Press, New York, NY.

Patel, B., and J. T. Eapen. 1989a. "Biochemical evaluation of naphthalene intoxication in the tropical acrid blood clam *Anadara granosa*." *Mar. Biol. 103*: 203-209.

Patel, B., and J. T. Eapen. 1989b. "Physiological evaluation of naphthalene intoxication in the tropical acrid clam *Anadara granosa*." *Mar. Biol. 103*: 193-202.

Péquignat, E. 1973. "A kinetic and autoradiographic study of the direct assimilation of amino acids and glucose by organs of the mussel *Mytilus edulis*." *Mar. Biol. 19*: 227-244.

Pütter, A. 1909. *Die Ernährung des Wassertieres und der Stoffhaushalt der Gewässer*. Gustav Fischer, Jena, Germany.

Reid, R. G. B. 1980. "Aspects of the biology of a gutless species of *Solemya* (Bivalvia: Protobranchia)." *Can. J. Zool. 58*: 386-393.

Reid, R. G. B., and F. R. Bernard. 1980. "Gutless bivalves." *Science 208*: 609-610.

Rice, M. A., K. Wallis, and G. C. Stephens. 1980. "Influx and net flux of amino acids into larval and juvenile flat oysters, *Ostrea edulis* (L.)." *J. Exp. Mar. Biol. Ecol. 48*: 51-59.

Riley, R. T., M. C. Mix, R. L. Schaffer, and D. L. Bunting. 1981. "Uptake and accumulation of naphthalene by oyster *Ostrea edulis*, in a flowthrough system." *Mar. Biol. 61*: 267-276.

Seibers, D., and A. Winkler. 1984. "Amino acid uptake by mussels, *Mytilus edulis*, from natural seawater in a flowthrough system." *Helgol. Meeresunters. 38*: 189-199.

Sindermann, C. J. 1985. "Keynote address: Notes of a pollution watcher." In F. J. Vernberg, F. P. Thurberg, A. Calabrese, and W. B. Vernberg (Eds.), *Marine Pollution & Physiology: Recent Advances*, pp. 11-30. Univ. of South Carolina Press, SC.

Somero, G. N., and R. D. Bowlus. 1983. "Osmolytes and metabolic end products of molluscs: The design of compatible solute systems." In P. W. Hochachka (Ed.), *The Mollusca, Volume 2: Environmental Biochemistry and Physiology*, pp. 77-100. Academic Press, New York, NY.

Stegeman, J. J. 1981. "Polynuclear aromatic hydrocarbons and their metabolism in the marine environment." In H. V. Gelboin, and P. O. P. Ts'o (Eds.), *Polycyclic Hydrocarbons and Cancer, Volume 3*, pp. 1-60. Academic Press, Inc., New York, NY.

Stegeman, J. J., and J. M. Teal. 1973. "Accumulation, release and retention of petroleum hydrocarbons by the oyster *Crassostrea virginica*." *Mar. Biol. 22*: 37-44.

Stewart, M. G. 1978. "Kinetics of neutral amino-acid transport by isolated gill tissue of the bivalve *Mya arenaria* (L.)." *J. Exp. Mar. Biol. Ecol. 32*: 39-52.

Stewart, M. G. 1979. "Absorption of dissolved organic nutrients by marine invertebrates." *Oceanogr. Mar. Biol. Ann. Rev. 17*: 163-192.

Stewart, M. G., and D. R. Bamford. 1975. "Kinetics of alanine uptake by the gills of the soft shelled clam *Mya arenaria*." *Comp. Biochem. Physiol. 52A*: 67-74.

Stewart, M. G., and D. R. Bamford. 1976. "The effect of environmental factors on the absorption of amino acids by isolated gill tissue of the bivalve, *Mya arenaria* (L.)." *J. Exp. Mar. Biol. Ecol. 24*: 205-212.

Stewart, M. G., and R. C. Dean. 1980. "Uptake and utilization of amino acids by the shipworm *Bankia gouldi*." *Comp. Biochem. Physiol. 66B*: 443-450.

Swinehart, J. H., J. H. Crowe, A. P. Giannini, and D. A. Rosenbaum. 1980. "Effect of divalent cations on amino acid and divalent cation fluxes in gills of the bivalve mollusc, *Mytilus californianus*." *J. Exp. Zool. 212*: 389-396.

Todd, D. K., R. M. Tinlin, K. D. Schmidt, and L. G. Everett. 1976. *Monitoring ground-water quality: Monitoring methodology*. U.S. Environmental Protection Agency, EPA-600/4-76-026, Las Vegas, NV.

U.S. Environmental Protection Agency. 1979. *Handbook for Analytical Quality Control in Water and Wastewater Laboratories*. U.S. Environmental Protection Agency - EMSL, EPA-600/4-79-19, Cincinnati, OH.

Wolf, C. R. 1982. "Oxidation of foreign compounds at carbon atoms." In W. B. Jakoby, J. R. Bend, and J. Caldwell (Eds.), *Metabolic Basis of Detoxication: Metabolism of Functional Groups*, pp. 5-28. Academic Press, New York, NY.

Worsey, M. J., and P. A. Williams. 1975. "Metabolism of toluene and xylenes by *Pseudomonas putida* (*arvilla*) mt-2: Evidence for a new function of the TOL plasmid." *J. Bacteriology 124*(1): 7-13.

Wright, S. H. 1979. "Effect of activity of lateral cilia on transport of amino acids in gills of *Mytilus californianus*." *J. Exp. Zool. 209*: 209-220.

Wright, S. H. 1982. "A nutritional role for amino acid transport in filter-feeding marine invertebrates." *Amer. Zool. 22*: 621-634.

Wright, S. H., S. A. Becker, and G. C. Stephens. 1980. "Influence of temperature and unstirred layers on the kinetics of glycine transport in isolated gills of *Mytilus californianus*." *J. Exp. Zool. 214*: 27-35.

Wright, S. H., T. L. Johnson, and J. H. Crowe. 1975. "Transport of amino acids by isolated gills of the mussel *Mytilus californianus* Conrad." *J. Exp. Biol. 62*: 313-325.

Wright, S. H., and G. C. Stephens. 1977. "Characteristics of influx and net flux of amino acids in *Mytilus californianus*." *Biol. Bull. 152*: 295-310.

Wright, S. H., and G. C. Stephens. 1978. "Removal of amino acid during a single passage of water across the gill of marine mussels." *J. Exp. Zool. 205*: 337-352.

HYDRAULIC FRACTURING TO ENHANCE IN SITU BIORECLAMATION OF SUBSURFACE SOILS

S. J. Vesper, M. Narayanaswamy, L. C. Murdoch, and W. J. Davis-Hoover

ABSTRACT

Hydraulic fracturing is a technique that creates permeable channelways for gaining access to low permeability subsurface soils. We have coupled hydraulic fracturing with in situ bioremediation to determine if contaminants can be degraded when they occur in fine-grained, overconsolidated soils. We have compared the performance of a well connected to a system of four hydraulic fractures to a conventional well during injection of hydrogen peroxide and nutrients to subsurface soils contaminated with gasoline. Increases in soil moisture, metabolic activity (measured by hydrolysis of fluorescein diacetate), and population of hydrocarbon-degrading microorganisms were observed in the vicinity of some of the fractures. No significant changes were observed in the vicinity of the conventional well. Hydrocarbon concentrations decreased in the soils around the hydraulically fractured well, whereas they remained nearly the same in the vicinity of the conventional well.

INTRODUCTION

Access to subsurface soil contamination is critical to the performance of any in situ remedial technology. This is especially true in low-permeability soils such as overconsolidated clays and silt. Hydraulic fracturing is a technique we developed to provide access to contaminated subsurface soils. Briefly, hydraulic fracturing (Murdoch et al. 1991; Davis-Hoover et al. 1991) involves injecting a sand-gel slurry into a borehole until the soil fractures and a pancake-shape layer is created. Several fractures approximately 0.5 cm thick and 10 or more meters in maximum dimension, can be created with as little as 20 cm of vertical spacing between.

Bioremediation uses microorganisms to degrade pollutants (Atlas & Pramer 1990) and in situ bioremediation is gaining in popularity because of its advantages over other remediation methods (Hinchee et al. 1991). Most organic contaminants are biodegradable under the appropriate conditions, but oxygen is often limiting (Floodgate 1984). The introduction of hydrogen peroxide is a common way to add oxygen to enhance in situ bioremediation (Riss & Schweisfurth 1985).

The Dayton Bioremediation site, a former fuel distribution and storage facility in Dayton, Ohio has subsurface soils contaminated with hydrocarbons caused by a leaking underground storage tank. The soil is silty-clay till, with low hydraulic conductivity. Bioremediation was selected by the on-site contractor to remediate the contamination. This paper will discuss some of the results from this research effort in which the bioremediation is enhanced by the addition of hydrogen peroxide and soluble nutrients through a well connected to four hydraulic fractures as compared to a conventional well.

MATERIALS AND METHODS

Site Description

The site, which is in Dayton, Ohio, is about 40 m by 40 m (Figure 1) with the main area of contamination about 10 m by 20 m east of the pit where the tanks were located. The surface of the site is covered with about 1 m of gravel. The subsurface soil is basically silty clay deposits (Figure 2). The ground slopes gently to the south-east. Limestone bedrock is at approximately 6 m depth and groundwater was absent from the till.

System Operation and Well Installation

In the area designated the hydraulic fracture well (FW), four hydraulic fractures were installed at 1.2 m, 1.8 m, 2.4 m, and 3 m depth. Each fracture received 270 to 400 kg of sand and were designed to be 5 to 6 m in diameter. A conventional well (CW) was installed 10 m north of FW. This well consisted of a 1.3-cm PVC pipe grouted in a 15-cm diameter boring to 4.6 m depth, which was filled with sand from 4.6 to 1.5-m. A capture trench was installed south and east of the wells with a return pump to recover water flowing from the site.

A remediation system was installed during the last week of November 1991 to deliver hydrogen peroxide (as the oxygen source), nutrients, and water by gravity flow from a tank located in a trailer on site. The system was designed, installed, and operated by the Foppe Thelen Group Inc., Cincinnati, Ohio. The details of the system and its operation are proprietary. The system commenced operation on December 6, 1991 (Day 0).

There were three periods of operation which were defined by the regularity of the system operation. System operation was intermittent during Period I (day 0 to 80) due to frozen pipes. It was in regular operation during Period II (day 81 to 202), and it was mostly inoperational during Period III (day 202 to 278). A flow totalizer was used to determined the total amount of injected liquid.

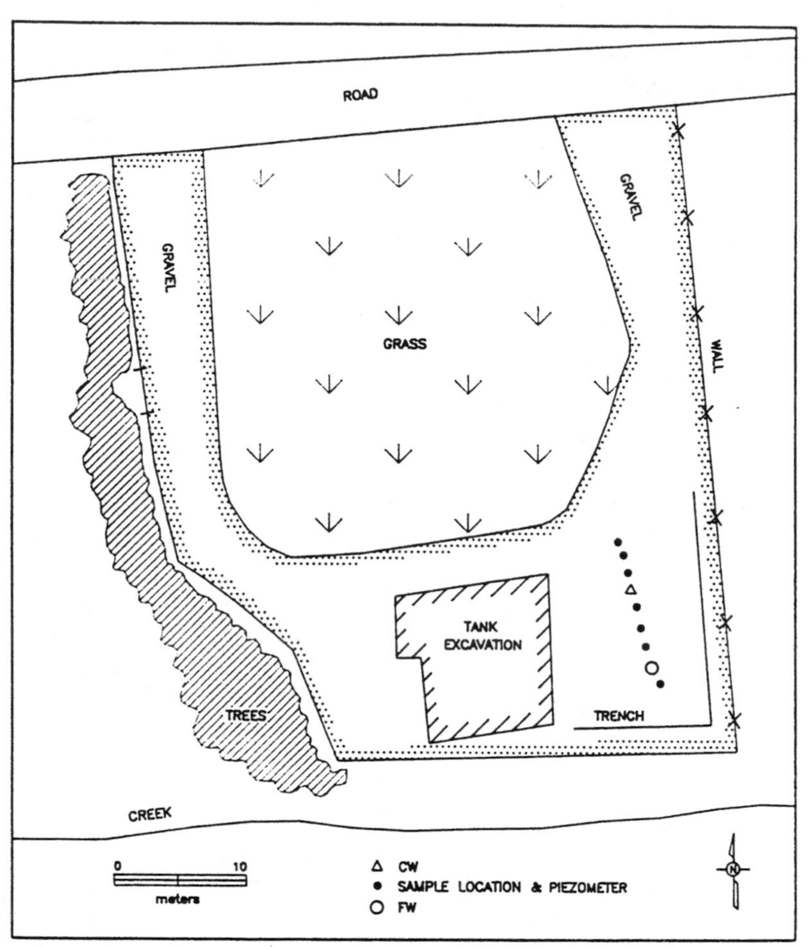

FIGURE 1. Map of Dayton bioremediation site.

Site Monitoring

Monitoring piezometers were installed at 1.5, 3, and 4.5 m north of each injection well, and 3 m south of FW. The monitoring piezometers around FW consisted of 4, 2.5 cm diameter PVC pipes positioned at the depth of each of the fractures but separated by a grout layer from each other. Around CW, three selected depths (1.5, 3, and 4 m) were chosen to install similar clusters of monitoring piezometers. Piezometers were monitored either by determining the elevation of water in the piezometers or, when they were full, by sealing the piezometers and measuring pressure head.

Injection rates were so low at this site that we were unable to find a flow meter that could accurately measure them. Therefore injection rates into the wells

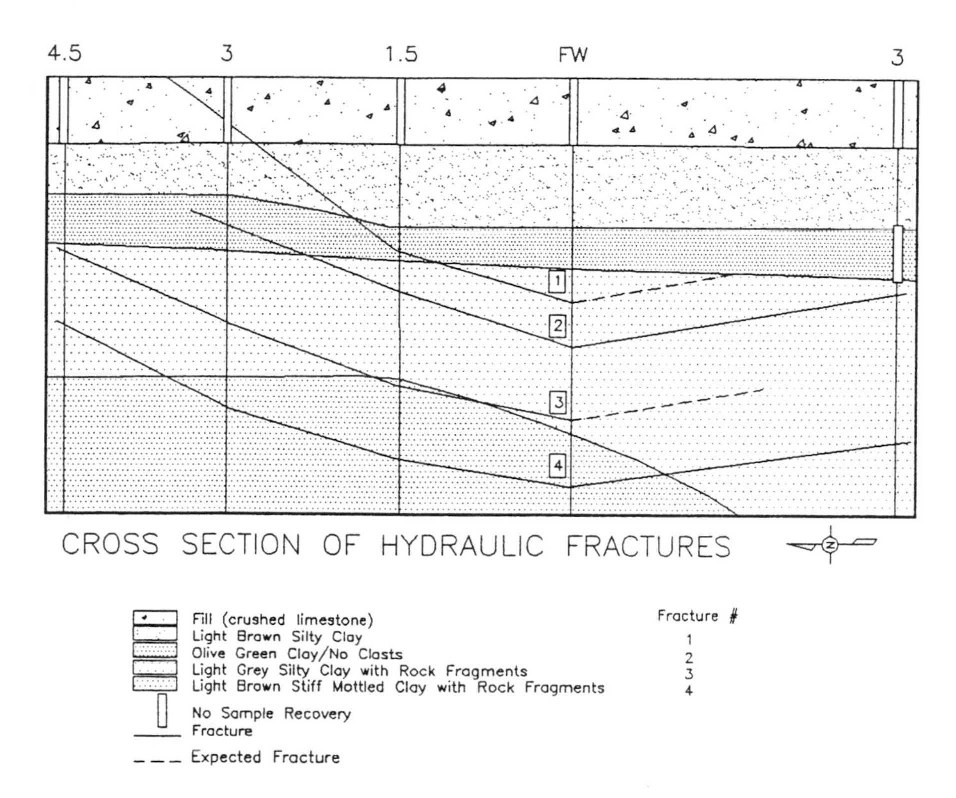

FIGURE 2. Cross section of Dayton bioremediation site showing stratigraphy and hydraulic fractures.

were determined by measuring the velocity of a tracer dye injected into the 2 m long transparent hose at the entrance to the well. Laboratory experiments were conducted to determine the relationship between the flowrate, determined using the dye method, and the average flowrate, as measured by collecting the water in a measuring cylinder in a given time. We were thus able to calibrate the flow measuring system.

Soil Sampling and Analyses

Soil samples were taken at the site by a commercial driller using a conventional split-spoon, 5-cm in diameter by 60-cm long. The initial samples were taken during the installation of the monitoring points (Figure 1). The second samples were taken on February 18, 1992 (Day 75), 30 cm due west of the first sample. The third sample was taken on July 24, 1992 (Day 233), 30 cm due north of the first sample. The cores were wrapped in plastic wrap, then sealed in plastic bags and returned to the laboratory under refrigerated conditions.

By examining the cores, the soil stratigraphy was determined and the position of each fracture in each core was mapped. In the laboratory, the cores were cut into 2.5-cm sections and subjected to a variety of analyses. The positions of the fractures were noted as they related to the samples. The uppermost section, 0 to 2.5 cm along each core, was taken for microbial population enumeration. This was done by aseptically shaving the outer layer from the section, then removing 1 g of soil and placing it in a 12-mL sterile tube with 5-ml of sterile water. The soil plus water was then sonicated for 5 min at 90% of power with the Heat Systems model W375 sonicator to release the microorganisms. To enumerate the total microbial population, the sonicate was plated on R2A medium using a Spiral Platter (Spiral Systems Inc.). The plates were incubated at 29°C until the number of colonies on the plate could be counted (usually in 48 hr). To enumerate the hydrocarbon degrader population, the sonicate was plated on a carbon-free salts medium (SM) (Dang et al. 1989) placed in sealed desiccators containing gasoline fumes. The colonies were counted after about 1 week.

The next lower section, from 2.5 to 5 cm, of the core was taken for moisture content (weight water/weight solids) analysis. The wet soil was weighed and then placed in a drying oven at 110°C. The dry soil was weighed, and moisture content determined by weight. The next lower core section, from 5 to 7.5 cm, was taken for fluorescein diacetate (FDA) hydrolysis analysis (Schnurer & Rosswall 1985), as a measure of microbial activity. This section was again aseptically shaved to remove the outer layer. Then 1 gram of the inner core was placed in a 125-mL sterile dilution bottle containing 100 mL of a 60-mM sodium phosphate buffer, pH 7.6, plus 0.5 mL of the stock FDA solution. The stock FDA solution was made by adding 2 mg of FDA (Sigma Chemical Co.) per mL of acetone. The bottles were mixed on a reciprocating shaker set at about 50 shakes per min. The incubation was done in a controlled-temperature room (Labline) at 12°C for 24 hr. After the incubation, samples were centrifuged in a microcentrifuge (Eppendorf) at 13,000 rpm for 1 min. The supernatant was then monitored at 490 nm using a Perkin Elmer spectrophotometer. The shavings from the sections were analyzed to determine the soil pH by adding 5 g of soil to 10 mL of deionized water, mixing for one-half hour, then letting this settle. The pH was measured with a soil pH probe (Orion) using a Corning pH meter.

The sequence of taking sample sections for each test was repeated throughout the entire length of each split spoon sample (except for the section taken for chemical analysis-see below). Thus six to eight samples could be taken for each split spoon. For this report, we have identified the three or four sections nearest to the fracture (N) and the three or four samples from between fractures (B).

For the statistical analysis of these results, the means of the moisture levels, relative microbial activity measurements, and microbial populations between FW and CW and between the initial and later samples, and the differences between soils near fractures (N) versus between fracture (B) were compared using a Student T-test with a 95% level of confidence (Tables 1, 2, 3, and 4). All the means are statistically different unless followed by the same letter (Gilbert 1987).

TABLE 1. Soil moisture content (%), microbial activity (OD 490), and hydrocarbon degrader population (cfu/mL) in the vicinity of FW and CW at 1.5 m from the injection well with time; uppermost fracture compared to similar control depth. The average (AVG) mean[a] from the soils is compared to soils near (N) the fracture and between (B) the fracture.

Well Type	FW (Frac. No. 1)					CW		
Sampling Time	Day 0	Day 75		Day 233		Day 0	Day 75	Day 233
Location	AVG	N	B	N	B	AVG	AVG	AVG
Moisture	10.5 A	31.2	22.1 B	22.3 B	18.1 B	10.1 A	11.5 A	10.9 A
Activity	.220 A	.269	.103 B	.191 A	.141 B	.102 B	.092 B	.020
Degraders	3.1×10^4 A	1.5×10^6 B	2.0×10^6 B	9.8×10^5 B	5.2×10^5 B	2.6×10^3 A	2.7×10^6 B	6.0×10^3 A

(a) All means are significantly different at the 95% level (Student T-test) unless followed by the same letter.

TABLE 2. Soil moisture content (%), microbial activity (OD 490), and hydrocarbon degrader population (cfu/g) in the vicinity of FW and CW at 1.5 m from the injection well; second fracture compared to similar control depth. The average (AVG) mean[a] from the soils is compared to soils near (N) the fracture and between (B) the fracture.

Well Type	FW (Frac. No. 2)					CW		
Sampling Time	Day 0	Day 75		Day 233		Day 0	Day 75	Day 233
Location	AVG	N	B	N	B	AVG	AVG	AVG
Moisture	10.5 A	19.5 B	14.6 A	20.0 B	11.9 A	10.0 A	12.1 A	10.0 A
Activity	.200	.145 A	.139 A	.145 A	.069 B	.102 B	.087 B	.022
Degraders	3.1×10^4 A	1.0×10^6 B	1.4×10^6 B	2.3×10^5 B	2.7×10^5 B	2.6×10^3 A	2.5×10^4 A	5.7×10^3 A

(a) All means are significantly different at the 95% level (Student T-test) unless followed by the same letter.

TABLE 3. Soil moisture content (%), microbial activity (OD 490), and hydrocarbon degrader population (cfu/g) in the vicinity of FW and CW at 1.5 m from the injection well; third fracture compared to similar control depth. The average (AVG) mean[a] from the soils is compared to soils near (N) the fracture and between (B) the fracture.

Well Type	FW (Frac. No. 3)					CW		
Sampling Time	Day 0	Day 75		Day 233		Day 0	Day 75	Day 233
Location	AVG	N	B	N	B	AVG	AVG	AVG
Moisture	10.5 A	21.8	12.0 A	10.8 A	11.1 A	10.1 A	12.0 A	10.1 A
Activity	.169	.223	.028 A	.039 A	.042 A	.102 B	.092 B	.020 A
Degraders	5.8×10^4 A	3.0×10^6	1.6×10^5 B	1.4×10^5 B	3.1×10^3 A	4.8×10^4 A	6.3×10^4 A	9.0×10^3 A

(a) All means are significantly different at the 95% (Student T-test) level unless followed by the same letter.

TABLE 4. Soil moisture content (%), microbial activity (OD 490), and hydrocarbon degrader population (cfu/g) in the vicinity of FW and CW at 1.5 m from the injection well; lowest fracture compared to similar control depth. The average (AVG) mean[a] from the soils is compared to soils near (N) the fracture and between (B) the fracture.

Well Type	FW (Frac. No. 4)					CW		
Sampling Time	Day 0	Day 75		Day 233		Day 0	Day 75	Day 233
Location	AVG	N	B	N	B	AVG	AVG	AVG
Moisture	10.5 A	12.2 A	11.1 A	11.3 A	11.4 A	10.1 A	11.5 A	9.5 A
Activity	.169 A	.059 BC	.036 B	.040 BC	.027 B	.135 A	.077 C	.065 BC
Degraders	5.8×10^4 A	4.7×10^5 A	1.4×10^5 A	1.5×10^5 A	1.5×10^4 A	4.8×10^4 A	6.3×10^4 A	3.9×10^4 A

(a) All means are significantly different at the 95% level (Student T-test) unless followed by the same letter.

Chemical Analyses

The cores taken for this investigation were analyzed for either benzene, toluene, and ethylbenzene (BTE fractions), or total petroleum hydrocarbons (TPH). The samples taken for BTE quantification were obtained by taking the lower 15 cm from each split-spoon sample and placing it quickly into 300 mL of methanol in U.S. Environmental Protection Agency-type jars. The samples were then tested according to EPA Standard Method SW-846.

The TPH samples were taken in the laboratory during the microbial analysis preparation. The core sections were shaved to remove the outer layer of soil, and the residuals from each core were combined and frozen at −50°C. The TPH level for each sample was determined using EPA Method 45.1.

RESULTS

Site Operation

The fractures had trajectories that climbed gently toward the surface (Figure 2). During the three periods of operation, the flow to the site varied (Figure 3). The rate of injection into FW varied from 2.5 L/min to 4.6 L/min and into the from 0.02 to 0.08 L/min. This resulted in 619 m^3 of water being injected into FW compared to 8 m^3 for CW.

The head pressures at the wellhead, as per the design, were similar at both FW and CW (Figure 4). However, at the pressure monitoring points away from the wells, the values in the vicinity of CW were always less than values observed on some days at FW. The overall head pressure distribution indicates higher head values in the vicinity of FW than CW. Figure 5 gives the head distribution in the vicinity of wells on a typical day (Day 201).

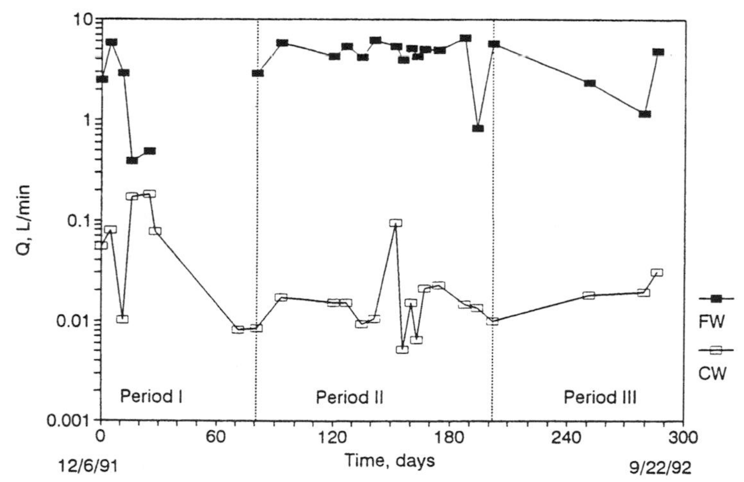

FIGURE 3. Flowrates into FW and CW during the remediation program.

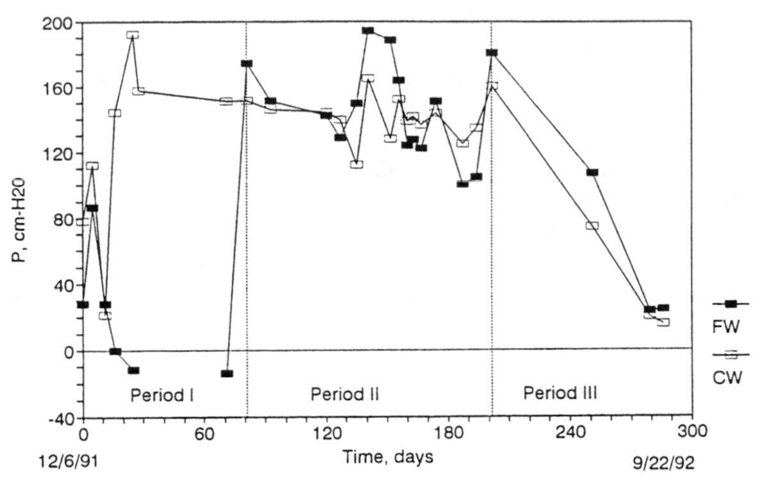

FIGURE 4. Pressures at FW and CW wellheads during the remediation program.

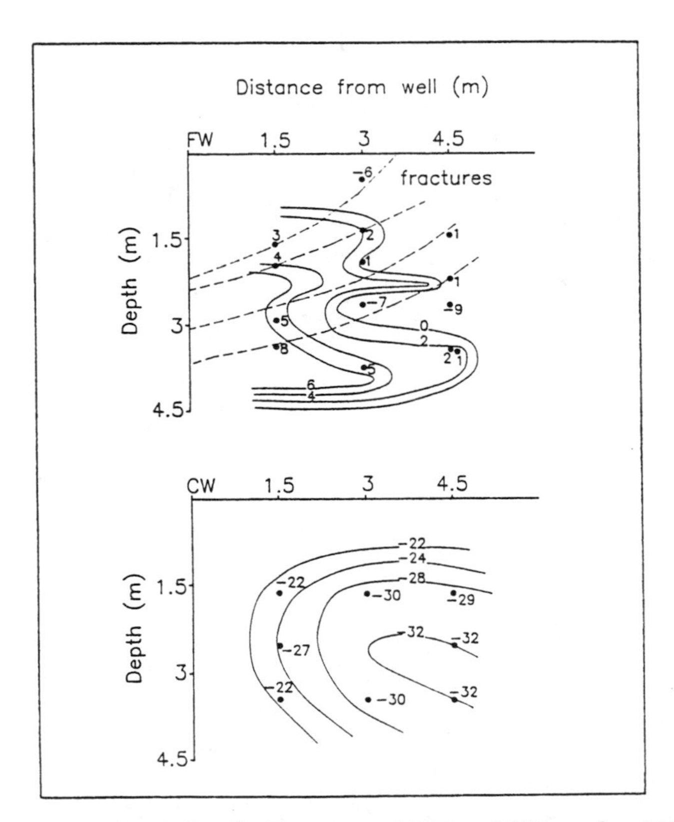

FIGURE 5. Typical head distribution around FW and CW on day 201. Heads are in cm water relative to data at the ground surface.

In Situ Bioremediation

Only the soils within the 1.5-m sampling area around both FW and CW have similar contamination. The areas further north of CW have minor contamination (data not presented). The data from the areas 1.5 m from FW and CW (Tables 1, 2, 3, and 4) give the results for moisture, microbial activity, and hydrocarbon degrader populations near and between each fracture compared to and for similar depths in the vicinity of CW.

The moisture content of the soil over the whole site at the time of the initial sampling was about 10%. At Day 75, the moisture content of the soil near the fractures had increased significantly to more than 20% compared to the moisture content in the initial soil samples and the soil between the fractures. The increase was greatest near the uppermost fracture, and the magnitude of the effect decreased with depth. No significant increase in moisture was observed in the soil around CW at the same depth. By Day 233, the moisture contents in the vicinity of the fractures had decreased compared to Day 75 but, in many cases, were still significantly greater than the initial contents. The moisture content measured around CW at Day 233 remained the same as earlier.

The initial microbial activity, as estimated by FDA analysis, was about twice as high around the upper two fractures compared to corresponding areas around CW, but in the area of the lower fractures, the differences were not as great. At Day 75, the microbial activity had increased significantly in the soil around fractures 1 and 3 but actually declined in the soil between these fractures in some cases compared to the initial activities. By Day 233, microbial activity had declined at all locations at the site compared to the initial samples. However, the microbial activity in the soil near fractures 1 and 2 was still significantly greater than that in the soil between fractures. The microbial activity declined around CW throughout the duration of the test.

The number of hydrocarbon degraders initially found in the soil around FW and CW were similar at about 10^4 cfu/g (with the exception of the uppermost soils around CW with 10^3). These populations increased to 10^5 and 10^6 cfu/g around and between the fractures by Day 75. Around CW, these populations only increased slightly with the exception of the shallowest depth where they increased 3 orders of magnitude (Table 1). By Day 233, the hydrocarbon degrader populations declined across the site, generally by about one order of magnitude. The total number of microorganisms in these soils parallelled the number of degraders and are not presented at this time.

The pH of the soil was between 7.2 and 8.4 in all samples. The more basic soils were generally found at the lowest depths. No changes in pH were observed around FW or CW as a result of the remediation effort (data not presented).

Contaminant Analysis

A summary of the chemical analyses at the 1.5-m sampling points for FW and CW is given in Table 5. The reductions reported in Table 5 are based on total BTE and total petroleum hydrocarbon concentrations. The totals from the

TABLE 5. Disappearance of BTE and TPH contaminants around FW and CW at 1.5 m from each injection well.

Well Type	FW		CW	
Distance from Well (m)	1.5		1.5	
BTE Disappearance				
Day 0 - Day 75	14.4 mg/kg	75%	0 mg/kg	0%
Day 0 - Day 233	11.4 mg/kg	59%	1.3 mg/kg	10%
TPH Disappearance				
Day 0 - Day 75	378 mg/kg	77%	0 mg/kg	0%
Day 0 - Day 233	347 mg/kg	71%	126 mg/kg	55%

2nd and 3rd determinations were subtracted from the initial totals to calculate the amount and percent disappearance by Days 75 and 233.

In the area around FW well, we saw what appears to be a 75% reduction in the BTE concentration by Day 75 but only a 59% reduction at Day 233. There was essentially no change in BTE concentrations around CW. The concentration of total petroleum hydrocarbons appears to have declined 71 to 77% near FW and 0 to 55% near CW (Table 5).

DISCUSSION

Site Operation

Weather and mechanical problems interfered with consistent operation of the Dayton Remediation site. The delivery system provided similar head pressures to each well, though the heads vary with time. Some of the variations are attributed to the filling and emptying of the feed tank. The fact that substantially more water was introduced into FW compared to CW indicates that the fractures acted as channels to increase flow. This was reflected in measurements at the monitoring points indicating relatively high heads in the soil surrounding FW, particularly in the vicinity of the fractures. Higher heads near the injection well are expected as the water flows from the well to the fractures. (Normally the on-site contractor injects fluid into a CW, whereas this system relies on gravity feed. It may be that, if fluid had been injected into the CW, the soil would have responded differently.)

In Situ Bioremediation

Although this field study continues, some observations can be made. The fractures have increased delivery of water to a tight clay, compared to FW. Observations that the upper fractures seemed to deliver more water than the lower fractures may simply be due a decrease in permeability with depth.

The increased microbial activity in the soil (FDA results) in the vicinity of the fractures indicates that oxygen and nutrients are flowing from the fractures into the soil. Generally, the area around the uppermost fracture seemed to respond with more activity than the lower fractures. It is possible that the higher organic contaminant concentrations around the upper fracture (data not given) provide more substrate for the microbes. The reason for the decline in metabolic activity by Day 233 may be the poor flow to the site immediately before the final sampling.

We found that the population of microbes increased significantly around FW. This increase was observed not only near the fractures but generally in the soil around FW. One would expect that, as bioremediation began, the population of the contaminant degrading microbes would increase. This apparently happened at the site in the vicinity of FW. In two cases, the population around CW increased between the Day 0 and Day 75 but, by Day 233, the population had returned to its initial population levels. By contrast, the populations around FW, remained significantly higher than background throughout the test.

Surprisingly, there was not always a good correlation between high metabolic activity and the population of degrader microorganisms. When the conditions appropriate for biodegradation become available, the existing population of organisms present in the soil may simply begin degrading the contaminant. Then there might be no net gain in population. On the other hand, the population may increase to respond to the new conditions. It may be that both phenomena are occurring at different locations in this soil.

Contaminant Disappearance

The results from the contaminant analyses indicate that more organic contaminants were eliminated in the soil around FW than around CW. The increases in metabolic activity and degrader populations suggest that the change in concentration resulted from biodegradation, although other processes may have contributed. Flushing of the soil by injected fluids, release of contaminants due to microbial surfactant production, and oxidation of organics by the peroxide may also have resulted in some decrease of contaminant concentration.

These disappearance data are based on limited sampling and, thus, can only be considered indicative of what might be happening in the soil. One would expect that contaminant concentrations would decline or, at least, stay the same over time at a particular location. However, in some cases, the contaminant concentration appears to increase with time at some locations. There could be many explanations for this but it seems most likely that contaminant concentrations could be highly variable even over the relatively short distances of the offset used between sampling dates. We have tried to limit this variation by taking samples very close together. Adequate sampling remains a major limitation to evaluating in situ remediation efforts.

ACKNOWLEDGMENT

This project was funded by the U.S. EPA's Risk Reduction Engineering Laboratory in Cincinnati, Ohio, under contract #68-C9-0031.

REFERENCES

Atlas, R. M., and D. Pramer. 1990. "Focus on Bioremediation." *ASM News 56*:7-9.

Dang, J. S., D. M. Harvey, A. Jaobbagy, and C. P. L. Grady, Jr. 1989. "Evaluation of Biodegradation Kinetics with Respirometric Data." *J. Water Pollut. Control Fed. 61*:1711-1921.

Davis-Hoover, W. J., L. C. Murdoch, S. J. Vesper, H. R. Pahren, O. L. Sprockel, C. L. Chang, A. Hussain, and W. A. Ritschel. 1991. "Hydraulic Fracturing to Improve Nutrient and Oxygen Delivery for *In Situ* Bioreclamation." In R. E. Hinchee and R. F. Olfenbuttel (Eds.), *In Situ Bioreclamation: Applications and Investigations for Hydrocarbon and Contaminated Site Remediation*, pp. 67-82. Butterworth-Heinemann, Stoneham, MA.

Floodgate, G. 1984. "Microbial Degradation of Oil Pollutants." In R. M. Atlas (Ed.), *Petroleum Microbiology*. Macmillan, New York, NY. pp. 355-398.

Gilbert, R. O. 1987. *Statistical Methods for Environmental Pollution Monitoring*. Van Nostrand Reinhold, New York, NY.

Hinchee, R. E., D. C. Downey, R. R. Dupont, P. Aggarwal, and R. E. Miller. 1991. "Enhancing Biodegradation of Petroleum Hydrocarbons through Soil Venting." *J. of Hazardous Materials 28*:3-11.

Murdoch, L. C., G. Losonsky, P. Cluxton, B. Patterson, I. Klich, and B. Braswell. 1991. *The Feasibility of Hydraulic Fracturing of Soil to Improve Remedial Actions*. Final Report USEPA 600/2-91-012. NTIS Report PB91-181818. p. 298.

Riss, A. and R. Schweisfurth. 1985. "Hydrogen Peroxide to Enhance *In Situ* Bioremediation." *Water Supply* 2:27-34.

Schnurer, J., and T. Rosswall. 1985. "Fluorescein Diacetate Hydrolysis as a Measure of Total Microbial Activity in Soil and Litter." *Appl. Environ. Microbiol.* 43:1256-1261.

OPTIMIZING OZONATION AND MICROBIAL PROCESSES TO REMEDIATE ATRAZINE-LADEN WASTE

C. J. Hapeman, D. R. Shelton, and A. Leeson

ABSTRACT

Development of remediation techniques for pesticide wastes, unusable application equipment rinsates, and registration-suspended pesticide stocks has led to a binary scheme involving ozonation followed by biomineralization of the resultant oxidized pesticides. Preliminary field tests indicated that the *s*-triazines were more recalcitrant. Ozonation of *s*-triazine gave rise to a mixture of products whose structure and abundance was dependent on the duration of ozonation. Oxidation of the *N*-alkyl to the *N*-acetyl and/or removal of the amino substituents occurred in all the primary and secondary ozonation products. Dechlorination and ring opening were not observed. Reaction of the amide was not found in the presence of an alkyl functionality. The final ozonation products can be utilized by microorganisms only as nitrogen sources; however, ammonia fertilizers present in pesticide wastes can be used preferentially. Attempts to increase the rate of biodegradation employing an *s*-triazine degrading pseudomonad (strain A) were unsuccessful; whereas, *Klebsiella terragena* (strain DRS-1) was found to mineralize the ozonation products in the presence of ammonia with the addition of a carbon source. Growth of DRS-1 cultures and *s*-triazine degradation were demonstrated in bench-scale reactors as well as the resiliency of DRS-1 to fluctuations in pH, substrate, and nutrient concentrations.

INTRODUCTION

Remediation of pesticide wastes, application equipment rinsate, and registration-suspended pesticides is of great concern to pesticide applicators and the agricultural community. Applicators are encouraged to minimize waste by collecting and reusing rinsates in subsequent applications; however, this is not always possible, particularly in areas were multiple crops are grown and a variety of pesticides are used. In the past, this material often was deposited on soil to evaporate and presumably degrade. This method, however, has proven to be inadequate as evidenced by the contamination of nearby groundwater supplies, farm wells (Aharonson 1987; Parsons & Witt 1988), and pesticide operational areas that have been in use for years without adequate ground protection (Winterlin et al. 1989).

Laboratory and field studies have shown that chemical pretreatment of pesticides enhances the rate of microbial mineralization, i.e., breakdown to CO_2, H_2O, NH_3 or NO_3^-, and inorganic salts (Hapeman-Somich 1991 and ref. therein). Evaluation of a treatment sequence of ozonation and subsequent biodegradation by indigenous soil microflora demonstrated the potential usefulness of this approach. The results further indicated that the *s*-triazine herbicides, such as the most widely used *s*-triazine, atrazine (2-chloro-4-ethylamino-6-isopropylamino-*s*-triazine), were somewhat recalcitrant to this treatment compared to the other pesticides (Somich et al. 1990). Further development and optimization of this binary remediation process (Figure 1) requires a thorough understanding of the chemical and biological processes and identification of the limiting parameters, particularly for the *s*-triazines.

Products from organic contaminants have been identified in some aqueous ozonation studies (Glaze 1986, Peyton et al. 1989) although most experiments have focused on total organic carbon loss or disappearance of parent material (Glaze 1987, Hoigne 1988). Other investigations have clearly demonstrated that ozone is not the only oxidizing species present under most aqueous ozonation conditions (Glaze 1987, Hoigne 1988, Peyton & Glaze 1987). The fate of the organic species is critical to understanding the overall ozonation process and in determining the active species and their respective roles in the degradation of the

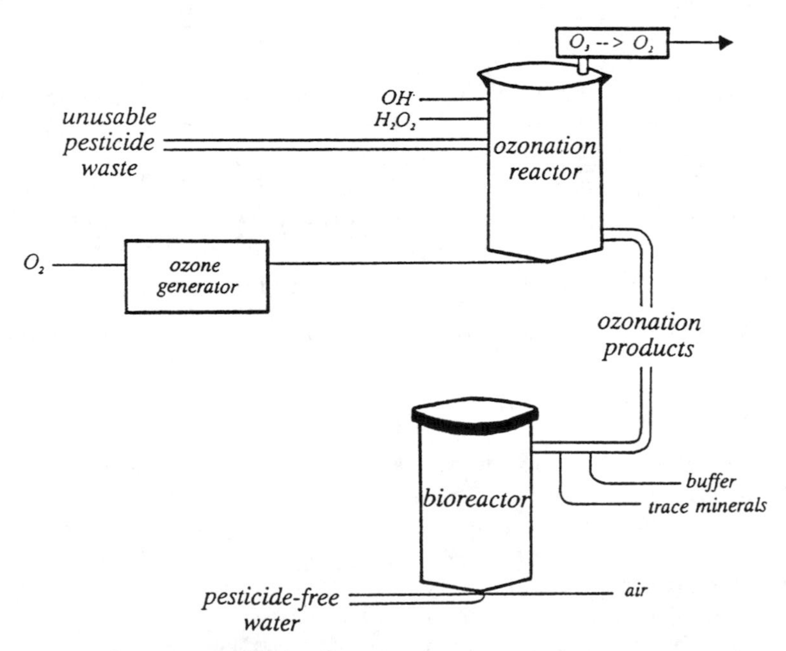

FIGURE 1. Binary remediation process.

organic compound. These demand, then, the isolation and characterization of the atrazine ozonation products and elucidation of the mechanisms involved in their formation and subsequent degradation.

Studies by Cook and Hutter (1981) led to the isolation of a pseudomonad (strain A) (PSA) that could utilize some *s*-triazines as sole nitrogen sources. This organism was found to be more effective in mineralizing a mixture of atrazine ozonation products than were indigenous soil microflora (Kearney et al. 1988), but degradation of the *s*-triazine ozonation products was completely inhibited in farm-generated wastes (Somich et al. 1990). Pesticide rinsate typically consists of a variety of formulating agents, surfactants, emulsifiers, fertilizers, and pesticides at concentrations less than 200 ppm. Preliminary experiments indicated that high ammonia concentrations (ca. 1%) were responsible for this inhibition. This result led to the isolation of a *Klebsiella terragena* (strain DRS-1) which was found to degrade the simple *s*-triazine, 2-chloro-4,6-diamino-*s*-triazine (Hapeman & Shelton 1993). Thus, another objective of the present study was to examine the effectiveness of DRS-1 in mineralizing atrazine ozonation products under simulated field conditions and to determine the limitations and requirements for scale-up.

MATERIALS AND METHODS

Chemicals

For convenience, the nomenclature system used by Cook (1987) is used here: A = amino, C = chloro, E = ethylamino, I = isopropylamino, O = hydroxy, and T = triazine ring. Several ozonation products contained an acetamido group, and in keeping with this nomenclature, D has been added to denote this moiety (Table 1).

The following were gifts from CIBA-GEIGY (Greensboro, North Carolina):

- Atrazine (2-chloro-4-ethylamino-6-isopropylamino-*s*-triazine)
- Formulated atrazine (Aatrex Nine-O, 85.5% atrazine and 4.5% other *s*-triazines)

TABLE 1. Atrazine and its ozonation products.

Abbreviation	Chemical Name
CIET	atrazine (2-chloro-4-ethylamino-6-isopropylamino-*s*-triazine)
CIAT	6-amino-2-chloro-4-isopropylamino-*s*-triazine
CEAT	6-amino-2-chloro-4-ethylamino-*s*-triazine
CDIT	4-acetamido-2-chloro-6-isopropylamino-*s*-triazine
CDET	4-acetamido-2-chloro-6-ethylamino-*s*-triazine
CDDT	2-chloro-4,6-diacetamido-*s*-triazine
CDAT	6-amino-4-acetamido-2-chloro-*s*-triazine
CAAT	2-chloro-4,6-diamino-*s*-triazine

- Simazine (2-chloro-4,6-diethylamino-*s*-triazine)
- Propazine (2-chloro-4,6-diisopropylamino-*s*-triazine)
- 6-amino-2-chloro-4-isopropylamino-*s*-triazine
- 6-amino-2-chloro-4-ethylamino-*s*-triazine
- 2-chloro-4,6-diamino-*s*-triazine
- 2-chloro-4,6-diamino-*s*-triazine-U-ring-^{14}C

Ozonation Procedure

Ozonation experiments were carried out in one of three previously described reactors:

1. Studies involving the isolation of intermediates and elucidation of the *s*-triazine degradation pathway were conducted using a custom-designed glass reactor where ozone/oxygen gas was passed over the top of the solution to decrease the ozone contact time with the solution, thus decreasing the rate of reaction (Hapeman-Somich et al. 1992).
2. A standard photoreactor retrofitted with a bottom feed sintered glass frit was used for all kinetic experiments (Somich et al. 1988).
3. A 9-L glass bottle equipped with a stainless steel airstone on a Teflon™ feed tube was used to generate large quantities of ozonated atrazine for microbial studies (Leeson et al. 1993). Ozone was generated using a PCI Ozone Generator Model GL-1B (PCI Ozone Corporation, West Caldwell, New Jersey) with oxygen feed. Oxygen/ozone was delivered to the reactor at 1 L/min with an ozone concentration between 0.2 and 1.0% as determined using a PCI Ozone Monitor Model HC.

Sample Analysis

Concentrations of atrazine and its degradation products were determined using response factors or concentration standard curves. Samples were taken at appropriate time intervals during the reaction and analyzed directly using a 0 to 50% acetonitrile/phosphoric acid buffer (pH 2) exponential gradient on a C-18 (ODS, 5 µm) column at a flowrate of 2 mL/min. Further details can be found in Hapeman-Somich et al. (1992), Leeson et al. (1993), and Hapeman (1993).

Ammonia concentrations were determined colorimetrically using a Technicon Autoanalyzer II, Model SPR-2431 (Technicon Instrument Corporation, Terrytown, New York). Samples were diluted with distilled water, and values were assigned relative to ammonium chloride standards.

Product Isolation

The ozonation reaction was monitored by HPLC as described above and carried out until all starting material was depleted or the maximum concentration of product was obtained. The reaction mixture was extracted with 50-mL ethyl acetate (3X); the extract was dried over Na_2SO_4 and concentrated in vacuo to ca. 5 mL and then to dryness using nitrogen. The residue was redissolved in several milliliters of acetonitrile and separation of the reaction products achieved by semi-preparative high pressure liquid chromatography (HPLC) (Hapeman-Somich et al. 1992).

Microbial Incubations

DRS-1 was isolated as described previously (Hapeman & Shelton 1993). PSA (NRRLB-12227) was obtained gratis from CIBA-GEIGY (Greensboro, North Carolina) and was originally isolated by Cook (1987). Microbial incubations with PSA and DRS-1 were conducted using sterile 250-mL Erlenmeyer flasks containing 50 mL of medium consisting of ozonated Aatrex Nine-O amended with 0 to 10% corn syrup (Tru-Sweet 42, American Fructose Corp., Decatur, Alabama), 20 mM phosphate buffer (pH 7), trace metals solution (Shelton & Somich 1988), and varying concentrations of ammonium sulfate (0 to 0.8 M ammonia). Flasks (3 replicates per experimental treatment) were incubated on a reciprocal shaker table (ca. 120 rpm) at 25°C. Metabolism was assessed by monitoring increases in turbidity (Klett-Summerson photoelectric colorimeter) and/or product dissipation.

Bench-Scale Reactor Studies

Two reactor configurations were constructed: a continuous-flow stirred tank reactor (CFSTR) and an upflow fixed-film column reactor. The column reactor was filled with Celite Biocatalyst Carrier R-625 (Manville, Denver, Colorado). Each reactor was inoculated with a dense DRS-1 culture, and then a solution of corn syrup added via one line and ozonated Aatrex Nine-O solution, phosphate buffer, and trace minerals through a second to prevent microbial growth in the tubing (Leeson et al. 1993). Ozonation product concentrations were monitored periodically in both the effluent and influent.

RESULTS

Ozonation of Atrazine and Its Degradation Products

Ozonation of atrazine gave rise to a complex mixture consisting of four primary products (6-amino-2-chloro-4-isopropylamino-*s*-triazine [CIAT], 6-amino-2-chloro-4-ethylamino-*s*-triazine [CEAT], 4-acetamido-2-chloro-6-isopropylamino-*s*-triazine [CDIT], and 4-acetamido-2-chloro-6-ethylamino-*s*-triazine [CDET]) which were subsequently degraded to three secondary products (2-chloro-4,6-diacetamido-*s*-triazine [CDDT], 6-amino-4-acetamido-2-chloro-*s*-triazine [CDAT], and 2-chloro-4,6-diamino-*s*-triazine [CAAT]) (Table 1 and Figure 2). Structural identification of CAAT, CEAT, and CIAT was established by comparison with authentic samples. Syntheses and nuclear magnetic resonance (NMR) and mass spectrometry provided structural verification of the remaining products (CDAT, CDDT, CDET, and CDIT) (Hapeman-Somich et al. 1992).

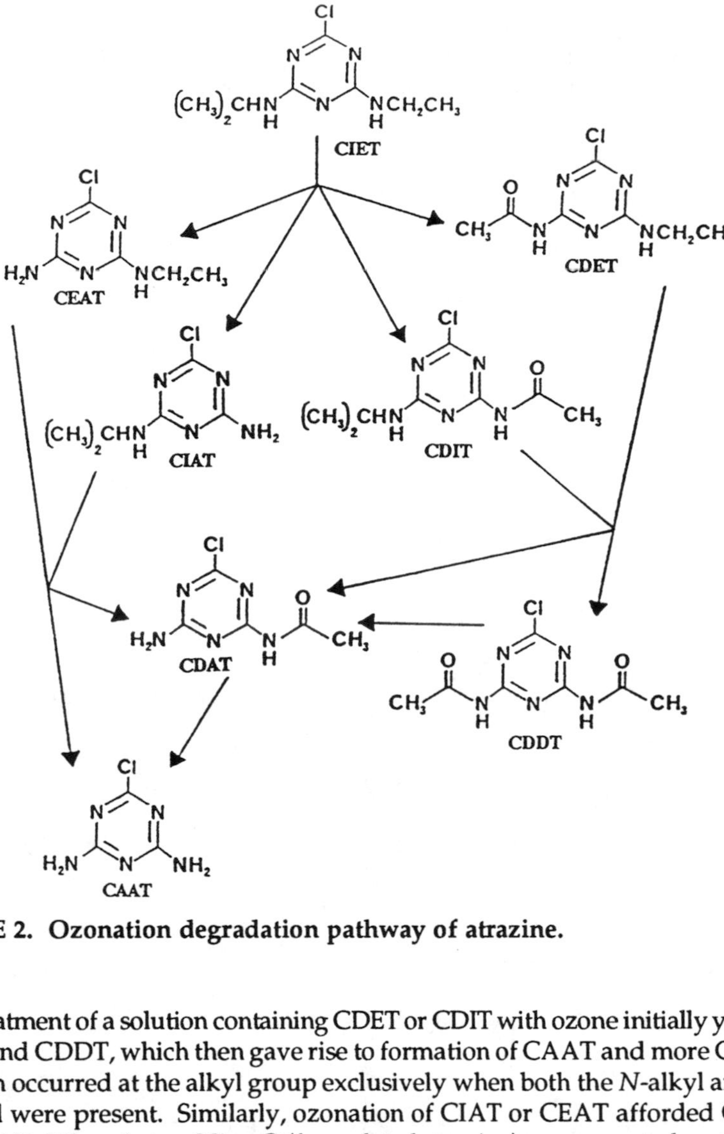

FIGURE 2. Ozonation degradation pathway of atrazine.

Treatment of a solution containing CDET or CDIT with ozone initially yielded CDAT and CDDT, which then gave rise to formation of CAAT and more CDAT. Reaction occurred at the alkyl group exclusively when both the *N*-alkyl and the *N*-acetyl were present. Similarly, ozonation of CIAT or CEAT afforded CAAT and CDAT. When Aatrex Nine-O (formulated atrazine) was ozonated, a mixture of CDAT and CAAT was formed. Further oxidation beyond CAAT did not occur.

Relative Ozonation Rates of Products

The ratio of the atrazine ozonation products changed as the reaction proceeded because their relative degradation rates differed. The significance of these compounds is reflected, therefore, in their relative ratio with respect to the disappearance of parent material, atrazine. At 35% loss of atrazine, the product ratios were CDIT 19%, CIAT 51%, CDET 1%, CEAT 12%, CDDT 2%, CDAT 4%, and CAAT 11%. Accordingly, the concentration of the primary products decreased and the secondary products increased as the reaction proceeded. At 90% depletion of an initial concentration of 0.153 mM of atrazine, the following concentrations were observed: CIET 15 μM, CDIT 22 μM, CIAT 30 μM, CDET 0.6 μM, CEAT 3.2 μM, CDDT 1.4 μM, CDAT 11 μM, and CAAT 17 μM (Figure 3) (Hapeman-Somich et al. 1992).

Mineralization of CAAT

The ability of PSA and DRS-1 to use CAAT as a nitrogen source for growth in the presence of additional nitrogen and/or carbon was examined (Figure 4). PSA was unable to degrade CAAT in the presence of ammonia (8 mM); concentrations of CAAT increased as CDAT (initial concentration = ca. 15 mg/L) was hydrolyzed to CAAT. However, PSA did degrade CAAT in the absence of ammonia. In contrast, CAAT was rapidly degraded by DRS-1 when both carbon (0.5% v/v) and 8 mM ammonia were present. No apparent CAAT degradation was observed in the absence of additional carbon (corn syrup), indicating that no significant concentrations of other utilizable carbon sources, such as formulations and surfactants, were present in the ozonated Aatrex.

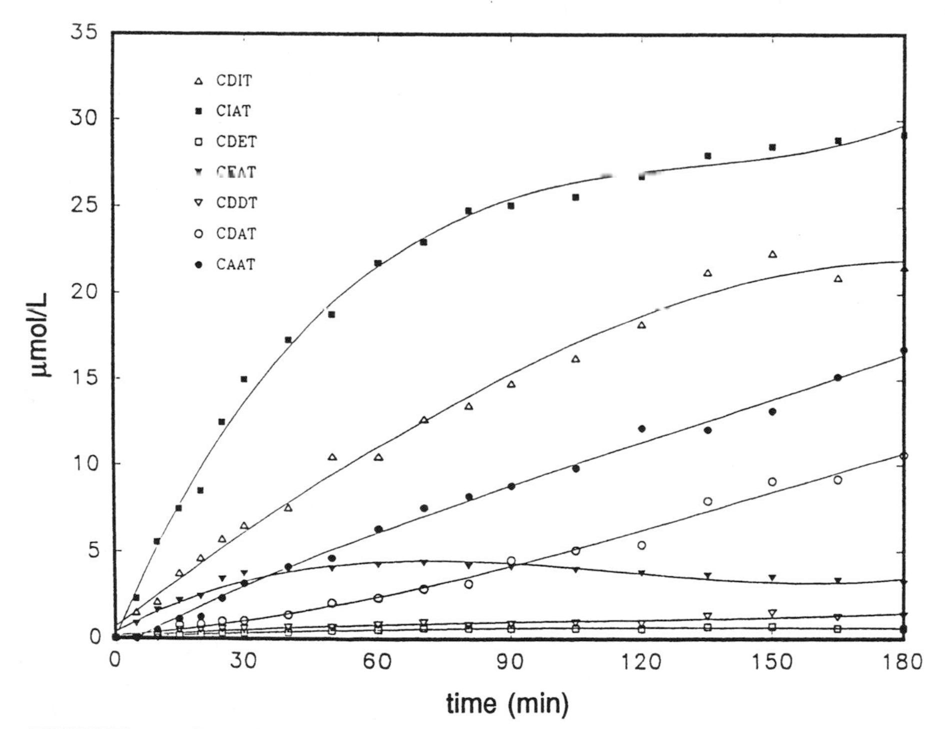

FIGURE 3. Atrazine reaction profile (formation of ozonation products; [atrazine]$_0$ = 153 μM).

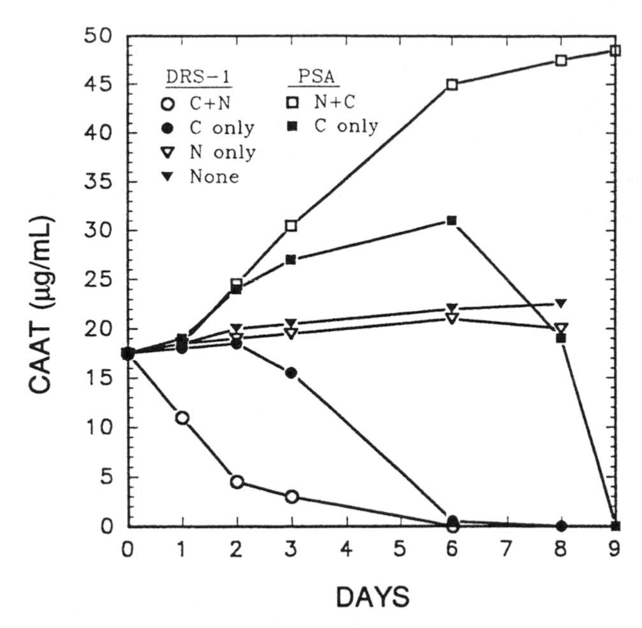

FIGURE 4. Degradation of CAAT by DRS-1 and PSA.

DRS-1 was observed to degrade CAAT over a wide range of ammonia concentrations. Microbial growth was not inhibited even at high ammonia concentrations (800 mM), as indicated by comparable increases in turbidity, although rates of CAAT degradation decreased somewhat as a function of increasing ammonia concentration (Leeson et al. 1993). In a separate experiment, varying concentrations of carbon (0, 0.1, 0.25, 1.0, 5.0, and 10% v/v corn syrup) produced virtually no differences in CAAT degradation. CAAT concentrations decreased from 18 to <1 mg/L after 3 days for all runs except 0% where ca. 30% of CAAT was degraded within 10 days (Leeson et al. 1993).

Radiolabeled CAAT-U-ring-^{14}C was added to DRS-1 cultures in biometer flasks, and $^{14}CO_2$ was trapped in NaOH to determine if CAAT dissipation was due to transformation or due to complete mineralization. No radioactivity was present in the NaOH solution on day 0 or day 1; however, 24%, 57%, and 77% of the initial activity was recovered as carbon dioxide after 5, 7, and 9 days, respectively.

Bench-Scale Reactor Studies

DRS-1 grew readily and degraded CAAT in a continuous flow stirred tank reactor (CFSTR), from which kinetic constants were determined (Leeson et al. 1993). DRS-1 also exhibited biofilm growth on the Celite solid support in an upflow column reactor. Column influent and effluent samples were monitored regularly (Figure 5). Significant CAAT removal was observed across the column when influent concentrations ranged from 15 to 30 mg/L. Breakthrough of CAAT occurred at day 3 due to a fluctuation in pH and whenever influent CAAT concentrations exceeded 30 mg/L.

DISCUSSION

Ozonation

Analysis of the aqueous ozonation of atrazine and its degradation products demonstrated that amino alkyl groups were the first site of attack. The *N*-alkyl group was either removed or converted to the *N*-acetyl, the *s*-triazine ring remained intact, and the chlorine was not displaced. Furthermore, the isopropyl group was not converted to the ethyl group nor were the ethyl or isopropyl moieties converted to an aldehyde (Hapeman-Somich et al. 1992). The ozonation reactions of CDET and CDIT clearly demonstrated that the *N*-alkyl group was far more reactive than the *N*-acetyl moiety. The reaction profile data (Figure 3) and additional experiments (Hapeman 1993) also showed that reaction at the *N*-ethyl moiety was much preferred over the *N*-isopropyl group.

Microbial Degradation

CAAT was consistently degraded by DRS-1 when a carbon source was provided, even in the presence of ammonia; however, significant CAAT degradation was not observed in the absence of an extraneous carbon source. Extensive

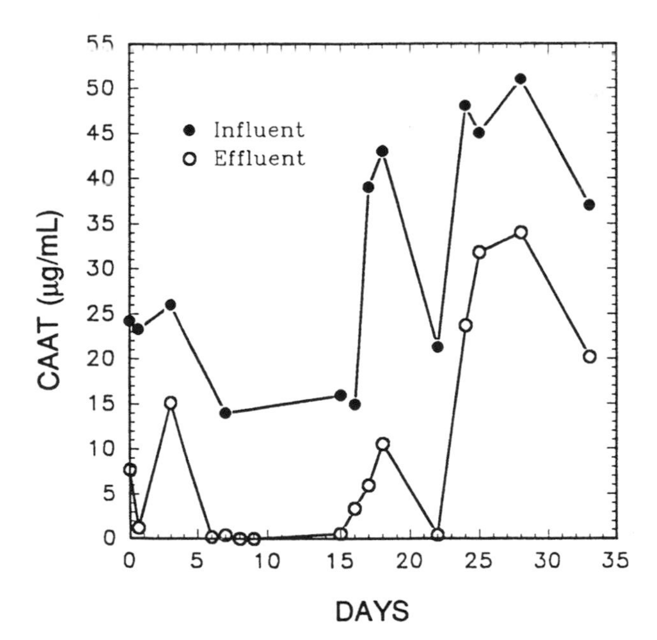

FIGURE 5. Degradation of CAAT in fixed-film column reactor.

mineralization of radiolabeled CAAT-U-ring-^{14}C to $^{14}CO_2$ indicated that both the amino substituents of CAAT and the nitrogen atoms of the *s*-triazine ring were used as a nitrogen source. In contrast to PSA, high concentrations of ammonia were not completely inhibitory to DRS-1 growth. Further experiments suggested that DRS-1 had a strong preference for CAAT compared to ammonia as the nitrogen source (Leeson et al. 1993), although the relative rate was dependent on the ammonia concentration. In addition, the ability of DRS-1 to grow in solutions containing ozonated Aatrex Nine-O indicated that the presence of formulations in the waste should not inhibit growth. Thus, the ability of DRS-1 to degrade CAAT in the presence of ammonia and formulation should improve the removal of *s*-triazines from the pesticide wastes.

Bioreactors

The results from the bench scale bioreactors indicated that both the CFSTR and the fixed-film reactor could support growth of DRS-1 and CAAT degradation. However, organisms in biofilms typically are more resistant to sudden fluctuations of nutrient supply or toxins in the reactor feed solution and as such are more practical for field operations. Although breakthrough of CAAT occurred at day 3 due to a fluctuation in pH, and when CAAT concentrations exceeded 30 mg/L, these problems were remedied by closer monitoring of pH and by decreasing the influent concentration, respectively. Alternatively, increasing the detention time also would cause the CAAT effluent concentration to approach zero. It is important to note that the biofilm culture was relatively tolerant of adverse conditions, because the system recovered from both perturbations.

CONCLUSION

In summary, these results have demonstrated the complete mineralization of formulated atrazine using the binary process of ozonation and biodegradation, and have provided some of the necessary details to proceed in the optimization of *s*-triazine remediation under simulated field conditions. Investigations are continuing to discern the extent of the various oxidizing species in the ozonation of atrazine and to optimize rates of degradation in conjunction with design and fabrication of pilot-scale reactors. With the appearance of *s*-triazine residues not only in agricultural sites but in ground and surface waters as well, these results should be useful to waste remediation investigators and may also be helpful to those developing methods for triazine residue removal in water treatment.

ACKNOWLEDGMENTS

The authors wish to thank Eton Codling, Carmen Gilotte, Julie C. Lin, Joseph J. O'Connell, and April Talcott Danchik for their technical assistance.

Mention of specific products or suppliers is for identification and does not imply endorsement by the U.S. Department of Agriculture to the exclusion of other suitable products or suppliers.

REFERENCES

Aharonson, N. 1987. "Potential Contamination of Groundwater by Pesticides." *Pure Appl. Chem.* 59: 1419.

Cook, A. M. 1987. "Biodegradation of *s*-Triazine Xenobiotics." *FEMS Microbiol. Rev. 46*: 93.

Cook, A. M., and R. J. Hutter. 1981. "*s*-Triazines as Nitrogen Sources for Bacteria." *J. Agric. Food Chem. 29*: 1135.

Glaze, W. H. 1986. "Reaction Products of Ozone: A Review." *Environ. Health Perspectives 69*: 151.

Glaze, W. H. 1987. "Drinking-Water Treatment with Ozone." *Environ. Sci. Technol. 21*: 224.

Hapeman, C. J. 1993. "Oxidation of *s*-Triazine Pesticides." Submitted to *Symposium Proceedings of Emerging Technologies in Hazardous Waste Management*. Atlanta, GA, American Chemical Society, Washington, DC.

Hapeman, C. J., and D. R. Shelton. 1993. "Field Studies Using Ozonation and Biomineralization as a Disposal Methodology for Pesticide Wastes." (submitted to *Chemosphere*.)

Hapeman-Somich, C. J. 1991. "Mineralization of Pesticide Degradation Products." *ACS Symp. Ser. 459*:133.

Hapeman-Somich, C. J., G.-M. Zong, W. R. Lusby, M. T. Muldoon, and R. Waters. 1992. "Aqueous Ozonation of Atrazine. Product Identification and Description of Degradation Pathway." *J. Agric. Food Chem. 40*:2294.

Hoigne, J. 1988. "The Chemistry of Ozone in Water". In S. Stucki (Ed.), *Process Technologies for Water Treatment*, pp. 121-243. Plenum Publishing Corp., New York, NY.

Kearney, P. C., M. T. Muldoon, C. J. Somich, J. M. Ruth, and D. J. Voaden. 1988. "Biodegradation of Ozonated Atrazine as a Wastewater Disposal System." *J. Agric. Food Chem. 36*: 1301.

Leeson, A., C. J. Hapeman, and D. R. Shelton. 1993. "Biomineralization of Atrazine Ozonation Products. Application to the Development of a Pesticide Waste Disposal System." (submitted to *J. Agric. Food Chem.*)

Parsons, D. W. and J. M. Witt. 1988. *Pesticides in Groundwater in the United States of America. A Report of a 1988 Survey of State Lead Agencies.* Oregon State University Extension Service, Corvallis, OR.

Peyton, G. R., C. S. Gee, M. A. Smith, J. Brady, and S. W. Maloney. 1989. "By-Products from Ozonation and Photolytic Ozonation of Organic Pollutants in Water: Preliminary Observations." In R. A. Larson (Ed.), *Biohazards of Drinking Water Treatment*, pp. 185-200. Lewis Publishers, Chelsea, MI.

Peyton, G. R., and W. H. Glaze. 1987. "Mechanism of Photolytic Ozonation." *ACS Symp. Ser.* 327: 76.

Shelton, D. R., and C. J. Somich. 1988. "Isolation and Characterization of Coumaphos-Metabolizing Bacteria from Cattle Dip." *Appl. Environ. Microbiol. 54*: 2566.

Somich, C. J., P. C. Kearney, M. T. Muldoon, and S. Elsasser. 1988. "Enhanced Soil Degradation of Alachlor by Treatment with Ultraviolet Light and Ozone." *J. Agric. Food Chem. 36*: 1322.

Somich, C. J., M. T. Muldoon, and P. C. Kearney. 1990. "On-Site Treatment of Pesticide Waste and Rinsate Using Ozone and Biologically Active Soil." *Environ. Sci. Technol. 24*:745.

Winterlin, W., J. N. Seiber, A. Craigmill, T. Baier, J. Woodrow, and G. Walker. 1989. "Degradation of Pesticide Waste Taken from a Highly Contaminated Soil Evaporation Pit in California." *Arch. Environ. Contam. Toxicol. 18*: 734.

SURVIVAL OF A CATABOLIC PLASMID INDEPENDENT OF ITS INTRODUCED HOST IN A FRESHWATER ECOSYSTEM

R. R. Fulthorpe and R. C. Wyndham

ABSTRACT

A 3-chlorobenzoate (3Cba) degrading organism, *Alcaligenes* sp. st. BR60, was introduced to nonsterile flowthrough lake ecosystem microcosms to monitor its survival and the survival of its catabolic plasmid, pBRC60. Microcosms were dosed with 3Cba from 10 nM to 50 µM in the incoming feed water. The introduced host-plasmid association — BR60(pBRC60) — was monitored using selective plate counts, while the total population of culturable cells carrying the catabolic plasmid was monitored with DNA probes specific for the catabolic genes carried on the plasmid. The 3Cba dosing experiments were repeated over three summers, and in each case, the BR60 populations did not consistently correlate to either the 3Cba dose or the 3Cba uptake rate in the microcosms, whereas these parameters did correlate well to the total pBRC60-carrying population. Indigenous recipients of pBRC60 were isolated and characterized, proving to be various strains members of four distinctly different species of which at least two were dominant hosts for this plasmid in these systems. Evidence for genetic instability of plasmid pBRC60 was found in a fifth species that played a role in both 3Cba and chloroaniline metabolism in these microcosms.

INTRODUCTION

Our understanding of the potential for inoculant or catabolic gene survival in natural systems is poor. In contaminated natural environments, chlorinated organics may occur at micromolar levels or less and although biodegradation of synthetic organics does occur at low concentrations (Lewis et al. 1988), studies suggest that the organisms responsible differ from those operating at higher concentrations (Rubin et al. 1982). The usefulness of laboratory-selected strains as inocula in natural situations has been demonstrated in only a few cases (Brunner et al. 1985, Edgehill & Finn 1983, Focht & Shelton 1987) and there are reports of inocula that failed to enhance biodegradation of their substrate in situ (Goldstein et al. 1985, McClure et al. 1989, Ramadan et al. 1990). Natural conditions present a number of problems for laboratory selected contaminant degraders reintroduced into the environment. Low substrate concentrations may not support the growth of these specialized strains, or the introduced organisms may abandon metabolism of the contaminant in favor of other substrates (Goldstein et al. 1985). Naturally occurring toxins, predators, and interspecific competition for nutrients may all reduce the survival of the introduced strain (Liang et al. 1982, Scheuerman et al. 1988). On the other hand, alternative carbon sources available in natural ecosystems may enhance the survival of introduced strains and hence the degradation of low level contaminants (Schmidt & Alexander 1985).

There is abundant evidence that bacterial genetic exchange processes occur in nature, including bacterial conjugation, phage transduction, and natural transformation (Carlson et al. 1985, Morrison et al. 1978, Stewart & Sinigalliano, 1990, Stotzky & Babich 1986, Trevors et al. 1987). However the rate and the importance of these processes to the degradation of xenobiotic chemicals are unknown.

This study examined the survival of a 3Cba-degrading bacterium in nonsterile flowthrough lake ecosystem microcosms contaminated with 3Cba and related chemicals, and assessed the horizontal transfer of the plasmid-borne catabolic genes. *Alcaligenes* sp. BR60 is capable of growth on 3Cba because of genes found within a transposable DNA element, Tn5271, carried by conjugative 85 kb plasmid, pBRC60 (Nakatsu et al. 1991, Wyndham et al. 1988). This paper summarizes the results of 3 years of release experiments that show that the original inoculum and the catabolic plasmid experienced differential survival in the microcosms. More complete details are published elsewhere, but here we present summary data from experiments carried out over a 3-year period on the relationship between 3Cba dosage and inoculum and plasmid survival, and the evidence for and scope of horizontal transfer of the plasmid. Work with chlorinated aromatics other than 3Cba showed that Tn5271 plays a role in the degradation of chlorobiphenyl and chloroaniline in conjunction with alternative hosts.

METHODS AND MATERIALS

Bacterial Strains

Alcaligenes sp. BR60 is described in Wyndham et al. (1988) and Wyndham and Straus (1988). The catabolic transposon is described in Nakatsu et al. (1991). The following strains were purchased from the American Type Culture Collection: *Pseudomonas acidovorans* (ATCC 15668), *P. alcaligenes* (ATCC 14909), *P. cepacia* (ATCC 25416), *P. fluorescens* (ATCC 13525), *P. pseudoflava* (ATCC 33668), *P. palleronii* (ATCC 17724), *P. putida* Biotype A (ATCC 12633), and *Chromobacterium violaceum* (ATCC 12472). *Alcaligenes* sp. H850 was provided by General Electric Company in Schenectady, New York. *Acinetobacter calcoaceticus* DON2, an aniline-degrading isolate from the Don River, was described by Wyndham (1986). *Moraxella* sp. G. was provided by Dr. Josef Zeyer of Geneva, Switzerland (Zeyer et al. 1985). *P. putida* PRS2015 was provided by Dr. Chakrabarty (Chatterjee & Chakrabarty 1983). General media used for strain maintenance and/or purification included 0.1% TYE agar (0.1% tryptone, 0.1% yeast extract, 1.8% agar), CPS agar (0.5 g casitone, 0.5 g peptone, 0.5 g soluble starch, 1 g glycerol, 0.04 $MgSO_4.7H_2O$, 0.2 g K_2HPO_4, 1 g agar in 1 L water) or nutrient agar. Selective media for isolation or maintenance of catabolic strains are based on a minimal medium (Medium A, see Wyndham et al. 1988) supplemented with 2 mM of the chloroaromatic compound and 5 ppm yeast extract.

Microcosm Design

Twelve microcosms representative of an epilimnetic sediment/water interface were set up at Queen's University Biological Station on Lake Opinicon near Elgin, Ontario, Canada. An extensive sampling of surficial waters and sediments in the area in early 1987 failed to turn up 3Cba-degrading organisms that hybridized with pBRC60, so the area was considered clean of pBRC60 prior to the release of BR60 (Fulthorpe 1991). The microcosms were large aquaria (72 cm × 40 cm × 22 cm high) containing surficial sediments from Lake Opinicon, complete with small macrophytes, detritus, and snails, taken directly from a shallow bay of the lake. A 5-cm layer of epilimnetic sediment was allowed to settle and was not subsequently disturbed. The 50 L of overlying water was aerated gently with porous stones. Epilimnetic lake water was continuously pumped from an intake pipe in the middle of an adjacent 2-m-deep bay into an overflowing header tank before delivery to the various microcosms at a controlled dilution rate. Sterile aqueous solutions of the chloroaromatic treatment chemical also were delivered at fixed rates to selected tanks. Overflow water from experimental tanks was passed through a carbon filter before entering a seepage bed. No nutrients were added, and all experiments took place at ambient temperatures.

In 1987, lake water was siphoned from the header tank to the microcosms using aquarium air tubing constricted with plastic clamps such that the dilution rate approximated 0.05 hr^{-1}. The 3Cba was delivered from sterile 4-mM reservoirs using commercially available intravenous apparatus. Water and 3Cba delivery systems were altered in 1988 to allow finer control on the feed concentration of 3Cba. Prefiltered (170 µM mesh) lake water flowed through 1-mL plastic syringes capped with upturned 20-gauge stainless steel needles, 3 cm below the water surface, producing a microcosm dilution rates of 0.01 hr^{-1}. Chemicals were delivered from sterile reservoirs to the water delivery lines via a 10-rpm 10-channel peristaltic pump (Watson-Marlow 501U) at the rate of 0.008 mL/min. Aquaria were sealed with plexiglass lids containing input and output tubes sealed in place so as to eliminate the transfer of aerosols or water spray between microcosms.

Experimental Treatments. The effect of 3Cba dose on BR60 and pBRC60 survival was investigated in 1987 and 1988. In 1987, microcosms were dosed with 3Cba concentrations in the feedwater that approximated 1 µM, 30 µM, and 0 (controls). BR60 grown on 3Cba was inoculated, and the survival of the BR60 and pBRC60 was monitored over a 56-day period. In 1988, microcosm 3Cba concentrations were set at 11 concentrations between 10 nM up to 25 µM, and zero dosage controls were included. Microcosms were inoculated with BR60 and tracked for 28 days. In 1989, the effect of alternative chloroaromatic substrates on the survival of pBRC60 was investigated; microcosms were dosed with various levels of 4-chloroaniline, 2-4-dichlorophenoxyacetic acid, or chlorobiphenyl and tracked for 100 days (Fulthorpe & Wyndham 1992). Controls for this experiment included a 25-µM 3Cba-dosed microcosm and two zero dose control microcosms that were later dosed with 50 µM 3Cba. The data obtained from these 3Cba-dosed microcosms are included in the overall 3Cba response summary that follows.

Enumeration of BR60 and Tn5271 Homologous ("Probe Positive") Cells. 3Cba agar plates supplemented with streptomycin (25 µg/mL) were used to enumerate populations of BR60, which were easily recognizable by their rapid growth and colony morphology. 3Cba agar without antibiotics was used to estimate the total culturable 3Cba-utilizing population as well as to isolate putative pBRC60 recipients. To quantify total culturable cells carrying DNA homologous to the 3Cba genes, cloned fragments of Tn5271 were used to make a P-32 labeled DNA probe specific for the 3Cba catabolic genes as described in Fulthorpe and Wyndham (1989). A most-probable-number DNA hybridization technique adapted from Fredrickson et al. (1988) was used to obtain an estimate of the number of "probe positive" (Tn5271 homologous) cells in microcosm water and sediments.

Uptake Rates of 3Cba in Microcosms. The degradation rates of the treatment chemicals in microcosms waters and sediments were determined using ^{14}C-labeled substrates obtained from Sigma radiochemicals as described in Fulthorpe and Wyndham (1989).

Isolate Characterization

Unique non-BR60 like colonies appearing on 3Cba were isolated by purification on fresh 3Cba agar or liquid medium. Throughout the 1989 field season, microcosm waters and sediments were plated on minimal medium agar supplemented with 3Cba, chlorobiphenyl, 4-chloroaniline/5 ppm yeast extract, or 2,4-D, depending on the treatment the microcosms received. Any colonies showing signs of chloroaromatic degradation were purified for future study. BIOLOG GN ID kits from BIOLOG Inc. (Hayward, California) were used to characterize Gram-negative isolates of interest. Representative isolates from distinct types were further characterized using standard tests (Gram stain, oxidase reaction, catalase activity, motility, poly-ß-hydroxybutyric acid accumulation, pigment production) as described by Smibert and Kreig (1981).

Plasmid and total genomic DNA was extracted as described in Fulthorpe and Wyndham (1992). Restriction digestions were carried out on plasmids to confirm their similarity to pBRC60.

Laboratory pBRC60 Host Range Experiments

To test the host range of the pBRC60 replicon, we used a Tn5 (kanamycin resistant - Km^r)-labeled derivative of pBRC60 that lacked the catabolic transposon, because stable maintenance of Tn5 and Tn5271 proved difficult and expression of Tn5271 in various strains was an unreliable marker of plasmid presence. Details on the plasmid construct and mating procedure are in Fulthorpe and Wyndham (1991).

RESULTS

Survival of BR60 and pBRC60 in 3Cba-Dosed Microcosms

Using the MPN-DNA hybridization technique, the theoretical lowest detectable level of probe-positive organisms was 8 cells/mL in the waters or 80 cells/mL in the sediments. In 1987, probe-positive organisms persisted in water and sediments of all 3Cba-dosed microcosms for the full 56 days of the experiment. In microcosms lacking 3Cba inputs, probe-positive counts in the sediments dropped to undetectable levels after 41 days. Throughout the experiments, MPN-DNA hybridization estimates of probe-positive cells were higher when a non-selective (0.1% TYE) medium was used in the MPN plates rather than a highly selective 3Cba broth media. We suspected, but could not prove at the time, that organisms other than BR60 (which grew well in the selective MPN plates) were carrying pBRC60 or Tn5271 in these systems.

In the 1988 experiments, when BR60 populations were more carefully monitored, a dramatic difference in the behavior of BR60 versus the behavior of probe-positive organisms was noted. Three days after inoculation, the BR60 populations averaged 1.0×10^4 colony-forming units (CFU)/mL in microcosm waters, and 1.1×10^5 CFU/mL in the surficial sediments. The highest populations were found in those microcosms receiving 10 nM, 40 nM, 700 nM, 10 μM, and 20 μM 3Cba, exhibiting no correlation with 3Cba concentration. After 14 days, populations had dropped dramatically to averages of 68 CFU/mL in the waters; 4.0×10^3 CFU/mL in the sediments. No colony-forming BR60 were detected in microcosm waters after 32 days, and average sediment populations had dropped to 1.7×10^3 CFU/mL. The probe-positive populations, i.e., those cells carrying Tn5271, were often orders of magnitude higher than the viable counts of BR60, and the mean log estimates of the probe-positive populations were significantly related to the log of the 3Cba feed concentration. Probe-positive cells averaged 2.35×10^3 cells/nmole of 3Cba in the microcosm waters and 1.07×10^4 cells/nmole of 3Cba in the sediments. For comparison, in continuous culture of BR60 in sterile creek water containing 200 μM 3Cba the yield averages 4.7×10^4 cells/nmole (Wyndham & Straus, 1988).

In 1989, the third summer of experiments, microcosms were left to run for a 100 day period. In Figure 1, the survival of BR60 and probe-positive cells is shown for two microcosms not receiving any 3Cba; one microcosm was not inoculated with BR60 and the other was. In the uninoculated microcosm, probe-positive organisms remain undetectable, although there is a transient population of BR60 detected, perhaps from the previous years experiments. In the inoculated microcosms, BR60 declines in both the water and the sediments, and the probe-positive

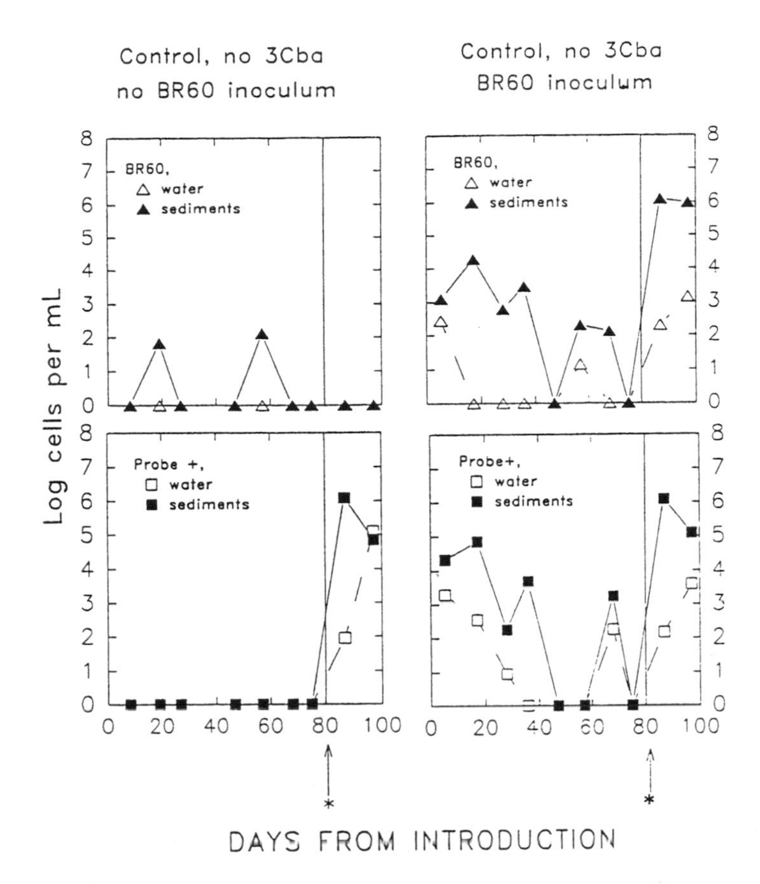

FIGURE 1. Populations of BR60 and probe-positive cells (number of cells hybridizing to a Tn5271 specific probe) versus time for two microcosms: on the left, results for a microcosm not dosed with 3Cba and not inoculated with BR60; on the right, results for a microcosm inoculated with BR60 but not dosed with 3Cba. Note that at day 80, the lake water feed to both microcosms was supplemented with 50 μM 3Cba.

population mirrors this decline. Toward the end of the experiment, both microcosms were dosed with 50 μM 3Cba, and in both cases probe-positive populations responded dramatically to this input and increased over several orders of magnitude. Curiously, in the uninoculated microcosm, BR60 was not responsible for this increase, while in the inoculated microcosm, the probe-positive population was comprised entirely of BR60 cells. This result emphasizes the importance of microcosm-specific factors in determining the ability of BR60 to respond to 3Cba inputs. In one case it was able to exploit the 3Cba to the exclusion of

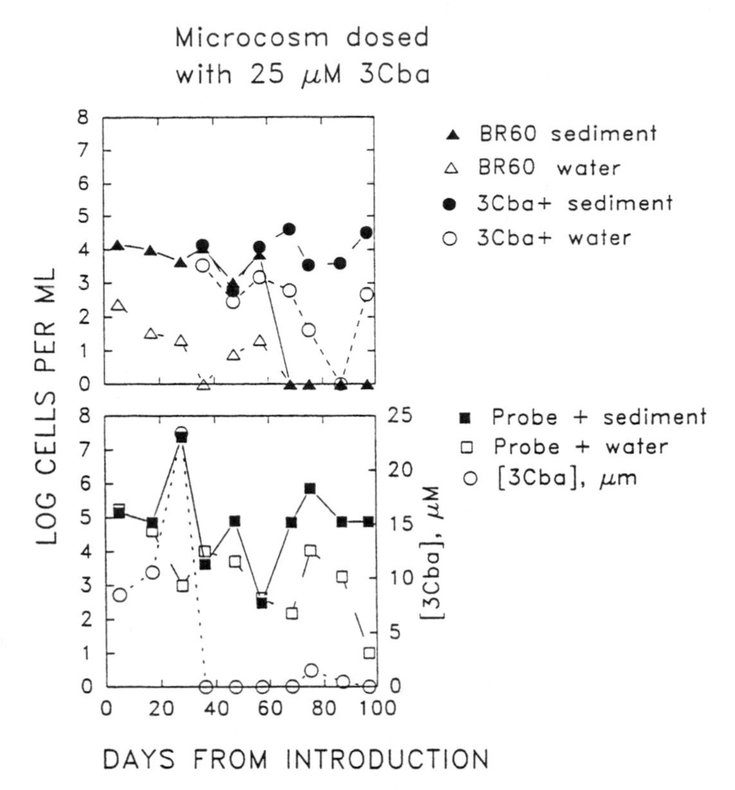

FIGURE 2. Populations of BR60, other 3Cba degraders, and probe-positive cells versus time for a microcosm dosed with 25 μM 3Cba over a 100-day period. The concentration of 3Cba in the microcosm is shown in the bottom graph. Note that 3Cba in the feed is fully degraded after about 38 days. Probe-positive populations persist in this microcosm, but BR60 populations are completely replaced by other 3Cba degraders after 60 days.

other degraders, in another, it was not, even though we see it is present in the sediments of the microcosm. Figure 2 shows a typical response to long-term exposure to 3Cba dosage, where BR60 populations survive for a time but eventually decrease in population, becoming all but undetectable in the final 4 weeks of the experiments. At the same time, there is little change in the probe-positive population and in the total number of 3Cba-degrading organisms. The BR60 population appears to have functioned in the system for a time, co-existing with other 3Cba degraders (note the populations in the microcosm waters) only to be completely replaced by alternative pBRC60 carriers toward the end of the experiment. The survival of BR60 observed in these 1989 experiments was actually better than the survival observed in the 1988 experiments, where replacement of BR60 by other 3Cba-degrading pBRC60 carriers took place more rapidly.

Correlations Between 3Cba Dosage Uptake in BR60 and Probe-Positive Populations

As mentioned earlier, 3Cba-dosed microcosms were set up, inoculated with BR60, and monitored over several weeks in each of three summers, 1987, 1988, and 1989. During each of these years, three key relationships were always observed:

1. The probe-positive populations was consistently higher than the BR60 population in 3Cba-dosed microcosms.
2. The probe-positive population was more strongly related to the 3Cba dosage than was the BR60 population.
3. The 3Cba uptake rate was significantly correlated to the probe-positive population, but not to the BR60 population.

Details on the specific relationships for each year are given in Fulthorpe and Wyndham (1989), but Figures 3 and 4 summarize data from all 3 years together. Figure 3 illustrates the dependence of the probe-positive and BR60 populations on the 3Cba concentration in the microcosm feedwater. Figures 4 illustrates the degree of dependence of 3Cba uptake rate in the microcosm waters on the probe-positive and BR60 populations. These highly repeatable relationships, combined with the seasonal data, are compelling evidence that the plasmid pBRC60 is the unit of interest in these systems, not the original host-plasmid association.

Nature of Probe-Positive Isolates

It was clear from the early 1987 experiments that probe-positive organisms other than BR60 were taking advantage of the 3Cba inputs to these microcosms. During the subsequent release experiments in 1988 and 1989, a concerted effort was made to isolate and purify bacteria from colonies on 3Cba agar and on other media, and to confirm their catabolic capabilities and genetic makeup. The 3Cba degraders were screened for the presence of pBRC60 by probing restricted total genomic or plasmid enriched DNA with a fragment from Tn5271 (Fulthorpe & Wyndham 1991, 1992). Here we summarize the various taxonomic groups that harbored pBRC60 in these experimental microcosms. Some characteristics of these strains are given in Table 1.

Yellow-Pigmented Group III Pseudomonads "PR117." In 1988, a total of eight 3Cba degraders were found to be of this group, and they were isolated from microcosms receiving a broad range of 3Cba dosages. In 1989, several members were again isolated, including one strain that was isolated as the 3Cba-degrading partner of a 3-chlorobiphenyl degrading co-culture derived from a 3-chloro-biphenyl dosed microcosm in a separate line of investigation (Fulthorpe & Wyndham 1992). All these isolates were motile, aerobic, slightly curved Gram-negative, narrow rods bearing multiple polar or peritrichous flagella. Colonies developed a nonfluorescent yellow pigment on CPS and KING's B agar, and to a lesser degree, on nutrient agar. Cells accumulated poly-ß-hydroxybutyric acid

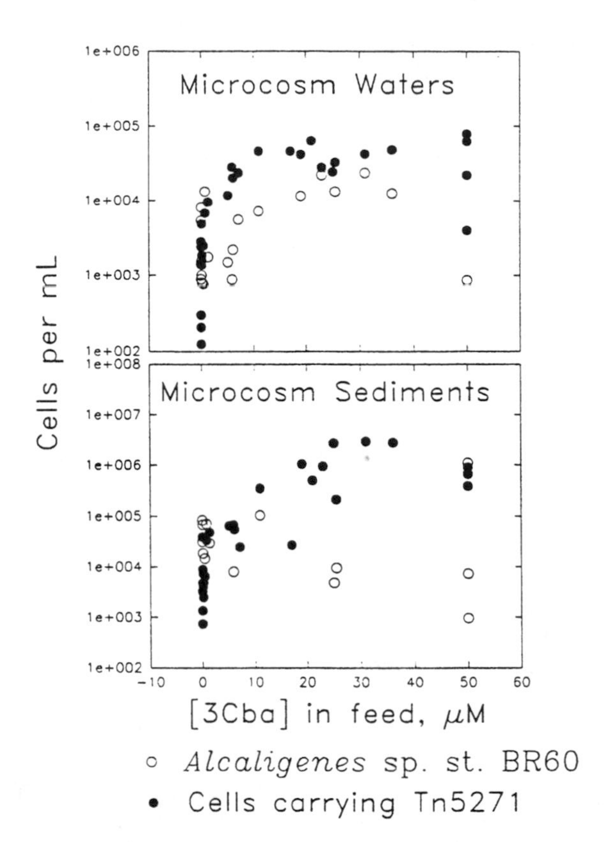

FIGURE 3. Average populations of BR60 and probe-positive cells versus 3Cba concentration in the lake water feed for all 3Cba-dosed and control (zero dose) experimental microcosms monitored in 1987, 1988, and 1989. Note that probe-positive populations generally are higher and more closely related to 3Cba inputs than are BR60 populations. Anomalously low probe-positive population in microcosm waters dosed at 50 μM was taken only 2 weeks after dosing began, and populations may not have had time to establish. All other populations are averages taken over at least 4 weeks.

as determined by the formation of sudan black crystals. The BIOLOG system could not identify these isolates, giving them a "poor" identification of *Alcaligenes faecalis* with a similarity of 0.49. All these isolates are tentatively identified as unknown members of the Group III pseudomonads, differing from all known yellow-pigmented strains in this group with respect to their use of carbohydrates.

Yellow-Pigmented Carbohydrate-Utilizing Group III Pseudomonad. PR42A grew rapidly on 3Cba plates, carried PBRC60, and was indistinguishable from the PR117 group in its colony morphology and growth on general media and on 3Cba

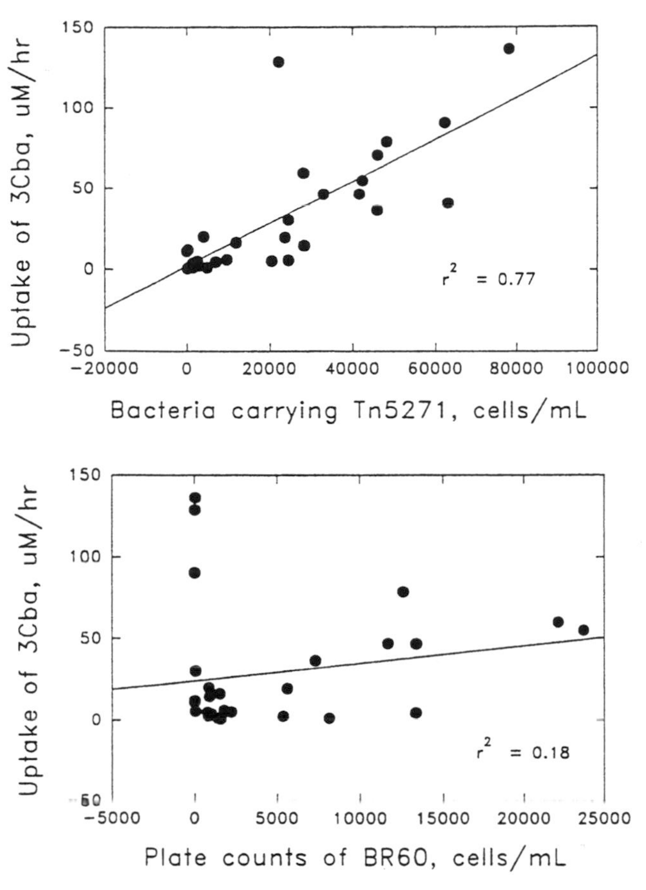

FIGURE 4. Chlorobenzoate uptake rates in microcosm waters versus probe-positive populations (top graph) and BR60 populations (bottom graph) present at time of uptake measurement. For top graph, linear regression is y = 0.00130x – 0.914; for bottom graph regression is not significant.

media. Its substrate utilization pattern was quite different from that of the PR117 group and from known Group III pseudomonads, but its GN ID gave it a poor identification of *Comomonas acidovorans*, so it could well be another member of the Group III pseudomonads. Again, no strains similar to this were re-isolated in 1989.

Non-Pigmented Unknown "PR120" Group. Only one member from this group was isolated in 1988, and it was from the microcosm receiving the highest dose of 3Cba that year (20 µM). However, it was repeatedly isolated from 25-µM and 50-µM dosed microcosms in 1989, perhaps indicative of a preference for higher 3Cba concentrations. These bacteria were motile, aerobic Gram-negative

TABLE 1. Characteristics of PBRC60 carrying strains isolated from flowthrough nonsterile microcosms.

	BR60	PR117	PR42A	PR24B	PR120	PR63
pigment	–	+	+	–	–	–
oxidase	+	+	+	+	+	+
catalase	+	+	+	+	+	+
flagellar type[(a)]	(pl)	per	per	pm	pm	pm
3-chlorobenzoate	+	+	+	+	+	–
chlorocatechol accum.	–	–	–	+	+	–
distance[(b)]	0.45	10.4	11.4	40.3	16.9	21.4

(a) pl = polar maltitrichous, variable; per = peritrichous; pm = polar monotrichous
(b) Distance from *Comomonas testosteroni* based on BIOLOG substrate use profile; units are average number of reactions not in common.

oval-shaped rods with polar monotrichous flagella forming small colorless colonies on general media. It could not be identified by its BIOLOG GN ID profile.

Fluorescent Pseudomonads. Most strains in this group lost the 3Cba+ phenotype after initial isolation. Only strain PR24B, isolated from a microcosm receiving only 0.12 µM 3Cba, retained full 3Cba metabolism and was found to carry pBRC60. It was identified as *Pseudomonas fluorescens* Biovar IV by its BIOLOG GN ID fingerprint and confirmed by its motility, polar monotrichous flagella, and the production of a fluorescent yellow-green pigment on KING's B agar. No members of this group were re-isolated in 1989 when no low-dose 3Cba experiments were carried out.

Acidovorax Species. PR63 was isolated from a microcosm receiving 10 µM 3Cba. It is also a motile aerobic, Gram-negative rod with polar monotrichous flagella. On first isolation, it exhibited scant but positive growth on 3Cba agar. This property was difficult to reestablish, nevertheless the organism hybridized to a pBRC60 probe, and extraction of the plasmid revealed that a rearrangement had taken place (see Fulthorpe & Wyndham 1991). The BIOLOG GN ID profile suggests that it is related to *Acidovorax delafieldii*; and it shares the characteristics of this genus, also a member of the Group III pseudomonads. Several similar strains isolated from a microcosm inoculated with BR60 and dosed with 50 µM 4-chloroaniline in a separate line of investigation (see Fulthorpe & Wyndham 1992) proved to be related to PR63. These strains were large, motile rods with single polar flagella identified by the BIOLOG GN ID system as *Acidovorax delafieldii*. Plasmid enriched preparations of their DNA were probed with fragments of the catabolic transposon. The strains were all carrying harbored Tn5271's flanking insertion sequence IS1071 on a plasmid unrelated to the pBRC60. These results suggest that pBRC60 and Tn5271 were unusually unstable in this particular group.

Laboratory Host Range. The results of the filter matings are shown in Table 2, where it can be seen that pBRC60 does not transfer well into known members of the Group III pseudomonads, nor into *Pseudomonas fluorescens*, at least in laboratory filter matings.

DISCUSSION

This work demonstrates some important things. First and foremost, it demonstrates that a 3-chlorobenzoate degradative element can be maintained in a "new" environment after introduction and that it can effect 3Cba degradation in those systems. It was clear after the first few weeks of release experiments that the species actually inoculated into the microcosms, BR60, was not necessarily the one that continued to carry and express the degradative genes. The important element in this study was not the nonindigenous bacteria; it was the nonindigenous plasmid.

Plasmid transfer has been demonstrated in natural aquatic environments. Bale et al. (1987) demonstrated transfer of a natural large plasmid between *Pseudomonas aeruginosa* strains on the surface of stones in a river bed at rates of 5×10^{-6} transconjugants/donor O'Morchoe et al. (1988) documented transfer of plasmids between *P. aeruginosa* strains in lake water at rates ranging from 10^{-6} to 10^{-4} transconjugants/donor, even though donor and recipient densities were low (10^3 to 10^4 cells/mL). In the present study, a comparable conjugation rate of 5×10^{-4} transconjugants per donor is calculated. The microcosm environment was conducive to plasmid transfer: that is temperatures were above 20°C and the sediment layer and natural lake water provided an abundance of solid surfaces known to enhance conjugative transfer rates (Genther et al. 1988). BR60 died off in these microcosms, but Wellington et al. (1990), O'Morchoe et al. (1988), and

TABLE 2. Results of filter mating between BR6025(pBRC40::Tn5) and various potential recipients.

Species	ATCC No.	Ratio[a]	Rate[b]
Acinetobacter sp. DON2		9	6.2×10^{-5}
Moraxella sp. G.		19	4.5×10^{-4}
P. acidovorans	15668	0.2	5.4×10^{-2}
P. cepacia	25416	82	2.2×10^{-3}
P. putida Biotype A	12633	44	5.5×10^{-4}
Chromobacterium violaceum	12472	200	nd
P. fluorescens	13525	100	nd
P. palleronii	17724	2,000	nd
P. pseudoflava	33668	6.5	nd

(a) Ratio of donor cells to recipients.
(b) Transconjugant cells per donor.
nd = transconjugants not detected.

Gealt et al. (1985) have all documented plasmid transfer from donors that did not actively grow in the test medium. Genther et al. (1988) have demonstrated that a large proportion of freshwater Gram-negative bacteria are "recipient-active" for a broad-host-range plasmid.

In this work, specific groups of pBRC60 carriers were important in these microcosms year after year. Judging by frequency of isolation alone, the yellow-pigmented PR117 and the nonpigmented PR120 group, and even BR60 itself were the important 3Cba degraders in these systems. The factors that determined the dominance of one over the other in a particular microcosm at a particular time were not clear, but were clearly variable. Under laboratory conditions, none of the recipients were efficient degraders of 3Cba, even at low concentrations (Fulthorpe & Wyndham 1991), so factors such as the nature of the organic matter in the sediments, the predatory community and the ability of the organism to attach to the surface of particles are probably of key importance. In spite of the dominance of the indigenous pBRC60 carriers in these systems, we cannot say that BR60 was an unimportant member of the 3Cba-degrading community because it showed fairly good survival for a time in 1989 and completely dominated one microcosm after 3Cba was added. We would not have predicted this from results from earlier experiments. There did seem to be some relationship between the microcosm 3Cba-dose and the type of pBRC60 carrier isolated, for instance only PR117 groups members and a *Pseudomonas fluorescens* pBRC60 carrier were isolated from very-low-dose microcosms, but the data are insufficient to generalize on this point.

What is most striking about the host range of pBRC60 in these systems is its lack of similarity to the host range as determined in the laboratory. It did not transfer well into two well-known Group III pseudomonads, nor into a *Pseudomonas fluorescens* strain, and yet Group III pseudomonads, albeit unknown members of this group, dominated the indigenous recipients, and a *Pseudomonas fluorescens* pBRC60 carrier was found. These facts emphasize gaps in our knowledge about plasmid behavior after introduction. Prediction of host range is complicated by our limited catalogue of nonpathogenic bacteria and problems of culturability and counterselection inherent in the methods used to study conjugation. A particular catabolic genotype may do much better than expected if its survival characteristics are predicted from the survival of its host and its apparent host range in the laboratory.

ACKNOWLEDGMENTS

This project was funded by the Natural Science and Engineering Research Council of Canada, and we thank the Queen's University Biology Station for the use of their facilities.

REFERENCES

Bale, M. J., J. C. Fry, and M. J. Day. 1987. "Plasmid transfer between strains of *Pseudomonas aeruginosa* on membrane filters attached to river stones." *J. Gen. Microbiol.* *133*: 3099-3107.

Brunner, W., F. H. Sutherland, and D. D. Focht. 1985. "Enhanced biodegradation of polychlorinated biphenyls in soil by analog enrichment and bacterial inoculation." *J. Environ. Qual.* *14*: 324-328.

Carlson, C., S. Steenberg, and J. Ingraham. 1985. "Natural transformation of *Pseudomonas stutzeri* by plasmids that contain cloned fragments of chromosomal DNA." *Arch. Microbiol. 140*: 134-138.

Chatterjee, D. K., and A. M. Chakrabarty. 1983. "Genetic homology between independently isolated chlorobenzoate-degradative plasmids." *J. Bacteriol. 153*: 532-534.

Edgehill, R. U., and R. K. Finn. 1983. "Microbial treatment of soil to remove pentachlorophenol." *Appl. Environ. Microbiol. 45*: 1122-1125.

Focht, D. D., and D. Shelton. 1987. "Growth kinetics of *Pseudomonas alcaligenes* C-0 relative to inoculation and 3-chlorobenzoate metabolism in soil." *Appl. Environ. Microbiol. 53*: 1846-1849.

Fredrickson, J. K., D. F. Bezdicek, F. J. Brockman, and S. W. Li. 1988. "Enumeration of Tn5 mutant bacteria in soil using a most probable number - DNA hybridization procedure and antibiotic resistance." *Appl. Environ. Microbiol. 54*: 446.

Fulthorpe, R. R. 1991. "Survival and transfer of a catabolic transposon in a freshwater ecosystem." PhD. Thesis. Carleton University, Ottawa, Ontario, Canada.

Fulthorpe, R. R., and R. C. Wyndham. 1992. "Involvement of a chlorobenzoate catabolic transposon, Tn5271, in community adaptation to chlorobiphenyl, chloroaniline, and 2,4 dichlorophenoxyacetic acid in a freshwater ecosystem." *Appl. Environ. Microbiol. 58*: 314-325.

Fulthorpe, R. R., and R. C. Wyndham. 1991. "Transfer and expression of the catabolic plasmid pBRC60 in wild bacterial recipients in a freshwater ecosystem." *Appl. Environ. Microbiol. 57*: 1546-1553.

Fulthorpe, R. R., and R. C. Wyndham. 1989. "Survival and activity of a 3-chlorobenzoate catabolic genotype in a natural system." *Appl. Environ. Microbiol. 55*: 1584-1590.

Fulthorpe, R. R., N. A. Straus, and R. C. Wyndham. 1989. "Bacterial adaptation to chlorobenzoate contamination in the Niagara region investigated by DNA:DNA colony hybridization." In G. W. Suter and M. A. Lewis (Eds.), *Aquatic Toxicology and Environmental Fate: Eleventh Volume*, pp. 59-71. ASTM STP 1007. American Society for Testing and Materials, Philadelphia, PA.

Gealt, M. A., M. D. Chai, K. B. Alpert, and J. C. Boyer. 1985. "Transfer of plasmids pBR322 and pBR325 in wastewater from laboratory strains of *Escherichia coli* to bacteria indigenous to the waste disposal system." *Appl. Environ. Microbiol. 49*: 836-841.

Genther, F. J., P. Chatterjee, T. Barkay, and A. W. Bourquin. 1988. "Capacity of aquatic bacteria to act as recipients of plasmid DNA." *Appl. Environ. Microbiol. 54*: 115-117.

Goldstein, M., L. M. Mallory, and M. Alexander. 1985. "Reasons for possible failure of inoculation to enhance biodegradation." *Appl. Environ. Microbiol. 50*: 977-983.

Lewis, D. L., R. E. Hodson, and H.-M. Hwang. 1988. "Kinetics of mixed microbial assemblages enhance removal of highly dilute organic substrates." *Appl. Environ. Microbiol. 54*: 2054-2057.

Liang, L. N., J. L. Sinclair, L. M. Mallow, and M. Alexander. 1982. "Fate in model ecosystems of microbial species of potential use in genetic engineering." *Appl. Environ. Microbiol. 44*: 707-714.

Morrison, W. D., R. V. Miller and G. S. Sayler. 1978. "Frequency of F116-mediated transduction of *Pseudomonas aeruginosa* in a freshwater environment." *Appl. Environ. Microbiol. 36*: 724-730.

McClure, N. C., A. J. Weightman, and J. C. Fry. 1989. "Survival of *Pseudomonas putida* UWC1 containing cloned catabolic genes in a model activated sludge unit." *Appl Environ. Microbiol. 55*: 2627-2634.

Nakatsu, C., J. Ng, R. Singh, N. Straus, and C. Wyndham. 1991. "Chlorobenzoate catabolic transposon Tn5271 is a composite class I element with flanking class II insertion sequences." *Proc. Nat. Acad. Sci. 88*: 8312-8316.

O'Morchoe, S. B., O. Ogunseitan, G. S. Sayler, and R. V. Miller. 1988. "Conjugal transfer of R68.45 and FP5 between *Pseudomonas aeruginosa* strains in a freshwater environment." *Appl. Environ. Microbiol. 54*: 1923-1929.

Ramadan, M. A, O. M. El-Tayeb, and M. Alexander. 1990. "Inoculum size as a factor limiting success of inoculation for biodegradation." *Appl. Environ. Microbiol. 56*: 1392-1396.

Rubin, H.E., Subba-Rao, R.V., and M. Alexander. 1982. "Rates of mineralization of trace concentrations of aromatic compounds in lake water and sewage samples." *Appl. Environ. Microbiol. 43*: 1133-1138.

Scheuerman, P. R., J. P. Schmidt, and M. Alexander. 1988. "Factors affecting the survival and growth of bacteria introduced into lake water." *Arch. Microbiol. 150*: 320-325.

Schmidt, S. K., and M. Alexander. 1985. "Effects of dissolved organic carbon and second substrates on the biodegradation of organic compounds at low concentrations." *Appl. Environ. Microbiol. 49*:822-827.

Smibert, R. M., and N. R. Kreig. 1981. "Systematics: General characterization." In P. Gerhardt, R. G. E. Murray, R. N. Costilow, E. W. Nester, W. A. Wood, N. R. Krieg, and G. B. Phillips (Eds.), *Manual of Methods for General Bacteriology*, pp. 409-443. American Society of Microbiology, Washington, D.C.

Stewart, G. J., and C. D. Sinigalliano. 1990. "Detection of horizontal gene transfer by natural transformation in native and introduced species of bacteria in marine and synthetic sediments." *Appl. Environ. Microbiol. 56*: 1818-1824.

Stotzky, F., and H. Babich. 1986. "Survival of and genetic transfer by genetically engineered bacteria in natural environments." *Adv. Appl. Microbiol. 31*: 93-138.

Trevors, J. T., T. Barkay, and A. W. Bourquin. 1987. "Gene transfer among bacteria in soil and aquatic environments: A review." *Can. J. Microbiol. 33*: 191-198.

Wellington, E. M., N. Cresswell, and V. Saunders. 1990. "Growth and survival of streptomycete inoculants and extent of plasmid transfer in sterile and nonsterile soil." *Appl. Environ. Microbiol. 56*: 1413-1419.

Wyndham, R. C., and N. A. Straus. 1988. "Chlorobenzoate catabolism and interactions between *Alcaligenes* and *Pseudomonas* from Bloody Run Creek." *Arch. Microbiol. 150*: 230-236.

Wyndham, R. C., R. K. Singh, and N. A. Straus. 1988. "Catabolic instability, plasmid gene deletion and recombination in *Alcaligenes* sp. BR60." *Arch. Microbiol. 150*: 237-243.

Zeyer, J., A. Wasserfallen, and K. N. Timmis. 1985. "Microbial mineralization of ring-substituted anilines through an ortho-cleavage pathway." *Appl. Environ. Microbiol. 55*:447-453.

EFFECT OF SALINITY, OIL TYPE, AND INCUBATION TEMPERATURE ON OIL DEGRADATION

J. R. Haines, M. Kadkhodayan, D. J. Mocsny, C. A. Jones, M. Islam, and A. D. Venosa

ABSTRACT

The effects of salinity, oil type, and temperature on crude oil biodegradation were studied using mixed cultures obtained from beaches in Prince William Sound, Alaska. The rate and extent of biodegradation were monitored by oxygen consumption and carbon dioxide production. Hydrocarbon degradation was confirmed by GC-MS. Salinity (between zero and 30 g NaCl/L) inhibited oil biodegradation: the lag times increased and oil biodegradation rate and extent decreased as the salinity increased. The effects of temperature were mixed. In all cases, the lag period decreased as temperature increased. The O_2 uptake rates (mg/L-day) increased between 12°C and 20°C, but were unaffected between 20° and 28°C. The maximum amount of O_2 consumed and CO_2 produced was different for each oil studied. Analysis of residual oil showed greater degradation of alkanes and aromatics from the crude oils than from number 2 fuel (F2) oil. Up to 96% of the alkanes were removed from the crude oils, whereas only 79% were removed from F2. Aromatic hydrocarbon degradation was more variable, ranging between a minimum of about 20% removal from North Slope (NS) crude oil to a maximum of nearly 70% from light Arabian (LA) crude oil.

INTRODUCTION

Spills of crude oil and refined petroleum products are common, because of the scale on which petroleum products are produced, transported, and consumed in modern society. A variety of technologies can be applied to clean up spilled petroleum. Physical-chemical treatment methods, such as absorbents, can effectively remove petroleum from some contaminated environments. Absorbents, however, are frequently landfilled or incinerated, which can generate air pollutants. The grounding of the *Exxon Valdez* focused attention on the use of bioremediation as an alternative cleanup tool. Biological treatment processes can convert petroleum to relatively innocuous products such as biomass, CO_2, and water. Although treatment rates can be slow and not all components are degradable, microbial degradation has been shown to reduce the quantity of polluting oil and the overt toxicity of petroleum components (U.S. EPA, 1991).

The biodegradation of oil has been studied and reviewed extensively, and the effects of environmental factors on petroleum degradation have been detailed (Atlas 1981,1984; Leahy & Colwell 1990). Although the interactions of different environmental factors have seldom been examined quantitatively, several broad generalizations can be made regarding the factors that affect the rate and extent of petroleum biodegradation. First, degradation is often slower at lower temperatures (Atlas 1986; Haines & Atlas 1982, 1983). Second, nitrogen, phosphorus, and iron are frequently necessary to enhance degradation (Dibble & Bartha 1976, 1979; Moucawi et al. 1981). Molecular oxygen is usually required to support hydrocarbon degradation. Finally, hydrocarbon-degrading bacteria are quite common in the environment. Most work has focused on one oil type (Ward & Brock 1976), or the effect of temperature has been studied for only one set of variables (kerosene spiked with tracer compounds, with one added nutrient mixture) (Cooney et al. 1985).

This study was undertaken as part of a research program to develop protocols that will provide a unified framework for testing and evaluating products designed to stimulate oil biodegradation (Venosa et al. 1991, Venosa et al. 1992). This bioremediation product testing protocol will incorporate and standardize the most important environmental variables that control petroleum degradation. During the protocol development effort, selected environmental factors will be varied to determine their effects on performance of either indigenous or inoculated microorganisms degrading several different kinds of oil. We have varied temperature, salinity, and oil type and have quantified their effects on the rate and extent of oil biodegradation by enrichment cultures. This paper summarizes the results of this work.

MATERIALS AND METHODS

Cultures

Oil-degrading consortia were originally derived from sediment and water samples from different sites in Prince William Sound, Alaska. Small quantities (10 g) of sediment were added to flasks containing 100 mL of modified Bushnell-Haas (BH) salts medium (Difco Laboratories, Detroit, Michigan) and 1 g/L North Slope (NS) crude oil. The BH medium was modified by adding more nitrogen (7.25 g/L KNO_3) and 20 g/L NaCl. The flasks were incubated on a shaker (200 rpm) for 14 d at 20°C. Transfers of cultures were made every three weeks thereafter into fresh BH medium with the same quantity and type of oil.

Respirometry

Oxygen uptake and CO_2 production were measured using N-CON model WB512 respirometers (N-CON Instruments, Larchmont, New York). The instrument uses a system of sensitive pressure sensors and calibrated solenoid valves to quantify and track the addition of oxygen to reaction flasks. Consumption of oxygen by the active biomass lowers the pressure in the reaction flask, and the sensors detect the change. A calibrated solenoid valve then delivers enough O_2 to balance the system. A computer records and calculates cumulative O_2 consumption as mg/L and O_2 uptake rate as mg/L/hr.

Each respirometer flask is 500 mL in volume and is sealed with a gastight cap containing a KOH trap for scrubbing CO_2 from the reactor headspace. The KOH trap is a polymethylpentene centrifuge tube fitted with a stainless steel cannula extending to the bottom of the tube. A syringe valve connected to the top of the cannula preserves the pressure integrity of the system and permits removal of the KOH for CO_2 measurements. The KOH solutions were added or withdrawn from the traps with a gastight syringe. The need to change the KOH trapping solution was determined by the color change observed in the KOH. The color change was indicated by 100 mg/L Alizarin red S dye dissolved in the KOH. Alizarin red S changes from purple to orange between pH 11 and 10.

CO_2 produced was quantified by measuring the pH of the KOH solution. The mass of CO_2 was calculated using a computer program written in our laboratory (D. Mocsny, personal communication). This program requires as inputs the normality, volume, temperature, and measured pH of the KOH in the CO_2 traps. The validity of this program was evaluated by using primary-standard Na_2CO_3 to produce known amounts of CO_2, which was subsequently trapped in KOH. The trapping efficiency was determined by precipitating the CO_2 as $BaCO_3$ with a $BaCl_2$ solution. The mass of precipitated $BaCO_3$ was determined gravimetrically, and the results were compared with those obtained using the pH measurement and calculation program. The methods agreed within 5%. The program is available from J. R. Haines.

Oil Chemistry

The oil remaining in the reaction flasks at the conclusion of each experiment was extracted into hexane (100 mL). At the beginning of extraction, 1 mL of a 400 ng/μL d_{10}-phenanthrene recovery standard was added to each flask. The hexane extract was separated from the aqueous phase and dried by passing a subsample through Na_2SO_4, 30 mL of the dried extract was collected for analysis. Prior to analysis, an internal standard mixture was added to each sample. The mixture contained d_8-naphthalene, d_{10}-anthracene, d_{12}-chrysene, and d_{12}-perylene in methylene chloride at 500 ng/μL. Each sample was analyzed using a Hewlett Packard 5890/5971 gas chromatograph-mass spectrometer (GC-MS). The operating conditions were column, DB-5; 30 m, 0.25 mm ID, 0.25 μm film thickness; splitless injection from an autosampler; injector temperature, 290°C; initial temperature, 55°C for 3 min; increase by 5°C/min to 280°C; then increase by 3°C/min to 310°C and hold for 10 min for a total run time of 73 min. The transfer line was held at 320°C. The GC-MS was operated in the selective ion monitoring (SIM) mode at 1.54 scans per second. Prior to analysis of samples a 5-point calibration curve was established. The relative response factors for a series of alkane and aromatic hydrocarbons were measured and used to calculate subsequent data. The QA/QC standards for our laboratory were derived from EPA method 8270. Hydrocarbon analysis data were reported as nanograms of analyte per milligram of original oil rather than quantity of analyte per quantity of water.

Salinity

The effect of salinity on oxygen consumption by oil-degrading organisms was evaluated in BH broth by varying the NaCl concentration between 0 and 30 g/L. Each respirometer flask contained 250 mL BH broth, 1,250 mg NS crude oil, and 1 mL of inoculum (10^6 cells/mL final). The inoculum was a mixed culture grown on NS oil as described above. The respirometer flasks were incubated at 20°C, and oxygen uptake was monitored for 30 days. CO_2 evolution was measured as described above. Oil analysis was not done for this series of experiments.

Incubation Temperature and Oil Type

The effect of incubation temperature on oxygen consumption, CO_2 evolution, and oil degradation was measured in a separate experiment. Respirometer flasks were prepared with 250 mL BH broth containing 20 g/L NaCl. Replicate flasks received 1,250 mg of South Louisiana (SL) crude, NS crude, light Arabian (LA) crude, and No. 2 (F2) fuel oil. Flasks were incubated at 12°C, 20°C, and 28°C. The CO_2 production and O_2 consumption were monitored as previously described. At the end of the incubation period, the residual oil was extracted with hexane and analyzed as described above.

RESULTS

Salinity

The effect of increasing NaCl concentration on the O_2 uptake and CO_2 production of NS crude oil degrading microorganisms is summarized in Figure 1. The flasks without added NaCl showed the most rapid onset of O_2 uptake, followed in order by flasks containing increasing concentrations of NaCl. The maximum rate of O_2 uptake was observed in cultures without added NaCl. The oxygen uptake rates followed the same pattern as the onset of O_2 uptake. The observed cumulative quantity of O_2 consumed was unaffected by the salinity, ranging from 2,000 to 2,800 mg O_2/L. Production of CO_2 closely tracked O_2 consumption.

Temperature Effects on Biodegradation of Different Oils

Replicate cultures of microorganisms were incubated concurrently at 12, 20, and 28°C to determine if incubation temperature affected the extent or selectivity of oil degradation. Four oils were used in these experiments: the three crude

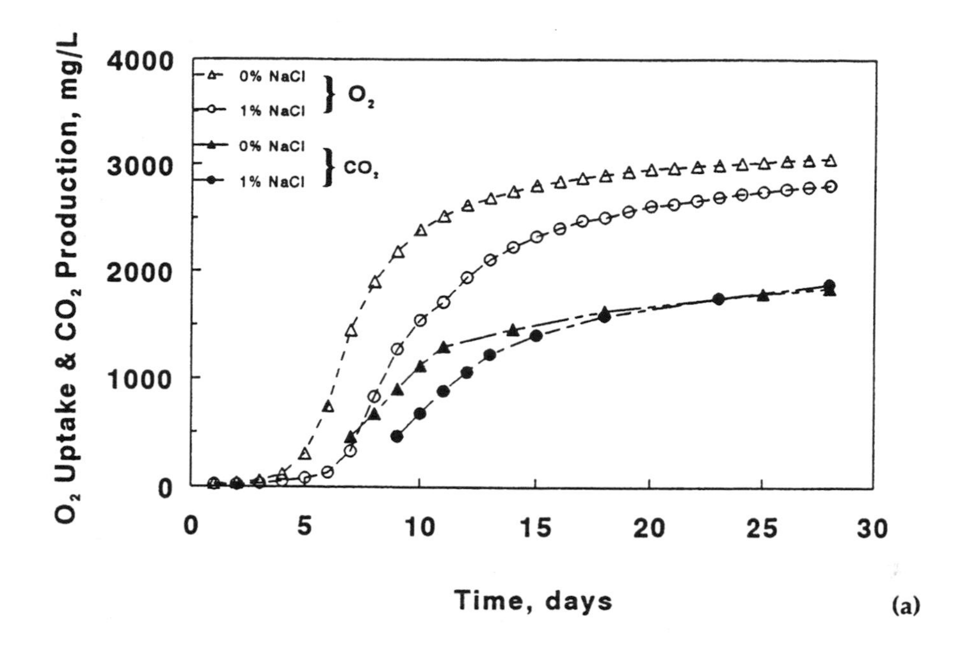

(a)

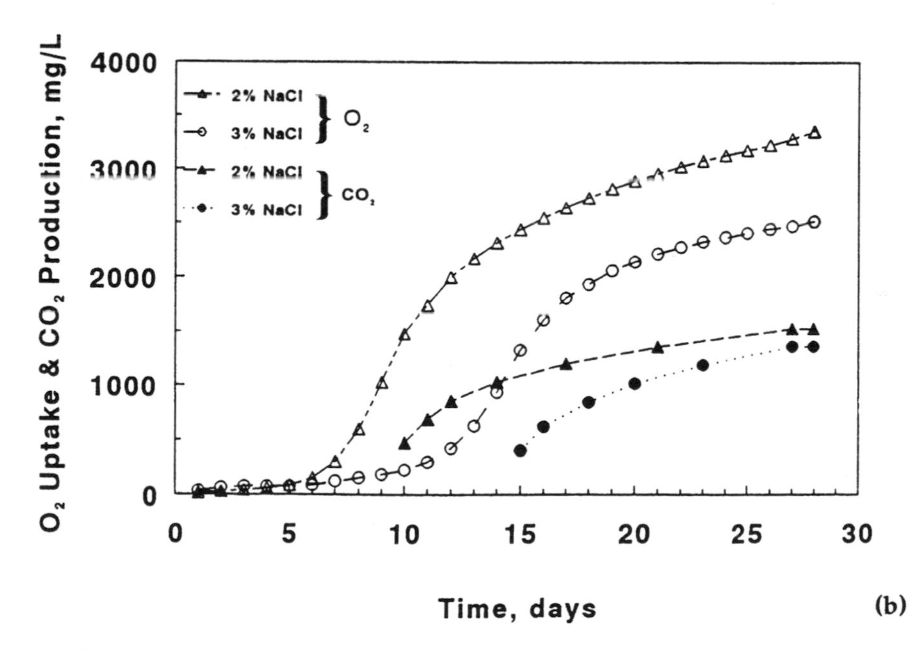

(b)

FIGURE 1. O_2 uptake and CO_2 production by microorganisms growing on North Slope crude oil at different NaCl concentrations. (a) 0 and 1% NaCl, (b) 2 and 3% NaCl.

oils (LA, NS, SL) and the one refined fuel (F2). The O_2 uptake, CO_2 production, and oil degradation were monitored for up to 40 days of incubation. Table 1 summarizes the results of these experiments in terms of O_2 uptake and CO_2 production. Only SL and NS crude oils yielded a difference in ultimate oxygen consumption at different temperatures. At 28°C, both NS and SL consumed more than 4,500 mg/L O_2. At lower temperatures, O_2 uptake was 4,000 mg/L or less for both oils. CO_2 production did not track O_2 consumption as a function of temperature in a predictable manner, implying that temperature might have affected the flow of carbon to biomass and CO_2 differently in the enrichment cultures. Both the LA and F2 oils produced relatively higher cumulative CO_2 production than did the NS and SL oils.

In all cases, increasing the incubation temperature from 12 to 20°C caused an increase in the rate of O_2 uptake and CO_2 production, as shown in Figure 2. No further change in the rate of O_2 uptake was observed in three of the four oil types when temperature was increased from 20 to 28°C. The changes in the target resolvable hydrocarbons over the period of the experiment are summarized in Table 2. Both aliphatic and aromatic hydrocarbons were highly degraded in all treatments, and the extent of degradation was about the same at all incubation temperatures. Alkanes were degraded by more than 90% in the three crude oils, but only by 70 to 80% in the No. 2 fuel oil. Aromatics were degraded almost equally among all four oil types. The highest losses of aromatic hydrocarbons were associated with incubation at 28°C for each of the crude oils.

TABLE 1. Effect of incubation temperature on O_2 uptake and CO_2 production from No. 2 fuel, Light Arabian, North Slope, and South Louisiana crude oils.

Oil Type	Incubation Temperature	O_2 Uptake, Cumulative mg/L	CO_2 Production, Cumulative mg/L
No. 2 Fuel Oil	12	2880[a]	2192[a]
	20	3050	2510
	28	3039	2193
Light Arabian	12	3300	2185
	20	3762	2856
	28	3340	2884
North Slope	12	2932	1879
	20	2172	2172
	28	5154	1656
South Louisiana	12	3745	2497
	20	4009	2343
	28	4573	3000

(a) Values are the mean of three replicates.

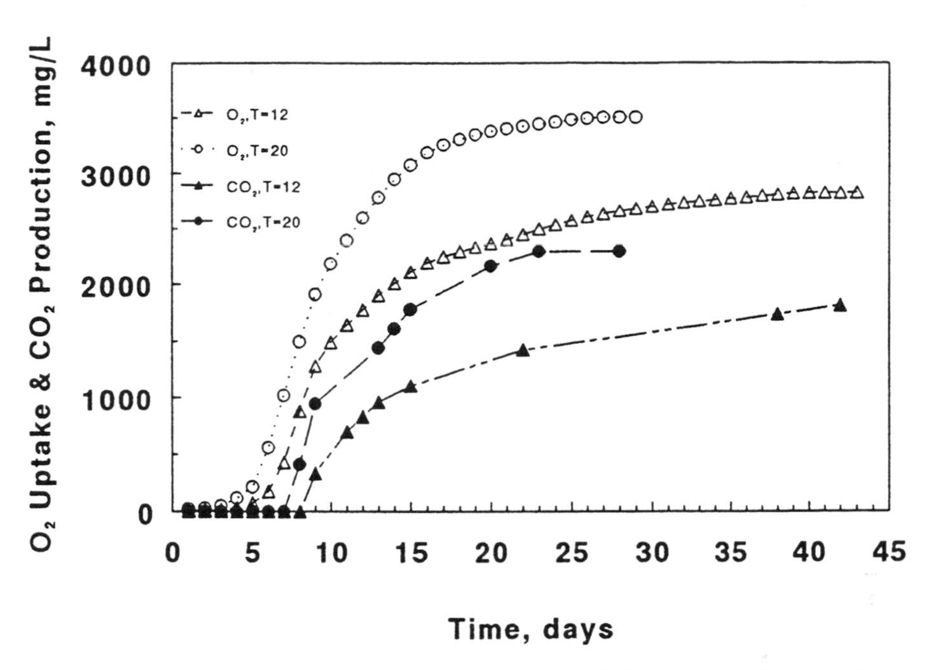

FIGURE 2. Effect of temperature on O_2 uptake and CO_2 production by microorganisms growing on light Arabian crude oil.

DISCUSSION

The active organisms in these experiments were isolated from a marine environment in which salinity ranges from near 0 to 30+ parts per thousand twice daily. These extreme changes occur due to the tidal flux of water into and out of beaches in Prince William Sound. Due to heavy rainfall there is also a constant input of fresh water into the body of most beaches in the Sound. In our experiments, the best degradation of oil occurred at the lowest NaCl content. An implication of this result is that removal of spilled oil from a saline environment and transfer to a lower salinity treatment area may enhance the ultimate biodegradation of the oil. If native marine organisms are restricted in their ability to utilize oil simply because they exist in a saline environment, the development of remediation products specifically adapted to the marine environment would be beneficial. Oil-degrading halophilic microorganisms would be potentially useful in marine environments where halotolerant organisms may only survive but may not be active in degrading oil.

The effect of temperature on oil degradation appears to be less important than other factors. Temperature changes from 20 to 28°C had little effect on the degradation of some oils. Increasing the temperature from 12 to 20°C had a more pronounced effect. The most likely reason for the change lies in the relative

TABLE 2. Biodegradation of alkanes and aromatic hydrocarbons at 12, 20, and 28°C.

Oil Type	$T^{o(a)}$ Alkanes[b]	$T^{o(a)}$ Aromatics[b]	Temperature °C	$T_f^{(a)}$ Alkanes	T_f^{a} Aromatics
Light Arabian	127,515	10,678	12	4,197	3,274
			20	4,731	7,614
			28	3,854	4,246
No. 2 Fuel Oil	116,938	40,098	12	25,021	21,139
			20	36,197	25,205
			28	27,763	21,322
North Slope	71,141	11,195	12	2,914	8,776
			20	2,711	7,586
			28	2,664	7,278
South Louisiana	72,582	7,370	12	5,729	4,835
			20	3,496	3,377
			28	3,565	2,672

(a) T_0 = time zero; T_f = final time of the experiment.
(b) All quantities are expressed as ng analyte/mg total oil.

fluidity of the oil and the corresponding availability of various oil components to active cells. One aspect of availability is the solubility of aromatic hydrocarbons as reported by Stucki and Alexander (1987). As a class, aromatic hydrocarbons are more resistant to biodegradation than the alkanes. Resistance is variable depending on prior exposure of the environment to hydrocarbons, as shown by Heitkamp et al. (1987). The three different types of crude oil were relatively equal in their susceptibility to biodegradation. The refined product, No. 2 fuel oil, was slightly more resistant to biodegradation with respect to the aliphatics fraction. Future experiments will attempt to identify the metabolic products that accumulate as a result of biodegradation in the reaction flasks. This should give us a better insight on the factors that lead to the differences observed.

The effects of several environmental variables on oil degradation have been demonstrated in this work. The results suggest how the performance of microbial bioremediation products might be improved. Although some factors are not amenable to manipulation, those environmental factors that are amenable should be carefully examined. The salinity of large bodies of water cannot be changed, but organisms can be selected to optimize performance in a saline environment. Similarly, a series of products could be developed with the prevailing temperature regimes of potential spill areas as the dominant selective factor. Products may be developed that are specific for classes of petroleum such as fuels, refined oils, and crudes.

Clearly, no one product or evaluation protocol will serve as a universal solution to the bioremediation of petroleum spills. A series of products will probably be needed, as will a series of protocols for testing those products. Our goal is to develop rational, equitable protocols for evaluating the potential effectiveness of products under a wide variety of conditions. This will ensure that cost-effective treatment options will be available to officials in charge of cleanup operations.

REFERENCES

Atlas, R. M. 1981. "Microbial Degradation of Petroleum Hydrocarbons: an Environmental Perspective." *Microbiol. Rev. 45*: 180-209.

Atlas, R. M. (ed.). 1984. *Petroleum Microbiology*. Macmillan Publ. Co., New York, NY.

Atlas, R. M. 1986. "Fate of Petroleum Pollutants in Arctic Ecosystems." *Water Sci. Technol. 18*: 59-67.

Cooney, J. J., S. A. Silver, and E. A. Beck. 1985. "Factors Influencing Hydrocarbon Degradation in Three Freshwater Lakes." *Microbial Ecol. 11*: 127-137.

Dibble, J. D., and R. Bartha. 1976. "Effect of Iron on the Biodegradation of Petroleum in Seawater." *Appl. Environ. Microbiol. 31*: 544-550.

Dibble, J. D., and R. Bartha. 1979. "Effect of Environmental Parameters on the Degradation of Oil Sludge." *Appl. Environ. Microbiol. 37*: 729-739.

Haines, J. R., and R. M. Atlas. 1982. "*In Situ* Microbial Degradation of Prudhoe Bay Crude Oil in Beaufort Sea Sediments." *Mar. Environ. Res. 7*: 91-102.

Haines, J. R., and R. M. Atlas. 1983. "Biodegradation of Petroleum Hydrocarbons in Continental Shelf Regions of the Bering Sea." *Oil Petrochem. Pollut. 1*: 85-96.

Heitkamp, M. A., J. P. Freeman, and C. E. Cerniglia. 1987. "Naphthalene Biodegradation in Environmental Microcosms: Estimates of Degradation Rates and Characterization of Metabolites." *Appl. Environ. Microbiol 53*: 129-136.

Leahy, J. G., and R. R. Colwell. 1990. "Microbial Degradation of Hydrocarbons in the Environment." *Microbiol. Rev. 54*: 305-315.

Moucawi, J., E. Fustec, P. Jambu, A. Ambles, and R. Jacquesy. 1981. "Biooxidation of Added and Natural Hydrocarbons in Soils: Effect of Iron." *Soil Biol. Biochem. 13*: 335-342.

Stucki, G., and M. Alexander. 1987. "Role of Dissolution Rate and Solubility in Biodegradation of Aromatic Compounds." *Appl. Environ. Microbiol. 53*: 293-297.

United States Environmental Protection Agency. 1991. *Alaska Oil Spill Bioremediation Project.* EPA/600/9-91/046a. Gulf Breeze, FL.

Venosa, A. D., J. R. Haines, W. Nisamaneepong, R. Govind, S. Pradhan, and B. Siddique. 1991. "Screening of Commercial Inocula for Efficacy in Stimulating Oil Biodegradation in Closed Laboratory System." *J. Haz. Mater. 28*: 131-144.

Venosa, A. D., J. R. Haines, and D. M. Allen. 1992. "Efficacy of Commercial Inocula in Enhancing Biodegradation of Weathered Crude Oil Contaminating a Prince William Sound Beach." *J. Ind. Microbiol. 10*: 1-11.

Ward, D. M., and T. D. Brock. 1976. "Environmental Factors Influencing the Rate of Hydrocarbon Oxidation in Temperate Lakes." *Appl. Environ. Microbiol. 31*: 764-772.

CLEANING OF RESIDUAL CONCENTRATIONS WITH AN EXTENSIVE FORM OF LANDFARMING

J. Harmsen, H. J. Velthorst, and I. P. A. M. Bennehey

ABSTRACT

Biological soil treatment by landfarming is a relatively simple and inexpensive method of cleaning soil contaminated by different organic compounds. Although microorganisms are able to degrade a large part of the pollution, residual concentration is left that is slowly or not noticeably degraded by microorganisms present in the soil. This non-bioavailable fraction of the pollutant is recognized by its slow transport from micropores in the soil in which the pollutant is adsorbed to the microorganism. Because it is difficult to increase the bioavailability of the residual concentration during the short term, it is necessary to extend the treatment time to give the pollutants time to desorb and become bioavailable. To prevent excessive costs, this extra time must be coupled with a decrease in activities on the landfarm. Therefore an intensive treatment in which the bioavailable part is removed must be followed by an extensive form of landfarming. The boundary conditions of this second cleaning step, extensive landfarming, are discussed.

INTRODUCTION

Soil polluted with organic contaminants can be treated using landfarming, a relative simple biological soil-cleaning technique, that is inexpensive. The polluted soil is spread out on a specially constructed site, where microbiological degradation is stimulated by soil cultivation and the addition of nutrients. Biological techniques are important, because several functions can be given to the cleaned soil while it is still bioactive, in contrast, for example, to thermal treatments. However, not all of the pollution can be removed with landfarming treatment. Depending on the properties of the polluting substance, soil characteristics and pollution age, 60 to almost 100% can be degraded by microorganisms (Harmsen 1991). The residual concentration thatis left, can be low, but often exceeds the standards for clean soil set by the government. It is said that the residual concentration is nonbioavailable.

Bioavailability is hard to define. A useful definition is the fraction that is in equilibrium with pore water, including microorganisms. This definition, however, takes no account of the time period in which the equilibrium should be achieved. In a landfarm months or years are available to clean the soil, whereas in a bioreactor the treatment must be restricted to days to prevent excessive costs. This paper discusses bioavailability relative to the possibilities of landfarming.

FACTORS INFLUENCING BIOAVAILABILITY

Organic Matter

Organic pollutants are adsorbed to the soil. The organic matter fraction is the most important fraction describing the partition of nonionic organic contaminants between soil particles and pore water. The adsorption to organic matter can be related to solubility; the lower the solubility, the stronger the adsorption (large partition coefficient). Partition is based on equilibrium, but this is not always achieved in soil. The value of the partition coefficient increases with longer contact periods (Boesten & van der Pas 1983).

Brusseau and Rao (1989) related the sorption rate constant, which is a measure for the time in which equilibrium can be achieved, to the partition coefficient. Using desorption experiments it was shown that soil the first desorbed fraction from a polluted soil follows equilibrium and that the desorption rate of the second fraction decreases with increasing partition coefficient (Harmsen 1991). The nonequilibrium behavior in soil can be made clear by the characteristics of the organic matter. This is not a smooth surface on which reversible sorption is possible. It has micropores that the pollutant must enter by diffusion. In the pores, the diffusion rate is retarded due to sorption to the organic matter. A component with or with little adsorption diffuses quickly into the pores. A strongly adsorbed component is retarded, and therefore the diffusion rate is much smaller (Wu & Gschwend 1986). Another field and laboratory experiment (Pignatello et al. 1990) indicates that slow desorption of pollutants results from molecular diffusion in organic matter and in intraparticle micropores.

Intraparticle Micropores

In addition to adsorption, slow desorption is often caused by entrapment in intraparticle micropores. Steinberg et al. (1987) presented results of 1,2-dibromoethane, a compound that only slightly adsorbs to organic matter and therefore should have a high desorption rate. Part of 1,2-dibromoethane, however, desorbs slowly, which has been attributed to diffusion into very small micropores. The cross section of these pores is of the same magnitude as the size of the molecules, and diffusion out of the pores is retarded by steric hindrance. This may be the reason for the lower effectiveness of in situ soil remediation techniques such as pump-and-treat and soil-air-venting technologies (Grathwohl et al. 1990). After a decrease in concentration in the water phase during treatment, the concentration increases again when the treatment is stopped.

In practice adsorption in organic matter and entrapment in micropores will occur simultaneously, and the distribution over both adsorption sites will depend on the sorption characteristics of the pollutant and the soil.

Time

It must be realized that in soil remediation practice, desorption always follows adsorption. The pollutant has been exposed in the soil for a longer period. During that period the fraction adsorbed on the adsorption sites in the micropores is increasing and the bioavailable fraction is decreasing. A recent pollution is adsorbed on the surface of a soil particle or soil aggregate and is attainable for microorganisms, and consequently a large fraction will be degraded. After adsorption on the surface, redistribution of the pollution over the entire soil particle takes place. The pollutant diffuses into the micropores. The bioavailable fraction decreases and the nonbioavailable fraction is not degraded during the treatment period. To be degraded, the nonbioavailable pollutant must shift from the micropores to larger pores where microorganisms can be present. This transport is a diffusion process in which time is an important parameter.

As a result of the diffusion process, landfarming treatment of a recent and an old pollution in soil gives different cleaning rates. A recent pollution is adsorbed on the surface and will be removed almost completely during biological treatment (Figures 1A and 1C) (Harmsen 1993). An old pollution is also adsorbed inside soil particles or aggregates, and the same biological treatment gives a lower cleaning rate (Figures 1B and 1D).

A TWO-STEP CLEANING APPROACH

Intensive Landfarming

Using the traditional way of landfarming, a degradation curve as given in Figure 2A is obtained. The term "intensive landfarming" is used to distinguish this form of landfarming from "extensive landfarming" (see next section). Landfarming starts with creating more or less optimal conditions for microbiological activity. Important are aerobic conditions, sufficient nutrients and a temperature high enough to support active microorganisms. After an adaptation period, which is often very short in a soil polluted years ago, the bioavailable part is degraded relatively quickly. This may be weeks for pentachlorophenol (Figure 2A) to years for a complex pollution such as crude oil (Harmsen 1991). The treatment period can be shortened by optimizing the process, for example by increasing the temperature by working indoors. It should be realized that too much optimization leads to a technique comparable to a bioreactor, and consequently costs more; landfarming must be an inexpensive soil remediation method.

Optimization, however, does not give lower residual concentrations. To understand the slow degradation of the residual concentration, the degradation

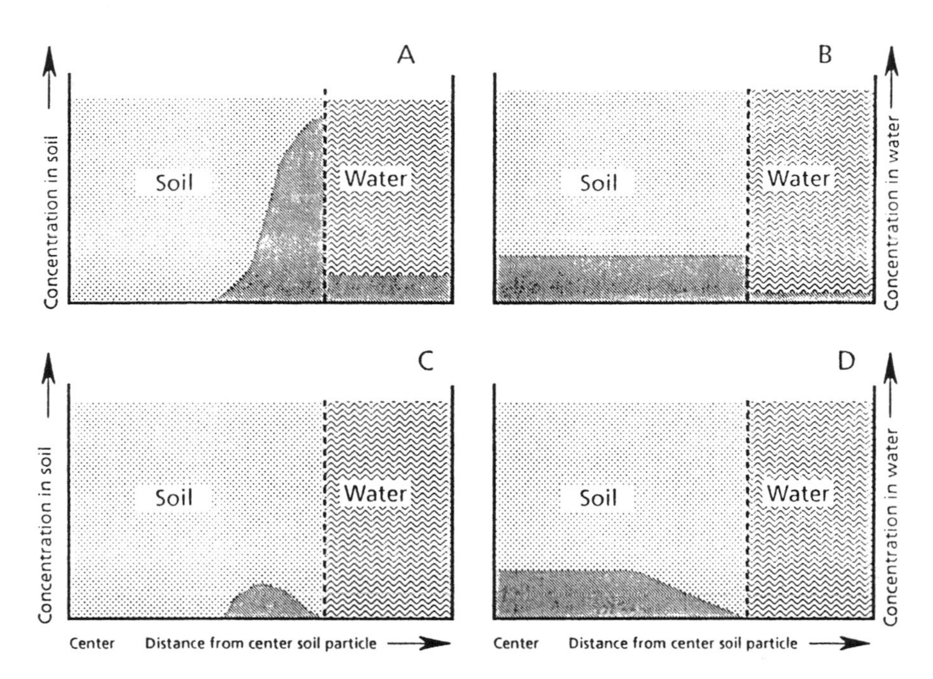

FIGURE 1. Distribution of a pollutant in one soil particle or aggregate. The concentration in the soil particle and the surrounding water phase is given as a function of the distance to the center of the soil particle, which is assumed to be spherical. Both vertical scales are different, concentration in soil >> concentration in water. A: Recent pollution before treatment. B: Old pollution before treatment (the pollutant has been in contact with the soil for a long time). C: Recent pollution after biological treatment. D: Old pollution after the same biological treatment. (From: Harmsen 1993, reprinted by permission of Kluwer Academic Publishers, Dordrecht, The Netherlands).

curve (Figure 2A) can be compared with a desorption curve (Figure 2B). The first curve was the result of a landfarming experiment with pentachlorophenol. The latter curve was obtained by eluting a soil column with an aqueous solution of pentachlorophenol (1 mg/L). After 2 weeks of adsorption, the desorption was started by elution of the column with clean water. At first the pentachlorophenol adsorbed in the macropores was leached out of the column. This will also be the bioavailable fraction, which was degraded immediately after start of the treatment (compare the quick decrease of the pollutant in the beginning of the curve in Figure 2A). After the release of the bioavailable part, the pentachlorophenol in the micropores diffuses to the larger pores and is leached out of the column in a low concentration. This pentachlorophenol is now available for degradation, resulting in a very slow decrease in the degradation curve.

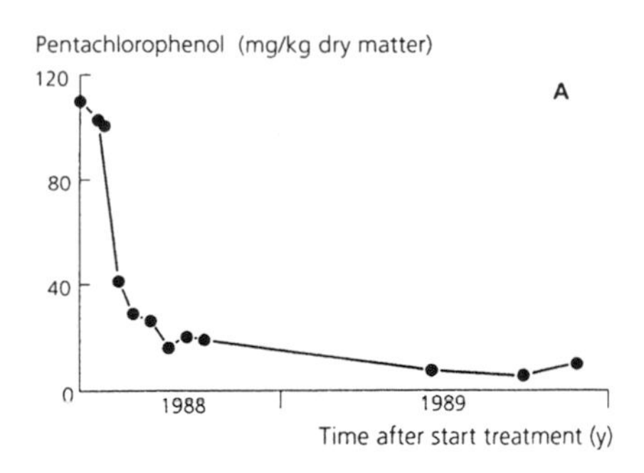

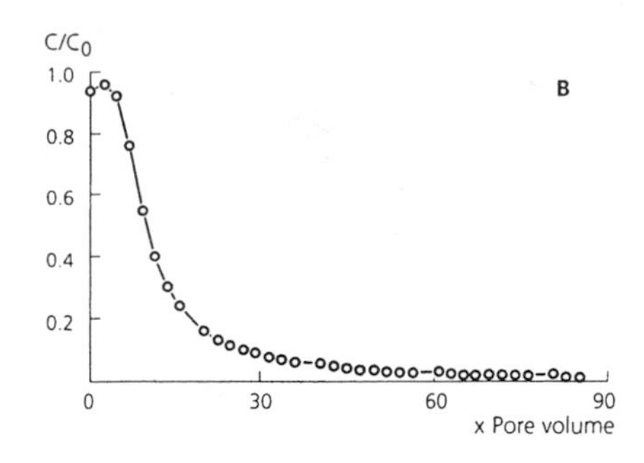

FIGURE 2. A: Degradation of pentachlorophenol in an earlier polluted soil using landfarming treatment. B: Leaching of a soil column saturated with an aqueous solution of pentachlorophenol for 2 weeks. The ratio between the concentration in the effluent and the saturation solution (C/C_0) has been related to the number of pore volumes leached through the column. The influent did not contain pentachlorophenol.

Extensive Landfarming

The nonbioavailable fraction of the pollution remains on sites that cannot be reached by microorganisms. To become bioavailable, this fraction must be transported from adsorption sites in micropores to larger pores in the soil. In this respect bioavailability of the residual concentration can be converted to a transport problem. For transport, a medium, a carrier, and time are necessary. Only the first two items can be influenced by human activities. We investigated both, by varying moisture content and using detergents and found that we were not able to increase the transport rate (Harmsen 1992). Time is the only factor left. Therefore, biological cleaning of polluted soils requires a new approach. As mentioned in the previous section, the bioavailable fraction can be removed quickly using already existing techniques such as intensive landfarming, but also using bioreactors. For the nonbioavailable fraction, an additional time-consuming method is needed to give the pollutant time to desorb and diffuse to microorganisms. An extensive form of landfarming could be possible (van den Bosch & Harmsen 1992).

A soil can be treated via extensive landfarming when the bioavailable fraction has already been removed by intensive landfarming or by another biological method. Extensive landfarming starts when cultivation or other activities have no noticeable effect on the degradation rate and continues until a concentration accepted by local authorities is reached (Figure 3). To be sure that degradation can continue, the soil is treated in such a way that, over the long term, aerobic circumstances are maintained. It may be necessary to add structure improvers such as compost and nutrients or to use a certain vegetation in order to fixate or aerate the soil. Cultivation to improve the soil structure must be minimized because it is expensive.

The main requirements for extensive landfarming are space and time. A location for extensive landfarming should be carefully selected. At the start the soil is not clean. This means that the soil cannot be used for all applications, i.e., it is not multifunctional. The soil should have a limited function, however with the knowledge that it will become clean or multifunctional in time. Extensive landfarming could be applied, for example on the top of landfills, on the bermes of roads at industrial complexes, at large civil works, at military complexes, but also in areas with a limited agricultural function such as forestry.

Time will be necessary to clean the soil. Figure 4A shows the reduction of a residual concentration of pentachlorophenol in a soil we have landfarmed under "optimal" conditions. Linear extrapolation in time of the last part of the curve (zero-order decrease) gave a cleaning period of about 2 years. A more probable nonlinear first-order decrease (see also Figure 2A) will lead to a longer period. A residual concentration of polycyclic aromatic hydrocarbons (PAHs) was present in the same soil. A decrease in concentration could hardly be seen (Figure 4B). A cleaning period of 10 years should be necessary using linear extrapolation in time. This is to be expected because, compared to pentachlorophenol, PAHs are more strongly adsorbed (larger partition coefficient soil/water) and will therefore diffuse more slowly.

CONCLUSIONS

At the end of a biological treatment procedure, the bioavailable part has been removed and a residual concentration has been left in the soil. This fraction, which is nonbioavailable, remains in soil pores devoid of living microorganisms. It may

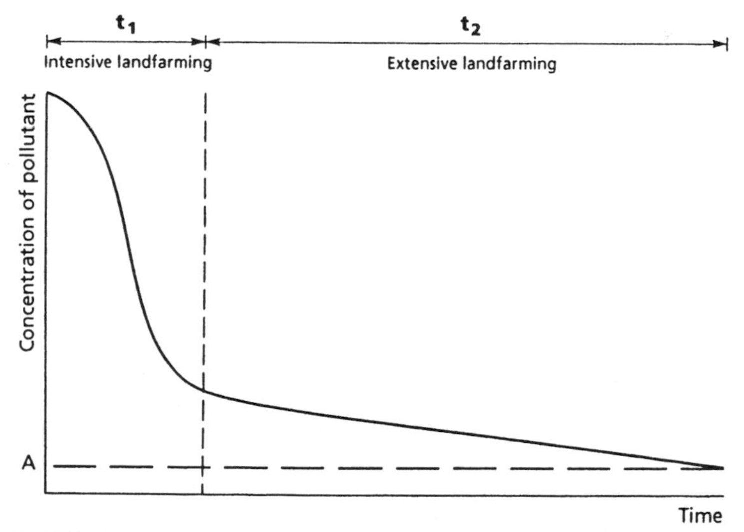

FIGURE 3. Biological treatment of a polluted soil using intensive and extensive treatment. $t_2 >> t_1$; A = concentration accepted by local authorities.

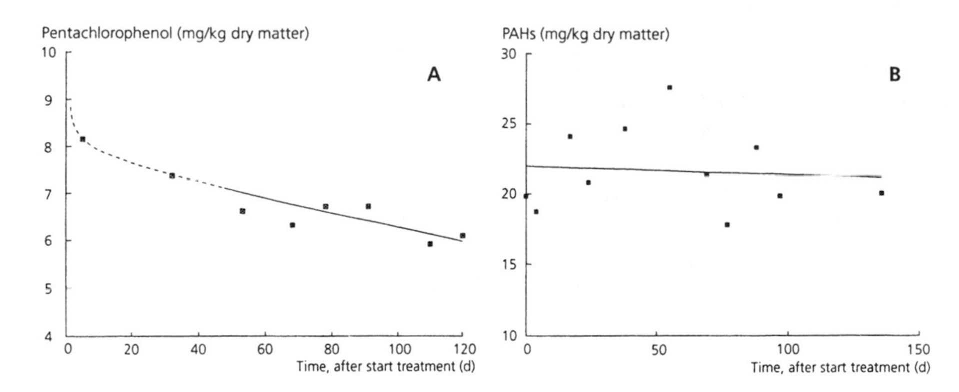

FIGURE 4. Microbiological degradation of a residual concentration under aerobic conditions and with a temperature of 20°C. Tenfold determination on each day, excluding statistical outliers. A: Pentachlorophenol. B: Polycyclic aromatic hydrocarbons (PAHs) (EPA list).

become bioavailable for degradation by a diffusion-limited desorption process, which is a very slow process. Therefore, biological cleaning of polluted soil requires a new approach. The bioavailable fraction can be removed quickly using already existing techniques (bioreactor, landfarming). For the nonbioavailable fraction an additional time-consuming method is needed to give the pollutant time to desorb. Use of the extensive form of landfarming is one possibility. It should be realized that desorption follows adsorption. The amount of time required for this extensive form of soil cleaning will be longer, if the contact period of the pollutant with the soil was longer. With longer contact time, the fraction adsorbed in the micropores is larger. Pollutants with a larger partition coefficient between soil and water also need a longer treatment period.

ACKNOWLEDGMENT

This research is part of The Netherlands Integrated Soil Research Programme.

REFERENCES

Boesten, J. J. T. I., and L. J. T. van der Pas. 1983. "Test of some aspects of a model for the adsorption/desorption of herbicides in field soil." *Aspects of Applied Biology* 4: 495-501.

Brusseau, M. L., and P. S. C. Rao. 1989. "The influence of sorbate-organic matter interactions on sorption nonequilibrium." *Chemosphere. 18* (9/10): 1691-1706.

Grathwohl, P., J. Farrell, and M. Reinhard. 1990. "Desorption kinetics of volatile organic contaminants from aquifer material." In F. Arendt, M. Hinsenveld, and W. J. van den Brink (Eds.), *Contaminated Soil '90*, pp. 343-350. Kluwer Academic Publisher, Dordrecht, The Netherlands.

Harmsen, J. 1991. "Possibilities and limitations of landfarming for cleaning contaminated soils." In R. E. Hinchee and R. F. Olfenbuttel (Eds.), *On-Site Bioreclamation: Processes for Xenobiotics and Hydrocarbon Treatment*, pp. 255-272. Butterworth-Heinemann, Stoneham, MA, USA.

Harmsen, J. 1993. "Managing bio-availability: An effective element in the improvement of biological soil cleaning?" In H. Eijsackers and T. Hamers (Eds.), *Integrated Soil and Sediment Research: A Basis for Proper Protection* (in press). Kluwer Academic Publishers, Dordrecht, The Netherlands.

Pignatello, J. J., C. R, Frink, P. A. Martin, and E. X. Droste. 1990. "Field-observed ethylene dibromide in an aquifer after two decades." *J. of Cont. Hydrol. 5*: 195-214.

Steinberg, S. M., J. J. Pignatello, and B. L. Sawhney. 1987. "Persistence of 1,2-dibromoethane in soils: Entrapment in intraparticle micropores." *Environ. Sci. Technol. 21* (12): 1201-1208.

van den Bosch, H., and J. Harmsen. 1992. "Report workshop Landfarming, Bilthoven, The Netherlands, October 7th 1991." RIVM report nr. 736101013. Bilthoven, The Netherlands (in Dutch).

Wu, S., and P. M. Gschwend. 1986. "Sorption kinetics of hydrophobic organic compounds to natural sediments and soils." *Environ. Sci. Technol. 20*: 717-725.

BIOSCRUBBING OF BUTANOL AND 2-BUTOXYETHANOL FROM SIMULATED CAN COATING OVEN EXHAUST

P. H. LeFevre and J. Croonenberghs

ABSTRACT

The Environmental Engineering Department of Coors Brewing Company designed and operated a pilot-scale bioscrubber used to investigate the destruction of butanol and 2-butoxyethanol from a production oven output and to determine production design parameters. The design is based on chemical engineering principles of a water scrubber and a packed-bed flooded column using microbiological cultures to support the VOC pollutant destruction capabilities. The design included limiting the amount of water used and increasing the throughput and the carbon input capabilities. The results of the simulated production runs include the increase from 85% to a range of 85 to 95% (100 to 125 g/m^3 per hour load rate) destruction efficiency and a minimal amount of water required to operate the unit. We believe that as a result of this project, this technique has wide application in the control of water-soluble, biodegradable volatile organic compounds (VOCs).

INTRODUCTION

Coors Brewing Company operates an aluminum can manufacturing facility to provide containers for the brewery at the rate of 4 billion cans per year. Air emissions containing butanol and 2-butoxyethanol are produced during the application of the internal coating that protects beer quality by separating the beer from the aluminum can. The emissions amount to several hundred tons per year.

The Environmental Engineering Department personnel used chemical engineering principles to design a water scrubber and enhance the pollutant destruction capabilities with microbiological cultures. The scrubber was built as a pilot-scale device to determine certain production design parameters that included using plastic (nondegradable) packing and a flooded-bed packed column.

The traditional means of destroying butanol and 2-butoxyethanol emissions involves the use of thermal oxidizers. However, thermal oxidizers are high-capital and high-maintenance equipment. We decided to investigate the possibility of using a microbial process to destroy the organics captured by a water scrubber. This idea was based on extending the work done in Europe and the United States on odor scrubbers, soil venting vapor scrubbers, and a can plant VOC scrubber (Ottengraf et al. 1986, Ottengraf 1987, Leson et al. 1991). Our improvement would be a low-pressure-drop, high-load-rate packed-bed with recirculated water and biomass. This configuration would be a high-efficiency water scrubber, loaded with microbes that destroy the VOCs as they are absorbed.

The design provided for a much smaller unit than those used at conventional odor sites. Such units often require the use of a separate building for the bed and considerable amounts of high-temperature ductwork to transport the polluted effluent for treatment.

THE EXPERIMENT

Experimental Apparatus

The apparatus we designed to test our concept (Figure 1) consisted of a blower to provide airflow, a steam injector operated by a programmable logic controller to control temperature, a humidifier, a metering solvent pump to provide a known concentration of solvent, a manifold to control airflow (up, down, and bypass), a packed-bed (1 m high by 1 m diameter) with rotating distributor arm, an aerated sump with recirculation pump, and monitoring instrumentation.

The bed was a 946-L (250-gal) high-density polyethylene tank filled with 1-inch Jaeger spherical polypropylene packing. The rotating distributor arm at the top of the tank was used to continuously feed the aqueous nutrient formula. The recirculation system consisted of a 200-L (53-gal) aerated tank and a recirculation pump to provide sump fluid to the distributor arm. The recirculation rate was 18.9 Lpm (5 gpm).

Bioscrubber Microbiology

The system was seeded with waste treatment plant return-activated sludge from the Coors waste processing plant and consisted of a complex consortium of protozoa, fungi, and bacteria. As the microbes adapted to the new environment, the higher forms were eliminated and only the bacteria remained. No attempts were made to characterize the microbial population, but periodic microscopic examinations were made to determine general population shifts.

Because this was a pilot environment, the unit was seeded several times, however, operation of the system under normal conditions would require only one seeding. Generally, within 1 week after seeding, microbial populations had stabilized.

Nutrient additions are necessary in this type of scrubber and consist of a synthetic, liquid medium, containing all macro- and micronutrients required to sustain the bacteria. In this instance, carbon was excluded from the nutrient because it was supplied in the solvent stream. The nutrient was developed

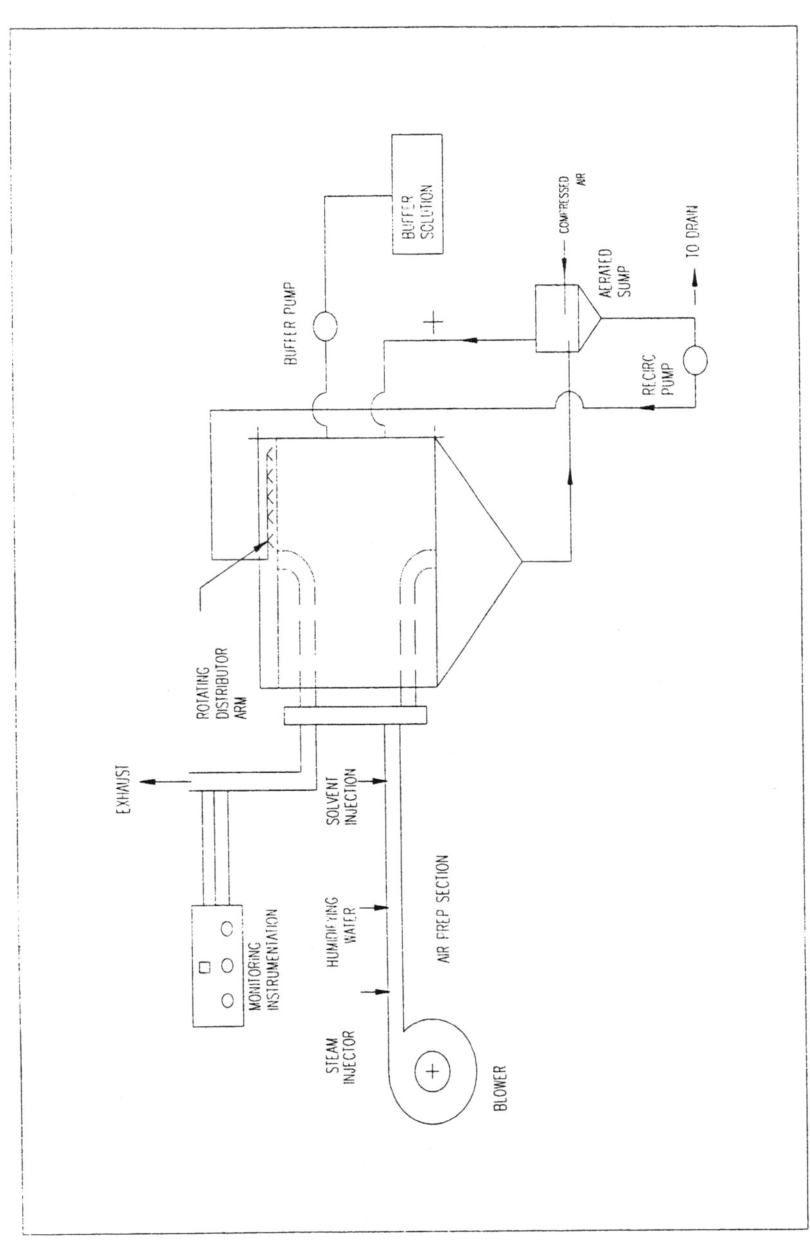

FIGURE 1. Bioscrubber equipment diagram.

specifically for this process and was batch-fed into the system daily. Buffering to maintain pH balance could not be accomplished by daily batch feeding, but was achieved by continuous addition of a phosphate buffer.

Daily Operation

The THC analyzer and the solvent pump were calibrated daily to ensure accuracy in the calculated VOC destruction figures. The aeration sump was 90% drained to allow for removal of biomass waste and to replenish the nutrient. A dedicated computer automatically downloaded raw data from the datalogger, performed the necessary calculations, and graphed the desired parameters.

The data collected included sump pH, inlet and outlet air temperature, pressure drop, airflow and total hydrocarbon (THC) concentration in the exhaust gasses. Data were collected at 15-minute intervals by a datalogger and downloaded daily to a computer that used spreadsheet software to calculate and plot pH (Figure 2), temperature, and airflow. THC in and out, percentage VOC destruction, and bed loading rate were also plotted.

Results

Experimental runs were made without microbes in the system to determine the scrubbing efficiency of the system as a water scrubber, using once-through water at a rate of 18.9 Lpm (5 gpm), and the results are shown in Figure 3. The airstream flow was 8 m^3/min at 37° containing 275 ppmv butanol and 2-butoxy-ethanol at 3:1. Without biomass present, the maximum scrubbing efficiency was 85%. With biomass present an 85 to 95% (100 to 125 g/m^3 per hour load rate) destruction efficiency was obtained. Because some parameters were difficult to maintain in the pilot version of this equipment, the increase of efficiency may appear minimal. However, with tighter process control, such as a manufacturing environment might provide, we project a sustained efficiency rate of 95% or better.

Typical production simulation runs were made under the same conditions and Figure 3 shows the results of a 4-day simulation run with biomass present. The initial increase in efficiency is due to biomass buildup and spikes represent monitoring equipment calibration. During the third day, a recirculation pump failed, which led to the fall off in efficiency. The fourth day we were still experiencing pump problems. The loss in efficiency illustrates the importance of having a recirculated sump.

Sump samples taken during the simulated run contained no solvents. Figure 3 shows the load rate, i.e., the amount of organics being treated by the scrubber during the 4-day run. Load rates in the range of 100 to 125 g/m^3 per hour were achieved.

CONCLUSIONS

A minimum boost in capture efficiency of 10% can be achieved by loading a wet scrubber with biomass. The microbes rapidly use the solvents as an energy

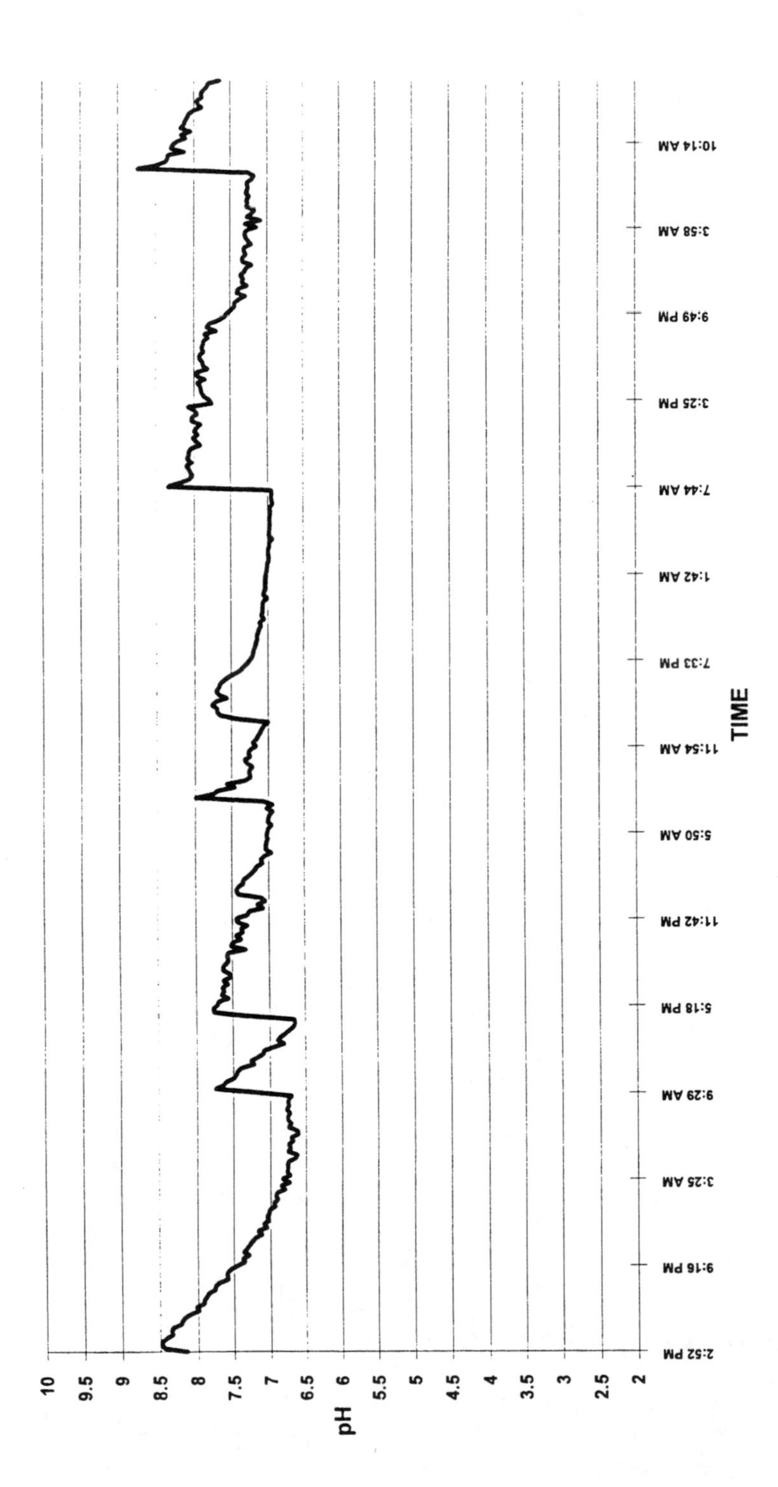

FIGURE 2. Bioscrubber pH.

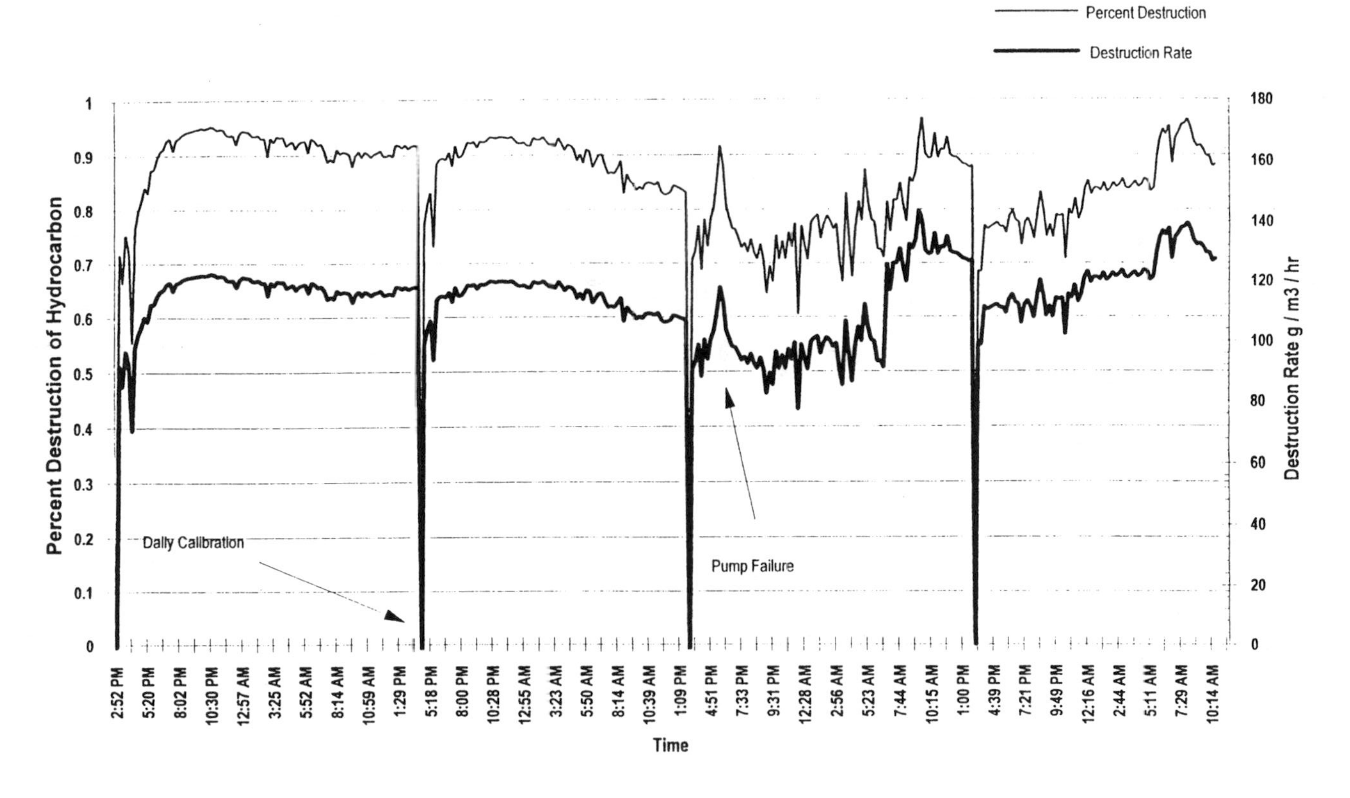

FIGURE 3. Bioscrubber percent destruction and load rate.

source, thus maintaining a very low solvent concentration throughout the bed and increasing the driving force for physical absorption. Our observations indicate that the suspended biomass in the recirculated sump fluid and the fixed biomass in the packing are both important to the overall capture efficiency. The maximum loading is a function of the biodegradability of the substrate and the ability of the bed to absorb a given chemical. These findings support the design of a wet scrubber that is significantly smaller than those based on traditional criteria. The scrubber requires very little water, about 0.5% of the once-through scrubber. The bioscrubber destroys contaminated chemicals at the point of capture rather than transferring them to another site for treatment.

REFERENCES

Leson, G., and A. M. Winer. 1991. "Biofiltration: An Innovative Air Pollution Control Technology for VOC Emissions." *J. Air & Waste Management Assoc. 41*(8): 1045-1053.

Ottengraf, S. P. P. 1987. "Biological Systems for Waste Gas Elimination." *TIBECH 5*: 132-136.

Ottengraf, S. P. P., J. J. P. Meeters, A. H. C. van den Oever, and H. R. Rozema. 1986. "Biological Elimination of Volatile Xenobiotic Compounds in Biofilters." *Bioprocess Engineering* 1: 61-69.

SLURRY BIOREMEDIATION OF POLYCYCLIC AROMATIC HYDROCARBONS IN SOIL WASH CONCENTRATES

F. J. Castaldi

ABSTRACT

Tests were conducted which evaluated the use of slurry bioremediation for treatment of polycyclic aromatic hydrocarbons (PAHs) in soil ash concentrates. The concentrates were obtained during operation of a soil-washing process. The concentrates were primarily silty-clay soils with 90 wt % of the particle sizes less than 30 μ. The major PAHs present in the concentrates were fluoranthene, pyrene, naphthalene, phenanthrene, and chrysene. The wastes also contained high concentrations of lead, zinc, copper, vanadium, chromium, and nickel. The soil wash concentrates were slurried to approximately 40 wt % solids in an airlift-loop reactor and inoculated with a mixed culture of hydrocarbon-degrading bacteria. After 60 days of batch slurry bioremediation, degradation was apparent with most soil-bound PAHs. A higher level of degradation was observed for the 2- through 4-ring PAHs than for the 5- to 6-ring PAHs. GC-MS analysis verified that degradation occurred for all PAHs during slurry bioremediation of the soil wash concentrates.

INTRODUCTION

Slurry bioremediation is a process to treat contaminated soils and sludges in a bioreactor. These waste materials are mixed with water to form a slurry, and if necessary, nutrients, microorganisms, and surfactants are added to enhance the biodegradation process. Slurry-phase bioremediation treats organic sludges and contaminated soils by extraction and biodegradation. The technique generally requires extensive, high-power mixing to suspend solids in the slurry and to maximize the mass transfer of the organic contaminants to the aqueous phase where biodegradation normally occurs.

Tests were conducted to evaluate the use of slurry bioremediation for treatment of polycyclic aromatic hydrocarbons (PAHs) in soil wash concentrates. The concentrates were obtained during operation of a proprietary soil-washing process developed in Germany. This soil-washing process is a two-step wet mechanical separation using water without the addition of detergents, solvents, acids, bases, or similar materials as the extracting agent.

The soil-washing technology concentrates pollutants in a froth discharged during flotation separation. The froth is then thickened and dewatered using conventional gravity thickeners and plate-filter presses. The pollutant concentrates used in this evaluation were generated from the treatment of contaminated soils obtained at a coke oven facility.

The fate of PAHs during slurry bioremediation is influenced by a number of factors which determine the degradation rate and extent of metabolism. These factors include the physicochemical properties of the PAH such as molecular size, water solubility, lipophilicity, volatility, concentration, and the presence of various substituents and functional groups. Environmental factors include temperature, pH, oxygen concentration, salinity, light intensity, sediment type, and the presence of growth-stimulating nutrient additives (e.g., cosubstrates and cometabolites). Microbial factors include the types, population, and distribution of microorganisms present in the bioreactor.

Much is known concerning the microbial metabolism of PAHs using both pure cultures and purified enzymes (Cerniglia 1984). Biodegradation of lower-molecular-weight PAHs by a variety of microorganisms has been demonstrated, and the biochemical pathways for a number of these compounds have been investigated. Bacteria oxygenate PAHs to form a dihydrodiol with a *cis* configuration. In this reaction, both atoms of molecular oxygen are incorporated into the PAH via a dioxygenase enzyme (Gibson et al. 1975, Jerina et al. 1976). The *cis*-dihydrodiols can undergo further metabolism via a pyridine nucleotide-dependent dehydrogenation reaction to yield catechols, which can act as substrates for ring cleavage enzymes with complete mineralization of the PAH (Heitkamp et al. 1988, Mahaffey et al. 1988). The genes for the initial oxidations of PAHs are localized on plasmids (Guerin & Jones 1988).

Although primary use of high-molecular-weight PAHs has not been demonstrated conclusively, biodegradation of these compounds under cometabolic growth conditions has been described (Mahaffey et al. 1988). In the presence of either naphthalene or phenanthrene, cometabolism of pyrene, 1,2-benzanthracene, 3,4-benzpyrene, and 1,2,5,6-dibenzanthracene by a mixed culture of flavobacteria and pseudomonads has been demonstrated (Cerniglia 1984). However, a strain of *Pseudomonas paucimobilis* has demonstrated a capacity to use fluoranthene, a high-molecular-weight PAH that is slowly degraded, as the sole source of carbon and energy for growth (Mueller et al. 1990). Therefore, it is reasonable to expect groups of pseudomonads to be effective against mixtures of PAHs such as are found at hazardous waste sites contaminated with creosote or coal tar.

A common degradative pathway demonstrated with another strain of *Pseudomonas paucimobilis* used the same ring cleavage enzymes for growth with naphthalene or biphenyl (Kuhm et al. 1991). Therefore, the ability to metabolize naphthalene and other aromatic hydrocarbons suggests that substituted derivatives also can be metabolized to a great extent by a single set of enzymes.

MATERIALS AND METHODS

The concentrates were composed of light material from gravimetric sizing (e.g., wood, tar fragments, coal, and coke); overflow from a hydrocyclone, essentially the fines portion of the soil with the adsorbed pollutants; and sludge from the flotation operation. The concentrates were primarily silty-clay soils with 90 wt % of the particle sizes less than 30 μ (Table 1). The organic contaminants present on the concentrates were primarily PAHs with fluoranthene, pyrene, naphthalene, phenanthrene, and chrysene being at the highest concentrations. The concentrates also contained elevated levels of benzo(a)anthracene, benzo(b)-fluoranthene, benzo(k)fluoranthene, and benzo(a)pyrene.

The pollutant concentrates were inoculated with a mixed culture of hydrocarbon-degrading bacteria belonging primarily to the genera *Pseudomonas* and *Acinetobacter*. Tap water, enriched with ammonium bicarbonate (NH_4HCO_3) and phosphoric acid (H_3PO_4) added to provide microbial macro-nutrients, was used to slurry the pollutant concentrates to approximately 40 wt % solids. The study was originally provided a sample of soil wash concentrate that had a moisture level of only 25 wt %, which is too dry for effective treatment in a slurry reactor. Consequently, the concentrates required additional moisture before they could be reacted in the slurry bioremediation process. The use of 40 wt % solids as the operational slurry level for the bioremediation test is considered the most cost-effective for a bioreactor treating contaminated soils.

Approximately 62 L of the soil wash concentrate bioslurry was placed in an airlift-loop reactor of the type manufactured by the EIMCO Process Equipment Company, Salt Lake City, Utah. This reactor is equipped with a bottom rake mechanism to move settled material to the airlift that then circulates the material to the slurry surface for re-aeration and mixing. The reactor also uses a low-shear impeller located on the central shaft to provide additional agitation. Aeration

TABLE 1. Soil wash concentrate grain-size distribution.[(a)]

Fraction (mm)	Weight (%)
> 0.125	0.0
0.125 > 0.064	0.7
0.064 > 0.032	8.7
0.032 > 0.016	17.9
0.016 > 0.008	22.5
0.008 > 0.004	17.8
0.004 > 0.002	14.9
0.002 > 0.001	8.7
<0.001	8.8

(a) Grain-size distribution obtained by hydrometer analysis.

and mixing were provided by a set of nitrile butadiene rubber membrane diffusers of the type manufactured by Gummi-Jager KG GmbH & Cie, Hannover, Germany. The diffusers are attached to the rake arm shaft near the bottom of the reactor.

The bioreactor was operated in the batch mode for 60 days. Reactor slurry was sampled at days 15, 30, and 60 and compared to the pollutant levels at the start of the experiment. Each sample was analyzed for PAHs by high-performance liquid chromatography (HPLC, EPA Method 8310); chemical oxygen demand (COD, U.S. Environmental Protection Agency [EPA] Method 410.1); biochemical oxygen demand (BOD, American Water Works Association [AWWA] Standard Method 5210); total solids (EPA Method 160.3); total volatile solids (EPA Method 160.4); oil and grease (O&G, EPA Method 413.1); oxygen uptake (AWWA Standard Method 2710B); dissolved phosphate (EPA Method 365.2); dissolved ammonia (EPA Method 350.2); dissolved Kjeldahl nitrogen (EPA Method 351.3); and metals by inductively coupled plasma atomic emission spectroscopy (EPA Method 6010). In addition, the initial and 60-day slurry samples were analyzed for the base/neutral and acid extractable organics by gas chromatography/mass spectrometry (GC-MS, EPA Method 8270) as a verification of the HPLC results for the PAH levels in the soil wash concentrate bioslurry.

Total plate counts of the heterotrophic microorganisms in the reactor were made at several points during the experiment using the AWWA Standard Method 9215. The relative toxicity of the slurry during the biotreatment test also was assessed with the Microtox Method, which uses a species of bioluminescent bacteria, *Photobacterium phosphoreum*, to measure the toxicity of a waste solution.

RESULTS

Data for slurry-phase contaminant levels in the batch slurry bioremediation process are presented in Table 2. These data include 16 PAHs presented by number of aromatic rings (i.e., 2- to 6-ring compounds) and molecular weight (i.e., 128 to 276) for naphthalene through indeno(1,2,3-cd)pyrene. Table 2 data show considerable variation in PAH concentration over the experimental reaction period. The apparent increases in the concentrations of soil-bound PAHs may reflect an increased PAH extraction efficiency of the analytical method rather than the unlikely production of soil-bound PAHs during the study. However, this same concentration increase also appeared with the slurry measurements of soluble COD, soluble BOD, and O&G. These data suggest a possible solubilizing effect that may have resulted from the bacterial inoculations of pollutant concentrates made at the start of the experiment. It also is apparent from these data that after 30 days of aeration and mixing, substantive biodegradation is occurring in the slurry bioremediation process. This may be verified by the increase in bacteria numbers from 1.0E+05 colony-forming units (CFU)/mL at the start of the experiment to 8.9E+07 CFU/mL after 60 days of aeration in the bioreactor. The apparent relative toxicity of the slurry, as measured with the Microtox method, also increased by a factor of 10 times the starting toxicity after 60 days of aeration, suggesting higher pollutant concentrations in the aqueous

TABLE 2. Slurry-phase contaminant levels during batch biodegradation.

Parameter	Reactor Slurry at Day 0	Reactor Slurry at Day 15	Reactor Slurry at Day 30	Reactor Slurry at Day 60
Naphthalene, µg/kg	3,000	10,100	16,500	9,100
Acenaphthylene, µg//kg	<2,300	<2,300	1,500	<12,000
Acenaphthene, µg/kg	5,000	<1,800	<1,800	<18,000
Fluorene, µg/kg	2,000	1,850	2,800	755
Phenanthrene, µg/kg	6,050	6,850	13,500	6,300
Anthracene, µg/kg	5,900	3,800	7,650	4,000
Fluoranthene, µg/kg	15,850	19,000	41,000	16,000
Pyrene, µg/kg	6,700	12,500	25,500	11,000
Benzo(a)anthracene, µg/kg	6,050	9,650	18,000	6,550
Chrysene, µg/kg	2,700	3,200	1,600	<750
Benzo(b)fluoranthene, µg/kg	8,450	12,950	17,500	6,800
Benzo(k)fluoranthene, µg/kg	3,900	5,550	8,550	3,750
Benzo(a)pyrene, µg/kg	6,300	9,700	21,000	7,950
Dibenz(a,h)anthracene, µg/kg	920	995	1,300	1,180
Benzo(g,h,i)perylene, µg/kg	5,700	5,200	14,000	10,450
Indeno(1,2,3-cd)pyrene, µg/kg	78	10,050	15,000	6,100
Total COD, mg/kg	96,304	96,266	98,131	97,980
Soluble COD, mg/L	103.7	971	1,338	1,938
Total BOD, mg/kg	563	675	638	60
Soluble BOD, mg/L	17.3	70.7	140	29
O&G, mg/kg	132	1,040	1,116	984
O_2 Uptake Rate, mg O_2/L-min	0.05	0.2	0.07	0.05

Reactor temperature averaged 25.3°C; reactor pH was maintained between pH 6.0 and 7.0; total solids in reactor averaged 411,900 mg/kg with total volatile solids of 66,800 mg/kg.

phase due to waste constituent dissolution. Therefore, the increase in apparent PAH concentration is probably real and not an artifact of the analytical method.

Solid-phase (i.e., dewatered reactor slurry) PAH concentrations, as determined by HPLC, at the start of the experiment and after 60 days of slurry bioremediation are presented in Table 3. Corresponding dissolved-fraction PAH levels at each sampling point also are presented in this table. These data show that degradation was apparent with most soil-bound PAHs over the experimental reaction period. A comparison of the removals of 2- to 4-ring PAHs with 5- to 6-ring PAHs indicates that a higher level of degradation of the 2- through 4-ring PAHs occurred during treatment in the slurry bioremediation process, i.e., 56% removal for the 2- to 4-ring PAHs versus 21.9% removal for the 5- to 6-ring PAHs.

Dissolved-fraction data presented in Table 3 indicate slightly higher levels of certain PAHs in the aqueous phase after 60 days of treatment than were found at the start of the experiment. These data suggest that waste constituent dissolution must be enhanced before biodegradation can occur.

TABLE 3. Solid-phase PAH levels after 60 days of slurry bioremediation.

	Soil Wash Concentrate		60-Day Reactor Solids	
Parameter	Solid Phase (μg/kg)	Dissolved Fraction (μg/L)	Solid Phase (μg/kg)	Dissolved Fraction (μg/L)
Naphthalene	35,000	<1.7	16,394	<1.7
Acenaphthylene	<12,000	<2.2	<12,000	7.6
Acenaphthene	<18,000	<1.7	<18,000	<1.7
Fluorene	5,100	<0.2	1,360	0.43
Phenanthrene	34,500	<0.61	11,350	<0.61
Anthracene	15,500	0.62	7,206	1.4
Fluoranthene	59,500	<0.2	28,825	<0.2
Pyrene	46,000	0.16	19,817	0.17
Benzo(a)anthracene	24,500	<0.012	11,800	<0.012
Chrysene	<750	<0.14	<750	<0.14
Benzo(b)fluoranthene	21,500	<0.017	12,250	<0.017
Benzo(k)fluoranthene	10,400	0.0093	6,756	<0.016
Benzo(a)pyrene	24,500	0.02	14,322	0.042
Dibenz(a,h)anthracene	1,210	<0.028	2,126	<0.028
Benzo(g,h,i)perylene	10,800	<0.072	18,826	<0.072
Indeno(1,2,3-cd)pyrene	15,150	<0.041	10,990	<0.041
% Moisture	25.8	NA	25.8	NA

All PAH analyses performed using HPLC chromatography (EPA Method 8310).
NA means not applicable.

Table 4 presents GC-MS solid-phase PAH concentration data at the start of the experiment and after 60 days of slurry bioremediation. Although reporting higher concentrations for most PAHs, the GC-MS analysis does verify the PAH removals seen using HPLC analytical measurements. The GC-MS data also indicate that degradation occurred for all PAHs, i.e., 2-ring through 6-ring PAHs were removed by the slurry bioremediation process. Apparent removals varied from a low of 47.1% for benzo(g,h,i)perylene to 72.9% for fluorene. In general, 2- to 4-ring PAHs were degraded to a higher level than 5- to 6-ring PAHs. The GC-MS data also show removal of dibenzofuran and 2-methylnaphthalene during slurry biotreatment.

Metals concentrations for soil wash concentrates and bioreactor slurries are presented in Table 5. Dissolved-fraction metals also are presented for two samples. Although there is consistency between samples with regard to the metals detected at each point in the treatment, there appears to be a loss of some metals in the process when all data are put on a common solids basis. This apparent loss is believed to be due to the addition of calcium carbonate and magnesium hydroxide to the reactor slurry for pH adjustment during the experiment. Both calcium and magnesium were seen to increase in concentration during the treatment period. Sodium hydroxide also was added to the reactor for pH adjustment during the experiment. It can be seen from the dissolved-fraction data that a higher level of sodium also was present in the bioreactor at the conclusion of the experiment.

DISCUSSION

Microbial growth on water-insoluble carbon sources such as PAHs often is accompanied by the appearance of surface-active compounds in the culture medium. The production of these surface-active compounds (i.e., biosurfactants) is connected mostly with growth limitations in the late-logarithmic and the stationary-growth phases. This occurs because the specific enzymes responsible for biosurfactant production are induced under growth-limiting conditions

TABLE 4. GC-MS verification of slurry bioremediation performance.

Parameter[a]	Soil Wash Concentrate	60-Day Reactor Solids	% Removal
Naphthalene	87	30.6	64.8
Acenaphthylene	13	4.5	65.4
Acenaphthene	9.2	2.7	70.65
Fluorene	14	3.8	72.9
Phenanthrene	77	21.6	71.95
Anthracene	35	11.9	66
Fluoranthene	130	46.8	64
Pyrene	98	32.4	66.9
Benzo(a)anthracene	68	25.2	62.9
Chrysene	81	32.4	60
Benzo(b)fluoranthene	120[b]	50.4[b]	58[b]
Benzo(k)fluoranthene	120[b]	50.4[b]	58[b]
Benzo(a)pyrene	56	25.2	55
Dibenz(a,h)anthracene	13	5.95	54.2
Benzo(g,h,i)perylene	34	18	47.1
Indeno(1,2,3-cd)pyrene	37	17.1	53.8
Dibenzofuran	27	9.4	65.2
2-Methylnaphthalene	16	5.4	66.3
Phenol	<10	<10	NA
Pentachlorophenol	<50	<50	NA
bis(2-Ethylhexyl)phthalate	3.7	4.3	0
% Moisture	25.8	25.8	NA

(a) All values are μg/g analyzed by GC-MS (EPA Method 8270).
(b) Benzo(b)fluoranthene coelutes with benzo(k)fluoranthene. The value reported is the sum of the two compounds.
NA means not applicable.

TABLE 5. Metals concentrations in soil wash concentrate and reactor slurry.

	Soil Wash Concentrate			60-Day Reactor Solids	
Parameter[(a,b)]	Solid Phase (mg/kg)	Dissolved Fraction (mg/L)	0-Day Reactor Solids (mg/kg)	Slurry Phase (mg/kg)	Dissolved Fraction (mg/L)
Aluminum	9,850	<0.2	5,450	3,300	<0.80
Antimony	<6.3	<0.1	<4.9	<6.3	<0.40
Arsenic	<19	<0.3	<15	<19	<1.2
Barium	160	0.14	80.5	41	0.69
Beryllium	0.56	<0.002	0.29	<0.13	<0.008
Boron	<38	<0.60	<30	<38	<2.4
Cadmium	0.35	<0.005	<0.25	<0.32	<0.02
Calcium	14,500	490	7,100	9,900	920
Chromium	45	<0.01	23.5	28	<0.04
Cobalt	7.4	0.012	4.0	2.7	0.096
Copper	125	<0.02	64	38	0.43
Iron	28,000	0.62	14,000	7,500	32
Lead	155	<0.05	76.5	42	<0.2
Magnesium	2,400	26	1,250	3,800	140
Manganese	420	1.3	210	130	<0.04
Molybdenum	4.0	<0.05	<2.5	<3.2	0.98
Nickel	22	<0.02	12	15	0.10
Potassium	2,300	6.6	1,300	820	93
Selenium	<19	<0.3	<15	<19	<1.2
Silicon	280	15	210	150	10
Silver	<0.63	<0.01	<0.49	<0.63	<0.04
Sodium	515	5.6	925	1,100	1,200
Strontium	76	0.90	37	24	1.6
Thallium	<6.3	<0.10	<4.9	<6.3	<0.4
Vanadium	48	<0.02	25.5	13	<0.08
Zinc	150	0.045	73.5	45	<0.08
Total Solids (%)	74.2	NA	44.8	37.0	NA

(a) All values analyzed by inductively coupled plasma emission spectroscopy (EPA Method 6010).
(b) Reactor pH was maintained between pH 6.0 and 7.0.
NA means not applicable.

(Hommel 1990, Oberbremer & Müller-Hurtig 1989). During the present experiment, slurry concentrations of soluble-fraction COD and BOD and freon-extractable oils increased over a portion of the reaction period by 1 order of magnitude. These data suggest that the release of surface-active compounds may have been occurring in the reactor, and that PAHs possibly were extracted from the soil wash concentrates into the aqueous phase, thus increasing their bioavailability. This increase suggests that PAH biodegradation in fine-grain soils will proceed only when the interfacial tension is lowered by the production of surface-active compounds by the microorganisms in the process.

Coincident with this phenomenon, the apparent amount of PAH biodegradation increased with the observed increases in conventional pollutant removals. This was expected because a portion of the COD is comprised of PAH compounds, especially those with 2 through 4 aromatic rings.

It also was observed during the experiment that 5- and 6-ring PAHs were degraded coincident with the 2- through 4-ring PAHs. The apparent levels of biodegradation of the high-ring PAHs were approximately half that of the low-ring compounds suggesting that the slow dissolution of these constituents may serve to effect a cometabolic condition in the bioreactor. The growth-stimulating carbon compounds in the bioreactor may have enhanced the biodegradation of the nongrowth (i.e., microbially persistent) compounds in the soil wash concentrates.

Although the soil wash concentrates contained relatively high concentrations of lead, zinc, copper, vanadium, chromium, and nickel in the solid phase, these same metals were at relatively low concentrations in the dissolved fraction both before and after slurry bioremediation. This may suggest that metal toxicity will not present a problem for slurry biotreatment of these wastes.

CONCLUSIONS

The study concluded that microbial degradation of PAHs in fine-grain soils can be accomplished in a slurry bioreactor at a 40 wt % slurry level without apparent substrate inhibition. The apparent removals of 2- to 4-ring PAHs compared with 5- to 6-ring PAHs indicates that a higher level of degradation of the 2- through 4-ring PAHs occurred during 60 days of treatment in the slurry bioremediation process, i.e., 56% removal for the 2- to 4-ring PAHs versus 21.9% removal for the 5- to 6-ring PAHs as determined by HPLC analysis of the slurry phase. Although the soil wash concentrates contained high concentrations of lead, zinc, copper, vanadium, chromium, and nickel, these metals did not present a toxicity problem for slurry bioremediation.

REFERENCES

Cerniglia, C. E. 1984. "Microbial Metabolism of Polycyclic Aromatic Hydrocarbons." *Advances in Applied Microbiology 30*:31-71.

Gibson, D. T., V. Mahadevan, D. M. Jerina, H. Yagi, and H. J. C. Yeh. 1975. "Oxidation of the Carcinogens Benzo(a)pyrene and Benzo(a)anthracene to Dihydrodiols by a Bacterium." *Science 189*:295.

Guerin, W. F., and G. E. Jones. 1988. "Mineralization of Phenanthrene by a Mycobacterium sp." *Applied and Environmental Microbiology 54*:937.

Heitkamp, M. A., J. P. Freeman, D. W. Miller, and C. E. Cerniglia. 1988. "Pyrene Degradation by a Mycobacterium sp.: Identification of Ring Oxidation and Ring Fission Products." *Applied and Environmental Microbiology 54*:2556.

Hommel, R. K. 1990. "Formation and Physiological Role of Biosurfactants Produced by Hydrocarbon-Utilizing Microorganisms." *Biodegradation 1*:107.

Jerina, D. M., H. Selander, H. Yagi, M. C. Wells, J. F. Davey, V. Mahadevan, and D. T. Gibson. 1976. "Dihydrodiols from Anthracene and Phenanthrene." *Journal of the American Chemical Society 98*:5988.

Kuhm, A. E., A. Stolz, and H. J. Knackmuss. 1991. "Metabolism of Naphthalene by the Biphenyl-Degrading Bacterium *Pseudomonas paucimobilis* Q1." *Biodegradation* 2:115.

Mahaffey, W. R., D. T. Gibson, and C. E. Cerniglia. 1988. "Bacterial Oxidation of Chemical Carcinogens: Formation of Polycyclic Aromatic Acids from Benz(a)anthracene." *Applied and Environmental Microbiology 54*:2415.

Mueller, J. G., P. J. Chapman, B. O. Blattmann, and P. H. Pritchard. 1990. "Isolation and Characterization of a Fluoranthene-Utilizing Strain of *Pseudomonas paucimobilis*." *Applied and Environmental Microbiology 56*:1079.

Oberbremer, A., and R. Müller-Hurtig. 1989. "Aerobic Stepwise Hydrocarbon Degradation and Formation of Biosurfactants by an Original Soil Population in a Stirred Reactor." *Applied Microbiology and Biotechnology 31*:582.

TREATMENT OF A COMPLEX LIQUID MATRIX USING POWDERED ACTIVATED CARBON TREATMENT

D. C. Erickson, K. A. DeFelice, N. J. Myers, and M. D. Klein

ABSTRACT

This paper presents the results of a treatability study that was conducted to demonstrate the effectiveness of Powdered Activated Carbon Treatment on a complex liquid matrix. The process known as the PACT® system combines physicochemical and biological treatment technologies for removing toxic and hazardous organic compounds from liquid wastestreams. The liquid matrix used for this study was collected from a landfill that formerly accepted municipal refuse, liquid industrial waste, and domestic sewage sludge. The effectiveness of the treatment was evaluated by the percent removals achieved for organic matter as measured by biochemical oxygen demand (BOD) and chemical oxygen demand (COD), ammonia, and priority and hazardous pollutants. Percent removals of BOD, ammonia, and priority and hazardous pollutants generally exceeded 95%. More than 90% of the COD and TOC present in the PACT® system influent was removed by the process. Emissions of volatile organic compounds (VOCs) during treatment were minimal.

INTRODUCTION

This paper presents the results of a pilot-scale treatability study that was conducted to demonstrate the effectiveness of the PACT® system for treating a complex liquid matrix. The liquid matrix used for this study was collected from the Lowry Landfill Superfund site in Arapahoe County, southeast of Denver, Colorado.

The Lowry Landfill formerly accepted municipal refuse, liquid industrial waste, and municipal sewage sludge. The disposal procedure used was known as codisposal. Codisposal consisted of excavating waste pits, filling them approximately three-quarters full with liquid, and then filling them with municipal refuse. Additional municipal refuse was then mounded several feet above the waste pits.

Environmental investigations began at Lowry Landfill in the mid-1970s and currently continue. The U.S. Environmental Protection Agency (EPA) became actively involved in monitoring activities in 1981, when Lowry Landfill was first considered for the National Priorities List (NPL). Lowry Landfill was placed on the NPL in 1984 and divided into operable units (OUs). The OUs are as follows: shallow groundwater and subsurface liquids (OU 1), landfill solids (OU 2), landfill gas (OU 3), soils (OU 4), surface water and sediment (OU 5), and deep groundwater (OU 6). The treatability study program for waste-pit liquid was conducted per the Administrative Order (U.S. EPA 1989) for OUs 1 and 6.

The liquid extracted from the waste pits has a high ammonia content (595 mg/L) organic content as measured by BOD (1690 mg/L) and COD (4840 mg/L). BOD refers to the amount of oxygen required for bacteria to aerobically stabilize degradable organic matter. COD refers to the amount of oxygen necessary to chemically oxidize the organic material to carbon dioxide (CO_2) and water.

The organic fraction of the waste-pit liquid includes VOCs, semivolatile organic compounds (SVOCs), pesticides, herbicides, dioxins, and furans. Radionuclides, metals, and a high total dissolved solids (TDS) content have also been detected in the waste-pit liquid.

During the treatability program, the PACT® system was selected as a representative biological treatment option for evaluating biological treatment of waste-pit liquid collected from the Lowry Landfill site. Options for disposal of effluent from an on-site full scale system include discharge to a publicly owned treatment works (POTW) or to a surface stream.

Biological processes can be used to remove hazardous and priority pollutants from a liquid matrix. Success of the biological treatment depends on a variety of factors including the biodegradability and toxicity of the compound, microbial nutrient requirements, pH, temperature, and dissolved oxygen concentrations. The biodegradability of organic compounds is of primary concern when considering biological treatment. The range of organic compounds that can be biologically degraded is large. However, complex water insoluble compounds such as 2,3,7,8-tetrachlorodibenzo-*p*-dioxin (2,3,7,8-TCDD) have been shown to resist biological treatment (Sferra 1989). Moreover, high concentrations of materials toxic to bacteria, such as metals in an influent waste stream, may inhibit the activity of a biological treatment system. Biological treatment of hazardous and priority pollutants in aqueous matrices has been reported in the literature. Hannah (1986) reported results of a study comparing the removals of toxic pollutants from wastewater treated using each of six physical or biological treatment processes. Results of the study showed that conventional activated sludge treatment provided superior removals of selected volatile and semivolatile organic compounds from wastewater when compared with either gravity separation, chemical precipitation, or biological lagoon treatments. Ying (1987) reported on the successful treatment of a chemical waste landfill leachate in sequencing batch bioreactors (SBRs). The SBRs operated on a fill and draw operating schedule with a hydraulic residence time (HRT) of 5 days. Near the end of the 5-day HRT, an addition of powdered activated carbon (PAC) was made to the biological reactor. The authors noted that effluent quality from the SBRs was improved with the addition of the PAC. The authors further noted that addition of the PAC improved treatment performance because of the concurrent dual organic removal mechanisms of adsorption and degradation of persistent compounds on PAC.

A subsequent report by Eckenfelder (1991) presented information supporting the addition of PAC to activated sludge systems to treat recalcitrant toxic organic chemicals. Reported results indicated removals of TOC and color from plastics, and dyestuff manufacturing wastewater increased with increasing PAC dosage.

The PACT® system was evaluated during this study because the addition of PAC to activated sludge systems has been shown to improve removals of hazardous substances. The PACT® system is a patented, proprietary process that is currently licensed by Zimpro Passavant Environmental Systems Inc., Rothschild, Wisconsin. The PACT® system involves the addition of PAC to the aeration basin of the activated sludge system for removing organic contaminants by adsorption and biodegradation. A two-stage continuous PACT® system was used for this study because (1) a high-quality effluent was required for the available effluent discharge options, (2) the influent wastewater contained high concentrations of organics and priority and hazardous pollutants, and (3) stripping of VOCs was a concern.

TEST OBJECTIVES

The primary objective of this paper is to provide data showing the effectiveness of the PACT® system in treating liquid matrices containing a high concentration of organic matter and priority pollutants. The criteria that were used to evaluate the PACT® system during the study are as follows:

- BOD, COD, priority pollutants removal
- Ammonia removal
- Physical removal of solids
- VOC emissions.

TESTING PROGRAM

This section provides a summary of the influent liquids, PACT® system operating conditions, and analytical testing procedures.

Influent Liquid

The PACT® system was evaluated during treatment of waste-pit liquid. The waste-pit liquid was placed in polyethylene 55-gallon U.S. Department of Transportation (DOT)-approved drums, and shipped to the Zimpro Passavant Environmental Systems facility. The waste-pit liquid was analyzed, and nutrients including nitrogen, phosphorus, and trace metals were added as necessary. The compounds that were added listed with their waters of hydration and their respective concentrations in the influent, after addition of the nutrient solution, are as follows:

Compound Added	Concentration (mg/L)
$MnSO_4{\cdot}H_2O$	1.12
$CuSO_4{\cdot}5H_2O$	1.68
$ZnSO_4{\cdot}7H_2O$	24.9
$CoCl_2{\cdot}6H_2O$	2.28
$FeCl_3{\cdot}6H_2O$	52.3
H_3PO_4	35.2

The pilot-scale PACT® system was seeded using activated sludge supplied by Zimpro Passavant. The seed consisted of activated sludge collected from the Wausau POTW, mixed with activated sludge collected from an ongoing PACT® system bench-scale study. PAC was added to the first- and second-stage aeration tanks before influent flow was initiated.

PACT® System

A diagram of the experimental pilot-scale PACT® system setup is shown in Figure 1. The pilot-scale PACT® system was assembled and operated by Zimpro Passavant. The major components of the pilot-scale system include first- and second-stage, stainless-steel, cylindrical, covered aeration tanks with mechanical mixers, external stainless-steel clarifiers with conical bottoms and motor-driven rakes, and peristaltic feed and mixed-liquor recycle pumps. Mixed-liquor dissolved oxygen (DO) was introduced to the aeration tanks using air diffusers located at the bottom of the aeration tanks. During treatment, the waste-pit influent was mixed with PAC and biological solids. The waste-pit liquid-carbon-biological solids mixture was continuously aerated to cause biological oxidation and adsorption of the organic contaminants. Virgin PAC was added to the second stage of the PACT® system, polishing the effluent before discharge. Following aeration, the mixture flowed to a clarifier and the powdered carbon and biological solids settled and were separated from the effluent liquid. Residual solids were dewatered using centrifugation and retained for analyses. For compounds that were slowly biodegraded but adsorbable on PAC, biological treatment may have been the ultimate means of removal because these materials were retained in the system longer than the liquid. In conventional activated sludge treatment, these same components would undergo biological treatment only during the HRT of the water because activated sludge has a limited adsorption capacity (Parker et al. 1993).

PACT® System Operating Conditions

PACT® system operating conditions were selected on the basis of influent chemical characteristics, desired effluent quality, and PACT® system operating

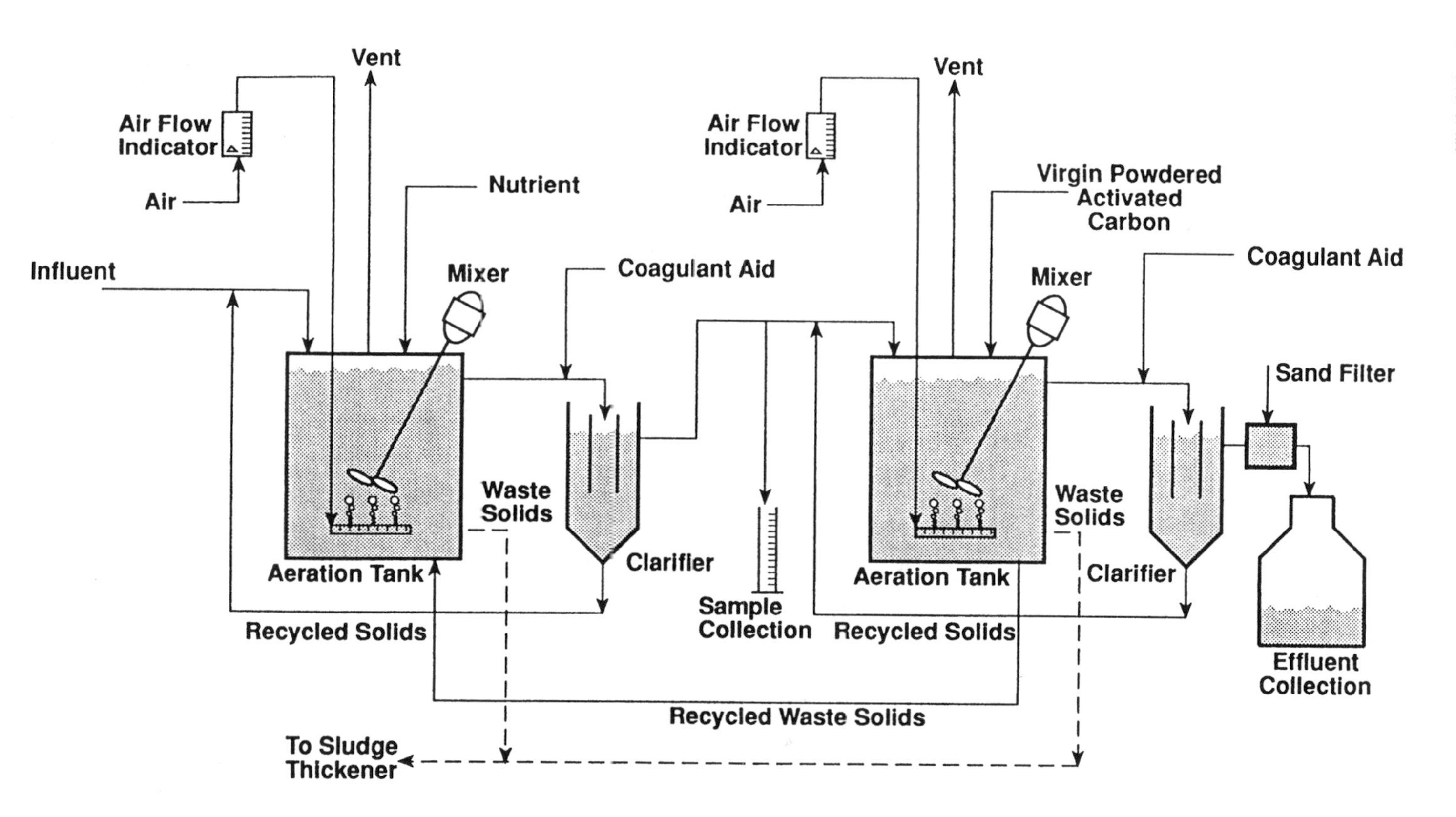

FIGURE 1. Schematic flow diagram of PACT® system two-stage pilot-scale system.

experience. The PACT® system operating conditions are summarized in Table 1. The range of HRT and solid retention time (SRT) values shown in Table 1 are typical values for PACT® system operations. Actual operating conditions used during waste-pit liquid treatment including HRT and SRT are proprietary and subject to Secrecy Agreements between parties involved in this project.

Mixed liquor was wasted directly from the first- and second-stage aeration tanks on a daily basis to maintain the desired solids retention time. Virgin PAC was added to the second stage of the PACT® system daily to maintain desired levels of PAC and biological solids.

The pH and dissolved oxygen levels in the aeration basin were monitored and adjusted as necessary using automatic controllers. The pH of the PACT® system was maintained between 6.5 and 8.0 using nitric acid for the duration of the study. DO levels in the first- and second-stage aeration basins were maintained above 2.0 and 3.0 mg/L respectively.

A granular media filter using silica sand was used after the second-stage clarifier to further reduce the amount of solids in the effluent from the second-stage clarifier. Media filtration alone did not result in sufficient solids removal, so a mixture of cationic and anionic polymers was added to the second-stage supernatant before media filtration to coagulate colloidal solids and to improve solids removal performance by filtration.

TABLE 1. Powdered Activated Carbon Treatment system operating conditions.

Parameter	Waste-Pit Liquid
Date	03/13/92 to 05/04/92
Influent flow	50 L/day
Oxygen uptake, mg/L/hr	52/14[a]
Mixed liquor DO, mg/L	3.9/5.7[a]
Mixed liquor pH	7.5/7.5[a]
Mixed liquor temperature, °C	21/21[a]
Hydraulic Retention Time (days)	0.2 - 5[b]
Solids Retention Time (days)	5 - 40[b]
Mixed liquor suspended solids, mg/L	11,400/8900[a]

(a) First stage/second stage

(b) The range is typical of values for the PACT® system. Actual values used for waste-pit liquid treatment are proprietary and subject to Secrecy Agreements between all parties involved in this project.

°C = degrees Celsius; DO = dissolved oxygen; mg/L = milligrams per liter; mg/L/hr = milligrams per liter per hour.

Steady-State Operation

Performance of the PACT® system was evaluated during steady-state operation. Steady-state operation commenced 31 days after PACT® system startup and was defined as operation after three SRTs elapsed during which (1) HRT and SRT were not adjusted, and (2) the mixed-liquor solids of each stage did not vary by more than 20% from the mean for that period for the respective stage. At the end of the period, the concentrations of the final effluent BOD and total suspended solids (TSS) values consistently were less than 30 mg/L.

Off-Gas Testing

The composition of organics in the aeration off-gas was evaluated by collecting and analyzing samples of the off-gas from the respective first- and second-stage aeration tanks. Aeration tanks were covered to control the release of the off-gases. A sample port was provided in the cover of each tank through which a sampling tube could be inserted. Off-gas samples were withdrawn and sampled using Tenax™-filled gas sampling tubes. The contents of individual tubes were thermally desorbed and were analyzed for volatile organic compounds using gas chromatography (GC) analytical methods.

Sampling and Analysis

Samples of the PACT® system influent and effluent were submitted for process analytical testing to assess removals of conventional analytes including BOD, COD, ammonia, and solids and for EPA Contract Laboratory Program (CLP) analysis to assess the removal of hazardous and priority pollutants. A summary of the process operational parameters testing schedule is presented in Table 2. Samples of the PACT® system influent and effluent were analyzed for VOCs, SVOCs, pesticides/polychlorinated biphenyls (PCBs), dioxins, metals, herbicides, radionuclides, and additional parameters.

RESULTS AND DISCUSSION

Table 3 presents a summary of the process analytical results for the PACT® system influent and final effluent. Figures 2 and 3 illustrate the respective PACT® system BOD and COD concentration data; Figure 4 illustrates the PACT® system ammonia concentration data. A summary of analytical results from the CLP characterization of the final effluent from the treatability study is presented in Table 4. The table includes concentration data for those organic and inorganic parameters detected in the PACT® system influent streams and includes analyte percent removals. Table 5 presents a summary of the PACT® system off-gas analytical data.

The concentrations of BOD in the influent and in the effluent are presented in Figure 2. The BOD values shown are for total BOD, which includes oxygen

TABLE 2. Process analytical program to evaluate Powdered Activated Carbon Treatment system process control.

	Number of Samples Per Week				
Type of Analysis Performed on a Weekly Basis	Influent	First-Stage Clarifier Effluent	Second-Stage Clarifier Effluent	First- and Second-Stage Aeration Mixed Liquor	Waste Solids
Biochemical oxygen demand	1	2	2	–	–
Chemical oxygen demand	1	2	2	–	–
Nonpurgeable organic carbon	1	2	2	–	–
Total suspended solids	1	2	2	–	–
Volatile suspended solids	1	–	2	–	–
Total Kjeldahl nitrogen	1	–	2	–	–
Ammonia nitrogen (as nitrogen)	1	2	2	–	–
Nitrate, nitrite (as nitrogen)	1	2	2	–	–
pH	2	2	2	Daily	–
Color	1	2	2	–	–
Total phosphorous	1	2	2	–	–
Mixed liquor volatile suspended solids	–	–	–	3	1
Mixed liquor suspended solids	–	–	–	3	–
Dissolved oxygen	–	–	–	Daily	–
Total solids	1	1	1	–	1
Oxygen uptake rates	–	–	–	3	–
Temperature	1	–	–	1	–
Microbial evaluation	–	–	–	1	–

– = no sample collection or analysis performed.

TABLE 3. Process analytical Powdered Activated Carbon Treatment system testing influent and effluent analyses results.

	PACT® System Influent												
Days Duration	BOD (mg/L)	COD (mg/L)	NPOC (mg/L)	TS (mg/L)	SS (mg/L)	SA (mg/L)	TKN (mg/L)	NH_3-N (mg/L)	NO_3-N (mg/L)	NO_2-N (mg/L)	P (mg/L)	Color (APHA)	pH (units)
1	1690	4840	1880	22,000	3600	3280	639	595	<0.50	<0.50	17.1	755	9.30
5	1890	5290	1740	27,100	5590	1970	737	590	<0.50	<0.50	16.5	864	8.90
12	2220	5660	2240	23,600	2540	1980	788	638	<0.50	<0.50	8.80	759	10.7
19	2520	5730	2480	21,100	184	140	–	638	<0.50	<0.50	–	791	10.4
26	2710	5400	2320	21,300	142	98.0	591	617	0.60	<0.50	1.90	681	10.6
33	2400	5360	–	21,300	194	138	684	592	<0.50	<0.50	54.0	623	10.7
40	2720	5490	1950	21,000	104	54.0	693	549	<0.50	<0.50	5.20	752	10.6
48	2830	5460	1850	21,000	250	158	724	638	<0.50	<0.50	38.0	648	10.5

(continues)

TABLE 3. (Continued).

Days Duration	PACT® System Effluent BOD (mg/L)	COD (mg/L)	NPOC (mg/L)	TS (mg/L)	SS (mg/L)	SA (mg/L)	TKN (mg/L)	NH_3-N (mg/L)	NO_3-N (mg/L)	NO_2-N (mg/L)	P (mg/L)	Color (APHA)	pH (units)
1	<6.00	138	72.0	6250	<5.00	<5.00	0.90	0.200	274	3.30	1.80	5.00	8.00
5	<6.00	181	53.0	11,600	7.00	<5.00	76.0	74.0	416	22.0	2.00	5.00	7.60
7	7.20	333	65.0	14,500	8.00	<5.00	139	133	500	19.0	2.10	5.00	7.90
12	<6.00	493	90.0	19,100	23.0	10.0	–	304	765	7.40	0.900	5.00	7.80
14	9.00	432	116	21,000	10.0	<5.00	301	301	978	6.90	1.10	15.0	7.60
19	23.0	560	134	22,700	30.0	<5.00	–	409	1110	15.0	1.90	74.0	7.80
21	25.0	376	147	21,900	42.0	18.0	–	361	1070	53.0	1.80	52.0	7.70
26	41.0	694	142	21,800	20.0	<5.00	16.0	12.0	1090	281	1.90	119	7.50
28	25.0	626	157	23,100	<5.00	<5.00	8.90	1.20	1060	245	2.00	94.0	8.10
33	10.1	483	–	21,000	<28.0	<28.0	10.0	1.10	1040	8.20	1.20	93.0	8.10
35	10.6	588	–	22,000	38.0	<10.0	9.80	1.00	1110	0.500	1.90	83.0	7.60
40	6.20	544	139	22,000	23.0	5.00	12.7	1.70	997	0.600	1.70	112	7.90
42	<6.00	509	123	21,000	–	–	10.0	–	1050	1.00	2.00	110	
48	<6.00	389	137	23,000	22.0	<10.0	14.0	0.400	1240	0.500	2.60	57.0	8.10
49	<6.00	511	127	–	<10.0	<10.0	12.0	1.20	1090	0.500		50.0	7.20

< = The target analyte was not detected at or above the specified quantitation limit.
– = data not available
APHA = American Public Health Association
BOD = biochemical oxygen demand; COD = chemical oxygen demand; mg/L = milligrams per liter; NH_3-N = ammonia nitrogen; NO_2-N = nitrite-nitrogen; NO_3-N = nitrate-nitrogen; NPOC = nonpurgeable organic carbon; P = phosphorus; SA = suspended ash; SS = suspended solids; TKN = total Kjeldahl nitrogen; TS = total solids.

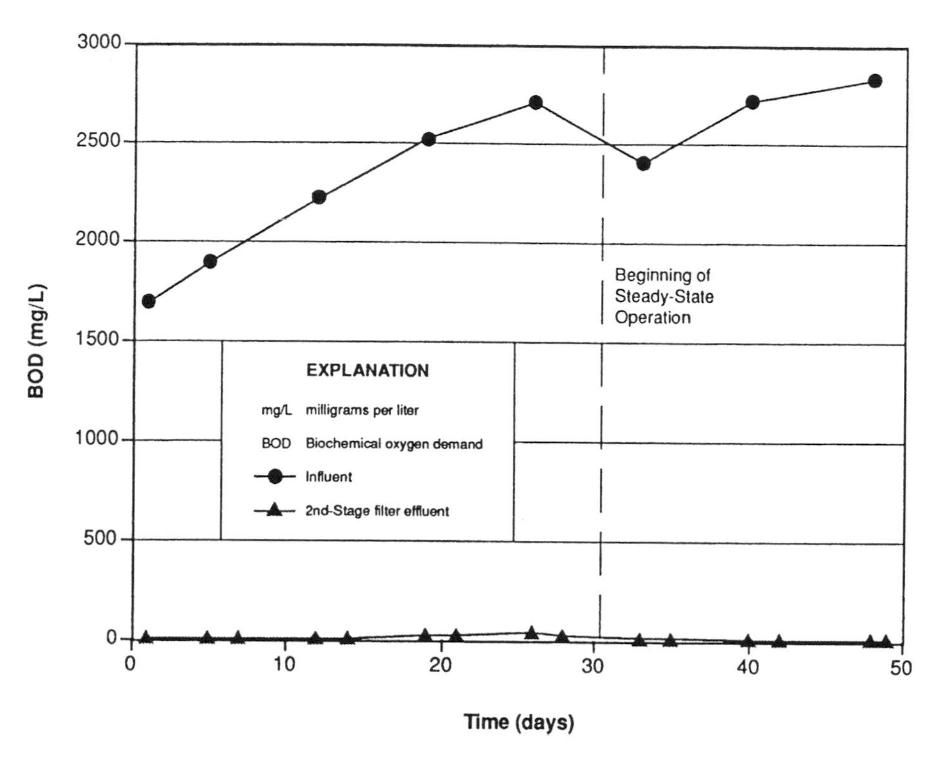

FIGURE 2. PACT® system influent and effluent BOD concentrations.

demand from oxidation of organic carbon and ammonia. The steady-state effluent BOD values may have been lower if the BOD test had been inhibited to prevent the extra oxygen demand of nitrifiers present in the BOD bottles. As illustrated in Figure 2, the PACT® system substantially reduced (by more than 90%) the influent waste-pit liquid BOD values to below 10 mg/L.

Figure 3 illustrates the results of the COD of the influent and effluent, respectively. The COD concentration of the effluent ranged between 100 and 700 mg/L.

The concentration of ammonia in the influent and the effluent are illustrated in Figure 4. Ammonia concentration in the effluent decreased after about 25 days. This decrease in the concentration of ammonia was caused by nitrification. Nitrification is the biologically mediated conversion of ammonia and organic nitrogen to the oxidized nitrogen forms (nitrate and nitrite). A significant amount of oxidation occurred in the first stage, reducing the ammonia nitrogen by about 50%. The second stage further reduced the ammonia concentration to an average of 1 mg/L during the steady-state period.

Information presented in Table 4 indicates that measurable amounts of VOCs, including such constituents as vinyl chloride, methylene chloride, acetone, 1,2-dichloroethene, methyl ethyl ketone, and benzene, were detected in waste-pit liquid influent. VOCs, however, were not detected in the PACT® system effluent

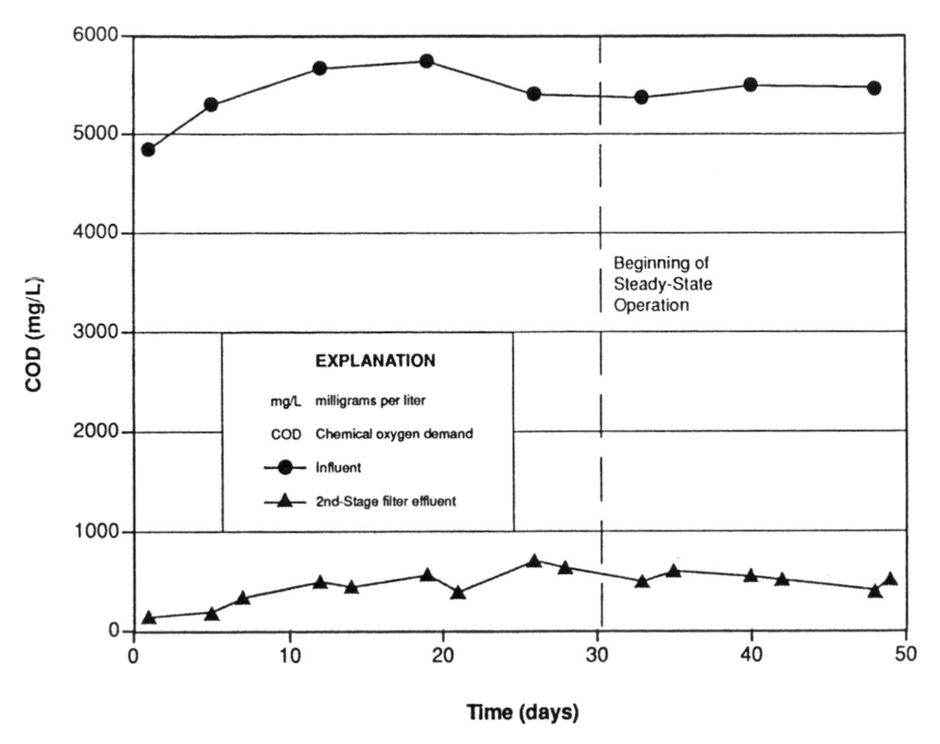

FIGURE 3. PACT® system influent and effluent COD concentrations.

stream. Phenol and 4-methyl phenol were the only SVOCs detected in the PACT® system influent. These compounds were removed to levels below detection limits during PACT® system treatment.

Dioxins and furans were analyzed and detected in the waste-pit liquid influent. Dioxin congeners were not detected in the PACT® system effluent. Concentration values for total dioxins and furans ranged from 34.5 picograms per liter (pg/L) (total PeCDF) to 89.4 pg/L (total HpCDF). The PACT® system effluent concentration values for total dioxins and furans were generally less than 10 pg/L. More than 89% of total dioxins and furans were removed.

Metals and inorganic parameters also were analyzed as part of the CLP analytical work. The PACT® system generally is not designed to remove metals from contaminated water; however, the concentration of some metals, including aluminum, arsenic, lead, and silver, decreased during PACT® system. Other metals, such as manganese and cobalt increased in concentration during PACT® system treatment.

This increase in metals concentration may have resulted from the addition of nutrients including metals, to the PACT® system influent. The analytical data presented in Table 4 also indicate that suspended solids were removed to a final concentration of less than 10 mg/L.

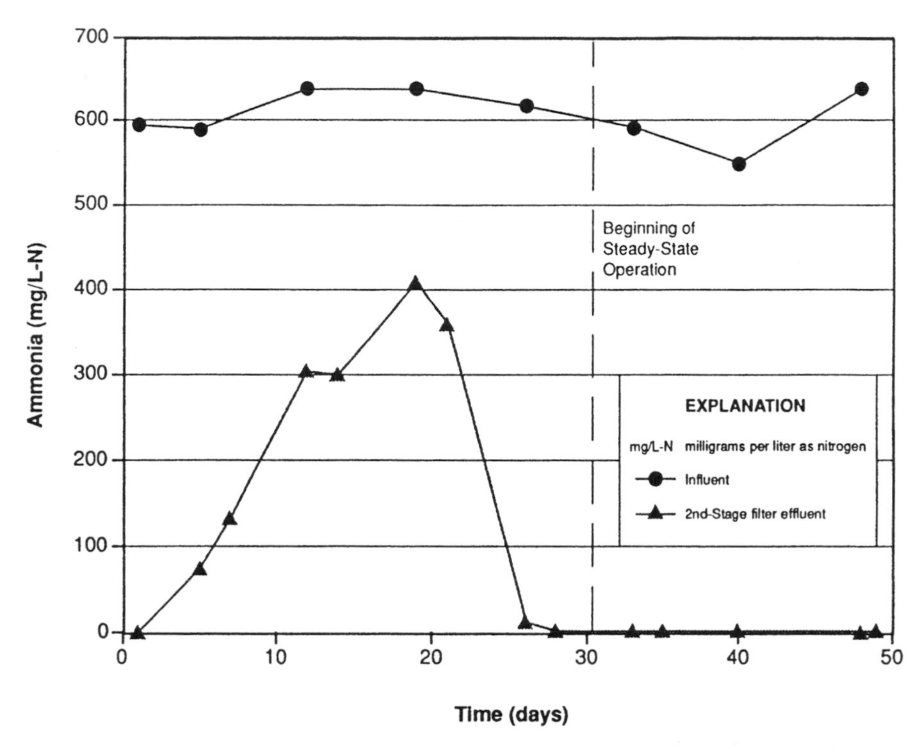

FIGURE 4. PACT® system influent and effluent ammonia concentrations.

Off-gas from the first-stage aeration tank contained dichloromethane, 1,1-dichloroethane, 1,2-dichloroethane, naphthalene, and *n*-propylbenzene (Table 5). The first three compounds were detected at low concentrations and were present in the waste-pit liquid influent to the PACT® system. The last two compounds, naphthalene and *n*-propylbenzene, were not detected in the influent liquid. Their presence in the off-gas is possibly an anomaly of the analytical method. The compounds elute late in the analysis; it is possible to mistake them for other compounds with similar retention time.

Losses of VOCs from the PACT® system were estimated by calculating a VOC mass balance for the waste-pit liquid influent and the aeration tank off-gas. Table 6 presents off-gas analytical results. More than 85% of the dichloromethane, 1,1-dichloroethane, and 1,2-dichloroethane were removed biologically or by adsorption; volatile emissions accounted for only a small fraction of the VOCs removed.

Dewatered residual solids from the PACT® system were analyzed for toxicity using the Toxicity Characteristic Leaching Procedure (TCLP) and other parameters. A summary of the analytical results is presented in Table 7. The concentrations of analytes in the dewatered solids were near or below quantitation limits and meet TCLP regulatory requirements for land disposal.

TABLE 4. Summary of percent removals achieved for waste-pit liquid.

Analyte	Influent Concentration	Effluent Concentration	Percent Removal
Volatile Organic Compounds (µg/L)			
Chloromethane	<10.0	<2.70	–
Vinyl chloride	120	<2.00	>98
Chloroethane	27.9	<5.00	>82
Methylene chloride	571	22.5 JB	96
Acetone	37,810	<3,500	>95
1,1-Dichloroethane	845	<0.500	>99
1,2-Dichloroethene (total)	132	<70.0	>46
Methyl ethyl ketone	12,200	<1,750 R	>85
Chloroform	11.3	<5.70	>49
1,2-Dichloroethane	11,200	<5.60	>99
1,1,1-Trichloroethane	120	<200	–
Trichloroethene	12.2	<5.00	–
Benzene	45.6	<5.00	–
4-Methyl-2-pentanone	1,840	<2.00	>99
2-Hexanone	112	<2.00	>98
Toluene	1,860	<2,420	–
Chlorobenzene	18.3	<300	–
Ethylbenzene	22.1	<680	–
Total xylenes	136	<440	–
Semivolatile Organic Compounds (µg/L)			
Phenol	1,488	<1.0	>99
4-Methylphenol	2,360	<1.0	>99
Metals (µg/L)			
Aluminum	283	138	51
Antimony	5.70	6.60	–
Arsenic	22.5	6.00	73
Barium	351	515	–
Beryllium	2.70	0.790	70
Boron	9,710	11,200	–
Cadmium	430	1.30	70
Calcium	13,600	69,000	–
Chromium	40.1	29.1	27
Cobalt	17.0	77.6	–
Copper	32.1	35.3	–
Iron	1,610	439	73
Lead	7.0	2.90	58
Magnesium	83,700	71,500	14
Manganese	50.9	332	–
Nickel	237	287	–
Potassium	424,000	387,000	8.7
Silver	2.30	2.80	–
Sodium	6,940,000	6,980,000	–
Vanadium	23.0	9.10	60
Zinc	68.3	306	–

TABLE 4. (Continued).

Analyte	Influent Concentration	Effluent Concentration	Percent Removal
Dioxins (pg/L)			
2378-TCDD	3.40	<2.80	>18
12378-PeCDD	5.50 JB	<4.50	>18
123478-HxCDD	9.70	<4.40	>54
123678-HxCDD	86.8 JB	<3.00	>96
123789-HxCDD	69.2 JB	<3.90	>94
1234678-HpCDD	3,010.0 JB	37.6 JB	98
OCDD	15,100.0 JB	<157 JB	>98
2378-TCDF	<5.80 JB	2.10	-
12378-PeCDF	2.80	<3.00	–
123478-HxCDF	18.2	<2.80	>84
234678-HxCDF	6.00 BJP	4.40	26
1234678-HpCDF	19.3	<2.10	>89
OCDF	38.9	<5.10	>86
Total TCDD	46.2	<2.80	>94
Total TCDF	48.1 JB	<2.10 JB	>95
Total PeCDF	34.5	<3.00	>91
Total HxCDF	74.6 JB	4.10	94
Total HpCDF	89.4 JB	<2.80	>96
Inorganic Parameters (mg/L)			
Alkalinity (as $CaCO_3$)	5,360	299	94
Orthophosphate	0.22	5.42	–
Total phosphorous	0.47	1.82	–
TOC	1,800	149	92
TOX	0.9	1.16	–
TSS	202	182	10
TDS	25,300	20,300	20
pH	9.56	6.51	–
Specific conductance (μmhos/cm)	26,000	26,500	–
Oil and grease	145	1.47	99
Turbidity (NTU)	31.5	6	81
TKN	488	8.6	98

– = Percent removal was negative or could not be calculated because the influent was less than the specified quantitation limit.
< = The analyte was not detected at or above the specified quantitation limit.
> = greater than.
BJP = Estimated maximum possible concentration; BOD = biochemical oxygen demand; $CaCO_3$ = calcium carbonate; COD = chemical oxygen demand; JB = The concentration is an estimate because the analyte was detected in the associated blank.
mg/L = milligrams per liter; NTU = nephlometric turbidity units; pg/L = picograms per liter; R = data are judged unusable; μg/L = micrograms per liter;
TKN = total Kjeldahl nitrogen.

TABLE 5. Powdered Activated Carbon Treatment off-gas analytical data.

	First-Stage Reactor Off-Gas	Second-Stage Reactor Off-Gas	First-Stage Reactor Off-Gas	Second-Stage Reactor Off-Gas
Sample Date:	04/28/92	04/28/92	04/29/92	04/29/92
Parameter				
1,1-Dichloro-ethane	0.238	<0.08	0.134	<0.08
1,2-Dichloro-ethane	0.397	<0.08	0.278	<0.08
Dichloro-methane	0.108	0.175	<0.08	0.129
Naphthalene	4.24[a]	4.58[a]	4.93[a]	4.49[a]
N-propyl-benzene	26.8[a]	19.6[a]	32.0[a]	22.4[a]
Sample Date:	04/30/92	04/30/92	05/01/92	05/01/92
Parameter				
1,1-Dichloro-ethane	0.332	0.081	0.322	<0.08
1,2-Dichloro-ethane	0.576	<0.08	0.299	<0.08
Dichloro-methane	0.254	0.195	0.125	<0.08
Naphthalene	3.26[a]	4.0[a]	4.0[a]	4.16[a]
N-propyl-benzene	16.8[a]	15.9[a]	12.7[a]	10.5[a]

Units are shown in micrograms per liter.
(a) 10-mL sample volume analyzed instead of the 250-mL sample volume analyzed for remaining analytes.
< = The target analyte was not detected at or above the specified quantitation limit.
– = data not available
mL = milliliter

TABLE 6. Off-gas volatile organic compound mass balance.

	Feed Concentration (µg/L)	Effluent Concentration (µg/L)	First-Stage Off-Gas Concentration (µg/L)	Second-Stage Off-Gas Concentration (µg/L)
Compound				
1,1-Dichloroethane	878	<5	0.238	<0.08
1,2-Dichloroethane	10370	<5	0.397	<0.08
Dichloromethane	1810	49	0.108	0.175

Mass Balance – Volatile Organic Compound
Liquid Flow = 28.51 First-Stage Gas Flow = 14,968 L/day
Second-Stage Gas Flow = 481 L/day

	Feed (mg)	Effluent (mg)	Gas (mg)	Removal (percent)
Compound				
1,1-Dichloroethane	25.0	<0.14	3.6	>85.0
1,2-Dichloroethane	295.5	<0.14	5.98	>97.9
Dichloromethane	51.6[(a)]	1.4	1.70	94.0

(a) From second volatile characterization.
< = The target analyte was not detected at or above the specified quantitation limit.
> = greater than.
L = liter; mg = milligram; µg/L = micrograms per liter.

CONCLUSIONS

Conclusions for the PACT® system study include the following:

- The PACT® system, as operated, provided effective treatment of organic material as measured by BOD and COD in the waste-pit liquid effluent.
- The PACT® system promoted the growth of nitrifying bacteria resulting in a reduction of ammonia concentrations to less than 2 mg/L in during steady-state operations.
- Concentrations of VOCs, including methylene chloride, methyl ethyl ketone, acetone, toluene, 2-hexanone, 1,1-dichloroethane, and 1,2-dichloroethane, were reduced to low (less than 10 µg/L) or nondetectable levels.

TABLE 7. Results of analysis of dewatered Powdered Activated Carbon Treatment residuals.

EPA Hazardous Waste Number	Parameter	CAS No.	Regulatory Level	TCLP Extract Analytical Data PACT® System Residuals
	Metals (mg/L)			
D011	Silver	7440-22-4	5.0	<0.01
D004	Arsenic	7440-38-2	5.0	<0.10
D005	Barium	7440-38-3	100.0	0.89
D006	Cadmium	7440-43-9	1.0	<0.01
D007	Chromium	7440-47-3	5.0	0.02
D009	Mercury	7439-87-3	0.2	<0.001
D008	Lead	7439-92-1	5.0	<0.06
D010	Selenium	7782-49-2	1.0	<0.10
	Pesticides (mg/L)			
D013	gamma-BHC (lindane)	58-89-0	0.4	<0.4
D012	Endrin	72-20-8	0.02	<0.020
D014	Methoxyclor	72-43-5	10.0	<10
D015	Toxaphene	8001-35-2	0.5	<0.5
D031	Heptachlor epoxide	76-44-8	0.008	<0.008
D020	Chlordane	57-75-9	0.03	<0.03
D031	Heptachlor	76-44-8	0.008	<0.008
	Herbicides (mg/L)			
D017	2,4,5-TP (Silvex)	83-72-1	1.0	<1
D016	2,4-D	94-75-7	10.0	<10
	Volatiles (mg/L)			
D043	Vinyl chloride	75-01-4	0.2	<0.025
D029	1,1-Dichloroethene	75-35-4	0.7	<0.025
D035	Methyl ethyl ketone	78-93-9	200.0	<0.25
D022	Chloroform	67-66-3	6.0	<0.025
D019	Carbon tetrachloride	56-23-5	0.5	<0.025
D018	Benzene	71-43-5	0.5	<0.025
D020	1,2-Dichloroethane	107-06-2	0.5	0.046
D040	Trichloroethylene	79-01-6	0.5	<0.025
D039	Tetrachloroethylene	127-18-4	0.7	<0.025
D021	Chlorobenzene	108-90-7	100.0	<0.025
D027	1,4-Dichlorobenzene	106-46-7	7.5	<0.025
	Semivolatiles (mg/L)			
D034	Hexachloroethane	67-72-1	3.0	<0.10
D036	Nitrobenzene	98-95-3	2.0	<0.10
D041	4,5-Trichlorophenol	95-95-4	400.0	<0.10
D024	Total cresols	95-48-7, 108-39-4	200.0	<0.10
D033	Hexachlorobutadiene	87-68-3	0.5	<0.10
D030	2,4-Dinitrotoluene	121-14-2	0.13	<0.10
D032	Hexachlorobenzene	118-74-1	0.13	<0.10

TABLE 7. (Continued).

EPA Hazardous Waste Number	Parameter	CAS No.	Regulatory Level	TCLP Extract Analytical Data PACT® System Residuals
	Semivolatiles (mg/L)			
D038	Pyridine	87-96-5	5.0	<0.10
D037	Pentachlorophenol	110-86-1	100.0	<0.10
D042	2,4,6-Trichlorophenol	88-06-02	2.0	<0.10
D041	2,4,5-Trichlorophenol	95-95-4	400.0	<0.10
	Miscellaneous			
	Corrosivity (pH)	Gross beta pCi/g		5.0
	Flash point (OC)	Gross beta pCi/g		>95
	Reactive cyanide (µg/g)	Gross beta pCi/g		<5.0
	Reactive sulfide (µg/g)	Gross beta pCi/g		100
	Total solids (%)	Gross beta pCi/g		20
	Gross alpha pCi/g	Gross beta pCi/g		1 ± 2
	Gross beta pCi/g	Gross beta pCi/g		0 ± 4

< = The target analyte was not detected at or above the reporting limit.
> = The target analyte was detected at or above the maximum reporting limit.
% = percent
± = plus or minus
°C = degrees Celsius; µg/g = micrograms per gram; CAS = Chemical Abstract Service; EPA = U.S. Environmental Protection Agency; mg/L = milligrams per liter; pCi/g = picocuries per gram.

- Organic compounds, including dioxins/furans, were removed by the PACT® system to concentrations that were generally near or less than the respective analytical quantitation limits.
- PAC addition, in conjunction with microbial activity, was adequate to minimize VOCs in the respective first- and second-stage aeration tank off-gas.
- Process residual solids meet regulatory TCLP requirements for land disposal.

ACKNOWLEDGMENTS

This study was funded by the Lowry Coalition in support of the Feasibility Study for OUs 1 and 6. The conclusions presented in this paper are not necessarily those of The Lowry Coalition. The authors wish to thank Jack Schuck and Tom Vollstedt of Zimpro Passavant Environmental Systems, Inc., Rothschild, Wisconsin, for their technical assistance during operation of the PACT® system.

REFERENCES

Eckenfelder, W. W. 1991. "Strategies for Toxicity Reduction in Industrial Wastewater." *Wat. Sci. Rech. 24*: 185.

Hannah, S. A., B. M. Austern, A. E. Eralp, and R. H. Wise. 1986. "Comparative Removal of Toxic Pollutants by Six Wastewater Treatment Processes." *J. Wat. Poll. Cont. Fed. 58*: 27.

Parker, W. J. Thompson, J. P. Bill, and H. Melcer. 1993. "Fate of Volatile Organic Compounds in Municipal Activated Sludge Plants." *Water Environ. Res. 65*: 58.

Sferra, P. R. 1989. "Biodegradation of Environmental Pollutants." In H. N. Freeman (Ed.), *Standard Handbook of Hazardous Waste Treatment and Disposal*, pp. 9.53-9.60. McGraw-Hill, Inc., New York, NY.

U.S. Environmental Protection Agency. 1989. "Second Amended and Restated Administrative Order on Consent and Conceptual Work Plan for The Remedial Investigation and Feasibility Study, Shallow Groundwater and Subsurface Liquids and Deep Groundwater Operable Units." Lowry Landfill, December 22.

Ying W., R. R. Bonk, and S. A. Stanley. 1987. "Treatment of a Landfill Leachate in Powdered Activated Carbon Enhanced Sequencing Batch Reactors." *Environ. Prog.* 6: 8.

ROTTING OF THERMOPLASTICS MADE FROM LIGNIN AND STYRENE BY WHITE-ROT BASIDIOMYCETES

O. Milstein, R. Gersonde, A. Huttermann, M. J. Chen, and J. J. Meister

ABSTRACT

White-rot basidiomycetes were able to biodegrade styrene (1-phenylethene) graft copolymers of lignin containing different proportions of lignin and polystyrene [poly(1-phenylethylene)]. The biodegradation tests were run on lignin/styrene copolymerization products that contained 10.3, 32.2, and 50.4 wt.% of lignin, respectively. The polymer samples were incubated with the white-rot fungi *Pleurotus ostreatus*, *Phanerochaete chrysosporium*, and *Trametes versicolor*, and the brown-rot fungus *Gleophyllum trabeum*. White-rot fungi degraded the plastic samples at a rate that increased with increasing lignin content in the copolymer sample. Both polystyrene and lignin components of the copolymer were readily degraded. Observation by scanning electron microscopy (SEM) of incubated copolymers showed a deterioration of the plastic surface. The brown-rot fungus did not affect any of these plastics, nor did any of the fungi degrade any of the pure polystyrene. White-rot fungi produced and secreted oxidative enzymes associated with lignin degradation in liquid media during incubation with lignopolystyrene copolymer. Fourier transform infrared (FTIR) spectra of the copolymers incubated with white-rot fungi have shown decreases of intensity in the whole range of absorbances characteristic of both lignin and polystyrene.

INTRODUCTION

Plastics contribute a significant and increasing portion by weight or by volume to the waste in municipal landfills and this plastic fraction is projected to increase. Since plastics became an integral part of contemporary life, opposition to placing plastics in landfills has grown because most synthetic polymers are resistant to biodegradation. The annual consumption of thermoplastic polystyrene has risen to 1.3×10^6 Mg in Western Europe and to 2.5×10^6 Mg in the United States (Society of the Plastic Industry 1991). This material is extremely recalcitrant to bioconversion, and opposition to incinerating plastics exists because of the potential of hazardous emissions. The degradability of plastics may be enhanced by linking selected, readily degradable substituents into the polymer chemical structure.

Products of starch, cellulose, and poly(hydroxybutric acid) blended into a synthetic polymer show appreciable biodegradability of the naturally occurring fraction of the plastic mixture but are not completely biodegradable. Lignin is the second most abundant biopolymer after cellulose but has not previously been used in these degradable plastics. Lignin occurs in the cell walls of all woody plants. About 50×10^6 Mg of lignin are released annually by the pulping industry. Most of this immense amount of biomass is discarded. However, the polyaromatic nature of lignin may represent an enormous supply of inexpensive, biodegradable chemicals. These could be used to produce an engineered material that would replace expensive petrochemicals with a renewable raw material of comparatively low cost. The increase in lignin use might be achieved by copolymerization of lignin with synthetic monomers (Meister 1991, Meister et al. 1991, Meister & Li 1992). Graft copolymer is formed by conducting a free radical polymerization with styrene in a nitrogen-saturated organic or aqueous/organic solvent containing lignin, calcium chloride, and a hydroperoxide.

The ultimate transformation of lignin in nature, its complete oxidation to CO_2, takes place primarily by the white-rot basidiomycetes. Brown-rot fungi, in contrast, leave the lignin essentially undegraded. In this paper we report how copolymerization of lignin and styrene monomer increases the susceptibility of the resulting lignopolystyrene product, and particularly its polystyrene moiety, to fungal degradation.

MATERIALS AND METHODS

Lignopolystyrene Complex: Lignin and Polystyrene Homopolymer

Most of the lignin used in these studies is a Kraft pine lignin prepared in "free acid" form with a number-average molecular weight (Mn) of 9,600, a weight-average molecular weight (Mw) of 22,000, and a polydispersity index of 2.29. The ash content of the lignin is 1.0 wt.% or less. The tested lignopolystyrene (LPS) copolymerizates were synthesized at the Department of Chemistry of the University of Detroit Mercy by the following method.

Pure styrene and a solution of lignin, calcium chloride, and dimethylsulfoxide were saturated with nitrogen (N_2) for 10 minutes. A 30 wt.% aqueous solution of H_2O_2 was added to the solution and, after 20 minutes, the monomer was added to the solution. The flask was stoppered and placed in a 30°C bath. The reaction mixture was stirred at 4 Hz for 48 h. This slurry then was added to 10 times its volume of acidified water (pH = 2) and the polymer was recovered by filtration.

This polymerization method was used to create lignin/styrene copolymerization products that contained 10.3 (LPS-10), 32.2 (LPS-32), and 50.4 (LPS-50) wt.% of lignin, respectively. The copolymerizates tested for biodegradation were supplied as a compression-molded sheet of 0.15-mm-thick by 5- to 7-cm-diameter circular plastic film. Polystyrene homopolymer, material RIPO, was used as received from Amoco Chemical Company, P.O. Box 400, Naperville, Illinois, USA. All compression moldings were done at 150°C and 192 kPa pressure for 1 min.

Organisms and Cultivation

The white-rot microorganisms used in this work were Basidiomycetes: white rot *Phanerochaete chrysosporium* Burdsall, *Trametes versicolor* I (L. ex Fr.) Quelet (ATCC 11235), and *Pleurotus ostreatus* v. florida (F6) (Jaquin ex Fr.), Kummer. The activities of the white-rot fungi were compared to those of the brown-rot fungus *Gleophyllum trabeum* (Pers. ex Fr.), Murrill.

The fungi were cultivated either on solid 2.5% agar medium for the study of copolymerizate biodegradation, or in liquid medium for the comparative study of the pattern of the analyzed oxidases. Both media contained the same concentrations of mineral salts, glucose, and a reduced content of nitrogen, as specified by Kirk et al. (1978) and Kern (1989).

The liquid medium (30 mL in 500-mL conical flasks) was placed in both control and treatment flasks (in triplicate). The test flasks also received 50 mg of either lignin, polystyrene, or the LPS complex. Mycelia grown for 9 days in the liquid medium without any added polymer, and then homogenized with Ultra-Thorax, were used as inoculum. Inoculated flasks were incubated in the dark as standing cultures for 3 weeks at 25°C. The inoculum for the plates was a piece of straw. The plates were placed at 25°C in a thermostated chamber at 100% humidity for 68 days and were aseptically opened to exchange the air every 7 days. Pressed plastic films measuring 0.4 cm^2 were placed directly on the surface of the solid medium near the inoculum.

Evaluation of Polymer Degradation

Biodegradation was tracked (1) by weight loss, in particular by decrease of the lignin and polystyrene components from the biodegraded complex, (2) by FTIR spectroscopy of the degraded samples, and (3) by SEM of the decayed polymer. Tested powdered copolymerizates were intimately bound with the growing fungi, thus, direct measurements of the loss of polymer weight were impossible. To evaluate loss of the tested copolymerizate, the nitrogen content of the aliquots of the dry material collected from the triplicates of inoculated plates and from the sterile control was measured. The amount of the nitrogen measured was extrapolated to the amount of fungal biomass by applying the same nitrogen-biomass ratio as found in the pure fungus from cultures of identical age and medium. The computed fungal biomass was subtracted from the recovered material in this way.

The nitrogen in the copolymerizates from inoculated plates and sterile controls was determined by elemental analysis after combustion of the dry sample in a quartz combustion reactor at 1020°C. The quantity of separate components, lignin and polystyrene, in the treated copolymer was analyzed by ultraviolet (UV) spectroscopy using multicomponent analysis methods.

Assay of Enzyme Activity

Lignin peroxidase (LiP) activity was measured through the rate of oxidation of veratryl alcohol (Tien & Kirt 1984). Laccase activity was measured by following color development of syringaldazine (4-hydroxy-3,5-dimethoxybenzaldehyde azine) at 525 nm (Harkin & Obst 1973). Mn(II) peroxidase activity was measured following oxidation of guajacol to tetraguajacol by monitoring visible light absorbance at 466 nm.

Scanning Electron Microscopy

The pieces of pressed copolymer (approximately 0.4 cm^2) from both the sterile control plates and from the plates with fungi were withdrawn after 68 days of incubation in a 25°C bath at 100% humidity. Withdrawn copolymerizates were then mounted on SEM stubs, sputter-coated with gold to thickness of about 10 nm, and observed and photographed using a Phillips SEM 515.

RESULTS AND DISCUSSION

Synthesis of Graft Copolymer

Data for reactions run to create samples for biodegradation are given in Table 1. All of these reactions were stirred at a rate of about 4 Hz throughout the synthesis. Reaction 1-14 produced LPS-10, reaction 1-34 produced LPS-32, and reaction 1-27 produced LPS-50. Proof of formation of the graft copolymer was obtained through mass balance of the fractionated reaction product, solubility tests, wetting tests, phase-partitioning tests, and FTIR analysis. These products have been shown to be poly(lignin-g-styrene)-containing materials by a series of solubility and extraction tests and are formed with almost 100% grafting efficiency for lignin. The copolymerization reaction of lignin and styrene is:

$$\text{Lignin} + n\ CH_2=\underset{\displaystyle C_6H_5}{CH} \longrightarrow \text{Lignin-}(CH_2\text{-}\underset{\displaystyle C_6H_5}{CH})_n\text{-}$$

SEM Visualization of Lignin/Styrene Copolymerizates Overgrown with Fungi

Fungal mycelia of all of the applied white-rot fungi had grown over the tested powdered LPS 4 or 5 days after inoculation of the plates. The most intensive enmeshing of the LPS copolymerizates was observed by the overgrown mycelial mat of *P. chrysosporium* and *T. versicolor*. Growth of *P. ostreatus* over the tested LPS was less intensive than growth of the other two white-rot fungi. All of the applied white-rot fungi and the brown rot *G. trabeum*, after 2 weeks of cultivation, completely overgrew the tested lignin powder. However, the brown rot

TABLE 1. Composition and yield of copolymerization reactions.

Sample Number	Composition (g)					Yield
	Lignin	Styrene	$CaCl_2$	H_2O_2 (mL)	Solvent	g/Wt. %
1-14	2.01	18.76	2.03	2.0	20.01	19.52/93.98
1-27	8.01	9.38	6.28	8.0	40.00	15.89/91.37
1-34	8.02	18.76	6.02	8.0	40.0	24.93/90.14

G. trabeum, even after 3 weeks of cultivation, colonized only external zones of the compact mass of LPS powder. Growth of both white-rot and brown-rot fungi was sporadic in and near the applied polystyrene.

The close encompassing of the particles of the tested LPS copolymerizates by white-rot mycelia was visualized clearly in the SEM, as shown in Figure 1. Moreover, mycelia of the white-rot fungi had produced capsular material outside the hyphae. Figure 1 also shows that this material engulfed particles of the degraded LPS, thus enhancing close contact between the fungi and the surface of the polymer complex.

The adhesion of microorganisms to surfaces of various composition is a decisive step in microbially induced corrosion (Whitekettle 1991). Presumably the active colonizers of polymer are able to adhere due to their ability to produce exocellular polymers composed primarily of nonionic and anionic polysaccharides. The formation of extracellular material that facilitated fungal adhesion on the surface of LPS copolymerizates was not observed in the tested brown rot *G. trabeum* although the hyphae of the fungus were found in the vicinity of the incubated polymer particles. Incubated lignin also was engulfed by the extracellular structures of white-rot fungi, similar to what was observed by Janshekar et al. (1982) during the degradation process caused by *P. chrysosporium*.

Incubation of the tested white-rot fungi with LPS copolymerizates that contained an increased wt.% of polystyrene (above 80%) caused a decrease of the production of the extracellular film-like material by the fungi.

Mass Reduction of Lignin/Polystyrene Constituents of the LPS

All of the white-rot fungi demonstrated an ability to decrease the weight of both constituents of LPS, regardless of the ratio of lignin and polystyrene (Figures 2B, 2C, and 2D). These white-rot basidiomycetes caused a range of weight loss of LPS copolymerizates that varied with the fungus with which the plastics were inoculated. The decomposing activity of *P. chrysosporium* and *T. versicolor* toward tested LPS copolymerizates exceeded the activity of *P. ostreatus* (Figures 2C, 2D, and 2B). All tested LPS showed insignificant weight loss of their constituents after incubation with the brown rot fungus *G. trabeum* (Figure 2A). However,

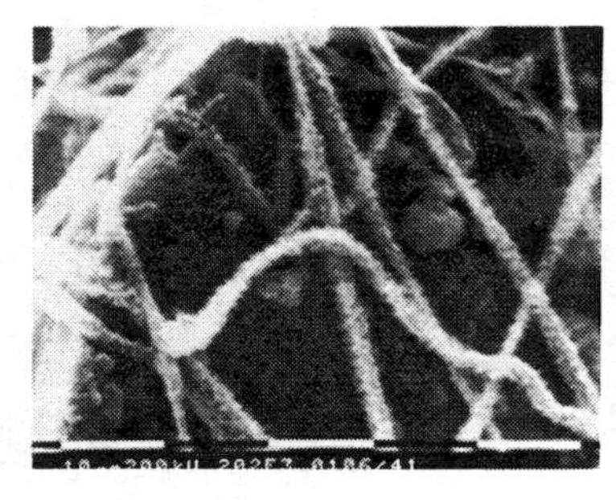

G. trabeum (A)

P. ostreatus (B)

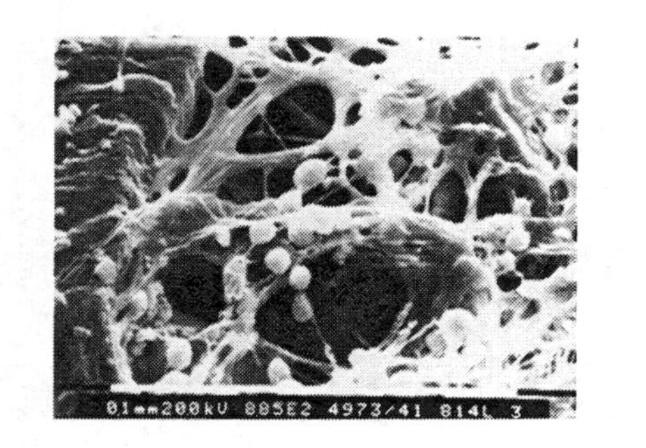

P. chrysosporium (C)

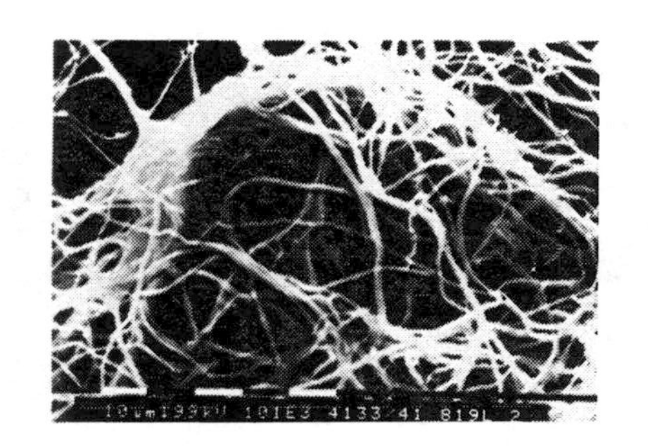

T. versicolor (D)

FIGURE 1. Scanning electron micrographs of powdered LPS-50 incubated for 30 days with *G. trabeum* (A), *P. ostreatus* (B), *P. chrysosporium* (C), and *T. versicolor* (D). Fungal hyphae have overgrown the polymerizate. The extracellular mucilage facilitates adhesion of hyphae and promotes efficient interaction with the plastic surface, as shown on B, C, and D. Bars, 10 µm.

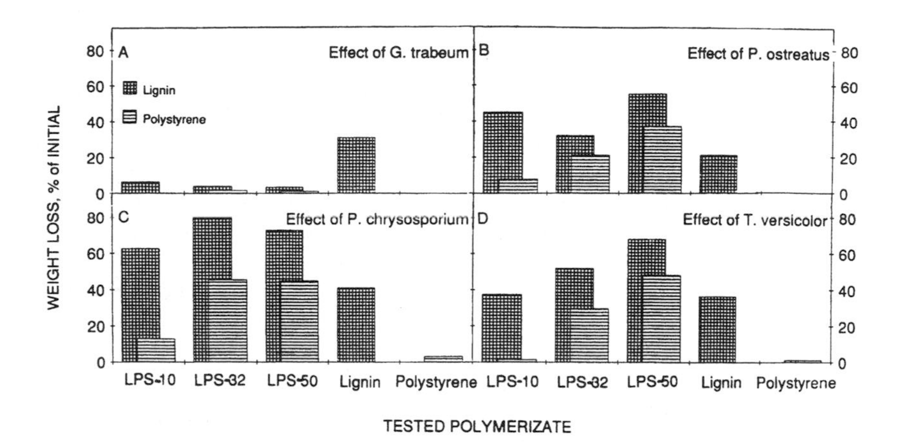

FIGURE 2. Mass loss of the constituents of lignopolystyrene graft copolymer (LPS) induced by fungal metabolism during 68 days of cultivation on solid media. LPS copolymerizates containing 10.3 (LPS-10), 32.2 (LPS-32) and 50.4 (LPS-50) wt.% of lignin, respectively, were incubated with *G. trabeum* (A), *P. ostreatus* (B), *P. chrysosporium* (C), and *T. versicolor* (D).

G. trabeum was able to deplete lignin applied as a natural polymer to an extent similar to the white-rot fungi. Decomposition of polystyrene incubated as a homopolymer was insignificant in all tested fungi.

The most efficient degradation of both constituents of LPS by white-rot fungi was observed with the plastics LPS-50 and LPS-32, containing 50.4 and 32.2 wt.% lignin, respectively. It appeared that the level of weight loss of polystyrene component from the incubated LPS was correlated with concentration of lignin in the copolymerizate. The measured weight loss of the LPS components could be due to mineralization as well as to modification followed by partial solubilization in a surrounding medium. This last type of conversion might be the cause of the biodegradation of lignin by the brown rot *G. trabeum*. Transformation of lignin caused by the brown-rot basidiomycete increases the number of polar groups in the lignin molecule after partial demethoxylation, hydroxylation, and less mineralization of lignin (Kirk 1975, Kirk & Farrell 1987).

The tested LPS copolymerizates, particularly their lignin and polystyrene components, were degraded by white-rot fungi in our experiments under conditions of solid-state fermentation. It appears that the conditions chosen for cultivation facilitate production of extracellular mucilage by the tested white-rot fungi. The extracellular capsular material, in turn, improves adhesion of hyphae on the plastic surface and intensifies the oxidative potential of the fungus. Moreover, the extracellular forms of the polysaccharides are not always present in a liquid culture of *P. chrysosporium* (Bes et al. 1987). It can be assumed that white-rot fungi in liquid media express their degradation potential toward incubated plastics less than do the same fungi cultivated in the solid state.

Pattern of Activity of Oxidative Enzymes in the Liquid Cultures

White-rot basidiomycetes, particularly *P. chrysosporium*, are responsible for the decomposition of the polymeric structure of lignin. During secondary metabolism, these fungi produce and secrete into the surrounding medium, two extracellular heme peroxidases, lignin peroxidase (LiP) and manganese peroxidase (MnP) (Kirk & Farrell 1987), that reportedly are associated with lignin degradation. However, many ligninolytic fungi do not produce detectable LiP. These white-rot fungi, particularly *T. versicolor*, produce one or more laccases in addition to MnP (Fahraeus & Reinhammar 1967). The pattern of oxidative activity secreted into the surrounding medium by white-rot fungi constitutes a unique combination of these enzymes that varies among strains and with the conditions of each organism's cultivation. The production of LiP, MnP, and laccase enzymatic activities in the liquid cultures of *P. chrysosporium*, *T. versicolor*, and *G. trabeum* supplemented with lignin or its copolymerizates with styrene is shown in Figures 3, 4, and 5, respectively.

LiP was found only in the culture medium of *P. chrysosporium*, and its variation with time is depicted in Figure 3, which shows a rapid increase of activity 10 days after inoculation. LiP activity in the culture medium with or without polymer addition reached a maximum level after 15 days followed by a decrease in enzymatic activity (Figure 3). Addition of either copolymerizate

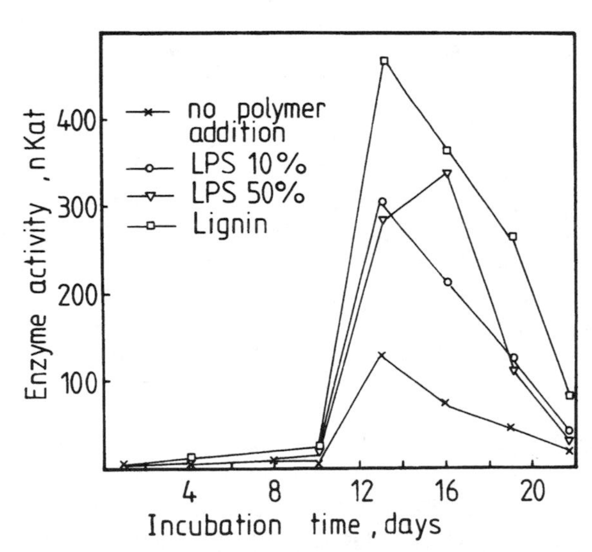

FIGURE 3. Production of extracellular LiP in standing liquid culture of *P. chrysosporium* supplemented with lignin (-□-), LPS-10 (-O-), LPS-50 (-▽-), or without addition (-x-) to the basal medium. The concentration unit nKat is billionths of the amount of enzyme needed to form 1 mole of product in 1 second.

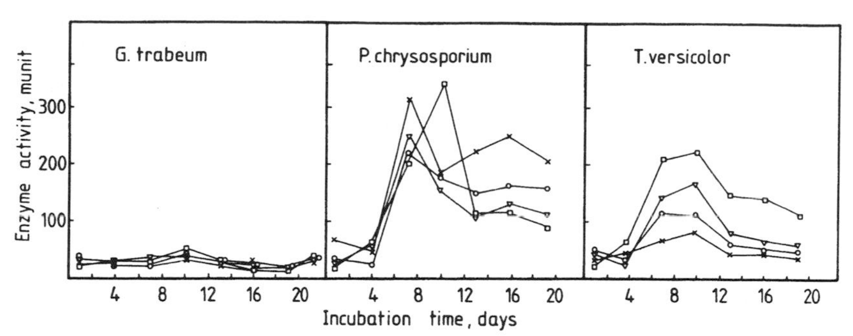

FIGURE 4. Production of extracellular MnP in standing liquid culture of *G. trabeum, P. chrysosporium,* and *T. versicolor* supplemented with lignin (-□-), LPS-10 (-○-), LPS-50 (-▽-), or without addition (-×-) to the basal medium. The concentration measure munit is one thousandth of the amount of enzyme needed to produce an initial increase of analyte absorbance at 465 nm of 1.00 per minute.

or lignin to the culture medium enhanced the level of LiP activity by almost 3 times when compared to the level of activity in culture media from control flasks.

MnP activity was detected in the culture medium of *P. chrysosporium* and *T. versicolor* and always appeared in the same period after inoculation as the LiP activity, as shown in Figure 4. No significant MnP activity was found in the culture medium of *G. trabeum*. The MnP activity of analyzed white-rot-fungi media reached a maximum after 10 to 12 days. Thereafter, the enzyme level began to decrease, as shown in Figure 4. However, in the late phase of cultivation, a second cycle of increase of MnP activity was observed in the culture media of both *P. chrysosporium* and *T. versicolor*. The level of MnP activity of the tested white-rot fungi was higher in the media supplemented with either lignin or its copolymerizates, as shown in Figure 4.

Figure 5 shows that significant levels of laccase activity were detected only in the culture medium of *T. versicolor*. Addition of lignin or LPS-50 to the medium led by a rapid increase of laccase activity. The level of enzymatic activity in the medium supplemented with lignin surpassed, by a factor of almost 5, the activity level of laccase in the medium with LPS-50. Under the test conditions, laccase activity in the culture medium of *T. versicolor* peaked twice during 22 days of cultivation, as shown in Figure 5.

Although the role of fungal extracellular enzymes in the degradation of lignin has been demonstrated (Tien & Kirk 1983), these enzymes do not seem to be prerequisite for lignin degradation in vivo (Sarkanen et al. 1991). However, LiP, MnP, and laccases can catalyze one-electron oxidation of phenolic and nonphenolic substrates producing cation-radical intermediates. The reactive phenoxy radicals, in turn, can mediate the oxidation of nonphenolic substrates (Kersten et al. 1990). The analyzed oxidative enzymes — LiP, MnP, and laccases — may modify the lignin macromolecule by introducing additional functional groups into its structure. These new functional groups render lignin and its copolymerizate with

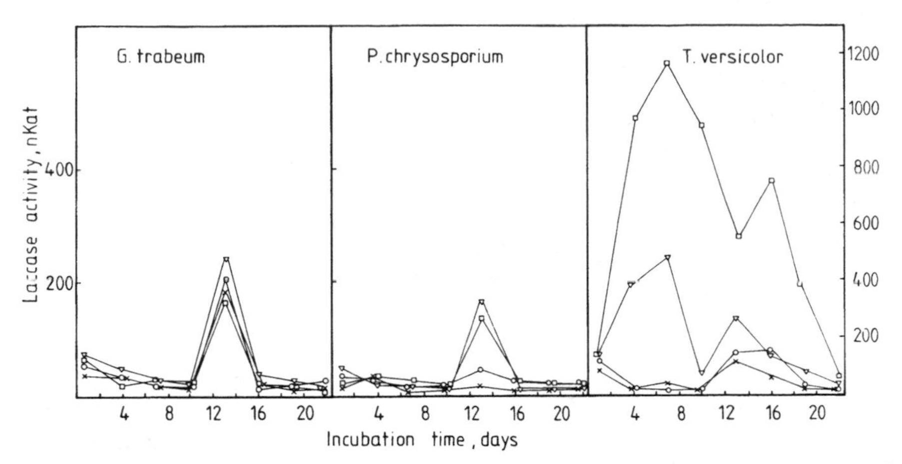

FIGURE 5. Production of extracellular laccase in standing liquid culture of *G. trabeum, P. chrysosporium,* and *T. versicolor* supplemented with lignin (-□-), LPS-10 (-○-), LPS-50 (-▽-), or without addition (-×-) to the basal medium. The concentration unit nKat is billionths of the amount of enzyme needed to form 1 mole of product in 1 second.

polystyrene more susceptible toward subsequent degradation by coordinated action of the enzymatic system of the whole organism.

Deterioration of the Plastic Surface

Additional evidence of bioconversion and degradation of the copolymers was obtained by SEM of the fungi-corroded surfaces of the plastics. SEM data for the LPS complex are shown in Figure 6. SEM data for the copolymer surface after hyphae from the fungus mycelia have grown over the surface show obvious traces of surface corrosion. The most common types of corrosion are striating, pitting, and occasional decay. Extensive pitting and striating were observed on the surfaces of plastics exposed to white-rot fungi, whereas very little deterioration of the surface could be seen on the plastic incubated with the brown rot *G. trabeum* or maintained on control plates.

Infrared Analysis

Differential, Fourier-transform, infrared spectroscopy (FTIR) was used to show the loss of functional groups from the copolymerizate. The equation for the calculation of the difference spectra is:

$$\begin{matrix}\text{Missing}\\\text{Infrared}\\\text{Absorbance}\end{matrix} = \left\{ \begin{matrix}\text{Absorbance}\\\text{Spectra of}\\\text{Copolymerizate}\\\text{from Control}\\\text{Flasks}\end{matrix} - \left[\begin{matrix}\text{Absorbance}\\\text{Spectra of}\\\text{Copolymerizate,}\\\text{Degraded}\\\text{Sample}\end{matrix} = \left(\begin{matrix}\text{Weight}\\\text{Percent}\\\text{Fungi in}\\\text{Sample}\end{matrix} \times \begin{matrix}\text{Absorbance}\\\text{Spectra of}\\\text{Fungal Mat}\end{matrix} \right) \right] \right\}$$

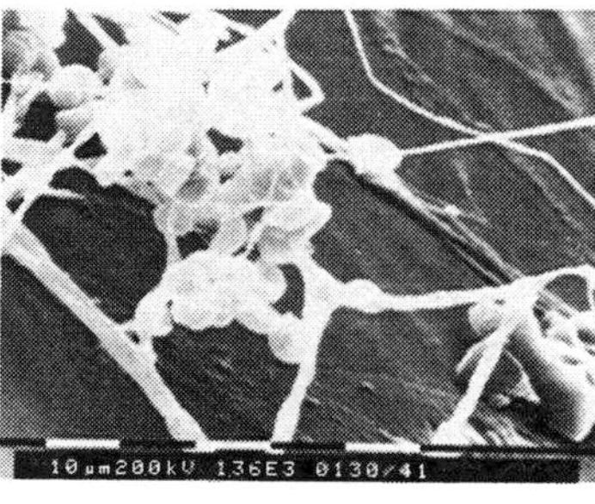

adhesion (A)

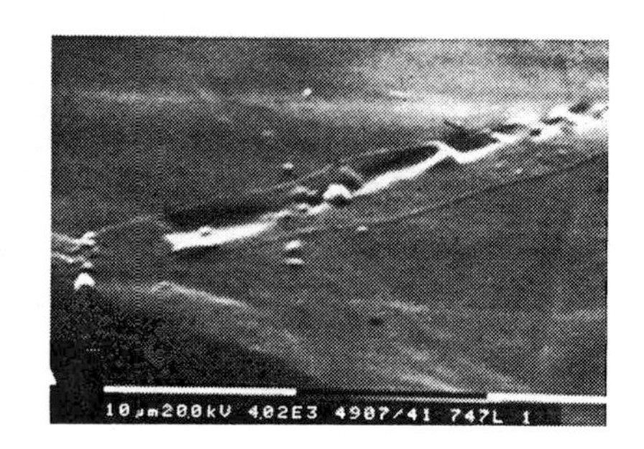

pitting (B)

striating (C)

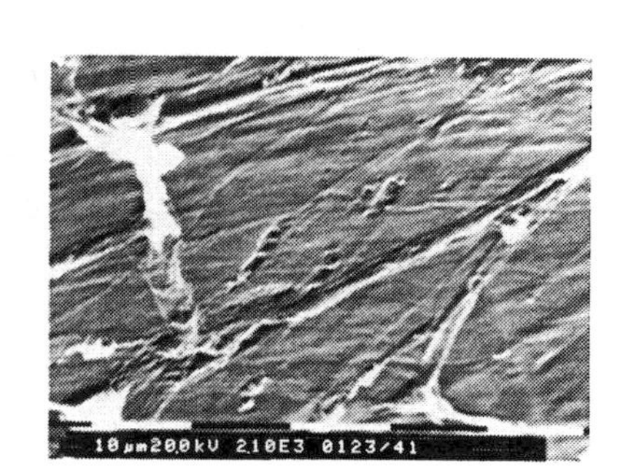

pitting and decay (D)

FIGURE 6. Scanning electron micrographs of pressed LPS incubated for 68 days showing adhesion (A) and different forms of surface deterioration caused by overgrown white-rot fungi. The surface deterioration is pitting (B), striating (C), and pitting and decay (D). Bars, 10 µm.

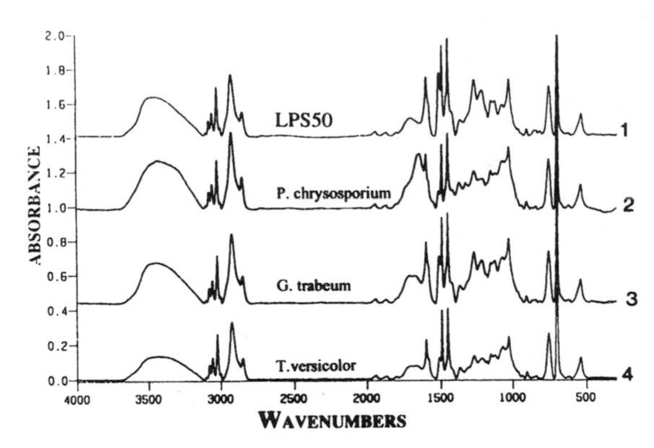

A = Fourier transform, infrared spectra of LPS-50

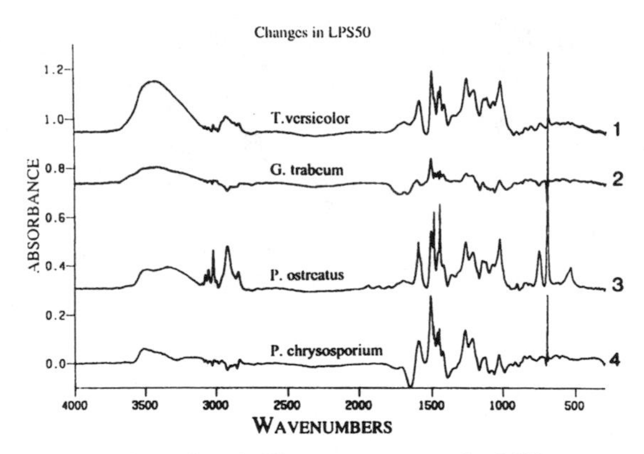

B = Calculated difference spectra for LPS-50 after 50 days of exposure to fungi

FIGURE 7. A = Fourier transform, infrared spectra of LPS-50 from sterile control plates and plates incubated for 50 days with *P. chrysosporium* (spectrum No. 2), *G. trabeum* (3), and *T. versicolor* (4). B = Calculated difference spectra for LPS-50 after 50 days of exposure to fungi. Difference spectra show significant losses of both components of the copolymerizate.

Figure 7 shows the FTIR spectra of copolymer recovered after 50 days of incubation with the four tested fungi. Lignin shows a small absorbance peak at a wavelength of 14.66 µm, whereas polystyrene shows a strong absorbance peak at 14.29 µm. The absorbance peaks, while proximate, are distinct and allow the materials to be differentiated. The height of these peaks is proportional to sample thickness and polymer concentration. Although this dependence on two properties of the analysis samples prohibits the use of the spectra to determine analytical concentrations, the data can be used to make qualitative comparisons between the incubated samples. These comparisons confirm the data obtained by the weight loss measurements. The 50 wt.% lignin sample shows significant peaks at both 14.29 and 14.66 μm, indicating significant loss of both lignin and polystyrene during incubation. The 32 and 10 wt.% lignin samples show proportionally smaller peaks at both wavelengths, indicating that smaller amounts of both components of the copolymerizate have been removed during incubation.

CONCLUSIONS

Bioconversion and degradation of lignin-styrene graft copolymer was verified by weight loss, SEM, oxidative enzyme production, and differential infrared spectroscopy. The most efficient degradation of lignin and polystyrene constituents of the copolymer by white-rot fungi was observed with the plastics with the highest lignin content, indicating that the level of weight loss of polystyrene component from the incubated LPS was correlated with concentration of lignin in the copolymerizate. Polystyrene frequently is used as a packaging material, and its use will probably increase because of growing concerns about polyvinylchloride in waste disposal streams. Currently, our society produces many commercial products of fully synthetic, recalcitrant materials. Copolymerization of synthetic sidechains onto naturally occurring backbones should be considered as a way of producing compounds that are more easily degraded in the environment. In particular, grafting of lignin with synthetic sidechains such as polystyrene will form a much more biodegradable material than synthesis of a polymer from pure, petroleum-derived hydrocarbons.

ACKNOWLEDGMENTS

Initial steps of the biodegradation study were supported in part by Grant 12-18938 A (Program: Nachwachsende Rohstoffe) from the Bunderministerium für Forschung und Technologie (BMFT) (Physikalisch-Technische Bundesanstalt (PTB), Jülich, Germany). Support of the copolymer physical property testing program by the U.S. Department of Agriculture under Grant No. 89-34158-4230, Agreement No. 71-2242B and Grant No. 90-34158-5004, Agreement No. 61-4053A, is gratefully acknowledged. We sincerely thank Konrad Wehr for his help and advice in SEM analysis.

REFERENCES

Bes, B., B. Pettersson, H. Lennholm, T. Iversen, and K. E. Eriksson. 1987. "Synthesis, Structure and Enzyme Degradation of an Extracellular Glucan Produced in Nitrogen-Starved Culture of the White Rot Fungus *Phanerochaete chrysosporium*." *Appl. Biochem. Biotechnol. 9*: 310-318.

Fahraeus, G., and B. Reinhammar. 1967. "Large-Scale Production and Purification of Laccase from Cultures of the Fungus *Polyporus versicolor* and Some Properties of Laccase A." *Acta Chem. Scand. 21*: 2367-2378.

Harkin, J. M., and J. R. Obst. 1973. "Syringaldazine, An Effective Reagent for Detecting Laccase and Peroxidase in Fungi." *Experientia 29*: 64-66.

Janshekar, H., C. Brown, T. Haltmeier, M. Leisola, and A. Fiechter. 1982. "Bioalteration of Kraft Pine Lignin by *P. chrysosporium*." *Arch. Microbiol. 132*: 14-21.

Kern, H. W. 1989. "Improvement in the Production of Extracellular Lignin Peroxidases by *Phanerochaete chrysosporium*: Effect of Solid Manganese (IV) Oxide." *Appl. Microbiol. Biotechnol. 32*: 223-234.

Kersten, P. J., B. Kalyanaraman, K. E. Hammel, B. Reinhammar, and T. K. Kirk. 1990. "Comparison of Lignin Peroxidase, Horseradish Peroxidase and Laccase in the Oxidation of Methoxybenzenes." *Biochem. J. 268*: 475-480.

Kirk, T. K. 1975. "Effect of a Brown-rot Fungus, *Lenzites trabea*, on Lignin in Spruce Wood." *Holzforschung 29*: 99-107.

Kirk, T. K., E. Shulz, W. J. Connors, L. F. Lorenz, and J. G. Zelkus. 1978. "Influence of Culture Parameters on Lignin Metabolism by *P. chrysosporium*." *Arch. Microbiol. 117*: 277-285.

Kirk, T. K., and R. L. Farrell. 1987. "Enzymatic 'Combustion': The Microbial Degradation of Lignin." *Ann. Rev. Microbiol. 41*: 465-505.

Meister, J. J. 1991. "Soluble or Crosslinked Graft Copolymers of Lignin, Acrylamide, and Hydroxylmethacrylate." U.S. Patent 5,037,931. Issued August 6, 1991.

Meister, J. J., A. Lathia, and F. F. Chang. 1991. "Solvent Effects, Species and Extraction Method Effects, and Coinitiator Effects in the Grafting of Lignin." *J. Poly. Chem. A. Poly. Chem. 29*: 1465-1473.

Meister, J. J., and C. T. Li. 1992. "Synthesis and Properties of Several Cationic Graft Copolymers of Lignin." *Macromolecules. 25*(1): 611-616.

Sarkanen, S., R. A. Razal, T. Piccariello, E. Yamamoto, and N. G. Lewis. 1991. "Lignin Peroxidase: Toward a Clarification of Its Role In Vivo." *J. Biol. Chem. 266*: 3636-3643.

Society of the Plastic Industry. 1991. *SPI Monthly Statistical Report on Resins*. 1990 Annual Summary as Compiled by Ernst and Young.

Tien, M., and Kirk, T. K. 1983. "Lignin-Degrading Enzyme from the Hymenomycete *Phanerochaete chrysosporium* Burds." *Science 221*: 661-662.

Tien, M., and T. K. Kirk. 1984. "Lignin-Degrading Enzyme from *Phanerochaete chrysosporium*: Purification, Characterization, and Catalytic Properties of a Unique H_2O_2-Requiring Oxygenase." *Proc. Natl. Acad. Sci., USA 81*: 2280-2284.

Whitekettle, W. K. 1991. "Effect of Surface-Active Chemicals on Microbial Adhesion." *J. of Industr. Microbiol. 7*: 105-116.

THE PHYSIOLOGY OF POLYCYCLIC AROMATIC HYDROCARBON BIODEGRADATION BY THE WHITE-ROT FUNGUS, *BJERKANDERA* SP. STRAIN BOS55

J. A. Field, E. Heessels, R. Wijngaarde,
M. Kotterman, E. de Jong, and J. A. M. de Bont

ABSTRACT

Bjerkandera sp. strain BOS55 is a promising polycyclic aromatic hydrocarbon (PAH)-degrading basidiomycete isolated from forest litter. In this study, the physiology of the fungus was investigated and the culture conditions for anthracene degradation were optimized. Unlike the model white-rot fungus, *Phanerochaete chrysosporium*, ligninolytic activity of the new isolate occurs during primary growth and it is not repressed by high nitrogen. Aeration studies showed that either a high level of dissolved oxygen or a high redox potential is required for rapid anthracene biodegradation. Optimal rates were obtained in an air atmosphere, by culturing the fungus in shallow media with an air-liquid interface of 1.1 $m^2\,L^{-1}$ or more. Under such conditions, 5-day-old cultures degraded 9 mg anthracene $L^{-1}\,d^{-1}$ when supplied with a starting concentration of 10 mg L^{-1}. The environmentally persistent PAH, benzo[a]pyrene, was degraded at a rate of 0.8 mg $kg^{-1}\,d^{-1}$ in a soil medium.

INTRODUCTION

Polycyclic aromatic hydrocarbons (PAHs) are important priority pollutants originating from coal gasification, coking, and wood preservation facilities. While low-molecular-weight (MW) PAHs are usually readily degraded, high-MW PAHs of five or more rings resist extensive bacterial degradation in soil and sediment media (Bossert & Bartha 1986, Heitkamp & Cerniglia 1987, Mueller et al. 1991). The recalcitrant behavior can be attributed to the limited bioavailability of PAHs strongly adsorbed onto soil organic matter (Manilal & Alexander 1991, Means et al. 1980, Volkering et al. 1992, Weissenfels et al. 1992).

White-rot fungi would be expected to have greater access to poorly bioavailable substrates because they secrete extracellular enzymes, such as lignin peroxidase (LiP) and manganese-dependent peroxidase (MnP), which are involved in the nonspecific oxidation of complex aromatic compounds such as lignin (Kirk & Farrell 1987). These enzymes could potentially penetrate deeper into soil fines than do microbial cells. MnP is peculiar because this enzymes simply oxidizes Mn^{2+} ions (Glenn et al. 1986). The resulting Mn^{3+} ions function as low MW mediators by initiating lignin oxidation at a distance from the enzyme (Lackner et al. 1991).

The extracellular peroxidases are directly implicated in the initial attack of numerous aromatic xenobiotic compounds including many PAH (Hammel 1992, Field et al. 1993). Purified LiP of the white-rot fungus, *Phanerochaete chrysosporium*, oxidizes PAH compounds to PAH quinones (Haemmerli et al. 1986, Hammel et al. 1986). Mn^{3+} ions in aqueous acetate likewise have been shown to oxidize PAH to acetoxy PAH derivatives and PAH quinones (Cremonesi et al. 1992, Cremonesi et al. 1989, Cavalieri & Rogan 1985). In fact, even horseradish peroxidase (HRP) of plant origin can oxidize some PAH compounds (Cavalieri et al. 1983). However, a few PAH compounds with a high ionization potential (IP) are not *in vitro* substrates of any peroxidase. A value of 7.5 eV (against a chloroanil electrode) has been established as the upper limit for LiP and Mn^{3+} (Cavalieri & Rogan 1985, Hammel et al. 1986).

Several studies have also shown that many PAHs are biodegraded *in vivo* by ligninolytic cultures of *P. chrysosporium* (Bumpus et al. 1985, Bumpus 1989, Dhawale et al. 1992, Field et al. 1992, Morgan et al. 1991, Sanglard et al. 1986). The list includes higher PAH compounds such as benzo[a]pyrene. Even compounds with IP values in excess of 7.5 eV (e.g., phenanthrene) are metabolized, indicating the presence of a yet unidentified enzyme. Several products, such as PAH quinones and dicarboxylic aromatics, temporarily accumulate in whole cultures before they are finally mineralized (Hammel et al. 1991, Hammel et al. 1992).

To date, most of the research concerning the degradation of xenobiotics by white-rot fungi has focused on the model organism *P. chrysosporium*. In a recent screening study, many newly isolated strains were found to degrade PAH as well or even better than *P. chrysosporium* (Field et al. 1992). An isolate from beech forest litter, *Bjerkandera* sp. strain BOS55, was clearly the best PAH degrader. This strain degraded 10 mg L^{-1} anthracene and benzo[a]pyrene by 99.2% and 83.1%, respectively, after 28 days. In this study, the physiology of PAH biodegradation by BOS55 was evaluated. Anthracene was used as a model substrate because anthraquinone, a known product of peroxidase, accumulated in high yields in the media (Field et al. 1992), thereby providing a measurable product of the initial ligninolytic attack.

MATERIALS AND METHODS

Bjerkandera sp. BOS55 was maintained on malt extract agar. Experiments were inoculated with an agar plug from the outer periphery of 7- to 10-day-old plates. Experiments were conducted with 5 mL of sterilized media, statically incubated in an air atmosphere (unless otherwise stated) at 30°C in loosely capped serum flasks. Two media were used, glucose-BIII consisting of 10 g L^{-1} glucose, BIII mineral medium (Tien & Kirk 1988) and 20 mM of 2,2-dimethylsuccinate (DMS) pH 4.5 buffer, whereas milled hemp stem wood (HSW) media (typically 30 g L^{-1}) contained no extra mineral salts and was buffered in 10 mM, pH 4.5 DMS. Then 10 mg L^{-1} PAH was provided by adding 0.05 mL of 1 g L^{-1} PAH in acetone. The experimental setup was based on sacrificing entire microcosms for PAH extraction (acetonitrile) and analysis (high-pressure liquid chromatography-diode array detector) according to Field et al. (1992). All experiments were conducted in triplicate. PAH incubated in uninoculated media or in NaN_3 (2 g L^{-1})-poisoned 12- to 14-day-old cultures served as controls. These were constantly monitored, and in general no significant anthracene elimination nor anthraquinone formation occurred. The killed fungi controls from glucose-BIII medium did, however, convert a small portion of the anthracene to anthraquinone; which perhaps can be attributed to traces of Mn^{3+} still present in the medium after azide poisoning. Poly R 478 dye (Sigma, St. Louis) was used as a ligninolytic indicator (Gold et al. 1988, Glenn & Gold 1983). Decolorization was monitored by the measuring absorbance ratio in the supernatant sampled at selected periods during the course of the experiment (De Jong et al. 1992b). At time 0, 0.8 g L^{-1} of dye was added. Poly R dye decolorization activity or anthracene-degrading activity was measured over 18 hours after adding dye or anthracene to whole cultures previously cultivated to a given age. MnP and manganese-inhibited peroxidase (MIP) activities were measured according to previously described procedures (De Jong et al. 1992a, De Jong et al. 1992b).

In order to test the effectiveness of *Bjerkandera* sp. BOS55 for the biodegradation of high MW PAH compounds in soil media, this fungus was incubated at 20°C with sterilized soil (sand = 89.5%, clay = 6.4%, and organic matter = 4.1%), which was artificially contaminated with 100 mg/kg benzo[a]pyrene supplied in acetone. The acetone was evaporated off in a drying oven prior to inoculation. The fungus was first grown out in a 5-L erlenmeyer flask containing 250-mL of 50 g L^{-1} HSW medium. On day 8, the culture was homogenized and transferred (1.2 mL) to the soil (2 g) placed inside a 250 mL serum flask. Autoclaved homogenized culture suspension added to soil served as the abiotic control. PAH was analyzed after 1 and 2 months incubation by extracting the soil with 4.25 mL of acetone (15 minutes of sonification followed by 1 hour on a shake table).

RESULTS AND DISCUSSION

Unlike *P. chrysosporium* (Kirk et al. 1978), *Bjerkandera* sp. strain BOS55 displayed robust ligninolytic activity during primary growth (Figure 1). Poly R dye decolorizing activities of the whole culture and extracellular peroxidase activities were already at high levels early on in the experiment when the fungus was still growing as judged by total CO_2 evolution. Anthracene (10 mg L^{-1}) was almost completely eliminated even before the lag phase was finished. Anthraquinone which initially accumulated was slowly being removed during the rest of the experiment.

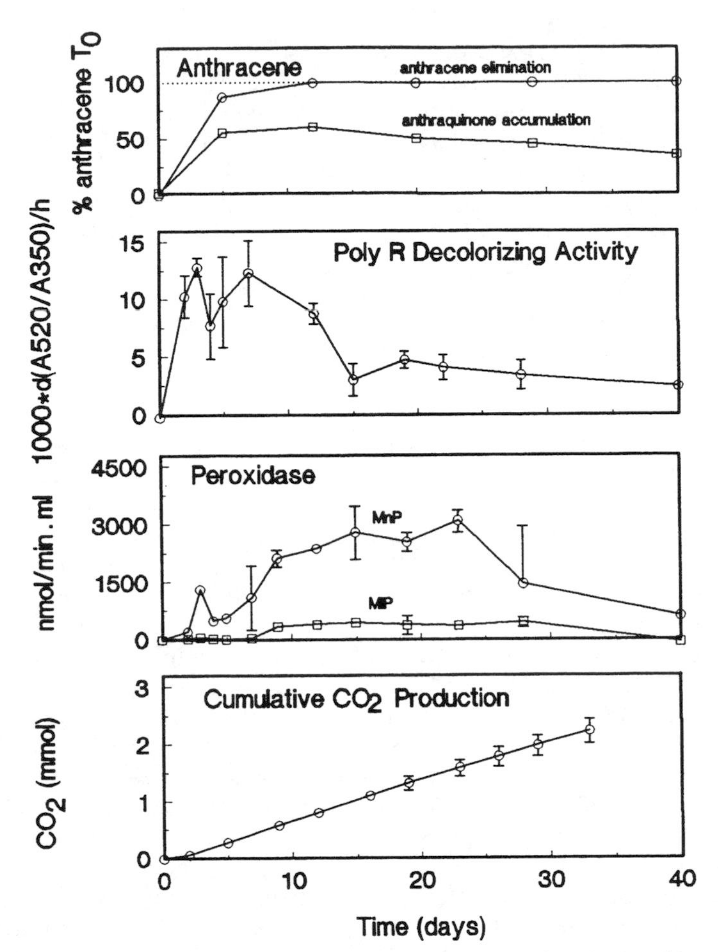

FIGURE 1. Biodegradation of anthracene, Poly R dye decolorization activity, peroxidase levels and cumulative CO_2 production during the growth of *Bjerkandera* sp. strain BOS55 (30 g L^{-1} HSW, air/liquid interface = 1.1 m^2 L^{-1}).

High N is known to repress Poly R dye decolorization by *P. chrysosporium* (Fenn et al. 1981, Glenn & Gold 1983). Neither anthracene biodegradation nor Poly R dye decolorization rates were repressed by high ammonia nitrogen levels (22 mM vs. 2.2 mM) when BOS55 was grown out on glucose-BIII medium (Figure 2), illustrating once again a distinct physiology compared to *P. chrysosporium*. High N stimulated the rate by which BOS55 degraded anthraquinone.

In experiments with small serum flasks (30 mL), anthracene biodegradation was repressed at cosubstrate concentrations in excess of 2 g L^{-1} HSW (Figure 3). However, increasing HSW concentration greatly enhanced both Poly R dye decolorization and extracellular peroxidase activity. Increasing HSW would increase the level of Mn^{2+} ions, which is known to stimulate MnP expression in

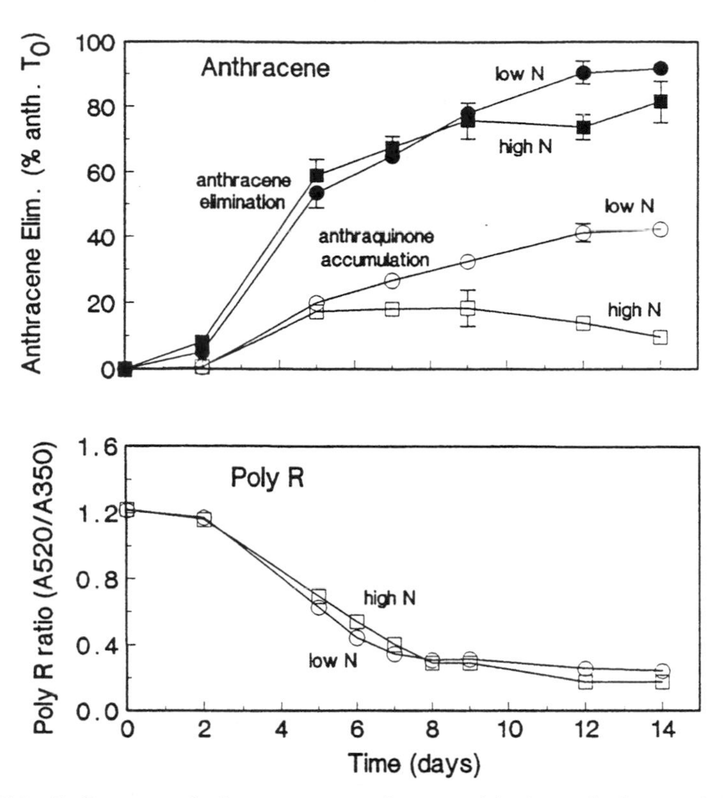

FIGURE 2. Influence of nitrogen on anthracene biodegradation and Poly R dye decolorization by *Bjerkandera* sp. strain BOS55 (10 g L^{-1} glucose in BIII basal media, air/liquid interface = 0.4 m^2 L^{-1}).

several white-rot fungi (Bonnarmc & Jeffries 1990, Brown et al. 1990, Michel et al. 1991, Périé & Gold 1991). In turn, MnP is known to be directly responsible for Poly R decolorization (Glenn & Gold 1985, Kuwahara et al. 1984). The small amount of cosubstrate carried over in the inoculum agar plug was sufficient to support some anthracene biodegradation. This suggests that only a minimal amount of enzyme activity is required to oxidize anthracene. Absence of anthracene degradation when using substrate-free inoculum confirms that at least some cosubstrate is required (results not shown). Diminished anthracene-degrading activity at high cosubstrate concentrations may reflect either target substrate competition between anthracene and lignin (in lignocellulose) or reduced oxygen supply due to faster growth (i.e., respiration) at higher cosubstrate levels.

In cultures with high cosubstrate concentrations, aeration stimulated anthracene biodegradation. The air-liquid surface area of the culture medium was varied by incubating the fungus in serum bottles of increasing diameter. Increasing the air-liquid surface dramatically improved anthracene biodegradation up to

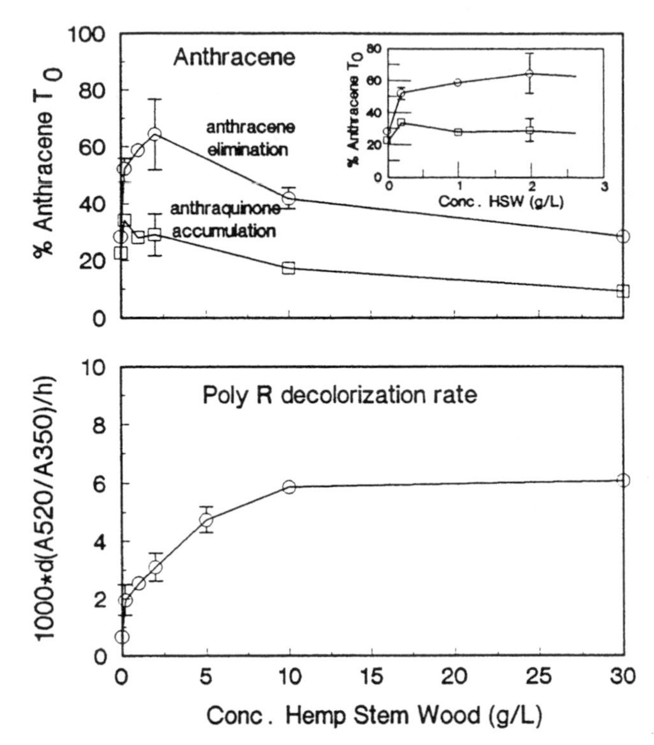

FIGURE 3. Role of HSW cosubstrate concentration on the biodegradation of anthracene and the rate of Poly R dye decolorization by *Bjerkandera* sp. strain BOS55 (air/liquid interface = 0.4 m^2 L^{-1}).

1.2 m^2/L (Figure 4), whereas only a minor effect on Poly R decolorization was observed. Anthracene (10 mg L^{-1}) was degraded at a rate of 8 to 9 mg L^{-1} d^{-1} in activity tests with 5- to 27-day-old shallow cultures. An 80% O_2 atmosphere compared to air stimulated anthracene biodegradation in cultures with limited air-liquid surface area (0.4 m^2/L), but no effect was evident with respect to Poly R dye and peroxidase activity. High O_2 had no effect on any parameter at optimal surface areas (results not shown), suggesting an adequate oxygen transfer in an air atmosphere. These results indicate that a higher aeration requirement exists for anthracene oxidation compared to Poly R oxidation and peroxidase secretion. This may result from an unusually high soluble O_2 or redox potential requirement for PAH degradation. Because the oxygen incorporated into PAH quinones by peroxidase comes from H_2O (Hammel et al. 1986), O_2 is not a likely requirement. Perhaps an indirect effect of O_2 is implicated, whereby O_2 maintains a high redox potential which in turn enhances the rate by which anthracene is oxidized by peroxidases.

Finally *Bjerkandera* sp. strain BOS55 was tested for the biodegradation of benzo[a]pyrene in soil media. After 56 days of incubation, BOS55 had removed 38.5% of the benzo[a]pyrene initially supplied at 100 mg/kg soil. The elimination

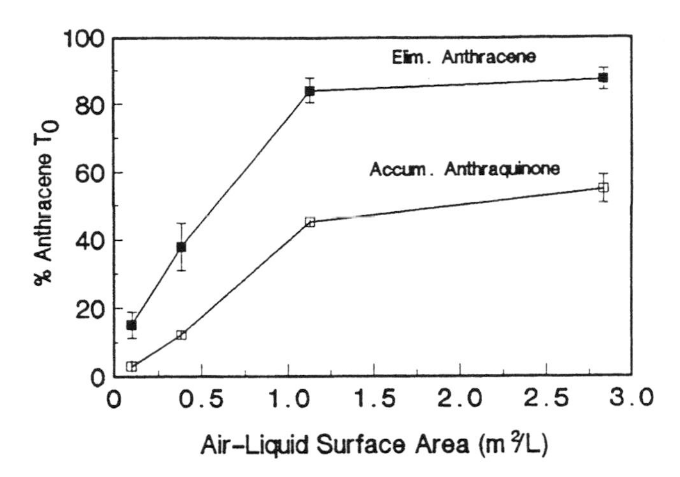

FIGURE 4. Effect of increasing culture air-liquid interface surface area on enhancing anthracene biodegradation by *Bjerkandera* sp. strain BOS55 (30 g L^{-1} HSW). Results are given for 5-day-old cultures.

was truly mediated by a biological reaction, because benzo[a]pyrene was recovered at 100% from the soil when incubated with the autoclaved fungus. The maximum rate of benzo[a]pyrene elimination corresponded to 0.8 mg per kg soil per day. This value is in close agreement with that found by Qiu and McFarland (1991), utilizing *P. chrysosporium* in a similar experiment but with a pure O_2 atmosphere.

REFERENCES

Bonnarme, P., and T. W. Jeffries. 1990. "Mn(II) regulation of lignin peroxidases and manganese-dependent peroxidases from lignin-degrading white-rot fungi." *Appl. Environ. Microbiol. 56*:210-217.

Bossert, I. D., and R. Bartha. 1986. "Structure-biodegradability relationships of polycyclic aromatic hydrocarbons in soil." *Bull. Environ. Contam. Toxicol. 37*:490-495.

Brown, J. A., J. K. Glenn, and M. H. Gold. 1990. "Manganese regulates expression of manganese peroxidase by *Phanerochaete chrysosporium*." *J. Bacteriol. 172*:3125-3130.

Bumpus, J. A., M. Tien, D. Wright, and S. D. Aust. 1985. "Oxidation of persistent environmental pollutants by a white rot fungus." *Science 228*:1434-1436.

Bumpus, J. A. 1989. "Biodegradation of polycyclic aromatic hydrocarbons by *Phanerochaete chrysosporium*." *Appl. Environ. Microbiol. 55*:154-158.

Cavalieri, E. L., E. G. Rogan, R. W. Roth, R. K. Saugier, and A. Hakam. 1983. "The relationship between ionization potential and horseradish peroxidase/hydrogen peroxide-catalyzed binding of aromatic hydrocarbons to DNA." *Chem. Biol. Interactions 47*:87-109.

Cavalieri, E., and E. Rogan. 1985. "Role of radical cations in aromatic hydrocarbon carcinogenesis." *Environ. Health Perspectives 64*:69-84.

Cremonesi, P., E. L. Cavalieri, and E. G. Rogan. 1989. "One-electron oxidation of 6-substituted benzo[a]pyrenes by manganic acetate. A model for metabolic activation." *J. Org. Chem. 54*:3561-3570.

Cremonesi, P., B. Hietbrink, E. G. Rogan, and E. L. Cavalieri. 1992. "One-electron oxidation of dibenzo[a]pyrenes by manganic acetate." *J. Org. Chem. 57*:3309-3312.

De Jong, E., J. A. Field, and J. A. M. de Bont. 1992a. "Evidence for a new extracellular peroxidase: Manganese inhibited peroxidase from the white-rot fungus *Bjerkandera* sp. Bos55." *FEBS Letters. 299*:107-110

De Jong, E., F. P. de Vries, J. A. Field, R. P. van der Zwan, and J. A. M. de Bont. 1992b. "Isolation and screening of basidiomycetes with high peroxidative activity." *Mycological Research 96*:1098-1104.

Dhawale, S. W., S. S. Dhawale, and D. Dean-Ross. 1992. "Degradation of phenanthrene by *Phanerochaete chrysosporium* occurs under ligninolytic as well as nonligninolytic conditions." *Appl. Environ. Microbiol. 58*:3000-3006.

Fenn, P., S. Choi, and T. K. Kirk. 1981. "Lignolytic activity of *Phanerochaete chrysosporium*: Physiology of suppression by NH4+ and L-glutamate." *Arch. Microbiol. 130*:66-71.

Field, J. A., E. de Jong, G. Feijoo Costa, and J. A. M. de Bont. 1992. "Biodegradation of polycyclic aromatic hydrocarbons by new isolates of white-rot fungi." *Appl. Environ. Microbiol. 58*:2219-2226.

Field, J. A., E. de Jong, G. Feijoo Costa, and J. A. M. de Bont. 1993. "Screening for xenobiotic degrading white-rot fungi." *Trends in Biotechnology. 11(2)*: 44-49.

Glenn, J. K., and M. H. Gold. 1983. "Decolorization of several polymeric dyes by the lignin-degrading basidiomycete *Phanerochaete chrysosporium*." *Appl. Environ. Microbiol. 45*:1741-1747.

Glenn, J. K., and M. H. Gold. 1985. "Purification and properties of an extracellular Mn(II)-dependent peroxidase from the lignin-degrading basidiomycete, *Phanerochaete chrysosporium*." *Arch. Biochem. Biophys. 242*:329-341.

Glenn, J. K., L. Akileswaran, and M. H. Gold. 1986. "Mn(II) oxidation is the principal function of the extracellular Mn-peroxidase from *Phanerochaete chrysosporium*." *Arch. Biochem. Biophys. 251*:688-696.

Gold, M. H., J. K. Glenn, and M. Alic. 1988. "Use of polymeric dyes in lignin biodegradation assays." In W. A. Wood and S. T. Kellog (Eds.), *Methods of Enzymology. Volume 161B. Lignin, Pectin and Chitin*, pp. 74-78. Academic Press Inc., San Diego.

Haemmerli, S. D., M. S. A. Leisola, D. Sanglard, and A. Fiechter. 1986. "Oxidation of benzo(a)-pyrene by extracellular ligninase of *Phanerochaete chrysosporium*." *J. Biol. Chem. 261*:6900-6903.

Hammel, K. E., B. Kalyanaraman, and T. K. Kirk. 1986. "Oxidation of polycyclic aromatic hydrocarbons and dibenzo[p]dioxins by *Phanerochaete chrysosporium* ligninase." *J. Biol. Chem. 261*:16948-16952.

Hammel, K. E., B. Green, and W. Z. Gai. 1991. "Ring fission of anthracene by a eukaryote." *Proc. Natl. Acad. Sci. USA 88*:10605-10608.

Hammel, K. E., W. Z. Gai, B. Green, and M. A. Moen. 1992. "Oxidative degradation of phenanthrene by the ligninolytic fungus *Phanerochaete chrysosporium*." *Appl. Environ. Microbiol. 58*:1832-1838.

Hammel, K. K. 1992. "Oxidation of aromatic pollutants by lignin degrading fungi and their extracellular enzymes." In H. Sigel and A. Sigel (Eds.), *Metal Ions in Biological Systems*, Vol. 28, pp. 41-60. Marcel Dekker, Inc., New York.

Heitkamp, M. A., and C. E. Cerniglia. 1987. "Effects of chemical structure and exposure on the microbial degradation of polycyclic aromatic hydrocarbons in freshwater and estuarine ecosystems." *Environ. Toxicol. Chem. 6*:535-546.

Kirk, T. K., E. Schultz, W. J. Connors, L. F. Lorenz, and J. G. Zeikus. 1978. "Influence of culture parameters on lignin metabolism by *Phanerochaete chrysosporium*." *Arch. Microbiol. 117*:227-285.

Kirk, T. K., and R. L. Farrell. 1987. "Enzymatic 'combustion': The microbial degradation of lignin." *Ann. Rev. Microbiol. 41*:465-505.

Kuwahara, M., J. K. Glenn, M. A. Morgan, and M. H. Gold. 1984. "Separation and characterization of two extracellular H_2O_2-dependent oxidases from ligninolytic cultures of *Phanerochaete chrysosporium.*" *FEBS Lett. 169*:247-250.

Lackner, R., E. Srebotnik, and K. Messner. 1991. "Oxidative degradation of high molecular weight chlorolignin by manganese peroxidase of *Phanerochaete chrysosporium.*" *Biochem. Biophys. Res. Comm. 178*:1092-1098.

Manilal, V. B., and M. Alexander. 1991. "Factors affecting the microbial degradation of phenanthrene in soil." *Appl. Microbiol Biotechnol. 35*:401-405.

Means, J. C., S. G. Wood, J. J. Hassett, and W. L. Banwart. 1980. "Sorption of polynuclear aromatic hydrocarbons by sediments and soils." *Environ. Sci. Technol. 14*:1525-1528.

Michel, Jr., F. C., S. B. Dass, E. A. Grulke, and C. A. Reddy. 1991. "Role of manganese peroxidase and lignin peroxidases of *Phanerochaete chrysosporium* in the decolorization of Kraft bleach plant effluent." *Appl. Environ. Microbiol. 57*:2368-2375.

Morgan, P., S. T. Lewis, and R. J. Watkinson. 1991. "Comparison of abilities of white-rot fungi to mineralize selected xenobiotic compounds." *Appl. Microbiol. Biotechnol. 34*:693-696.

Mueller, J. G., S. E. Lantz, B. O. Blattmann, and P. J. Chapman. 1991. "Bench-scale evaluation of alternative biological treatment processes for the remediation of pentachlorophenol- and creosote-contaminated materials: Solid phase bioremediation." *Environ. Sci. Technol. 25*:1045-1055.

Périé, F. H., and M. H. Gold. 1991. "Manganese regulation of manganese peroxidase expression and lignin degradation by the white-rot fungus *Dichomitus squalens.*" *Appl. Environ. Microbiol. 57*:2240-2245.

Qiu, X., and M. J. McFarland. 1991. "Bound residue formation in PAH contaminated soil composting using *Phanerochaete chrysosporium.*" *Hazardous Waste & Hazardous Materials. 8*:115-126.

Sanglard, D., M. S. A. Leisola, and A. Fiechter. 1986. "Role of extracellular ligninases in biodegradation of benzo[a]pyrene by *Phanerochaete chrysosporium.*" *Enzyme Microb. Technol. 8*:209-212.

Tien, M., and T. K. Kirk. 1988. "Lignin peroxidase of *Phanerochaete chrysosporium.*" In W. A. Wood and S. T. Kellog (Eds.), *Methods of Enzymology*. Volume 161B. *Lignin, Pectin and Chitin,* pp. 238-248. Academic Press Inc., San Diego, CA.

Volkering, F., A. Breure, A. Sterkenburg, and J. G. van Andel. 1992. "Microbial degradation of polycyclic aromatic hydrocarbons: Effect of substrate availability on bacterial growth kinetics." *Appl. Microbiol. Biotechnol. 36*:548-552.

Weissenfels, W. D., H-J. Klewer, and J. Langhoff. 1992. "Adsorption of polycyclic aromatic hydrocarbons (PAHs) by soil particles: Influence on biodegradability and biotoxicity." *Appl. Microbiol. Biotechnol. 36*:689-696.

DETERMINATION OF KINETIC PARAMETERS FOR A MULTISUBSTRATE INHIBITION MODEL

J. S. Bonner, R. L. Autenrieth, and B.-H. Bae

ABSTRACT

A deterministic model was developed to describe microbial growth rate as a function of multiple substrate concentrations for acclimated heterogeneous cultures. An experimental component accompanied the model development and the data were used for calibrating and testing. Those compounds evaluated include phenol with glucose and pentachlorophenol, and phenol with dichlorophenol and pentachlorophenol. Conditions were simulated to introduce a gaseous or liquid waste to a suspended culture medium of acclimated organisms. A series of experiments were conducted on batch and continuous flow cultures using initial rate experiments to evaluate growth and substrate utilization rates and yield. These were used in the model in conjunction with a parameter estimation routine to determine maximum growth rates, half-saturation constants, and inhibition constants. The model assumes that each substrate influences the degradation of other substrates in a multicomponent environment. This interactive model assumes that if three substrates are present in less than saturating conditions, they all must affect the overall growth rate of the organisms. Other model assumptions are that there is no substrate volatilization, biosorption, or inhibition due to pH fluctuations, and that there is a specific cell yield for each substrate. The equation used to model cell growth was based on the Monod and Haldane models, and on an enzymatic model describing two substrates reacting on a single enzyme. The model was successful at predicting interactive behavior and generating useful kinetic coefficients necessary for reactor design. Additionally, the coefficients indicated mechanisms of biodegradation including cometabolism, gratuitous metabolism, and cases where the compounds were surprisingly not interactive. The model revealed that the initial rate method was constrained by the physiological state of the microorganisms which is particularly important for more than two substrates. Most important is that growth under multiple substrate conditions resulted in kinetic coefficients significantly different from those obtained under single-substrate conditions. Considering the complexity of most environmental applications encountered, more than one possible carbon source always exists. Therefore, these interactions must be considered to accurately predict reactor performance in the multi-component environment.

INTRODUCTION

Wastewaters, particularly industrial wastewaters, typically contain a mixture of several noninhibitory and inhibitory substrates. To accurately mimic the characteristics of these wastewaters, biodegradation studies should be performed on reactors containing more than one substrate, but few such studies have been reported. Microbial growth on multiple substrates varies depending on culturing conditions. The type of substrate uptake can range from simultaneous utilization of all substrates to sequential utilization of substrates resulting in multi-exponential growth phases. In sequential growth on multiple substrates, cells first grow on the fastest assimilated substrate in the medium indicating that the microorganisms have an inherent ability to optimize their growth behavior according to the available substrates (Kompala et al. 1984), known as the diauxie phenomenon (Dixon & Webb 1979). Assuming that the growth rate of the organisms could be limited by only one substrate at a time, the overall growth rate would be equal to the lowest growth that would be predicted from the separate single-substrate Monod model. This type of sequential utilization of substrate was modeled assuming a low degree of interaction within the biochemical cell pathways (Bader 1976).

Grady (1990) notes diauxie growth curves are the exception rather than the rule in wastewater treatment systems. Substrates tend to be simultaneously used because the metabolic control mechanisms that govern sequential substrate removal do not operate efficiently due to the carbon-limited environments. Sequential substrate removal usually occurs in rapidly growing cultures, whereas simultaneous substrate removal prevails for longer solids retention time (SRT), resulting in slow-growth cultures (Grady 1990). Because simultaneous removal of substrates is prevalent in slow growth reactor systems, models can be developed that relate the effects of substrate interaction on cell growth rates. Bader (1976) defined a dual-substrate interactive model as one that assumes that if two substrates are present in less than saturating concentrations, then both must affect the overall growth rate of the organisms. It was found that different levels of interaction between two substrates may occur, depending on the relative values of their particular rate constants.

Limited information exists on the specific effects of multiple xenobiotic substrates on cell growth rates. However, the overall effects of some mixtures containing noninhibitory wastes and inhibitory wastes have been studied (Beltrame et al. 1984, Grady & Lim 1980, Machado 1985, Papanastasious 1982). Papanastasious (1982) studied the biodegradation of 2,4-dichlorophenoxyacetate (2,4-D) and glucose using a 2,4-D-acclimated batch culture. Biodegradation of glucose consistently followed Monod kinetics, whereas biodegradation of 2,4-D followed the Andrews substrate inhibition model (Papanastasious 1982). Because mutual inhibition occurred when both substrates were present, these results indicated the occurrence of concurrent growth forming additional cell mass. The high biomass caused an overall increase in the rate of the target substrate (2,4-D) utilization. From studies conducted on pure culture in continuous stirred tank reactors (CSTRs) fed 2-chlorophenol (2-CP), Machado (1985) concluded that the fate of a single organic component in a multicomponent feed is complicated by interactive effects.

Kinetic studies regarding multiple substrates are extremely limited. Yoon et al. (1977) developed a model for the growth of an organism on multiple substrates where a competitive inhibition effect occurs by one substrate on the other substrate's uptake. Bader (1978) proposed a double substrate interactive model in which substrates may or may not interact with each other. Recently, McCreary (1991) developed an inhibitory substrate interactive model based on the Monod equation, the Haldane equations, and Bader's (1978) noninhibitory substrate interactive model. The equation accurately predicted growth rates for supporting substrates (glucose and phenol) and pentachlorophenol (PCP) as a non-growth-supporting substrate in batch reactors.

The research presented here is part of a larger effort to characterize and understand microbial kinetics during multiple substrate degradation. This paper will discuss the derivation and application of a mathematical model developed to predict microbial degradation for acclimated populations fed multiple substrates. The experimental data indicated that kinetic information determined from individual substrates were inadequate for predicting microbial performance when exposed under multiple substrate conditions. For this reason, an interactive model was developed to describe microbial growth for multiple substrates that include inhibitory and noninhibitory sources.

MODEL DEVELOPMENT

A deterministic model to describe the microbial growth rate as a function of multiple substrate concentrations was generated. The developed model assumed that each substrate influences the degradation of other substrates in a multicomponent environment. This interactive model assumed that if three substrates are present in less than saturating conditions, they all must affect the overall growth rate of the organisms. Other model assumptions included no substrate volatilization or biosorption. The appropriateness of these assumptions was verified analytically and confirmed with literature-determined chemical characteristics. Both pH and temperature were held constant for the cultures and were not considered as variables in the observed response.

The equation used to model cell growth was based on the principles of the Monod model, the Haldane model, and an enzymatic model describing two substrates reacting on a single enzyme. The Monod model assumes that with increasing substrate concentrations the growth rate asymptotically approaches a maximum value. The Haldane model assumes that at some critical substrate concentration, cell growth reaches a maximum value, and further increases in substrate concentration inhibit the growth rate. The dual-substrate enzyme model described by Bailey and Ollis (1986) does not incorporate substrate inhibition effects. The equations for each model are given below:

Monod Equation:

$$\mu = \mu_{max} \left[\frac{S}{K_S + S} \right] \tag{1}$$

Haldane Equation:

$$\mu = \mu_{max} \left[\frac{S}{K_S + S + \dfrac{S^2}{K_i}} \right] \tag{2}$$

where μ = Cell growth rate coefficient (hr^{-1})
μ_{max} = Maximum cell growth rate coefficient (hr^{-1})
S = Substrate concentrations, respectively (mg/L)
K_s = Half-saturation constants (mg/L)
K_i = Inhibition constants (mg/L)

Dual-substrate enzyme equations (Bailey & Ollis 1986):

$$\frac{dS_1}{dt} - v_{\lambda_1} = v_{\lambda_{1max}} \left[\frac{\dfrac{S_1}{K_1}}{1 + \dfrac{S_1}{K_1} + \dfrac{S_2}{K_2}} \right] \tag{3}$$

$$\frac{dS_2}{dt} = v_{\lambda_2} = v_{\lambda_{2max}} \left[\frac{\dfrac{S_2}{K_2}}{1 + \dfrac{S_1}{K_1} + \dfrac{S_2}{K_2}} \right] \tag{4}$$

where v_{λ_1} = velocity of reaction for substrate, S_1
v_{λ_2} = velocity of reaction for substrate, S_2

Based on these models, the following multiple substrate interactive model has been developed. For cultures receiving growth-supporting substrates glucose (S_1), phenol (S_2), and non-growth-supporting substrate PCP (S_3), the following equation was used:

$$\mu = \mu_{max} \left[\frac{\frac{S_g}{Ks_g} + \frac{S_p}{Ks_p}}{1 + \frac{S_g}{Ks_g} + \frac{S_p}{Ks_p} + \frac{S_p^2}{Ki_p} + \frac{S_{PCP}^2}{Ki_{PCP}}} \right] \tag{5}$$

where S_g, S_p, S_{PCP} = Glucose, phenol, PCP concentrations, respectively (mg/L)
Ks_g, Ks_p = Half-saturation constants for glucose and phenol, respectively (mg/L)
Ki_p, Ki_{PCP} = Inhibition constants for phenol and PCP, respectively (mg/L)

Equation 5 is based on the following substrate removal rate equations and growth rate equation

$$\frac{dS_1}{dt} = \frac{1}{Y_1} \mu X \tag{6}$$

$$\frac{dS_2}{dt} = \frac{1}{Y_2} \mu X \tag{7}$$

$$\frac{dS_3}{dt} = \frac{1}{Y_3} \mu X \tag{8}$$

$$\frac{dX}{dt} = \mu X \tag{9}$$

where Y_1, Y_2, Y_3 = Cell yields for respective substrate
X = Total biomass concentration (mg/L)

In this formulation, the removal of each substrate (S_1, S_2, and S_3) contributes to the growth of total cell mass (X) with respective cell yield (Y_1, Y_2, and Y_3), not necessarily to the growth of specific cell mass (X_1, X_2, and X_3). The model allowed for a specific cell yield for each substrate removal equation. This approach was based on the fact that the equation does not take into account decay or substrate luxury uptake. Because these terms were not incorporated into the model equation, there cannot be a constant cell yield throughout the initial rate experiments due to reactor dynamics.

This model was applied to a set of experiments that assessed growth on glucose, phenol, and PCP. The acclimated culture could use glucose (S_g) and phenol (S_p) as primary substrates. It was determined that the bacteria could not use PCP as a growth-supporting substrate in batch-fed culture. Glucose cannot inhibit the cell growth rates at the concentrations used. At a critical concentration of phenol and beyond, inhibition of the cell growth rate would occur. Like 2,4-dinitrophenol (DNP) and other acidic aromatic compounds, PCP may acts as an uncoupler. An uncoupler dissipates the proton-motive force across the inner mitochondrial membrane interfering with ATP formation, but allows normal electron transport from NADH to O_2 (Stryer 1988). Therefore, PCP only consumes energy and does not support growth.

If only one primary substrate was present and there was no cell growth inhibition, the model equation (Equation 5) reduced to that of the Monod equation (Equation 1). If only one substrate was present and cell growth inhibition occurred, then the model equation reduced to a typical Haldane inhibition equation (Equation 2). This model was then incorporated into a parameter estimation algorithm developed by Ernest et al. (1991) to determine the maximum growth rate, saturation, and inhibition coefficients as independent parameters. For a given culture, all individual substrate concentrations and the corresponding growth rates for all kinetic experiments on that reactor were input into the parameter estimation algorithm to obtain the model-fitted kinetic values. This model was applied to single, dual, and triple substrate cultures to evaluate μ_{max}, effective half-saturation values, and inhibition constants. A summary of kinetic coefficients approximated by the model parameter estimation algorithm is presented in Table 1 for the batch-fed cultures.

For the CSTR cultures receiving phenol (S_p), 2,4-dichlorophenol (DCP, S_{DCP}), and PCP (S_{PCP}), a modified equation was used:

TABLE 1. Estimated model coefficients for single-, dual-, and triple-substrate conditions in batch reactors.

Reactor	μ_{max} (hr^{-1})	Ks_g (mg/L)	Ks_p (mg/L)	Ki_p (mg/L)	Ki_{PCP} (mg/L)
Batch Cultures					
50 mg/L phenol	1.23	—	44.69	$1.0 \cdot 10^6$ [a]	—
200 mg/L phenol	0.84	—	9.27	678	—
450 mg/L glucose + 50 mg/L phenol	0.73	165	6.13	511	—
300 mg/L glucose + 200 mg/L phenol	0.32	11.4	199	22,967	—
400 mg/L glucose + 100 mg/L phenol + 1 mg/L PCP	0.83	92.0	5.58	111	0.067

(a) High value of inhibition constant indicating negligible or no inhibition.

$$\mu = \mu_{max}\left[\frac{\frac{S_p}{Ks_p} + \frac{S_{DCP}}{Ks_{DCP}}}{1 + \frac{S_p}{Ks_p} + \frac{S_{DCP}}{Ks_{DCP}} + \frac{S_p^2}{Ki_p} + \frac{S_{DCP}^2}{Ki_{DCP}} + \frac{S_{PCP}^2}{Ki_{PCP}}}\right] \quad (10)$$

where S_p, S_{DCP}, S_{PCP} = Phenol, DCP, and PCP concentrations, respectively (mg/L)
Ks_p, Ks_{DCP} = Half-saturation constants for phenol and DCP respectively (mg/L)
Ki_p, Ki_{DCP}, Ki_{PCP} = Inhibition constants for phenol, DCP, and PCP, respectively (mg/L)

In this case, PCP was still considered a non-growth substrate. Phenol (S_p) and DCP (S_{DCP}) were inhibitory requiring the additional terms in the denominator of equation 10. Table 2 presents a summary of the kinetic coefficients approximated by the model parameter estimation algorithm.

MATERIALS AND METHODS

Long-term studies and initial rate experiments were performed on activated sludge cultures grown in pulse-fed batch reactors and CSTRs. For fed-batch

TABLE 2. Estimated model coefficients for single-, dual-, and triple-substrate conditions in continuous flow culture reactors.

Reactor	μ_{max} (hr^{-1})	Ks_g (mg/L)	Ks_p (mg/L)	Ki_p (mg/L)	Ki_{DCP} (mg/L)	Ki_{PCP} (mg/L)
Phenol Only Culture[(a)]						
	0.91	26.2	—	82.9	—	—
	0.96	74.57	—	28.85	—	—
	0.90	207.9	—	37.45	—	—
Phenol + DCP + PCP Culture						
PCP + phenol (92.7 mg/L)	0.73	9.26	—	1,553	—	0.20
DCP + phenol (108 mg/L)	0.81	5.30	0.63	109.2	1.30	—
PCP + DCP (11.3 mg/L) + phenol (111 mg/L)	0.80	9.16	7.80	$>1.0 \cdot 10^6$	0.80	0.32

(a) Phenol concentration varied up to 400 mg/L.

reactors, glucose, phenol, and PCP were chosen as the carbon sources and nine reactors were operated consisting of different combinations and concentrations of three carbon sources. For CSTRs, phenol, DCP, and PCP were selected as the carbon sources and three reactors with different combinations of substrates were operated. The activated sludge was obtained from two sources. The first was the Baytown Central Waste Treatment Plant aeration basin located in Bayfort Industrial District (Harris County, Texas) and the second source was the Texas A&M University wastewater treatment plant aeration basin (College Station, Texas). The phenolic compounds were analyzed using high-performance liquid chromatography or gas chromatography. Glucose was analyzed using an enzymatic assay kit (Sigma Co. St. Louis, Missouri). Dissolved-phase bulk substrate was monitored by total organic carbon. In initial rate experiments, suspended solids was measured by means of an electronic particle counter (Coulter Electronics, Louton, Beds, England) because the total suspended solids (TSS) method was not sensitive enough to detect the small change of biomass. An empirical relationship between floc counts and TSS was developed. Other culturing conditions are the same as those already reported (Bae et al. 1992, McCreary 1991).

MODEL PREDICTIONS

Single- and Dual-Substrate Feed

Kinetic data from previous work (Autenrieth et al. 1993) on single-substrate phenol and dual-substrate glucose and phenol also were modeled using the parameter estimation routine. The kinetic data from the dual-substrate phenol and glucose experiment were modeled by incorporating the three experiments (glucose, phenol, and glucose + phenol) into the parameter estimation algorithm simultaneously. The growth rate (μ) and substrate concentration (S_g, S_p, and/or S_{PCP}) values obtained from each initial rate experiment were input into the parameter estimation algorithm along with model state equations. For example, three initial rate experiments were performed on a dual-substrate reactor. The initial rate experiment generated a μ for each substrate concentration. The μ values and the corresponding substrate concentration(s) for all initial rate experiments on a dual-substrate reactor were input into the parameter estimation algorithm to estimate μ_{max} and each constant for the range of substrate concentrations. This was done so the model could predict growth rates independent of substrate concentration and combinations.

The model was successful in predicting growth rates for inhibitory and non-inhibitory single-substrate concentrations of phenol (Table 1). The Ki_p predicted for the 50 mg/L culture indicated that phenol was not inhibitory at this concentration. The Haldane equation can be reduced to the Monod equation because the S_p/Ki_p term became negligible. In the 200 mg/L phenol reactor, growth inhibition occurred at phenol concentrations greater than 40 mg/L. Typical coefficients for maximum growth rates of bacteria fed phenol range from 0.57 to 1.2 hr^{-1} (Hickman & Novak 1984, Klecka & Maier 1985). Half-saturation constants for phenol have been reported to range from 0.634 to 245 mg/L and inhibition constants vary from 7.5 to 700 mg/L (Beltrame et al. 1984, Powlowsky & Howell 1973, Rozich et al. 1983, Sokol & Howell 1981).

The model successfully predicted the microbial growth response for the dual-substrate reactor, 450 mg/L glucose and 50 mg/L phenol, with respect to varying glucose concentrations (Figure 1). However, the model underpredicted the growth response in terms of the phenol concentration for the 450:50 reactor (Figure 2). At this phenol:glucose ratio, the glucose concentration far exceeds that of phenol and this phenol concentration is considered noninhibitory. If glucose degradation were independent of phenol, there would be no interaction effects between the microbes degrading the two substrates. At low phenol concentrations and high glucose concentrations, the degradation does not appear to be interactive. Because the model assumes interaction effects between microbes, it cannot fit the data.

The model closely approximated microbial response for the 300 mg/L glucose and 200 mg/L phenol reactor (Figures 1 and 2). The generated inhibition constants for phenol (Ki_p) indicated greater phenol inhibition at its lower concentration, e.g., the 50:450 culture than 200:300 culture (Table 1). The generated phenol inhibition constant for the 200:300 culture was large and can be assumed to be infinity for practical purposes, making the S_p/Ki_p term negligible. Due to the inverse relationship between inhibition constants and the extent of inhibition, larger values of the inhibition constant indicate little inhibition. The model predictions help to expose a synergistic interaction between glucose and phenol at comparable molar concentrations possibly due to microbial species selectivity.

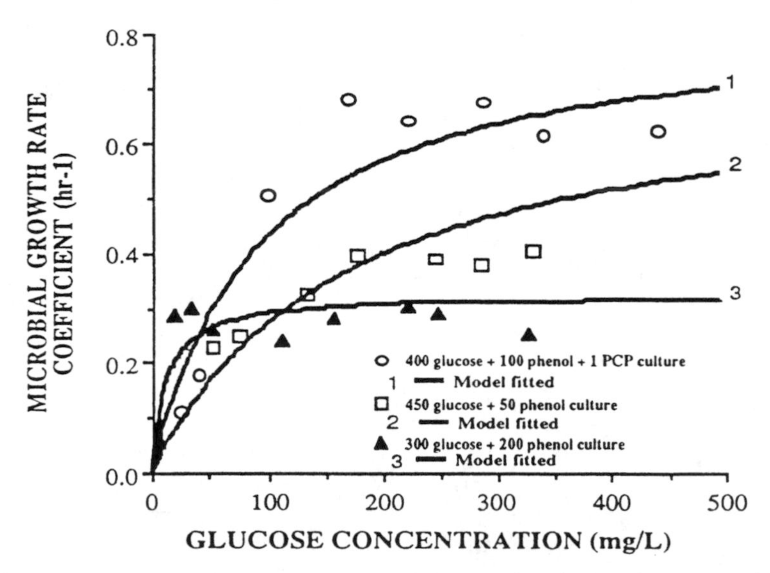

FIGURE 1. Model predicted behavior for growth rate as a function of glucose concentration compared to observed data for multiple-substrate batch-fed cultures.

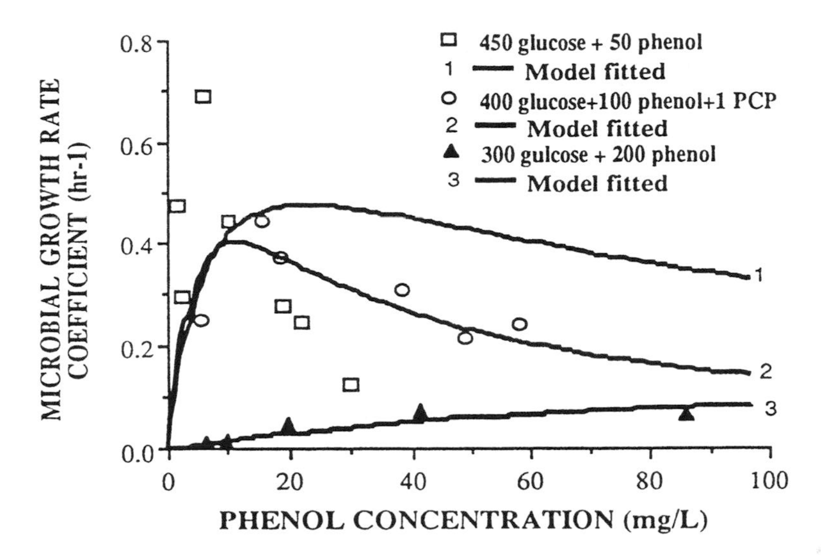

FIGURE 2. Model predicted behavior for growth rate as a function of phenol concentration compared to observed data for multiple-substrate batch-fed cultures.

Triple-Substrate Feed

The model also predicted the observed trends for noninhibitory and inhibitory growth observed for the batch-fed triple-substrate reactor with glucose, phenol, and PCP (Figures 1 and 2). When glucose was fed to these bacteria as the sole source of carbon, the predicted line was similar to a saturation curve (Figure 1). When phenol was fed to these bacteria as the sole source of carbon, the predicted line followed an inhibition function (Figure 2). Comparison of Ki_{PCP} for the fed-batch reactors (Table 1) and the CSTRs (Table 2) showed that the PCP was not as inhibitory in CSTR cultures where organisms were never exposed to glucose. In batch reactors, the phenol inhibition constant (Ki_p) was lower in the triple substrate reactor than in the dual-substrate reactors, indicating that phenol was more inhibitory under these conditions. Phenol may have been more inhibitory in the triple substrate reactor due to the combined impact of two inhibitory substrates. Comparing results from the 200:300 culture and the triple substrate reactor, the metabolic pathway for phenol degradation in the triple-substrate reactor may have been used for PCP decomposition. Figure 3 shows the degradation of PCP by the 400:100:1 culture in the presence of phenol. In this reactor, the degradation of phenol and PCP occurred simultaneously. Once phenol disappeared in the reactor, the PCP degradation slowed considerably. This suggests possible enzyme induction by phenol for PCP degradation or catalysis of PCP by phenol degrading enzyme because the ring-fission of PCP occurs before the complete dechlorination of PCP (Rochkind et al. 1986, Ruckdeschel et al. 1987).

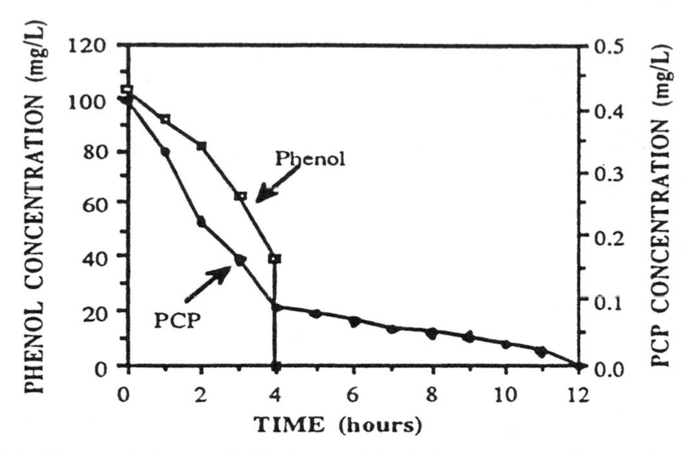

FIGURE 3. Degradation of PCP in a batch-fed reactor acclimated to 400 mg/L glucose, 100 mg/L phenol, and 1 mg/L PCP.

Also, this concentration of phenol could have been more inhibitory because the two chemicals were metabolized by one pathway, thus causing competition between the enzymes that catalyze the subsequent reactions. Another explanation for the greater phenol inhibition may be due to the occurrence of feedback inhibition in the metabolism of PCP. At some step in PCP degradation pathway, intermediate metabolites possibly chlorocatechols, may inhibit other enzymes in the metabolic pathway. The inhibited enzymes may have been the enzymes that catalyzed the reactions for phenol decomposition. Without the isolation of specific enzymes, these assumptions cannot be verified. However, the model has provided more evidence in evaluating degradation mechanisms in the absence of these analyses.

A separate set of experiments were conducted on microorganisms fed phenol with DCP and PCP in CSTRs. Figure 4 depicts a comparison between the observed

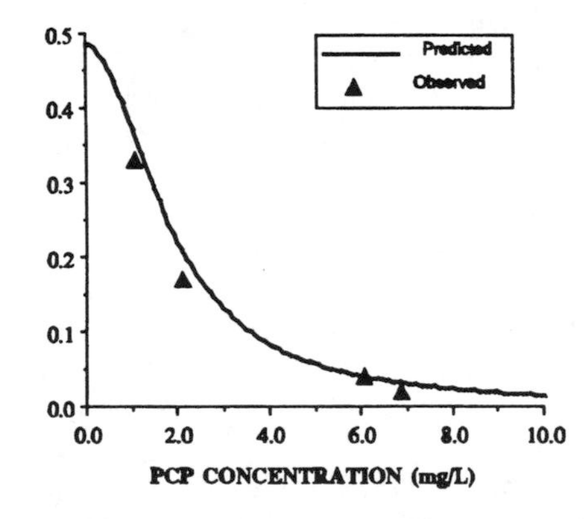

FIGURE 4. Model comparisons between predicted and observed growth rate data for CSTR cultures (phenol + DCP + PCP) as a function of PCP concentration in the presence of 92.7 mg/L phenol.

performance and that predicted by the interactive model as a function of PCP concentration in the presence of 92.7 mg/L phenol without any DCP. Similarly, Figure 5 compares microbial performance with model estimates as a function of DCP concentration in the presence of 108.0 mg/L phenol without any PCP. In both cases, with an increase in substrate concentration, growth is severely inhibited. The extent of this inhibition and an indication of substrate interaction effects are presented in Table 2, which provides the estimated kinetic coefficients for tested cultures. Because all combinations of substrates were tested, the substrate interactions can be more closely delineated. PCP is the most inhibitory substrate followed by DCP and phenol. An observed trend is that, as the number of inhibitory substrates is increased, the inhibition response (Ki) for the individual substrates is increased.

DISCUSSION

Table 1 summarizes the predicted coefficients from the model simulations for each reactor. Kinetic coefficients for single-substrate phenol (50 and 200 mg/L) and dual-substrate glucose and phenol experiments also were generated using the Monod and Haldane equations (Autenrieth et al. 1993). As expected, the kinetic coefficients generated for the single-substrate reactors were similar using the Monod, Haldane, and model equations. The μ_{max} values for the single-substrate 50 mg/L phenol reactor were 1.16 hr^{-1} using the Monod equation and 1.23 hr^{-1} using our interactive model equation. The μ_{max} values for the single-substrate 200 mg/L phenol reactor were 0.89 hr^{-1} using the Haldane equation and 0.84 hr^{-1} using our model equation. The closeness in these values verified

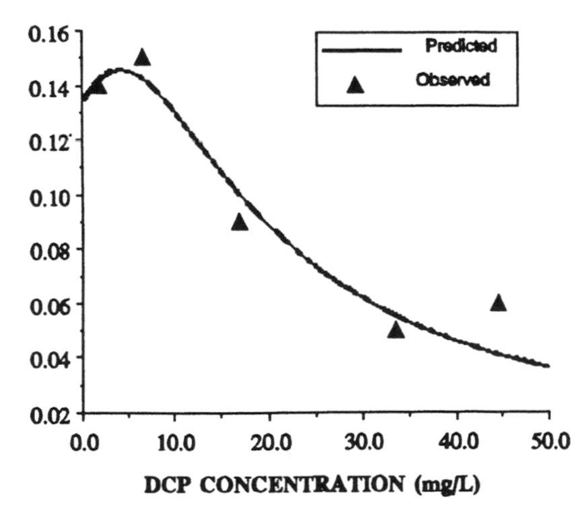

FIGURE 5. Model comparisons between predicted and observed growth rate data for CSTR cultures (phenol + DCP + PCP) as a function of DCP concentration in the presence of 108.0 mg/L phenol.

that our model equation could successfully model single-substrate kinetics. Unlike the Monod or Haldane equations, the interactive model assumed interactive effects of substrates and incorporated terms for these effects. For a dual-substrate reactor the kinetic coefficients based on Monod or Haldane kinetics had to be estimated for each substrate separately. This resulted in growth rates in excess of one for most combinations of the 50:450 culture due to the additional growth from the second substrate present.

The smallest maximum growth rate coefficient occurred in the 300 mg/L glucose and 200 mg/L phenol reactor. This reactor also had the largest values for the phenol saturation and inhibition constants for the dual-substrate reactors, making phenol least inhibitory in this reactor as verified by the Haldane kinetic analysis by Autenrieth et al. (1993). Although the growth rate was lower than the other reactors, there was effectively no cell growth inhibition due to phenol as indicated by the high Ki value (Table 1). The small growth rate could be due to interactive effects. At high phenol concentration, the mixed culture could not degrade phenol as efficiently as with both glucose and phenol present. At this phenol concentration range, microorganisms may need an alternative energy source for the synthesis of necessary enzymes. Therefore, phenol may tentatively suppressed the growth rate rather than inhibit it.

The smallest values for phenol saturation and inhibition constants occurred in the triple-substrate batch reactor (Tables 1). Phenol reached saturation very early, and it was most inhibitory in this reactor. PCP was less inhibitory in CSTRs than in batch reactors, as indicated by the high PCP inhibition constant in CSTR culture (Tables 1 and 2). These results may indicate a difference in culture acclimations to chlorinated phenols, e.g., DCP and PCP.

In general the CSTR cultures were more inhibited than the batch-fed cultures as can be seen by comparing the generated inhibition constants for the individual substrates except for PCP. Phenol, the primary growth-supporting substrate, in particular generated a much greater inhibition effect in CSTR cultures (Table 2) than was observed in the batch cultures (Table 1). Such results indicate that batch-fed culturing may be a more appropriate culturing technique, than the continual exposure of cultures to inhibitory substrates. This finding can be explained through the speculation of microbial ecology. In batch culture, only phenol degrading microorganisms grow rapidly at first, but phenol nondegrading microorganisms slowly recover their growth at the latter stage of reactor operation with the reduced toxicity of phenol and the presence of phenol degradation products. Therefore, all microbial species can be retained in the batch culture, resulting in a more diverse and stable system. However, in CSTR culture, when phenolic compounds are the primary substrates, the growth rate of phenol nondegrading microorganisms will be slow and eventually washed out if the growth rate is less than a critical value, leaving only phenol degrading organisms. Compared to batch culture, CSTR culture cannot retain all the microbial species which were present at first. This reduction in microbial diversity make the culture more susceptible to fluctuation in other environmental conditions. Therefore, monitoring and assessing microbial diversity and species dominance is particularly important in multiple substrates, mixed culture operations.

CONCLUSIONS

Our interactive model was successful in fitting the observed parameters and predicted parameters for one triple-substrate reactor and one dual-substrate reactor. However, the model was not successful in fitting the observed parameters for the dual-substrate 50 mg/L phenol and 450 mg/L glucose reactor. This suggests that other factors must be considered in the multiple substrate kinetic equations. The experiments indicated that the reactors were sensitive to changes in concentrations and type of substrate feed. There may be some cases where the particular substrate combinations do not have an interactive relationship, as was the case in the 50 mg/L phenol and 450 mg/L glucose reactor. However, when the substrate feeds remain the same (phenol and glucose) but the molar concentrations are similar, as in the 200 mg/L phenol and 300 mg/L glucose reactor, substrate interaction is indicated and the data can be successfully modeled. It appears that there was some critical inhibitory substrate concentration in which induction of enzyme systems occurs. Once enzyme induction commences, interactive effects in degradation become more significant. Below this critical substrate concentration there may be no interaction between substrates, and therefore other models may be more appropriate to describe cell growth and inhibition. Individual substrate inhibition was greater under the CSTR culturing conditions than those of a batch, indicating that batch culturing is more desirable for long-term treatment of multiple mixtures of inhibitory substrates.

REFERENCES

Autenrieth, R. L., J. S. Bonner, and E. McCreary. 1993. "Multiple Substrate Biodegradation Kinetics: Glucose, Phenol, and Pentachlorophenol." Submitted for publication.

Bader, F. G. 1976. "Analysis of Double-Substrate Limited Growth." *Biotechnology and Bioengineering 20*: 183-202.

Bae, B.-H., A. T. Ralph, R. L. Autenrieth, and J. S. Bonner. 1992. "Degradation Kinetics for Phenolic Compounds with Mixed Microbial Cultures in Continuous Flow Reactor." *Proceedings of Water Environmental Federation 65th Annual Conference, Research Symposia 1*, New Orleans, LA. pp. 321-331.

Bailey, J. E., and D. F. Ollis. 1986. *Biochemical Engineering Fundamentals.* McGraw-Hill Inc., New York, NY. p. 92.

Beltrame, P. 1984. "Inhibiting Action of Chloro- and Nitro-Phenols on Biodegradation of Phenol: A Structure-Toxicity Relationship." *Chemosphere 13*: 3-9.

Dixon, M., and E. C. Webb. 1986. *Enzymes*, 3rd ed., Academic Press, New York, NY, p. 72.

Ernest, A. N., J. S. Bonner, and R. L. Autenrieth. 1991. "Model Parameter Estimation for Particle Transport." *ASCE Journal of Environmental Engineering 117*: 573-595.

Grady, C. P. L., Jr. 1990. "Biodegradation of Toxic Organics: Status and Potential." *ASCE Journal of Environmental Engineering 116*: 805-821.

Grady, C. P. L., Jr., and H. C. Lim. 1980. "Biological Wastewater Treatment." In P.N. Chevemisinoff (Ed.), *Kinetics of Biochemical Systems*, Marcel Dekker Inc., New York, NY.

Hickman, G. T., and J. T. Novak. 1984. "Acclimation of Activated Sludge to Pentachlorophenol." *J. Water Pollut. Control Fed. 56*: 364-369.

Klecka, G. M., and W. J. Maier. 1985. "Kinetics of Microbial Growth on Pentachlorophenol." *Applied and Environmental Microbiology 49*: 46-53.

Kompala, D. S., D. Ramkrisshna, and G. T. Taso. 1984. "Cybernetic Modeling of Microbial Growth on Multiple Substrates." *Biotechnology and Bioengineering 26*: 1272-1281.
Machado, R. J. 1985. "Kinetics of Dual Substrate Removal by an Axenic Culture of Bacteria." M.E. Report, Clemson University, Clemson, SC.
McCreary, E. 1991. "Biodegradation of Multicomponent Hazardous Wastes Using Activated Sludge Bacteria." M. S. Thesis, Texas A&M University, College Station, TX.
Papanastasious, A. C. 1982. "Kinetics of Biodegradation of 2,4-Dichlorophenoxyacetate in the Presence of Glucose." *Biotechnology and Bioengineering 24*: 2001-2011.
Powlowsky, U., and J. A. Howell. 1973. "Mixed Culture Biooxidation of Phenol." *Biotechnol. and Bioeng. 15*: 889-896.
Rochkind, M. L., J. W. Blackburn, and G. S. Sayler. 1986. "*Microbial Decomposition of Chlorinated Aromatic Compounds.*" U.S. Environmental Protection Agency Technical Report, EPA-600/2-86/090, Cincinnati, OH.
Rozich, A. R., A. F. Gaudy, and P. D. D'Adamo. 1983. "Predictive Model for Treatment of Phenolic Wastes by Activated Sludge." *Water Research 17*: 1453-1466.
Ruckdeschel, G., G. Renner, and K. Schwarz. 1987. "Effects of Pentachlorophenol and Some of Its Known and Possible Metabolites on Different Species of Bacteria." *Applied and Environmental Microbiology 53*: 2689-2692.
Sokol, W., and J. A. Howell. 1981. "Kinetics of Phenol Oxidation by Washed Cells." *Biotechnol. and Bioeng. 23*: 2309-2049.
Stryer, L. 1988. *Biochemistry*. 3rd ed., p. 421. W. H. Freeman and Company, New York, NY.
Yoon, H., G. Klinzing, and H. W. Blanch. 1977. "Competition for Mixed Substrates by Microbial Populations." *Biotechnology and Bioengineering 19*: 1193-1210.

BIODEGRADATION OF *BIS*(2-ETHYLHEXYL)PHTHALATE, ETHYLBENZENE, AND XYLENES IN GROUNDWATER: TREATABILITY STUDY SUPPORTING IN SITU AQUIFER BIOREMEDIATION

D. A. Graves, C. A. Lang, and J. N. Rightmyer

ABSTRACT

Biotreatability studies were performed to evaluate the efficacy of in situ biological treatment of impacted groundwater at a Superfund site in New Jersey. The site once supported a manufacturer of wall coverings. Groundwater at the site is impacted with *bis*(2-ethylhexyl)phthalate (DEHP), ethylbenzene, and xylenes. Site characterization was performed to determine the potential for implementing a successful in situ bioremediation strategy on site. Biotreatability studies were conducted to determine the biodegradability of DEHP, ethylbenzene, and xylenes in site groundwater. Various treatments were established to evaluate target compound removal by biological and physical mechanisms. Biodegradation of target compounds was tracked by sampling reaction vessels at critical time intervals determined by respirometric evaluation of activity in each treatment. Microbial respiration was measured with a computerized respirometer. The biodegradation of ethylbenzene, *o*-xylene, and DEHP was demonstrated, and analysis of variance and comparison of means among treatments indicated that biological activity resulted in significantly lower levels of target compounds at the 95% confidence level. Groundwater was augmented with activated sludge and a bacterial strain capable of degrading DEHP. Augmented treatments effectively reduced the ethylbenzene and xylene concentrations. The DEHP degrader reduced the level of DEHP at a rate slightly greater than that observed with indigenous bacteria. The activated sludge treatment was less effective in removing DEHP, and it was also less efficient in coupling oxygen consumption with organic carbon mineralization.

INTRODUCTION

In situ bioremediation has been successfully applied to the remediation of aquifers impacted with hydrocarbons such as benzene, toluene, ethylbenzene, xylenes, lubricating oils, and fuels. In practice, mechanisms for the introduction of oxygen and nutrients and control of pH are employed to enhance the subsurface environment to stimulate biological activity. A variety of engineering designs may be employed to ensure proper control of environmental parameters. However, an a priori determination of the site characteristics and the biodegradability of target compounds is required before remedial designs are considered.

A site under consideration for in situ bioremediation once housed a production facility for the manufacture of vinyl wall coverings. During operation, process sludges were disposed in a surface impoundment. Additionally, the site contained numerous support facilities including underground storage tanks, an aboveground tank farm, various process waste tanks, cooling water equipment, and other processing equipment. Approximately 18,900 L of water-immiscible material consisting of DEHP, xylenes, lubricating oil, ethylbenzene, and naphtha have been pumped from the site subsurface since 1984.

The bioremediation of phthalates has not been extensively documented. Estimated half-lives in groundwater range from a few days to more than a year (Howard et al. 1991, Subba-Rao et al. 1982, Verschueren 1983). Thus, the biodegradation of this class of compounds cannot be confidently predicted, particularly in complex matrices such as soil or groundwater. Additionally, regulatory procedures often mandate demonstration of target compound biodegradation as part of technology screening or feasibility studies. The biodegradation of DEHP, ethylbenzene, and xylene in environmental samples was evaluated as part of a remedy screening study for a Superfund site in the Northeast United States.

The treatability study evaluated the potential applicability of bioremediation as a remedial technology for contaminant treatment in site groundwater. Chemical constituents targeted for monitoring throughout the remedy screening study were DEHP, ethylbenzene, and xylenes. The amount of each targeted chemical constituent removed, relative to the initial amount in groundwater, was used to evaluate the effectiveness of the technology. The main goals of the study were (1) to show the effectiveness of biological treatment strategies in reducing the concentrations of target compounds, and (2) to provide design information required for the next level of evaluation.

MATERIALS AND METHODS

Groundwater and Soil Sampling

Three representative well locations were selected for groundwater sampling. These wells were reported to provide a representative sampling of the full range of volatile organic compounds (VOC) and semivolatile organic compounds (SVOC) concentrations. VOCs ranged from nondetectable to 136,000 µg/L. SVOCs ranged from nondetectable to 67,898 µg/L. Soil samples were collected from an area having concentrations of DEHP and total targeted volatiles ranging from 3,100 to 30,000 mg/kg and 0.2 to 532 mg/kg, respectively.

Site Characterization for Bioremediation

Soil and groundwater samples were analyzed for pH, soil moisture content, residual nutrient content, and groundwater mineral content. The compatibility of nutrients with groundwater and the adsorption of nutrients by soil-groundwater slurries also were determined. The ability to transport oxygen was estimated by examining the stability of hydrogen peroxide in soil-groundwater slurries and the potential for sustaining a high dissolved oxygen content in the slurry.

Microbiological characteristics such as the population density of aerobic heterotrophic bacteria, hydrocarbon-degrading bacteria, and DEHP-degrading bacteria were quantified in each sample. The spread plate technique was used to determine the density of the microbial population in each sample. Total heterotrophic microbes were grown on dilute nutrient agar. Hydrocarbon-degrading microorganisms were cultured on a carbon-free, mineral salts, noble agar medium with the sole carbon and energy source being benzene, ethylbenzene, toluene, and xylene vapor. Phthalate degraders were cultured on the same medium containing DEHP as the sole carbon and energy source. The precision of the plate count method is approximately 0.5 order of magnitude. The response of bacteria to nutrient and oxygen stimulation also was evaluated.

Biotreatability Equipment and Materials

DEHP, ethylbenzene, and xylenes were analyzed using gas chromatography methods modified from U.S. Environmental Protection Agency (EPA) Method 8240 for VOCs and EPA Method 8270 for SVOCs. Total organic carbon (TOC) content was determined using a Dohrmann Total Carbon Analyzer. The pH of treatments was determined using a pH electrode. The nitrogen, as ammonia, and phosphate content of each treatment was determined using modified Standard Method 4500-NH_3 F and 4500-P E, respectively (Clesceri et al. 1989). Microbial respiration was quantified using computerized respirometers (N-Con Systems, Larchmont, New York).

Respirometric Biodegradation Studies. Batch biodegradation studies were performed on site groundwater samples. The studies were conducted in computerized respirometers, which measure oxygen consumption by microorganisms and replace the oxygen to avoid oxygen limitations. The total mass of CO_2 produced during the study was determined at the end of the study by measuring the inorganic carbon content of a potassium hydroxide (KOH) trapping solution in each respirometer vessel. This approach for monitoring microbial activity is nondisruptive to the sample vessel and is not labor intensive. Respirometric analysis of microbial activity also provided a real-time indication of contaminant biodegradation. Using this approach, sampling points were chosen based on microbial activity rather than on an arbitrary time schedule.

Treatments included nutrient-amended groundwater, nutrient- and activated sludge-amended groundwater, nutrient-amended groundwater augmented with 108 DEHP-degrading microbes/mL, nutrient- and mercuric chloride-amended groundwater, and untreated groundwater in a zero-headspace container. Each treatment was conducted in triplicate. This experimental design demonstrated the biodegradation of target compounds by contrasting untreated and biologically inhibited controls, accounted for mass balance among treatments, and indicated losses by biological and physical mechanisms.

In addition to evaluating biodegradation of DEHP by indigenous microorganisms at naturally occurring microbial densities, the benefit of augmenting groundwater with activated municipal sludge or DEHP-degrading microorganisms isolated from site soil samples was evaluated. DEHP degraders were grown in liquid mineral salts medium supplemented with 500 mg/L DEHP. Activated sludge was collected from a local publicly owned treatment works (POTW) in Knoxville, Tennessee, and washed to deplete the residual dissolved carbon in the sludge. Biological solids (activated sludge) were added to a final solids density of 860 mg/L. The use of alternative sources of microorganisms provided a means to increase the overall microbial density and diversity compared to groundwater alone.

A single groundwater composite (18 L) was prepared from which all treatments were established. Triplicate analyses were performed on this composite. The composite groundwater sample was found to have very low concentrations of DEHP, ethylbenzene, and xylene; therefore, 3 mL of a solution consisting of 1 part DEHP, 1 part ethylbenzene, and 1 part *o*-xylene was added to the 18-L composite sample and thoroughly mixed with the water to aid dissolution. An equal mass of water was removed from the bottom of the composite sample container and placed in each respirometer vessel. Nutrients and alternative microorganisms were added to appropriate treatments. Biologically inhibited controls were treated with mercuric chloride. The untreated controls were placed in zero-headspace vessels with no nutrient or microorganisms added and incubated in the dark at room temperature for the duration of the study. The remaining vessels were connected to the respirometer, maintained at constant temperature in the dark, and stirred continuously using a magnetic stir bar.

Oxygen consumption was continuously measured and recorded every 2 hours. CO_2 production, pH, residual nutrients, and microbial density were determined at the end of the study. Contaminant and TOC analyses were performed initially on the composite sample, on each vessel twice during the study as indicated by microbial respiration, and at the end of the study. The study was terminated when microbial respiration became asymptotic.

Statistical Methods. Reduction in target compound concentration at the end of the study was evaluated using single-classification Model I analysis of variance (ANOVA) (Sokal & Rolfe 1981). The mean residual concentrations of DEHP, ethylbenzene, and xylene were compared for each treatment to determine if a significant difference at the 95% confidence level could be attributed to treatment. The null hypothesis stated that no difference existed among means of the various treatments. Additionally, planned comparisons were conducted using the F-test to determine if significant differences at the 95% confidence limit occurred between treatments.

RESULTS AND DISCUSSION

Site Characterization for Bioremediation

Soil and groundwater samples were subjected to a battery of tests designed to evaluate the potential for implementing a successful in situ bioremediation treatment strategy.

Microbial Enumerations. Microbial enumerations are shown in Table 1. The indigenous heterotrophic, hydrocarbon-degrading, and phthalate-degrading microbial population densities observed in soil samples ranged from 9.2×10^5 to 3.0×10^7 colony-forming units per gram dry weight (CFU/g). The initial microbial density in groundwater also is indicated.

TABLE 1. Microbial enumerations and response of indigenous microorganisms to oxygenation and nutrient amendment.

Sample	Initial (CFU/g)[a]	Oxygen (CFU/g)[a]	Oxygen and Nutrients (CFU/g)[a]
Heterotrophs			
Soil A	3.0×10^7	5.2×10^7	1.4×10^8
Soil B	1.4×10^7	2.0×10^7	1.1×10^7
Soil C	3.3×10^6	3.0×10^7	9.4×10^7
Groundwater	3.0×10^5	8.8×10^6	3.5×10^7
Hydrocarbon degraders			
Soil A	1.6×10^7	6.9×10^7	3.5×10^8
Soil B	3.2×10^6	1.1×10^7	7.5×10^6
Soil C	4.0×10^6	1.3×10^7	3.1×10^8
Groundwater	4.7×10^4	4.2×10^6	3.6×10^7
Phthalate Degraders			
Soil A	1.4×10^7	1.7×10^7	1.3×10^7
Soil B	9.2×10^5	4.8×10^7	5.4×10^7
Soil C	1.5×10^6	6.6×10^6	9.2×10^6
Groundwater	4.7×10^4	1.1×10^6	7.5×10^6

[a] CFU = colony-forming unit

The response of native microorganisms to oxygenation and nutrient augmentation is an important factor in determining the likelihood of enhancing biodegradation. Table 1 also shows the results of microbial stimulation tests. Because stimulation was measured as growth of bacteria, the plate count method was used to quantitate the level of microbial activity induced by oxygenation and nutrient augmentation. Increase in microbial population density greater than 0.5 order of magnitude is considered to be a positive response. As indicated in Table 1, heterotrophs, hydrocarbon-degrading bacteria, and phthalate-degrading bacteria in most samples responded to oxygenation and nutrient amendment by increasing in cell density by greater than 0.5 order of magnitude. Heterotrophs in soil sample B failed to respond to treatment, and hydrocarbon degraders showed a weak response, but phthalate degraders responded very well to both treatments. Bacteria in groundwater (Table 1) responded favorably to stimulation as indicated by growth in the presence of oxygen and nutrients.

In general, the results of the microbial stimulation tests indicated that, under appropriate conditions, microbes present within the impacted soils and aquifers should respond favorably to nutrient and oxygen amendment. Nutrients enhanced the growth of bacteria in all cases except for heterotrophs and hydrocarbon degraders in soil sample B.

pH and Residual Nutrients. The pH of site samples is shown in Table 2. The optimal pH for bioremediation is generally accepted to be within the range of 6 to 8. A pH outside of this range may reduce microbial metabolism and biodegradation. The soil pH ranged from 5.6 to 7.7. All of the soil samples except one had a pH within the preferred range. The groundwater sample pH was 6.4.

The nitrogen, as ammonia, and orthophosphate content of site soil and groundwater samples is shown in Table 2. The nitrogen content of soil ranged from nondetectable to moderate. Ammoniacal nitrogen in groundwater was 4.5 mg/L. The phosphate content in the soil ranged from 80 to 221 mg/L but was nondetectable in the groundwater. Therefore, the low concentration of soluble

TABLE 2. Physical and chemical characteristics of site soil.

Sample	pH	Ammonium (ppm)[a]	Phosphate (ppm)	Soil Moisture (percent)	Ca^{++} (ppm)	Mg^{++} (ppm)	Cl^- (ppm)	Fe (ppm)
Soil A	5.6	<4	80	8	NA[b]	NA	NA	NA
Soil B	7.7	63	158	11	NA	NA	NA	NA
Soil C	7.3	4.7	221	13	NA	NA	NA	NA
Groundwater	6.4	4.7	<0.5	NA	61.6	33.3	45	427

(a) ppm, mg/kg for soil and mg/L for water.
(b) NA, not applicable.

nutrients and the microbial growth stimulated by the addition of nutrients indicates that nutrient additions will expedite successful bioremediation.

Total Organic Carbon Analysis of Groundwater. The total organic carbon content of the groundwater was 17 to 18 mg/L. This value represents relatively low organic carbon loading in the groundwater.

Mineral Analyses. The calcium, magnesium, and iron content of groundwater was evaluated because these elements can catalyze chemical reactions that may interfere with aquifer bioremediation. The divalent cations, calcium and magnesium, react with phosphate to form an insoluble precipitate. In the case of severe phosphate precipitation, wells and geological formations can become clogged.

Hydrogen peroxide is frequently used as an oxygen source for aquifer bioremediation. As hydrogen peroxide decomposes, oxygen is liberated. Soluble iron can react with hydrogen peroxide to form insoluble iron oxides and hydroxides. Because the rate at which peroxide decomposes is related to its ultimate efficiency in transporting oxygen through an aquifer, iron can limit oxygen movement. The calcium, magnesium, and iron content of site groundwater is shown in Table 2. The concentration of calcium and magnesium was moderate; the iron content was high.

Nutrient Compatibility. In order to determine the interaction between calcium, magnesium, and phosphate, a nutrient compatibility test was performed. The microbial nutrient blend Restore® 375 contains tripolyphosphates that act as calcium and magnesium chelators; therefore, by adding more Restore® 375, less precipitation occurs due to the increased concentration of the chelator.

The initial addition of Restore® 375 (10,000 mg/L) was based on the chelating power of the nutrient blend and the mineral content of the groundwater. The groundwater composite received the first nutrient amendment of 10,000 mg/L with no precipitation or cloudiness. These results indicate that nutrient amendment to the groundwater can be accomplished with no anticipated complications.

Nutrient Adsorption. Soils generally have an affinity for ammonium and phosphate. The binding capacity of a soil impacts the ability to transport nutrients away from the injection area. The data in Figure 1 show the amount of ammonia, phosphate, and chloride bound to a composite of the soil samples after each of three nutrient additions. The results indicated that after the initial addition of nutrients, approximately 40% of the ammonium, 80% of the added phosphate, and none of the chloride bound to the soil. The adsorption of ammonium and phosphate to the soil initially was high but showed of signs of saturation and reduced adsorption with each subsequent addition of nutrients.

The results of the nutrient adsorption test suggest that significant amounts of the added nutrients initially will bind to the soil. However, the binding capacity of the soil will saturate as nutrients are added, thus permitting nutrient transport through the aquifer.

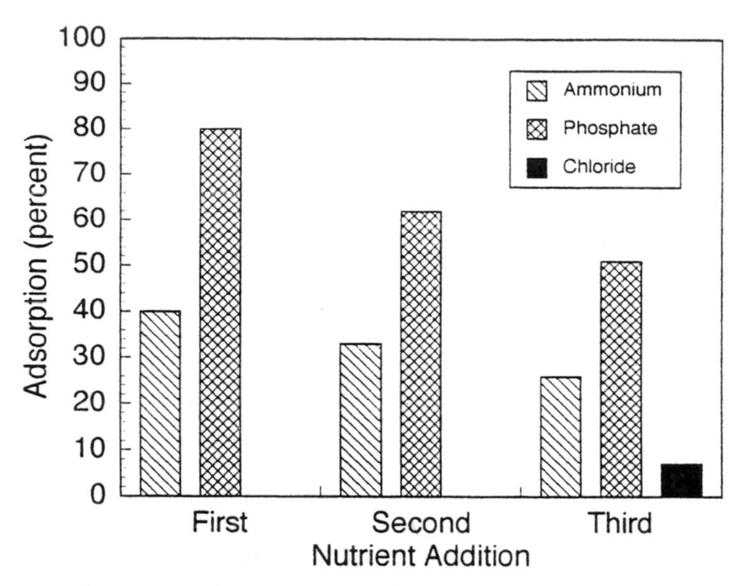

FIGURE 1. Adsorption of ammonium, orthophosphate, and chloride by a soil water slurry.

Hydrogen Peroxide Stability/Oxygenation Potential. Transport of oxygen through an aquifer usually is the limiting factor for in situ bioremediation. Oxygenation of other sites has been achieved by applying hydrogen peroxide. When hydrogen peroxide decomposes, the resulting oxygen may be present as dissolved oxygen or as free oxygen gas. The movement of oxygen through an aquifer depends on the persistence of hydrogen peroxide and the level of dissolved oxygen in the water over time.

Figure 2 tracks the persistence and concentration of dissolved oxygen in a 9 parts water to 1 part soil slurry composed of site samples. The slurry initially was amended with about 868 mg/L hydrogen peroxide. Oxygen concentration was quantitatively tracked until changes in the oxygen distribution between gaseous and dissolved phases became asymptotic. After about 4 hours, the dissolved oxygen content of the slurry was about 70 mg/L, and release of gaseous oxygen, as a result of hydrogen peroxide decomposition, was asymptotic.

A second application of 600 mg/L hydrogen peroxide was made. Decomposition progressed at about the same rate observed for the first application. Overall, the results indicated moderate hydrogen peroxide stability.

Biotreatability Study Results

Evaluation of site characteristics suggested that bioremediation was feasible. Therefore, a treatability test was conducted to determine the biodegradability

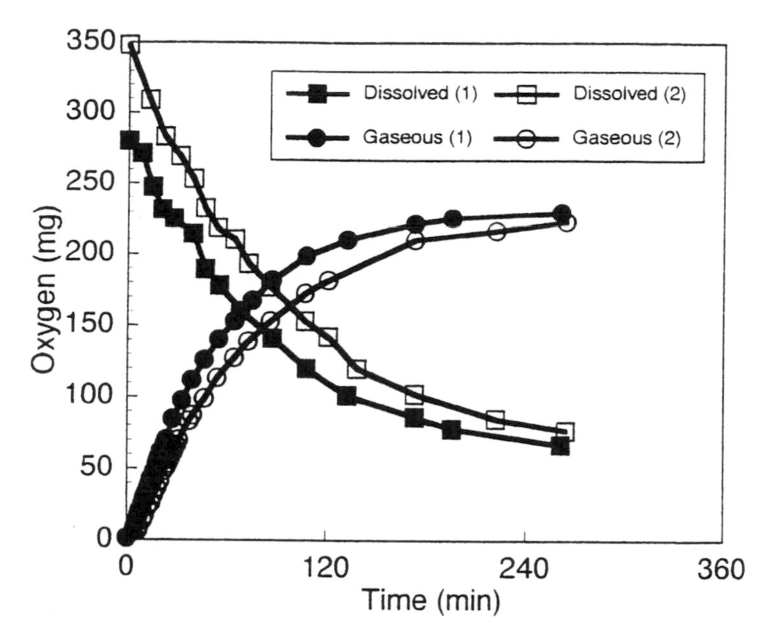

FIGURE 2. Oxygen delivery potential and hydrogen peroxide stability. Results indicate the concentration of total dissolved oxygen, which includes the oxygen available as hydrogen peroxide, and gaseous oxygen resulting from hydroxide peroxide decomposition.

of target compounds by native bacteria as well as by activated sludge and a selected strain capable of using DEHP as its sole carbon and energy source.

Reduction in Target Compound Concentration During Treatment. The loss of DEHP, ethylbenzene, and xylene was determined at four points during the batch groundwater biotreatability test (Figure 3). The results indicate that biological treatment contributed to the removal of target compounds from biologically active treatments. The biologically inhibited treatment (GW, HgCl, & Ntr) indicates the loss of target compounds through physical processes such as volatilization and adsorption to the reaction vessel. The vapor pressures of ethylbenzene and xylene are high enough that an appreciable amount of these compounds should volatilize and be lost when the treatment vessels were opened for sampling. The results shown in Figure 3 suggest this occurred. The vapor pressure of DEHP is so low that volatilization is not a major contributor to the abiotic loss of DEHP from the treatment vessels. The zero-headspace treatment represents the absolute change attributable to experimental and analytical artifacts in a completely untreated control. A 45% loss of ethylbenzene, 33% loss of xylene, and 18% loss of DEHP was observed in this control treatment. Ethylbenzene and xylene removal in all

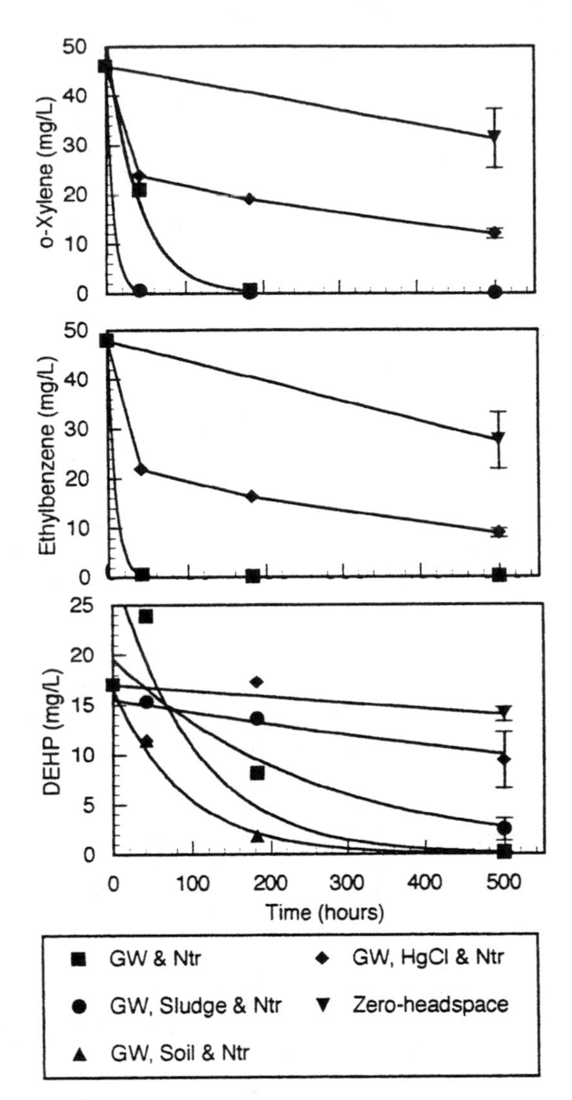

FIGURE 3. Target compound removal during groundwater batch treatment. Each point represents the mean value from the three treatments. The standard deviation of each treatment mean is indicated for the final data point.

biologically active controls was greater than 99.9%. Volatilization could account for 80% of the ethylbenzene loss and 75% of the xylene loss. DEHP removal in the two treatments using indigenous microbes was approximately 99.7%. DEHP removal by the activated sludge treatment was only 86%.

TABLE 3. Planned comparisons among treatments.

Source of Variation	Significance
Ethylbenzene	
Among Treatments	** [a]
Abiotic Vs. Biologically Active	**
Untreated Vs. Biologically Active	**
Sludge Vs. Indigenous Bacteria (groundwater and DEHP augmented)	ns [b]
Sludge Vs. Groundwater Bacteria	ns
Sludge Vs. DEHP degraders (augmented)	ns
Groundwater Bacteria Vs. Soil Bacteria (augmented)	ns
Untreated Vs. Abiotic	**
Xylene	
Among Treatments	**
Abiotic Vs. Biologically Active	**
Untreated Vs. Biologically Active	**
Sludge Vs. Indigenous Bacteria (groundwater and DEHP augmented)	ns
Sludge Vs. Groundwater Bacteria	ns
Sludge Vs. DEHP degraders (augmented)	ns
Groundwater Bacteria Vs. Soil Bacteria (augmented)	ns
Untreated Vs. Abiotic	**
DEHP	
Among Treatments	**
Abiotic Vs. Biologically Active	**
Untreated Vs. Biologically Active	**
Sludge Vs. Indigenous Bacteria (groundwater and DEHP augmented)	**
Sludge Vs. Groundwater Bacteria	**
Sludge Vs. DEHP degraders (augmented)	**
Groundwater Bacteria Vs. Soil Bacteria (augmented)	ns
Untreated Vs. Abiotic	**

(a) At the 95% confidence level, a significant variance component due to treatment existed.
(b) At the 95% confidence level, no significant variance component due to treatment was observed.

The significance of target compound removal was evaluated using ANOVA and the F-test. The results of the statistical comparisons of each treatment are shown in Table 3. For ethylbenzene and xylene, a significant variance was detected at the 95% confidence level between the biologically active treatments and the abiotic and the untreated conditions. There was also a significant variance between the abiotic treatment and the zero-headspace control. This variance is attributed to volatilization in the abiotic treatment because it was opened two more times than the zero-headspace treatment. No difference in the removal of ethylbenzene and xylene was observed among any of the biologically active treatments. These results strongly suggest that the biological component of the treatment process contributed to the removal of ethylbenzene and xylene.

The same statistical process was used to evaluate the removal of DEHP from each treatment. Biological treatment was shown to significantly contribute to DEHP removal at the 95% confidence level. A significantly greater DEHP removal was detected in groundwater treated with indigenous microbes (GW, Soil & Ntr in Figure 3) compared to the treatment augmented with activated sludge (GW, Sludge & Ntr). No difference was detected between the native groundwater and groundwater augmented with DEHP-degrading bacteria isolated from site soil.

Biodegradation Rate Constants and Half-Lives. The biodegradation rate constant was determined from the target compound removal data. Table 4 indicates the degradation rate constant and half-life for DEHP in each treatment. The biologically active treatments had higher rate constants and shorter half-lives than the abiotic control. The treatments containing indigenous microbes had much shorter half-lives for DEHP than did the activated sludge treatment.

The removal data for ethylbenzene and xylene were not appropriate for regression analysis of rate constants and half-lives; therefore, the initial concentration and the concentration at the first sampling point were used. In the case of ethylbenzene, the degradation rate exceeded the sampling frequency; therefore, the concentration was nondetectable at the first sampling point. The estimated biodegradation rate constant for ethylbenzene is ≥ -0.2 hr^{-1} with a half-life of ≤ 3.3 hr based on complete removal within 43 hours. The biodegradation rate constant and half-life were also calculated for xylene using the initial concentration and the concentration at the first sampling point. The rate constant for the activated sludge treatment was -0.2 hr^{-1} with a half-life of 3.3 hours. The rate constants for the two treatments containing indigenous microbes were essentially the same, with a rate constant of -0.02 hr^{-1} and a half-life of 37 to 39 hours.

TABLE 4. DEHP biodegradation rate constants and half-lives in each treatment.

Treatment	Biodegradation Rate Constant (k) (hrs^{-1})	Half-life (hours)
GW & Ntr	-9.9×10^{-3}	70
GW, Sludge & Ntr	-3.9×10^{-3}	176
GW, Soil & Ntr	-1.1×10^{-2}	62
GW, $HgCl_2$ & Ntr	-8.7×10^{-4}	800
Zero-headspace	0	infinite

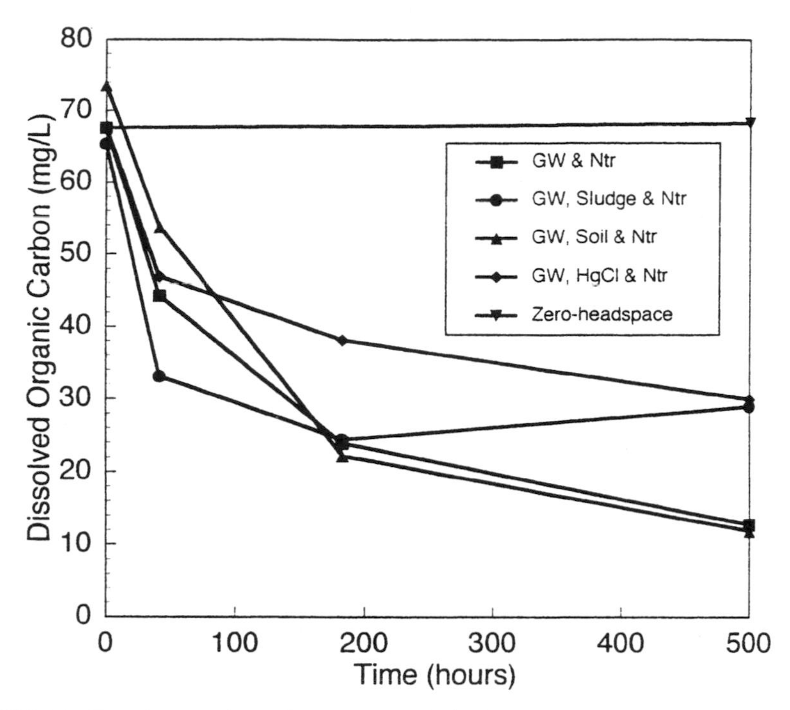

FIGURE 4. Total organic carbon removal during treatment.

Total Dissolved Organic Carbon Content. The reduction in TOC during the test period is shown for each treatment in Figure 4. The general trends in TOC removal are consistent with the removal of target compounds in Figure 3. The TOC in the activated sludge augmented treatment is higher than for other biologically active treatments due to the TOC contributed by decaying sludge. The TOC in the DEHP-degrading bacteria augmented treatment (GW, Soil, & Ntr) was slightly higher than the groundwater (GW & Ntr) treatment for the first two sampling points. At later points in the test, the TOC of these two treatments was essentially the same. The initially higher TOC in the GW, Soil & Ntr treatment was attributed to soluble organic carbon associated with the bacteria used to augment this treatment. Volatile suspended solids (VSS) analysis of these treatments supports this interpretation. As indicated in Figure 5, the volatile solids content of these two treatments quickly equalized to a similar value and remained very similar for the rest of the study. The microbial density of these treatments, reported in Figure 6, also supports population equilibration to the same level.

Microbial Respiration. Figure 7 tracks microbial respiration in each treatment except the zero-headspace control. The figure shows the mean ± one standard deviation at each point collected during the study. Because the data collection frequency was every 2 hours, data were plotted as a line rather than as individual

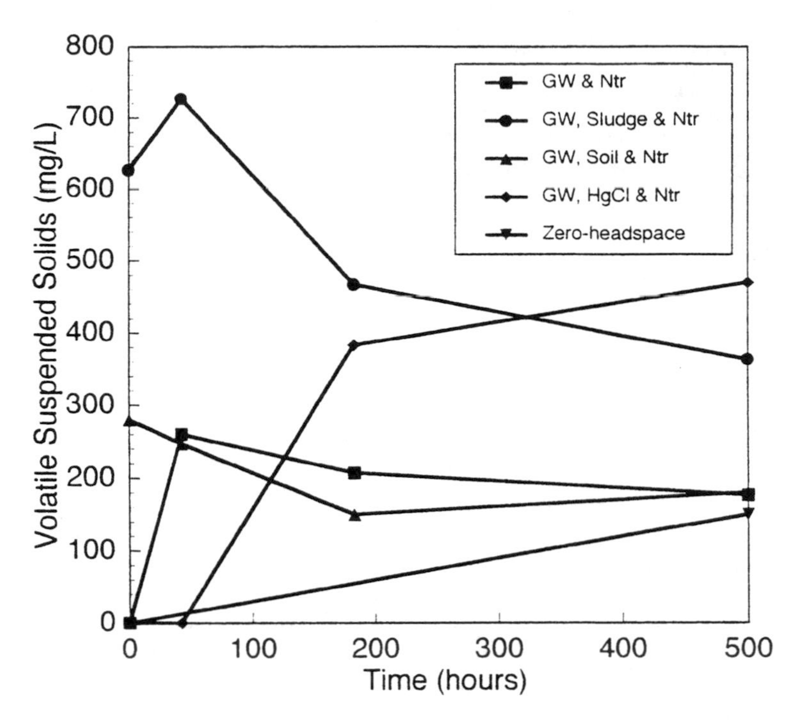

FIGURE 5. Change in volatile suspended solids content during treatment.

points. The results indicate that the activated sludge augmented treatment consumed more than two times more oxygen than the other biologically active treatments; however, target compound loss was not enhanced as shown in Figure 3 and Table 3. This treatment represents an unfavorable alternative because much more oxygen was required to achieve the same result. Oxygen consumption by the treatment augmented with DEHP degraders was slightly greater than the unaugmented groundwater treatment. The effectiveness of the mercury treatment for inhibiting biological activity was demonstrated by the virtual absence of oxygen consumption in this treatment.

Inorganic Carbon Production. The production of inorganic carbon (CO_2) complements the consumption of oxygen. However, absolute comparisons of oxygen consumed and CO_2 produced are difficult because some organic carbon is converted into biomass rather than being mineralized. Table 5 shows the amount of oxygen consumed, the total inorganic carbon (TIC) (CO_2) generated, and the TOC consumed during the study. Observed oxygen consumption was slightly higher than expected based on TIC produced and TOC consumed for the groundwater treatment and the DEHP degrader-augmented treatment; however, the slight difference in observed and expected oxygen consumption suggests that both treatments efficiently mineralized TOC. These results are in

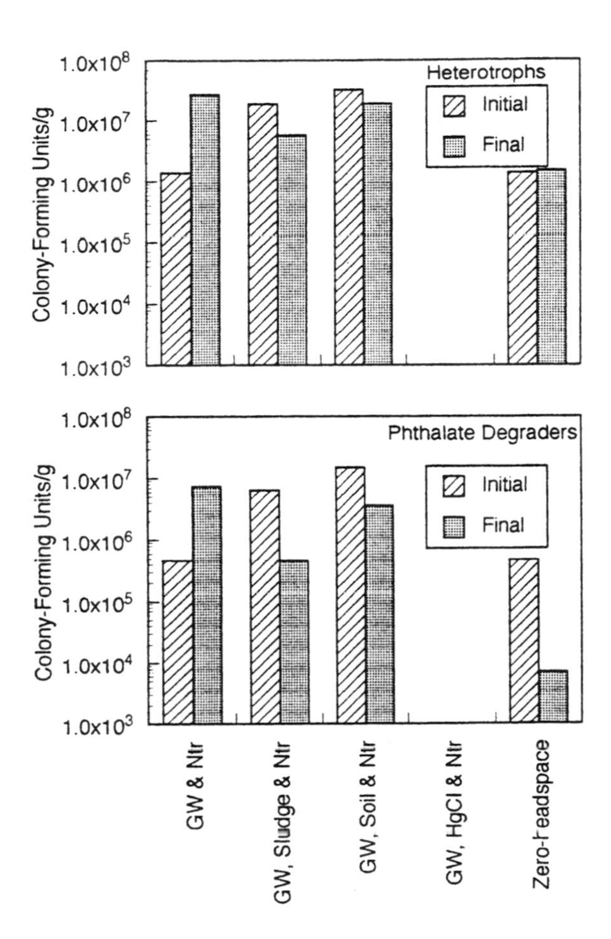

FIGURE 6. Change in microbial density in each treatment.

contrast to those observed for the activated sludge treatment which was much less efficient in mineralizing TOC.

Microbial Density. Figure 6 tracks the microbial density of heterotrophs and phthalate-degrading bacteria at the start of treatment and at the end. The microbial density in the groundwater and nutrient treatment increased during treatment. The microbial density of the activated sludge and the DEHP degrader-amended treatments decreased during treatment. The abiotic control contained no bacteria and the zero-headspace treatment had a low stable heterotroph population. The DEHP degrader density decreased during incubation.

Total suspended solids (TSS) and VSS analyses were performed on each treatment. The average VSS for each treatment is shown in Figure 5. TSS showed similar trends (data not shown). Solids analyses generally confirmed the microbial

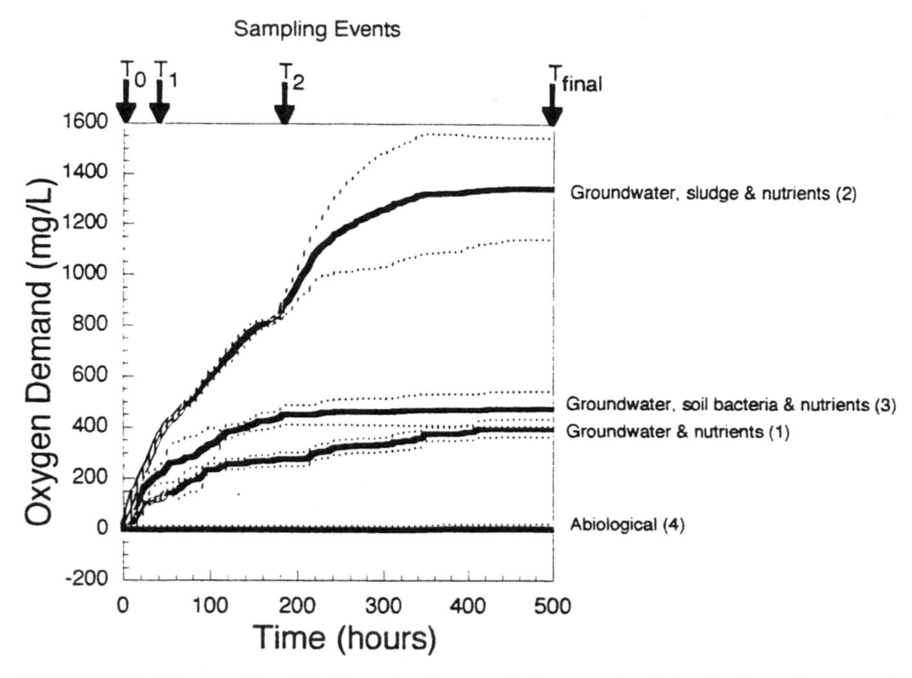

FIGURE 7. Mean microbial respiration and the standard deviation about each treatment. Mean values are indicated by heavy solid lines. Standard deviations are indicated by thin broken lines.

density determinations shown in Figure 6. However, the solids data for the abiotic control indicated an increase in TSS and VSS. This was due to a visibly noticeable precipitate caused by the presence of 500 mg/L $HgCl_2$. Precipitates were not observed in any other treatment.

Volatile solids for the groundwater and phthalate degrader-amended groundwater treatments had equalized by the end of the study. This observation suggests that the addition of extra bacteria will not have a long-term effect on the native population density. Similarly, the VSS content of the activated sludge treatment gradually fell after the first few hours of treatment.

Nutrient Utilization and pH Changes. Ammonia and orthophosphate concentrations in each treatment were examined. No appreciable reduction in nutrient concentration occurred during treatment.

Changes in the system pH were measured during treatment (data not shown). The abiotic control and the zero-headspace control had an essentially constant pH. The activated sludge treatment displayed an obvious acidification changing from an initial value of 6.8 to 5.2. Acidification may be a result of the extra organic carbon introduced into the system with the activated sludge. The other biologically active treatments became slightly alkaline.

TABLE 5. Respiration and mineralization of organic carbon.

Parameter	GW & Ntr	GW, Sludge & Ntr	GW, Soil & Ntr	GW, HgCl & Ntr	Zero -headspace
Total Oxygen Consumed (mg/L)	397	1344	474	13	ND
TIC Evolved (mg/L)	117	220	140	64	ND
TIC Initial (mg/L)	56	57	58	57	57
TIC Generated (mg/L)	61	164	82	7	ND
TOC Consumed (mg/L)	55	36	62	NA	0
Expected O2 based on TIC (mg/L)	325	874	437	25	ND
Expected O2 based on TOC (mg/L)	293	194	329	NA	ND

NA, not applicable

ND, not determined

CONCLUSIONS

The treatability study determined that native microbes possess the necessary metabolic capability to degrade DEHP, ethylbenzene, and xylene. The preliminary evaluation of site characteristics indicated that the site is acceptable for in situ aquifer bioremediation. Nutrient addition and oxygenation will be required. Nutrient compatibility and adsorption tests suggested marginal site characteristics for these parameters; however, since they have been identified, proper design of the treatment system can overcome potential problems before they cause operational difficulties in the field. Oxygenation of the aquifer will be challenging; however, it appears to be possible using hydrogen peroxide.

The groundwater biotreatability study indicated efficient biological removal of target compounds using indigenous microorganisms. The biodegradation of these compounds was demonstrated in bench-scale batch studies; however, biodegradation has not been determined under in situ conditions where complex interactions among water, soil, ions, organic compounds, and microbes exist. Results indicated successful biodegradation of target compounds. More than 99.9% of the ethylbenzene was removed from all biologically active treatments within 43 hours; more than 99.9% of the xylene was removed from the activated sludge treatment within 43 hours. Xylene removal from the two biologically active groundwater treatments proceeded at a slower rate compared to the sludge-amended treatment with approximately 50% removed after 43 hours. Xylene was nondetectable (<6.6 µg/L) in all biologically active treatments after 183 hours. Appreciable loss of ethylbenzene (approx. 80%) and xylene (approx. 75%) was observed in the abiotic control. This loss was attributed to volatilization.

Information on the biodegradation of DEHP in environmental samples is limited. This study provides degradation parameters that can be used as benchmarks for evaluating other treatability studies and potential field applications.

DEHP was biodegraded with a half-life of approximately 60 to 70 hours in the treatments containing site microbes. DEHP removal efficiency was 99.7%. Activated sludge was much less efficient in DEHP biodegradation with a calculated DEHP half-life of 176 hours and a removal efficiency of 86%. Virtually no DEHP was lost from the abiotic control. Microbial enumerations, VSS content, and target compound loss generally indicate that the addition of microbes to the groundwater does not provide a long-term benefit. These results suggest that DEHP can be biodegraded in situ using microorganisms indigenous to the site provided that nutrients and oxygen can be efficiently delivered to the impacted area of the aquifer.

REFERENCES

Clesceri, L. S., A. E. Greenberg, and R. R. Trussell (Eds.). 1989. *Standard Methods for the Examination of Water and Wastewater*, 17th ed. American Public Health Assoc., Washington, DC.

Howard, P. H., R. S. Boethling, W. F. Jarvis, W. M. Meylan, and E. M. Michelanko. 1991. *Handbook of Environmental Degradation Rates*, pp. 448-449. Lewis Publishers, Inc., Chelsea, MI.

Sokal, R. R., and F. J. Rolfe. 1981. *Biometry: The Principles and Practice of Statistics in Biological Research*, 2nd ed. W. H. Freeman and Co., San Francisco, CA.

Subba-Rao, R. V., H. E. Rubin, and M. Alexander. 1982. "Kinetics and Extent of Mineralization of Organic Chemicals at Trace Levels in Fresh Water and Sewage." *Appl. Environ. Microbiol.* *43*: 1139-1150.

Verschueren, K. 1983. *Handbook of Environmental Data on Organic Chemicals*, pp. 575-578. Van Nostrand Reinhold Company, Inc., New York, NY.

IMMUNOLOGICAL TECHNIQUES AS TOOLS TO CHARACTERIZE THE SUBSURFACE MICROBIAL COMMUNITY AT A TRICHLOROETHYLENE-CONTAMINATED SITE

C. B. Fliermans, J. M. Dougherty, M. M. Franck, P. C. McKinzey, J. E. Wear, and T. C. Hazen

ABSTRACT

Effective in situ bioremediation strategies require an understanding of the effects pollutants and remediation techniques have on subsurface microbial communities. Therefore, detailed characterization of a site's microbial community is important. Groundwater samples were collected from a trichloroethylene (TCE)-contaminated site before and after in situ air stripping, bioventing, and methane injection bioremediation techniques were used. Subsamples were processed for heterotrophic plate counts, acridine orange direct counts (AODC), community diversity, direct fluorescent antibody (DFA) enumeration for selected bacteria, and Biolog® evaluation of enzyme activity. The presented data describe the use of specific techniques to evaluate bacterial communities in groundwater from both a synecologically and an autecologically perspective with regard to the subsurface treatments during bioremediation. AODCs were orders of magnitude higher than plate counts and remained relatively constant with depth except for slight increases at the surface depths and the capillary fringe of the vadose zone. Nitrogen-transforming bacteria, as measured by serospecific DFA, were significantly affected both by the in situ air stripping and the methane injections. Microbial utilization of selected organic compounds was measured by Biolog® technology, and differed among the wells in relationship to their stimulation both by air and methane. The complexity of subsurface systems makes the use of selective monitoring tools essential.

BACKGROUND

The Savannah River Site (SRS) provides a unique opportunity to blend collaborative partners from industry, academia, and government in order to execute extensive research projects. The Gas Research Institute (GRI) in collaboration with Savannah River Technology Center (SRTC) has been funding research and development of a methanotrophic treatment process for TCE-contaminated groundwater for the past 4 years. During one such activity, the Integrated Demonstration Project, indigenous microorganisms were stimulated through the use of dual horizontal wells to degrade trichloroethylene (TCE), tetrachloroethylene (PCE), and their daughter products in situ by the addition of gaseous nutrients to the contaminated zone. Biodegradation is a highly attractive remediation strategy because contaminants are destroyed, not simply transferred to another location or immobilized. Bioremediation has been shown to be among the most cost effective technologies where applicable (Radian 1989; Legrand 1993). The application of horizontal well technology to bioremediation has formed the foundation of the SRS Integrated Demonstration Project and provided significant advantages over vertical wells and conventional bioremediation techniques. The increased surface area provided by the horizontal wells has allowed better delivery of nutrients and a more efficient recovery of gas and water, while minimizing formation plugging (Looney & Kaback 1991).

Because air/methane mixtures have been shown to stimulate selected members of the indigenous microbial community that have the capability to degrade TCE (Little et al. 1988; Vogel & McCarty 1985; Wilson & Wilson 1985), the principal nutrients supplied via the horizontal wells were methane (1 to 4%) and nitrogen in air. Although the lower horizontal well provided an efficient delivery of gas throughout the contaminated region, a vacuum was applied to the upper well located in the vadose zone. Such a combination of wells encouraged air/methane movement through the upper saturated and lower vadose zones while inhibiting the spreading of the organic plume (Looney & Kaback 1991). An extensive monitoring program using existing monitoring wells and soil borings has served to determine the biological response in the soil, sediment, and groundwater systems following the injection of air/methane. Data from Phase I (air injections), Phase II (1% and 4% methane injections), and Phase III (pulsing of 4% methane) of the Integrated Demonstration Project have illustrated the effectiveness of in situ bioventing and methane injections for the bioremediation of TCE and PCE (Hazen 1992).

This manuscript describes an affect of the injection perturbations on selected microbiological components in the groundwater influenced by the horizontal wells. The data are derived from the use of species-specific and serospecific fluorescent antibodies to detect and enumerate selected microbial populations. Additionally, the analyses included the use of Biolog® technology to determine the metabolic capability of the microbial populations present in the groundwater with respect to 95 different carbon and energy sources.

MATERIALS AND METHODS

Soil, sediment, and groundwater samples were collected aseptically as previously described (Fliermans & Balkwell 1989; Fliermans & Hazen 1990; Phelps et al. 1989). Groundwater samples were collected from 12 monitoring wells on a bimonthly schedule over a 12-month period. Wells were pumped according to established protocol (WSRC 1991) in order to stabilize chemical, physical, and biological parameters before a 4 liter samples were collected. Viable bacterial densities were determined by plating samples on laboratory medium as described by Balkwell and Ghiorse (1985). Total bacterial densities in groundwater samples were measured by direct epifluorescence microscopy using AODC techniques of Balkwell & Ghiorse (1985).

A modified direct immunofluorescent technique using serospecific polyvalent fluorescent antibodies (Fliermans et al. 1992) was employed to measure selected bacterial strains in groundwater samples which were concentrated 250-fold by centrifugation. These antibodies were prepared as described by Bohlool & Schmidt (1968) and Fliermans et al. (1974). Aliquots (10 μL) of the groundwater samples were heat-fixed onto toxoplasmosis slides, layered with blocking fluid, stained with the specific antibodies, washed in phosphate buffered saline, and viewed by epifluorescence microscopy.

Selected bacterial strains were chosen because they represented some of the microbial structural components of the ecosystem being investigated. Species-specific direct fluorescent antibodies were prepared against the major microbial communities involved in nitrogen transformation in soils and groundwater, i.e., *Azotobacter, Nitrosolobus, Nitrosomonas, Nitrobacter,* and *Bradyrhizobium,* as well as those species of *Nitrosomonas* involved in the production of enzymes capable of degrading TCE; a bacterium that is pathogenic, but widely distributed in nature (*Legionella pneumophila*); a widely distributed bacterium associated with our systems and involved in iron transformations (*Thiobacillus ferrooxidans*); and two organisms that have been involved in the destruction of TCE as well as transformation of methane in the laboratory (*SRL-MIIF, Methanobacterium*). Each of these organisms have been isolated from SRS habitats except for *Bradyrhizobium japonicum.*

More defined characterization for the bacteria are as follows:

- Azotobacter chroococcum is a free-living heterotrophic, nitrogen-fixer that lives under aerobic conditions.
- *Nitrosolobus multiformis* and *Nitrosomonas europea* are morphologically distinct chemolithotrophs that obtain their energy from oxidizing ammonia to nitrite while fixing CO_2 as their sole carbon source.
- *Nitrobacter agilis* and *Nitrobacter winogradskyi* are chemolithotrophs that secure their energy by oxidizing nitrite to nitrate while fixing CO_2 as their sole carbon source.
- *SRL-MIIF* is a Type II methanotroph isolated from the SRS with trichloroethylene (TCE)-degrading capabilities.
- *Methanobacterium formicicum* is a heterotrophic, strict anaerobe that produces CH_4 from organic compounds such as acetate and cysteine.
- *Thiobacillus ferrooxidans* grows as a chemolithotroph that gets its energy by oxidizing ferrous iron to ferric iron while fixing CO_2 as its carbon source.

- *Legionella pneumophila* Serogroup 1 is a human pathogen and the major etiologic agent of Legionnaires' Disease, as well as an aquatic and terrestrial bacterium.
- *Bradyrhizobium japonicum* is a symbiotic nitrogen-fixer that is present in association with soybean nodulation and occupies ecological niches near the soil surface.

Unconcentrated groundwater samples were pipetted (150 µL/well) into each microtiter well of triplicate GN Biolog® plates and incubated at the in situ groundwater temperature of 23°C. The optical density of each well was read at 590 nm after 3 weeks of incubation, and the reduced color in each of the wells was recorded. After incubation each plate was read using a Biolog® plate reader, and the optical density of the tetrazolium dye was recorded. The Biolog® automated plate reader was programmed to zero the instrument based on the optical density color of the reference well (A-1) in the Biolog® plate. Therefore, optical density readings determined at each inoculum pattern were significantly above the threshold level. The data were expressed as optical density for each of the 95 organic compounds tested, placed into reactive groups as previously described (Gorden et al., in press), and evaluated. The enzyme activity associated with the utilization of each of the compounds was expressed in terms of the location of the sampled well along with the in situ perturbations occurring at the time of sampling.

Data were analyzed by analysis of variance using JMP, version 2 (SAS Institute Inc., Cary, NC). Heterogeneity of variances was reduced using Log(Y+1) transformation. Pairwise comparisons were made using the Tukey method because it provided narrower confidence limits than methods used for general contrasts. The Kramer adjustment allowed for the unequal sample sizes tested. For all tests a probability level of 0.05 was assumed to be the critical level of significance.

RESULTS

The locations of the monitoring wells used in the Integrated Demonstration Project and sampled during this investigation are shown in Figure 1. Generally the lower numbered wells, MHT-1 through MHT-7, are more affected by the changes being incorporated into the horizontal wells, whereas the higher numbered wells, MHT-8 through MHT-11, are less affected. Moreover these higher numbered wells are affected at a latter time period as it took longer for the perturbation to reach these wells. Much of the physical/chemical data collected and analyzed from MHT-7 suggests that this well is clearly an outlier. The microbiological data also reflect this assessment. This may be a result of the way the well was physically developed rather than reflective of the groundwater at that location.

Data from the groundwater sampling reported here were acquired using the DFA and Biolog® techniques to assess selected and general bacterial dynamics in the groundwater, respectively. The DFA technique provided the collection of autecological data necessary to determine the population densities of selected nitrogen-cycling bacteria in the subsurface as well as other selected microbial

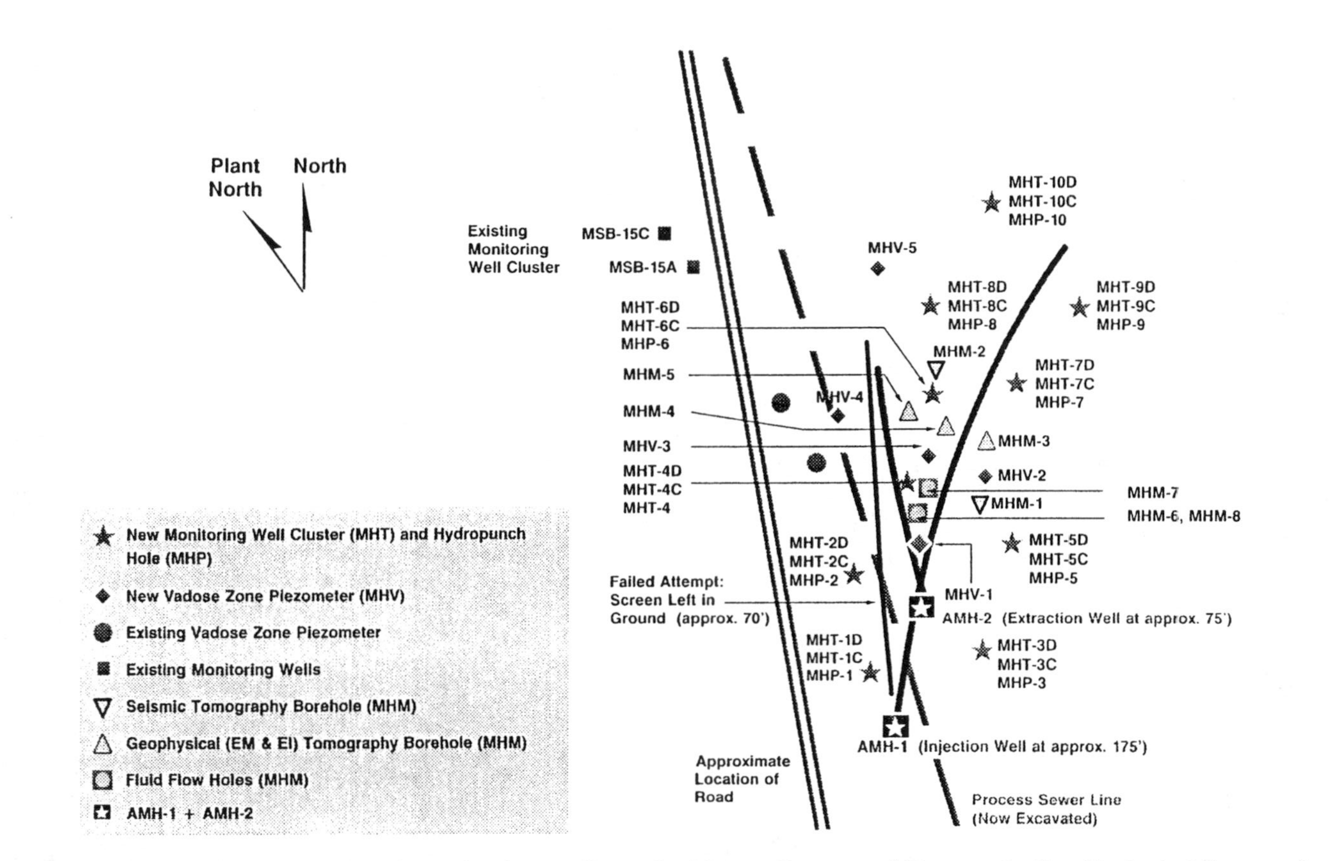

FIGURE 1. Location of horizontal and monitoring wells at the M-area Integrated Demonstration Project at Savannah River Site.

populations that were affected by the in situ bioventing and methane injection perturbations. Moreover, the Biolog® technique provided synecological data on the metabolic capability of the groundwater microbial community without regard to a particular organism.

The data in Figure 2 show the statistical analyses of DFA concentrations of *Azotobacter chroococcum* for all 12 wells during the perturbation regimes. Densities of *Azotobacter chroococcum* were significantly higher ($p<0.05$) during air and 1% methane injection than at any other time. The data indicate that in wells most affected by the perturbations the densities of *A. chroococcum* were greatest just after the beginning of the air injection and fell off rather dramatically after the start of the methane injections. Although the *Azotobacter* populations were highly variable, bioventing (which consisted of air injection in the bottom horizontal at 200 scfm along with vacuum extraction in the upper horizontal well at 240 scfm) provided a nitrogen source that appeared to encourage the growth of the free-living nitrogen-fixers. Such a stimulation is consistent with the DFA data. Subsequently the addition of 4% methane significantly reduced the densities of the *A. chroococcum* population, and those populations have been able to recover to the pretreatment levels during the pulsing regime of 4% methane and air.

The data in Figure 3 show the statistical analyses of DFA concentrations of *Nitrosomonas europea* for all 12 of the wells during the perturbation regimes. Densities of *Nitrosomonas europea* were significantly lower during injection of 4% methane than during other pretreatment or air injection. Once nitrogen is reduced during nitrogen fixation, the first step in the nitrification process is the oxidation of ammonia to nitrite. Thus if the population densities of the free-living nitrogen-fixers were to increase, it is reasonable to assume that the bacterial populations of the chemoautotrophic nitrifiers might increase as well. The data for the 12 sampled wells indicated that the densities of *N. europea* were greatest just after the start of air injection and fell off dramatically after the start of the

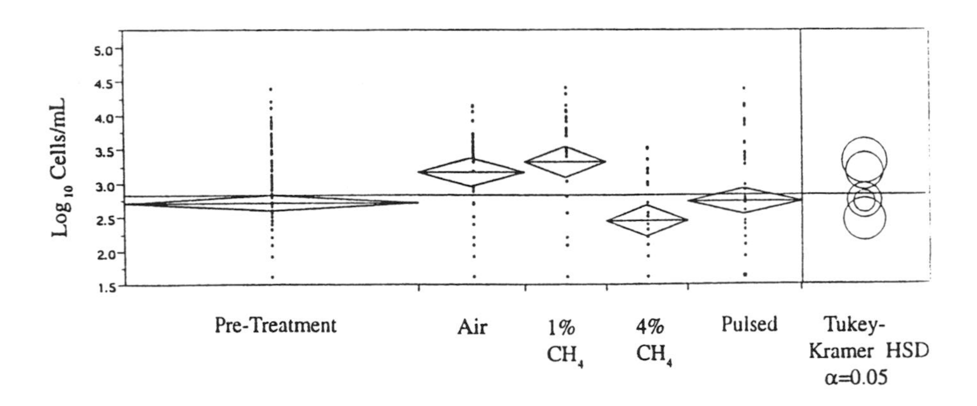

FIGURE 2. Density and statistical analyses of *Azotobacter chroococcum* in groundwater samples from 12 wells during in situ remedial perturbations.

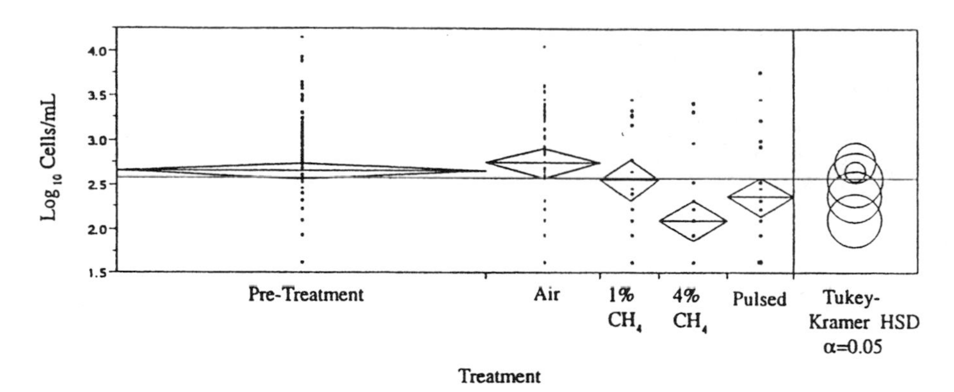

FIGURE 3. Density and statistical analyses of *Nitrosomonas europea* in groundwater samples from 12 wells during in situ remedial perturbations.

methane injections. The populations have as yet not been able to recover to the pretreatment levels, even with the pulsing regime.

Nitrosolobus multiformis, a lesser studied nitrifier, was never detected in very large concentrations in any of the samples and appears to play a rather minor role, numerically speaking, in these habitats. Densities of *N. multiformis* did not differ significantly between injection regimes.

The data in Figure 4 show the statistical analyses of DFA concentrations of *Nitrobacter agilis* and *Nitrobacter winogradskyi* for all 12 of the wells during the perturbation regimes. The DFA for these two bacteria were combined for the analyses, since they are so very much alike and may in fact be serotypes of *N. winogradskyi*. The DFA concentrations of *Nitrobacter agilis* and *Nitrobacter*

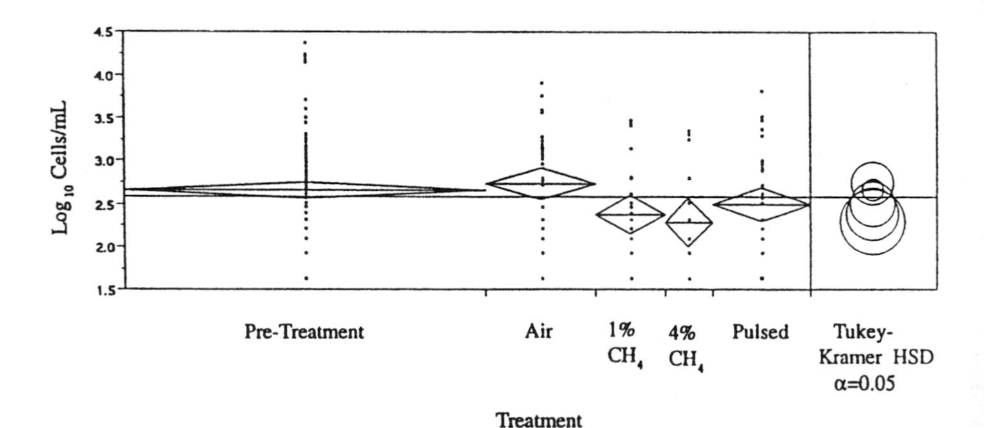

FIGURE 4. Density and statistical analyses of *Nitrobacter agilis* and *Nitrobacter winogradskyi* in groundwater samples from 12 wells during in situ remedial perturbations.

winogradskyi appear to follow a similar pattern as the ammonia-oxidizing bacteria. Once the methane injections began, the densities of the nitrite-oxidizing bacteria in these wells declined dramatically to the point where the concentrations of 70% of the samples were below detectable limits for *N. agilis* and *N. winogradskyi*. Densities of *N. agilis* and *N. winogradskyi* did not differ significantly between injection regimes when all the wells were grouped together for analyses at the $p<0.05$ level.

The data in Figure 5 show the statistical analyses of DFA concentrations of *SRL-MIIF* in each of the 12 wells. The data indicate that the bacteria were observed throughout the sampling perturbation period. Of the 12 samples collected before methane injection, only 25% of the samples showed detectable levels of the bacterium whereas 36% showed its presence after the methane injections had begun. Densities of *SRL-MIIF* were not significantly different between injection regimes when all the wells were grouped together for analyses at the $p<0.05$ level. In the overall analyses of the 12 wells it appears that this particular organism was not actively stimulated by the injection of methane or any of the perturbation regimes.

The data in Figure 6 show the statistical analyses of DFA concentrations of *Methanobacterium formicicum* in all 12 of the wells. Densities of *Methanobacterium formicicum* were significantly higher ($p<0.05$) prior to treatment and during air injection than during pulsed injection but were not significantly different from densities during 1% or 4% methane injection. There is a trend in the data that indicates that the injection of 1 and 4% methane decreased the densities of this bacterium. Statistically the densities of *Methanobacterium formicicum* were significantly lower ($p<0.05$) during pulsed injection of 4% methane than at any other time.

The data in Figure 7 demonstrate the statistical analyses of DFA concentrations of *Thiobacillus ferrooxidans* in the sampled wells. The data indicate that the densities

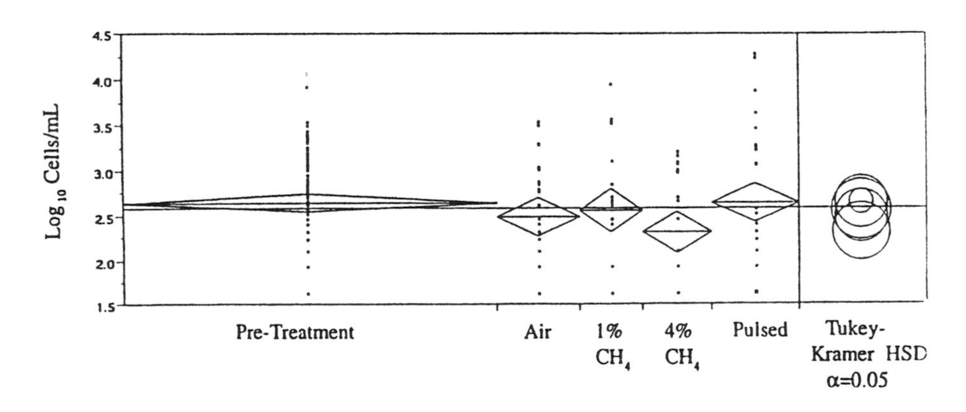

FIGURE 5. Density and statistical analyses of *SRL-MIFF* in groundwater samples from 12 wells during in situ remedial perturbations.

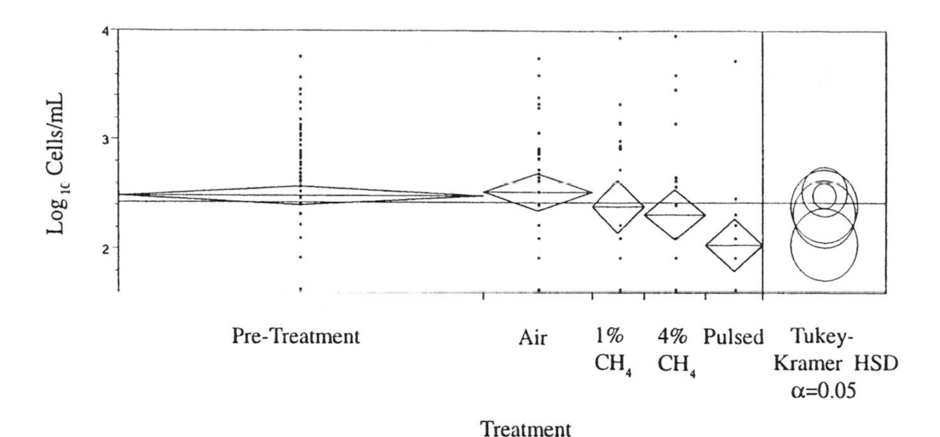

FIGURE 6. Density and statistical analyses of *Methanobacterium formicicum* in groundwater samples from 12 wells during in situ remedial perturbations.

of this particular strain of *T. ferrooxidans* were significantly higher ($p<0.05$) during 1% methane injection than at any other time. No significant difference was seen among the other treatments.

Finally, The data in Figure 8 show the statistical analyses of DFA concentrations of *Legionella pneumophila* Serogroup 1 in each of the 12 wells. The data indicate that the perturbations did not enhance the densities of *L. pneumophila* Serogroup 1, but infact the densities of *L. pneumophila* Serogroup 1 were significantly higher ($p<0.05$) prior to treatment than during 1% methane, 4% methane or pulsed perturbations.

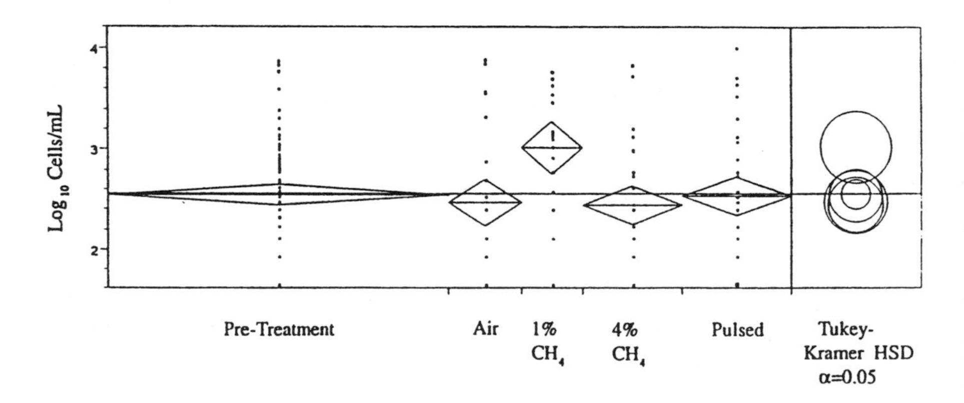

FIGURE 7. Density and statistical analyses of *Thiobacillus ferrooxidans* in groundwater samples from 12 wells during in situ remedial perturbations.

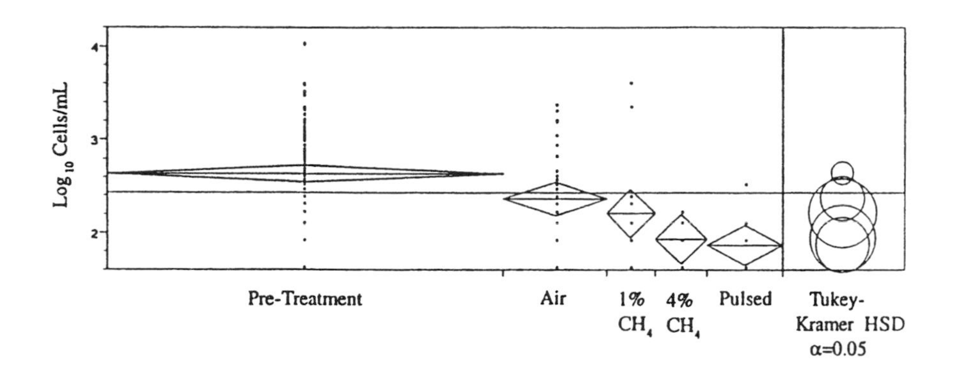

FIGURE 8. Density and statistical analyses of *Legionella pneumophila* serogroup 1 in groundwater samples from 12 wells during in situ remedial perturbations.

Bradyrhizobium japonicum was not found in any of the groundwater samples, although it was observed in the shallow soil samples from the site (data not shown).

Biolog® plates were used to determine the metabolic versatility and activity of the microbial consortia present in the groundwater samples collected from the wells. Groundwater samples from each sampling period were inoculated into triplicate GN Biolog® test plates. The optical density of the tetrazolium reactions, a indication of the ability of the groundwater microbial community to metabolize the 95 individual compounds listed in Table 1, were measured and plotted with respect to the perturbations made in the horizontal wells.

The tetrazolium reactions were grouped based on their chemical relationships and plotted. The data in Figure 9a and b show examples of the reaction of the microbial community in each of the 12 wells to the utilization of the tested carbohydrates (28) and phosphorylated compounds (3), respectively over the duration of the Integrated Demonstration Project. There is a clear difference in the ability of the microbial consortia among the 12 wells to use the various tested compounds. Each of the 95 individual compounds is currently being evaluated for significance of utilization both among and between wells and the various treatments.

DISCUSSION

Each of the two basic approaches to the study of microbial systems in nature, the synecological and the autecological approaches, has its merits and shortcomings. In the synecological approach, the entire microbial community is investigated in relationship to a given habitat. In this approach there is little concern about the type of microorganisms involved in the functioning of the habitat, but rather focus is given to the extent of microbial processing that occurs. This approach provides the investigator with process-oriented data and information

TABLE 1. Sole carbon sources present in Biolog® GN microtiter plates.

Carbohydrates	**Carboxylic Acids**	**Amino Acids**
N-Acetyl-D-galactosamine	Acetic acid	D-Alanine
N-Acetyl-D-glucosamine	*cis*-Aconitic acid	L-Alanine
Adonitol	Citric acid	L-Alanyl-glycine
L-Arabinose	Formic acid	L-Asparagine
D-Arabitol	D-Galactonic acid lactone	L-Aspartic acid
Cellobiose	D-Galacturonic acid	L-Glutamic acid
i-Erythritol	D-Gluconic acid	Glycyl-L-aspartic acid
D-Fructose	D-Glucosaminic acid	Glycyl-L-glutamic acid
L-Fucose	D-Glucoronic acid	L-Histidine
D-Galactose	α-Hydroxybutyric acid	Hydroxy-L-proline
Gentiobiose	β-Hydroxybutyric acid	L-Leucine
α-D-Glucose	γ-Hydroxybutyric acid	L-Ornithine
m-Inositol	ρ-Hydroxyphenylacetic acid	L-Phenylalanine
α-Lactose		L-Proline
Lactulose	Itaconic acid	L-Pyroglutamic acid
Maltose	α-Keto butyric acid	D-Serine
D-Mannitol	α-Keto glutaric acid	L-Serine
D-Mannose	α-Keto valeric acid	L-Threonine
D-Melibiose	D,L-Lactic acid	D,L-Carnitine
β-Methylglucoside	Malonic acid	γ-Aminobutyric acid
D-Psicose	Propionic acid	
D-Raffinose	Quinic acid	**Amides**
L-Rhamnose	D-Saccharic acid	Succinamic acid
D-Sorbitol	Sebacic acid	Glucuronamide
Sucrose	Succinic acid	Alaninamide
D-Trehalose		
Turanose	**Aromatic Chemicals**	**Amines**
Xylitol	Inosine	Phenylethylamine
	Urocanic acid	2-Aminoethanol
Esters	Thymidine	Putrescine
Mono-methylsuccinate	Uridine	
Methylpyruvate		**Alcohols**
	Polymers	2,3-Butanediol
Phosphorylated Chemicals	Glycogen	Glycerol
	α-Cyclodextrin	
D,L-α-Glycerol phosphate	Dextrin	**Brominated Chemicals**
Glucose-1-phosphate	Tween 80	Bromosuccinic acid
Glucose-6-phosphate	Tween 40	

for an overall picture of the functioning of the habitat. Such data are highlighted by information on the cycling of nutrients through an ecosystem and the transformation of chemical components of the system by microorganisms intrinsic to the ecosystem without regard to the microorganisms involved. Such investigations often rely on the mineralization or transformation of particular compounds which have been characteristically labeled, thus providing a marker for transformation assessment. The markers can thus be followed during various stages of

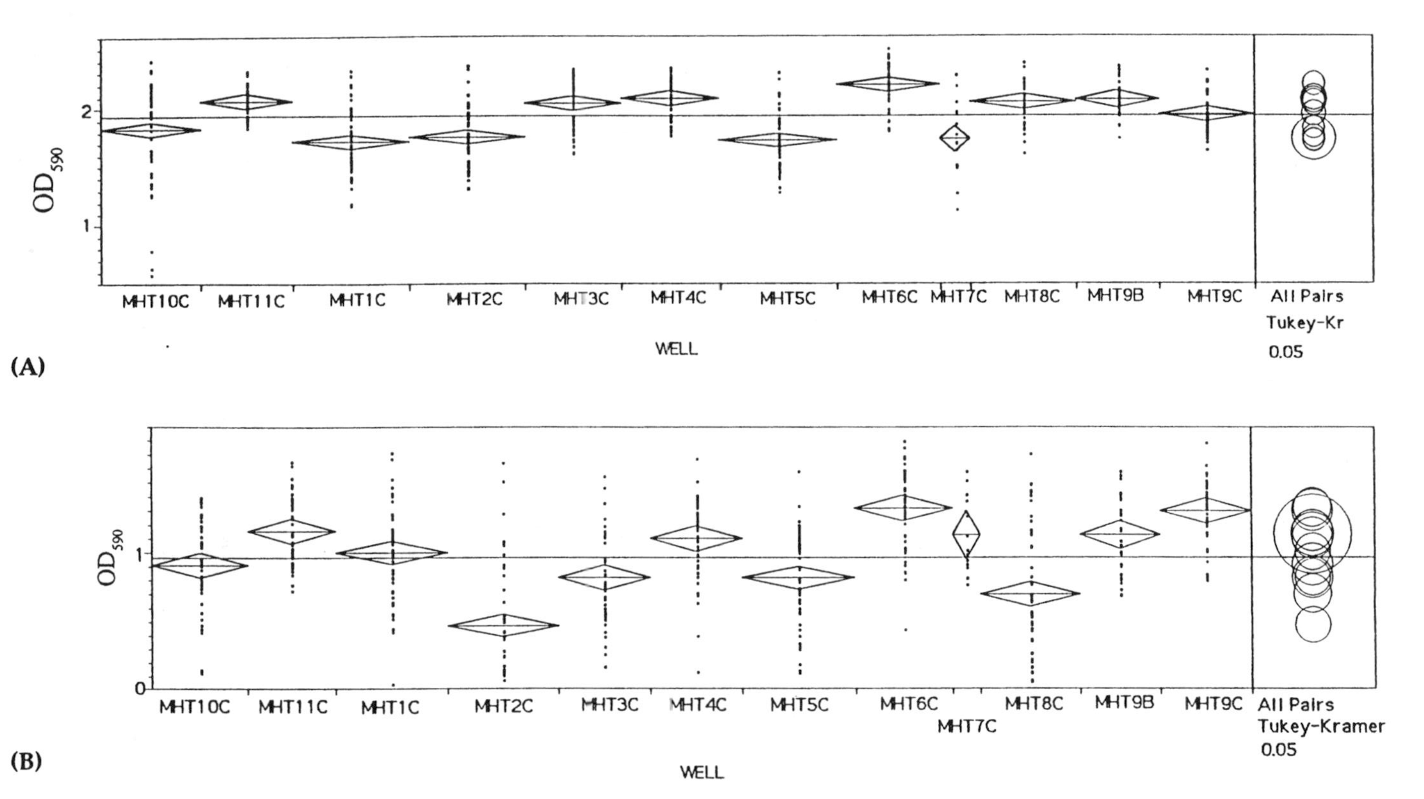

FIGURE 9. (A) Activity and statistical analyses of microbial community populations in the groundwater of all 12 wells with respect to 28 carbohydrate compounds. (B) Activity and statistical analyses of microbial community populations in the groundwater of all 12 wells with respect to 3 phosphorylated compounds.

incubation, degradation, and mineralization to enable measurements of the breakdown products, intermediates, and rates.

Furthermore, the autecological approach focuses on a particular microorganism and asks questions with regard to that organism and how it functions in the studied habitat. The data and information gained from this perspective are exceedingly valuable, particularly when dealing with an organism that is potentially pathogenic or is greatly involved in a functional aspect of the ecosystem. Such an approach allows one to pursue questions about the organism, its particular functioning in a given ecological niche, and the potential control of the organism that could not otherwise be made. The disadvantage of the autecological approach is that it requires specialized techniques for the identification and assessment of the bacterium. Few, if any, bacteria can be reliably characterized by their morphological structure even at the genus level. Many bacteria are pleomorphic, depending on their nutritional status, and thus morphological considerations are not acceptable criteria for assessment.

The apparent lack of techniques for autecological investigations is overcome through the use of specific serological and genetic probes for identifying the bacterium to a particular species and subgroup without culturing the organism. Under appropriate conditions the DFA technique can evaluate a bacterium as to its density, distribution, viability, state of physiological health, and transformation of selected isotopically labeled compounds using the techniques of epifluorescent microscopy, cytochrome activity (Fliermans et al. 1981), fluorescent probes (Rodriguez et al. 1992), and microautoradiography (Fliermans & Schmidt 1975), respectively.

Our discussions here center on the autecological approach to the ecology of selected bacteria. Specific polyvalent and/or monoclonal fluorescent antibody probes have been developed to view, enumerate, and assess the role of these bacteria in complex ecological environments that are neither sterile, monoculture, nor easily sampled. Immunofluorescent techniques initially were used in diagnostic medical microbiology, and only in the last few decades have they been employed in the areas of microbial ecology for agricultural, terrestrial, and aquatic microbiology investigations. More recently, DFA has been used to define the ecology of microorganisms associated with the degradation of toxic and hazardous wastes.

As with any technique there are positive and negative features. The DFA technique is capable of visualizing a specific bacterium in very complex habitats such as soils, metal corrosion piping, sediments, vascular plant material, coal, sewage sludge, or virtually any habitat. If one employs the DFA technique for any of the defined bacteria, as little as a single specific cell per milliliter of liquid sample can be visualized. This detection limit could be pushed even further given a larger sample size and a greater concentration factor, but the value of such is not apparent. DFA prepared with serospecific polyvalent antibodies against whole cells or bacterial cell walls are the most useful tools to date for quantifying and visualizing a bacterium in its natural habitat. Using polyvalent antibodies, the DFA can be made specific enough to react with single species or serogroups of any genus without cross-reacting with nonhomologous organisms outside the serogroup or with other species or genera of bacteria.

The DFA approach also can be used with monoclonal antibodies, although monoclonal antibodies have relatively little use in an in situ setting. This lack of functionality is owed to the high specificity of the antibody allowing only a very limited number of organisms in one serotype to react with the DFA prepared from monoclonal antibodies. Although another serotype of the same species may be present in the habitat, if it is not the one reacting with the specific monoclonal antibody, it will not be detected and thus the number of a particular bacteria will be greatly underestimated. Monoclonal applications are valuable in the typing of pure cultures of specific groups of bacteria within the genus, species, and serogroup of the homologous strain that may have come from a selected environment and caused a specific clinical or environmental response. In these circumstances monoclonals can be used to accurately determine the similarity of the organisms in both diagnostic and forensic investigations.

A second limitation of the DFA technique is that the DFA conjugate stains intact organisms, whether alive or dead. If the organism being detected were lysed either by natural events in their habitat or through human-induced events (i.e., addition of biocides), then cell wall debris caused by such events will stain and be observed microscopically. While at first this may seem to be a problem, in the real world it is less of a problem because bacteria are an excellent source of nutrients being at the lowest end of the food chain and, once dead, they are no longer capable of keeping the "wolves" from the door and are readily lysed and serve as food for other microbial populations.

Thus, in nonsterile aquatic or terrestrial habitats, the dead bacterium, is not readily detected by the DFA technique because the organism does not appear to remain intact for an extended period of time. This phenomenon has been observed in sterile soils and sterile aquatic systems. Bohlool & Schmidt (1968) placed killed DFA-positive organisms into a sterile soil system, and the DFA-positive bacteria remained intact for extended periods of time (days). In nonsterile systems, the DFA-positive organisms were observed as debris within a few hours. Thus, it is likely that the intact bacteria observed in the habitat are viable, but may be nonculturable as Hussong et al. (1987) have demonstrated for *Legionella*.

CONCLUSIONS

The data from these autecological investigations associated with the Integrated Demonstration Project indicate that selected microbiological communities can be efficiently, effectively, and accurately monitored using immunofluorescent techniques. It is necessary to note that, although these immunofluorescent probes are specific for the microbial constituents, they may or may not be representative of the most numerous or metabolically active bacteria that are present in these subsurface environments. Only one organism, *Bradyrhizobium japonicum*, has not been isolated from the Integrated Demonstration Site or the SRS complex. The other bacteria studied in this investigation are present at the site, are serologically homologous, and are likely to be the bacteria associated with the various transformations, particularly for nitrogen cycling.

The data for each of the bacteria studied provide some interesting observations. The data in Figure 2 for *Azotobacter chroococcum* indicate that the bacterium was stimulated by both air and 1% methane during the course of injections into the subsurface but was rather stongly inhibited by the 4% methane. During the pulsed perturbations the densities of *A. chroococcum* nearly recover to the pretreatment stages but not to those levels observed during the purging of air and 1% methane to the subsurface. This is probably somewhat reasonable since the bacterium is a free-living nitrogen-fixer and air contains high levels of nitrogen. Only when the methane concentration levels were enhanced to a constant 4% was *Azotobacter chroococcum* inhibited.

The data in Figure 3 for *Nitrosomonas europea* indicates that the organism was present more often before the methane injections than after. *N. europaea* is an ammonia-oxidizin g, Gram-negative chemolithotroph that produces nitrite from ammonia through a monooxygenase involvement. Suzuki et al. (1976) demonstrated that in *N. europaea*, ammonia oxidation was inhibited by CO, methane, and methanol. Thus the lowering of *N. europaea* densities during 4% methane injections would be compatible with the known physiology of the organism. Additionally, Hyman et al. (1988); Vannelli et al. (1990) and Rasche et al. (1991) have demonstrated the physiological versatility of *N. europaea* with its degradation of halogenated-aliphatic compounds, including trichloroethylene (TCE). Thus *N. europaea* may well be involved in situ with the degradation of TCE, but inhibited by the increase concentrations of methane as demonstrated by the DFA technique.

The data in Figure 4 represents the density and statistical analyses for *Nitrobacter agilis* and *Nitrobacter winogradskyi* for all 12 of the wells during the perturbation regimes. *Nitrobacter* spp. are Gram-negative organisms that grow as chemolithotrophs using CO_2 as the sole source of carbon while oxidizing nitrite to nitrate as an energy source. They represent the terminal step in nitrification while depending on the end product of *Nitrosomonas*. Thus it is not surprising that when *Nitrosomonas* declined during the methane perturbation, the decline of *Nitrobacter* would follow.

The response of *SRL-MIIF* to the remedial perturbations was rather uninteresting. *SRL-MIIF* was part of a microbial consortia that has been shown to actively degrade TCE aerobically (Fliermans et al. 1988), yet there was no significant change in the density of this organisms with any of the perturbations (Figure 5).

The data in Figure 6 represents the density and statistical analyses for *Methanobacterium formicicum*, which is a Gram-positive to variable bacterium strict anaerobe that oxidizes hydrogen and utilizes ammonia as its nitrogen source. Once the subsurface perturbations began, the densities of this particular organism declined rather precipitously. Although these subsurface systems are aerobic, anaerobic niches are readily maintained by the biological activity of heterotrophic organisms or the low diffusivity of oxygen through tight geological formations. The perturbations had a major component of air and may well have lowered the niches where *M. formicicum* was able to survive.

The data in Figure 7 represent the density and statistical analyses for *Thiobacillus ferrooxidans*, which is a Gram-negative bacterium capable of mixotrophic growth. It is rather common member of the soil bacteriology community. Only under perturbations of 1% methane did the densities of this organisms increase and then return to pretreatment levels after subsequent 4% and pulsed treatments.

Legionella pneumophila Serogroup 1 showed a considerable drop in cell density with all of the perturbations including air stripping alone. The data indicate that *L. pneumophila*, while part of the subsurface population, is not stimulated by any of the remediation techniques used in this investigation. Although *L. pneumophila* represents a very minor part of the pathogenic organisms in nature, it is necessary to establish that the in situ remediation techniques that were used do not stimulate this pathogenic portion of the microbial community.

The use of Biolog® GN plates during these remediation techniques provides the capability of evaluating microbial systems adapted to metabolizing selected chemicals as their sole carbon and/or energy source. In adapting the Biolog® system to groundwater sampling, the goal was not to provide a consistent level of microbial inoculum, but rather to provide a constant volume of groundwater with an indigenous bacterial density of ca. 10^4 to 10^5 cells of a variety of species to the Biolog® plates. Thus, the time of incubation of the Biolog® plates needs to be standardized for the particular groundwater, but generally it was 3 weeks at groundwater temperatures of 23°C.

Based on the results from the GN screening plates a variety of information was obtained. One is able to determine the best metabolite or group of metabolites that stimulate the microbial community present in the sampled groundwater. Because subsurface environments in a particular local are not microbiologically similar (Fliermans & Balkwell 1989), Biolog® information can be used to selectively stimulate the naturally occurring organisms at a particular location under the various perturbation regimes encountered.

Initial results from both the carbohydrate and phosphorylated compound utilization data demonstrated that each well had its own microbially active communities as measured by their utilization responses. The data from these two sets of compounds provide representation of the data currently being evaluated for each of the 95 compounds present in the GN Biolog® plates. Previous studies (Fliermans & Balkwell 1989; Fliermans & Hazen 1990) on the subsurface microbial populations at SRS have demonstrated the stimulation of microbial populations to nitrogen additions. This same phenomenon has been observed for selected amino acids, aromatics, esters, amides, amines, and polymers (in press).

Thus, these results suggest that Biolog® technology is a useful tool for screening bacterial isolates and consortia to determine their ability to survive, to metabolize, to degrade selected organic chemicals, and potentially to provide screening of groundwater systems for compounds that are useful in stimulating the microbial populations during bioremediation projects.

ACKNOWLEDGMENTS

The Gas Research Institute in collaboration with Savannah River Technology Center has been funding research and development of a methanotrophic treatment process for TCE-contaminated groundwater for the past 4 years. University and industry investigators funded by GRI have been integrated with the experience and expertise of several national laboratories (U.S. Department of Energy and U.S. Environmental Protection Agency) to provide the greatest experience and resource of bioremediation expertise ever assembled for any bioremediation demonstration.

REFERENCES

Balkwell, D. L., and W. C. Ghiorse. 1985. "Characterization of subsurface bacteria associated with shallow aquifers in Oklahoma." *Appl. Environ. Microbiol. 50*: 580-588.

Bohlool, B. B. and E. L. Schmidt. 1968. "Nonspecific staining: its control in immunofluorescent examination of soil." *Science 162*: 1012-1014.

Fliermans, C. B., and D. L. Balkwell. 1989. "Life in the terrestrial deep subsurface." *BioScience 39*: 370-377.

Fliermans, C. B., B. B. Bohlool, and E. L. Schmidt. 1974. "Detection of *Nitrobacter* in natural habitats using fluorescent antibodies." *Appl. Microbiol. 27*: 124-129.

Fliermans, C. B., and T. C. Hazen. 1990. "Serological specificity of *Aeromonas hydrophila* as measured by immunofluorescence photometric microscopy." *Canad. J. Microbiol. 26*: 161-168.

Fliermans, C. B., T. C. Hazen and R. L. Tyndall. 1992. "Modified direct fluorescent antibody technique as a monitoring tool for *Legionella*." In *The 4th International Symposium on Legionella*, Orlando, FL.

Fliermans, C. B., T. J. Phelps, D. Ringelberg, A. T. Mikell, and D. C. White. 1988. "Mineralization of trichloroethylene by heterotrophic enrichment cultures." *Appl. Environ. Microbiol. 54*: 1709-1714.

Fliermans, C. B., and E. L. Schmidt. 1975. "Autoradiography and immunofluorescence combined for autecological study of single cell activity with *Nitrobacter* as a model system." *Appl. Microbiol. 30*: 676-684.

Fliermans, C. B., R. J. Soracco, and D. H. Pope. 1981. "Measurement of *Legionella pneumophila* activity in situ." *Curr. Microbiol. 6*: 89-94.

Fliermans, C. B. and R. L. Tyndall. 1992. "Associations of *Legionella pneumophila* with natural ecosystems." In *The 1992 International Symposium on Legionella*, s p.30, Orlando, FL.

Gorden, R. W., T. C. Hazen and C. B. Fliermans. "Use of Biolog® technology for hazardous chemical screening." *Microbiological Techniques* (accepted for publication).

Hazen, T. C. 1992. "Test Plan or *In Situ* Bioremediation Demonstration of the Savannah River Integrated Demonstration Project." DOE/OTD TTP No. SR 0566-01. Westinghouse Savannah River Company. Aiken, SC.

Hussong, D., R. R. Colwell, M. O'Brien, E. Weiss, A. D. Pearson, R. M. Weiner, and W. D. Burge. 1987. "Viable *Legionella pneumophila* not detectable by culture on agar media." *Bio/Technology 5*: 947-950.

Hyman, M. R., I. B. Murton, and D. J. Arp. 1988. "Interaction of ammonia monooxygenase from *Nitrosomonas europaea* with alkanes, alkenes and alkynes." *Appl. Environ. Microbiol. 54*: 3187-3190.

Legrand, R. 1993. "Methanotrophic Treatment Technology." *Proceedings EMR Technical Interchange Symposium*. Salt Lake City, Utah. March, p. 1193.

Little, C. D., A. V. Palumbo, S. E. Herbes, M. E. Lindstrom, R. L. Tyndall, and P. J. Gilmer. 1988. "Trichloroethylene biodegradation by a methane-oxidizing bacterium." *Appl. Environ. Microbiol. 54*:951-956.

Looney, B. B., and D. S. Kaback. 1991. "Field demonstration of *in situ* air stripping using horizontal wells." *Proceedings of Waste Management '91*, Vol. 1, pp. 527-535. Cosponsored by the American Nuclear Society, American Society of Mechanical Engineers, U.S. Department of Energy, and the University of Arizona.

Phelps, T. J., C. B. Fliermans, T. Garland, S. M. Pfiffner, and D. C. White. 1989. "Recovery of deep subsurface material for microbiological studies." *J. Microbiol. Methods 9*: 267-279.

Rasche, M. E., M. R. Hyman, D. J. Arp. 1991. "Factors limiting aliphatic chlorocarbon degradation by *Nitrosomonas europaea* - cometabolic inactivation of ammonia monooxygenase and substrate-specificity." *Appl. Environ. Microbiol. 57*: 2986-2994.

Radian. 1989. *Phase 1 - Technical Feasibility of Methane Use for Water Treatment: Task 2 — Preliminary Economic Analysis.* Gas Research Institute, Chicago.

Rodriguez, G. G., D. Phipps, K. Ishiguro, and H. F. Ridgeay. 1992. "Use of fluorescent redox probe for direct visualization of actively respiring bacteria." *Appl. Environ. Microbiol. 58*: 1801-1808.

Suzuki, I., S.-C. Kwok, and U. Dular. 1976. "Competitive inhibition of ammonia oxidation in *Nitrosomonas europaea* by methane, carbon monoxide or methanol." *FEBS Lett. 72*: 117-120.

Vannelli, T., M. Logan, D. M. Arciero, and A. B. Hooper. 1990. "Degradation of halogenated aliphatic-compounds by the ammonia-oxidizing bacterium *Nitrosomonas europaea*." *Appl. Environ. Microbiol. 56*:1169-1171.

Vogel, T. M., and P. L. McCarty. 1985. "Biotransformation of tetrachloroethylene to trichloroethylene, dichloroethylene, vinyl-chloride, and carbon-dioxide under methanogenic conditions." *Appl. Environ. Microbiol. 49*:1080-1083.

Wilson, J. T., and B. H. Wilson. 1985. "Biotransformation of trichloroethylene in soil." *Appl. Environ. Microbiol. 49*:242-243.

WSRC. 1991. WSRC 3Q5. "Hydrogeologic Data Collection Methods, Procedures, and Specifications." Westinghouse Savannah River Company, Aiken, SC.

Zimmerman, R., R. Iturriaga, and J. Becker-Birck. 1978. "Simultaneous determination of the total number of aquatic bacteria and the number thereof involved in respiration." *Appl. Environ. Microbiol. 36*: 926-935.

IN SITU BIOREMEDIATION OF CREOSOTE-CONTAMINATED SOIL: COLUMN EXPERIMENTS

G. D. Breedveld and T. Briseid

ABSTRACT

The effectiveness of three in situ bioremediation techniques to clean soil highly contaminated with creosote has been studied in a laboratory column experiment. Sandy soil samples from a contaminated site containing 6.3 g/kg of total polycyclic aromatic hydrocarbons (PAHs)(16 EPA) were treated with air saturated water circulation with nutrient addition alone, forced air aeration alone, and forced air aeration combined with nutrient addition. After 170 days at 16°C only 14% of total PAH was removed in the water circulation treatment, whereas 38% was removed in the forced air aeration and 67% in the forced air aeration combined with nutrient addition. A reference column with free gas diffusion showed 34% PAH removal. Total organic carbon (TOC) and CO_2 respiration measurements indicate that leaching and evaporation have only a minor contribution to the removal of organic compounds.

INTRODUCTION

Creosote oil has been used extensively for wood preservation purposes in Norway. During the 1950s, the production of creosote-treated wood was approximately 130,000 m^3 per year. Since then the production has gone down to approx. 20,000 m^3 per year in 1989. Current wood-preserving techniques are mainly based on copper, chromium, and arsenic salts (CCA). At present only two creosote works are active in Norway. The risks for soil and groundwater pollution in relation to creosote preservation have long been underestimated. Although site investigations at only two former sites are completed, the results indicate extensive soil and groundwater pollution. To remediate creosote-contaminated sites, excavation and incineration are widely used in other European countries. Bioremediation would be a cost-effective alternative treatment method. In situ bioremediation has mainly been used to clean up sites polluted by oil and oil products (Staps 1990). Bioremediation based on water circulation focuses on pollution in the saturated zone. To clean up the unsaturated zone, bioventing technology can be used (van Eyk & Vreeken 1989, Hinchee & Miller 1990). Water circulation was shown to be able to clean artificial PAH-polluted soil to a large extent in laboratory experiments, while the method was less successful in field samples from a gaswork site (Werner 1991). Forced air aeration of coal tar-polluted soil was shown to initiate biodegradation of PAHs (Lund et al. 1991). In the present study, soil from the Lillestrøm site, a former creosote work site of the Norwegian State Railways and Norwegian Telecom, is used to evaluate the feasibility of in situ bioremediation techniques to clean highly creosote-contaminated soil. The following techniques are compared:

- Water circulation with air and nutrient addition to the water phase.
- Forced air aeration.
- Forced air aeration and addition of a nutrient solution.

MATERIALS AND METHODS

Soil samples were taken at the Lillestrøm site from the uppermost 0.5 m in the center of pollution. The sandy soil was sieved over a 4-mm sieve and homogenized. The soil material was packed in four 4 liter glass columns (Table 1). The columns were operated to simulate the in situ bioremediation techniques as follows:

- Column A, saturated, upflow circulation of water saturated with air oxygen and added nutrients (200 mg N/L and 50 mg P/L).
- Column B, unsaturated, downflow forced air aeration and downflow percolation of nutrient solution (200 mg N/L and 50 mg P/L).
- Column C, unsaturated, downflow forced air aeration (moisturized air).

TABLE 1. Installation and operation parameters of the column experiment.

Parameter	Description
Column	Glass column 10 cm diam. and 50 cm high. Stainless steel top and bottom with 100 µm screen. PTFE connections and tubing.
Filling material	Coarse sand (<4 mm) 5.6 kg dry weight, 16.0% water content, porosity n = 0.40
Temperature	15-18°C
Water circulation	Column A: 5-7 L/d Column B: 0.7-0.9 L/d
Nutrients	N: NH_4Cl, 200 mg N/L P: Na_2HPO_4, 50 mg P/L
Air aeration	Column B and C: 6-7 L/d

- Column D, reference column, unsaturated, no additions, loosely covered to reduce evaporation while allowing gas diffusion.

A schematic flowchart is shown in Figure 1. The operation parameters were kept constant during a 6-month period. The main installation and operation parameters are given in Table 1.

Columns B and C were monitored for CO_2 production in the gas phase using a CO_2 trap (NaOH). The water phase in columns A and B was monitored for pH, conductivity, oxygen, and nutrient content. Every third day the first 51 days and every week during the rest of the experiment, 2 L of the circulation solution in column A was replaced by fresh nutrient solution. After 3, 12, 45, 86, 128, and 170 days of operation, water samples were taken for analysis of the PAH content. Water samples for bacterial counts were taken after 3, 45, and 170 days.

After 170 days of operation, the soil material was extruded from the columns and divided into three samples (0-15, 15-30, and 30-47 cm depth). Each sample was homogenized. The soil samples were analyzed for PAH content (16 EPA) using high-pressure liquid chromatography (HPLC) with an ultraviolet-fluorescence detector. All measurements were done in duplicate. TOC was determined by CO_2 evolution after thermal combustion. The total number of microorganisms was determined by plate counts using nutrient agar. Mineral agar with creosote as the sole carbon source was used for the plate counts of the number of creosote degraders.

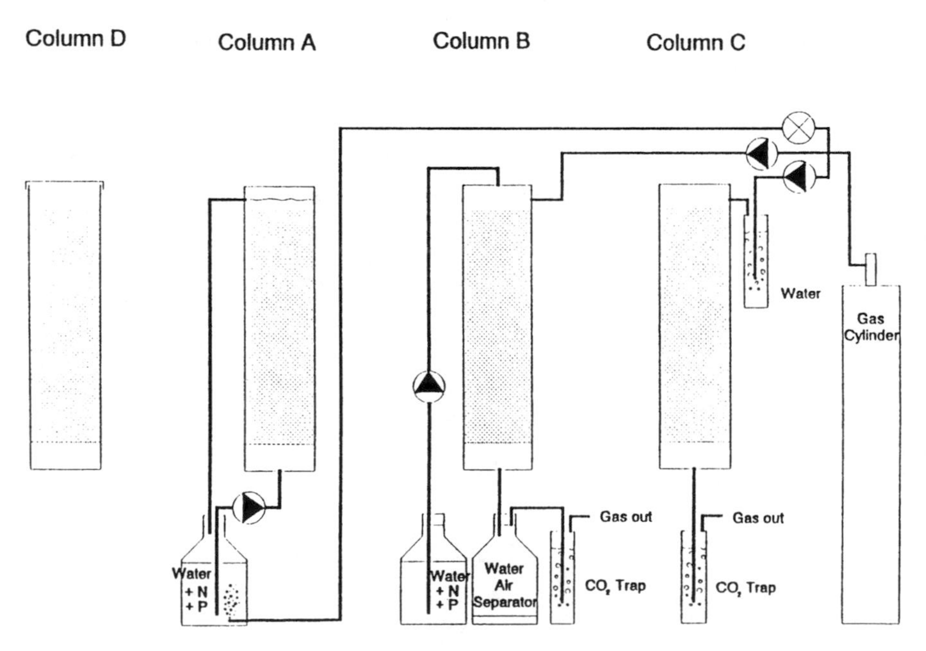

FIGURE 1. Schematic setup of the column experiment.

RESULTS AND DISCUSSION

Soil Characterization

The main characteristics of the soil material used in the column study are given in Table 2. The sandy material has a high carbon content consisting mainly of organic carbon. This organic carbon is believed to result from asphaltic creosote compounds in thc soil and not from soil organic matter. The soil contained a high number of creosote degraders. Approximately 50% of the microorganisms found in the plate counts could use creosote as the sole carbon source.

Gas Samples

The CO_2 production rate in columns B and C increased rapidly after the start of the experiment. After 6 days column C showed a maximal CO_2 production rate of 40 mg CO_2-C/kg/d, which declined slowly to 12 mg CO_2-C/kg/d after 170 days. Column B showed a maximal CO_2 production rate of 65 mg CO_2-C/kg/d after 18 days. After 170 days this had declined to 4 mg CO_2-C/kg/d. During the 170 days of the experiment columns B and C produced, respectively, 5.3 and 3.5 g CO_2-C/kg (Figure 2).

An attempt was made to measure the evaporation of PAH at the beginning of the experiment by activated carbon adsorption. The results were below detection limits (10 µg/L for naphthalene and 50 µg/L for total hydrocarbons).

Water Samples

The total PAH concentration in water samples increased the first 12 days to approximately 1800 µg/L in column A and 1100 µg/L in column B. After the

TABLE 2. Physical, chemical, and biological characterization of soil sample from the Lillestrøm site.

Parameter	Content
Sum 16 PAH (mg/kg)	6,300
Total carbon (%)	4.47
Total organic carbon (%)	4.17
pH (H_2O)	6.7
Total phosphorus (mg/kg)	443
Total nitrogen (mg/kg)	1,000
Total sulfur (mg/kg)	316
Total microbial count (CFU[a]/g)	2×10^7
No. of creosote degraders (CFU/g)	1×10^7

(a) CFU = colony-forming unit.

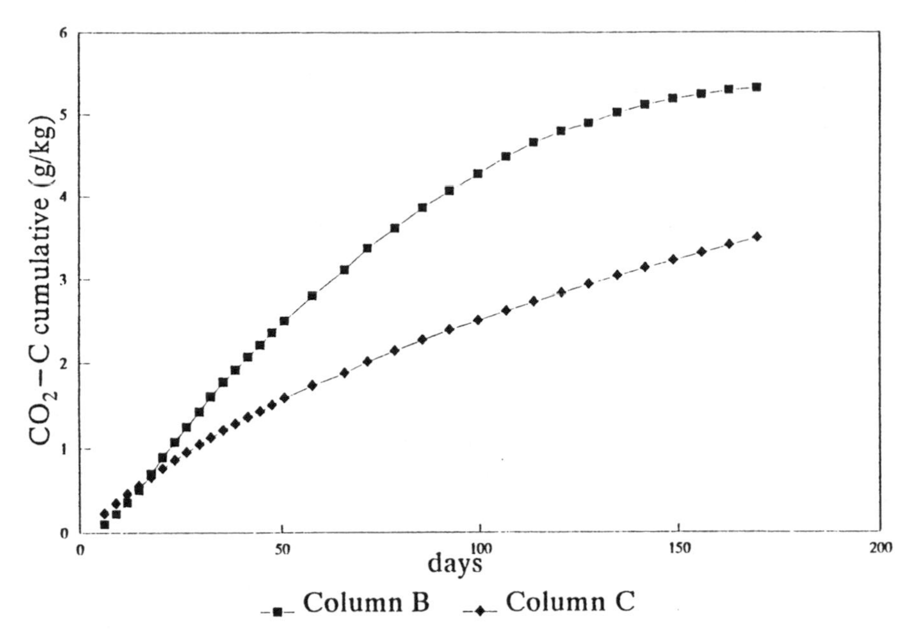

FIGURE 2. Cumulative carbon dioxide production in the forced air aeration columns with nutrient addition (B) and without nutrient addition (C).

12th day the concentrations decline rapidly and are below detection limit after 128 days (25 µg/L total PAH). After 170 days PAH concentration in leachate from column B has increased again to 200 µg/L (Figure 3).

The phosphate and ammonium concentrations in the leachate increase rapidly in the first 50 days, and level off to an almost constant concentration. In column A phosphate shows a total breakthrough at the end of the experiment. The ammonium concentrations in columns A and B show breakthrough after approximately 80 days. After 86 days the nitrate concentration increases in the leachate from column B, while the ammonium concentration declines, indicating nitrification.

The number of creosote-degrading microorganisms in the leachate declines rapidly from an initial 1×10^8 colony-forming units (CFU)/mL to 1×10^5 CFU/mL after 45 days. In column B, the decline continues to 5×10^4 CFU/mL after 170 days. In column A the number of creosote degraders had increased after 170 days to 2×10^6 CFU/mL.

Soil Samples

The total content of PAH (16 EPA) in soil samples at three depths after 170 days of treatment is given in Figure 4. The mean reduction of total PAH content is 14% in column A, 67% in column B, 38% in column C, and 34% in column D. The mean content of the 16 separate PAH compounds in the four columns shows

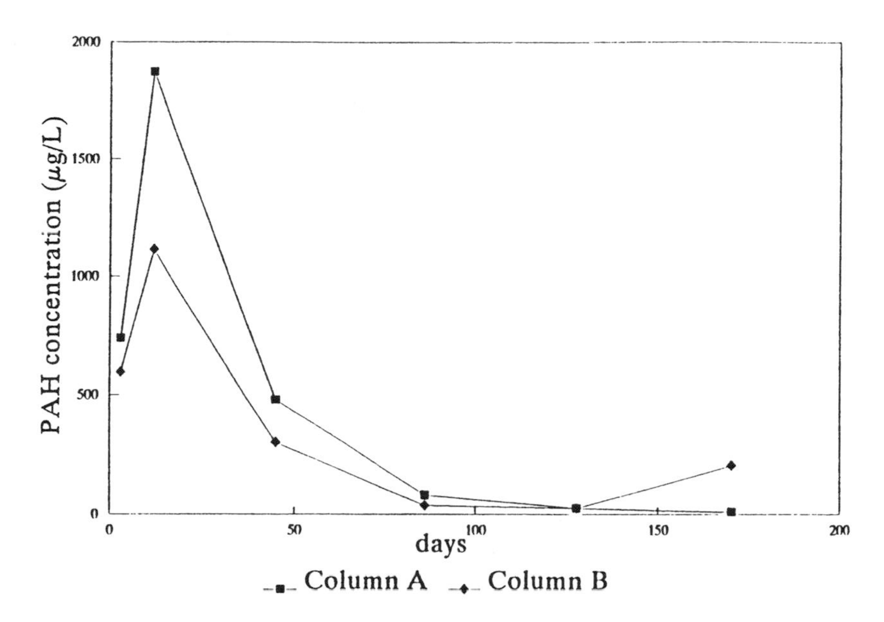

FIGURE 3. Total PAH (16 EPA) concentration in leachate from the water circulation column (A) and forced air aeration and nutrient addition column (B).

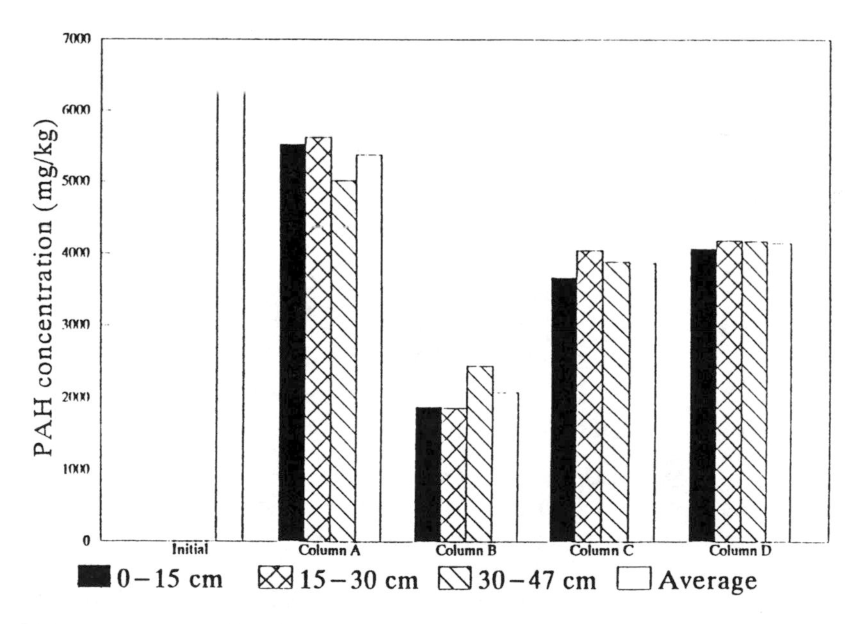

FIGURE 4. Total PAH (16 EPA) content in the soil columns at three different depths and the weighed average content after 170 days treatment.

that in column A only the naphthalene and anthracene contents are clearly reduced (Figure 5). Columns C and D show a decline in all the lighter 2- and 3-ring PAH compounds. Column B shows in addition a clear decline in the content of 4-ring PAHs, fluoranthene, pyrene, benzo(a)anthracene, and chrysene (Figure 5).

TOC content in the soil samples shows a mean reduction of 9% (3 g C/kg) in column A. Columns C and D show a decline in TOC content of 7% (3 g C/kg). TOC content in column B decreased by 16% (6.8 g C/kg).

The total number of microorganisms and the number of creosote degraders in the soil samples are shown in Figure 6. The figure shows a clear decrease in the number of creosote degraders in columns B, C, and D. In column A the number of creosote degraders increased, indicating that all the organisms found in plate counts can use creosote as the sole carbon source.

CONCLUSIONS

The soil material used in these experiments shows that a large population of creosote-degrading microorganisms is present in highly creosote-contaminated

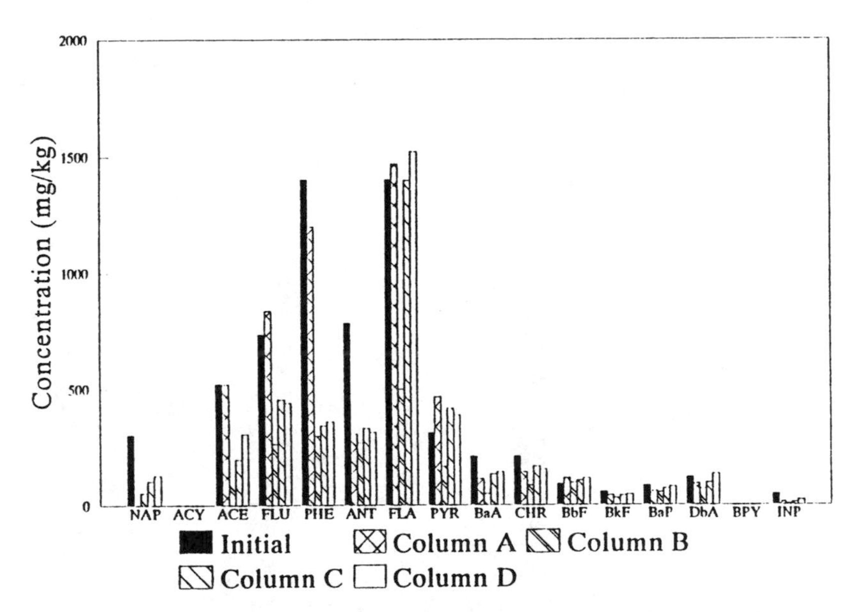

FIGURE 5. Mean concentration of single PAH compounds in the soil columns after 170 days treatment: NAP = naphthalene, ACY = acenaphthylene, ACE = acenaphthene, FLU = fluorene, PHE = phenanthrene, ANT = anthracene, FLA = fluoranthene, PYR = pyrene, BaA = benzo(a)anthracene, CHR = chrysene, BbF = benzo(b)fluoranthene, BkF = benzo(k)fluoranthene, BaP = benzo(a)pyrene, DbA = dibenzo(a,h)anthracene, BPY = benzo(ghi)-perylene, INP = indeno(1,2,4-cd)pyrene.

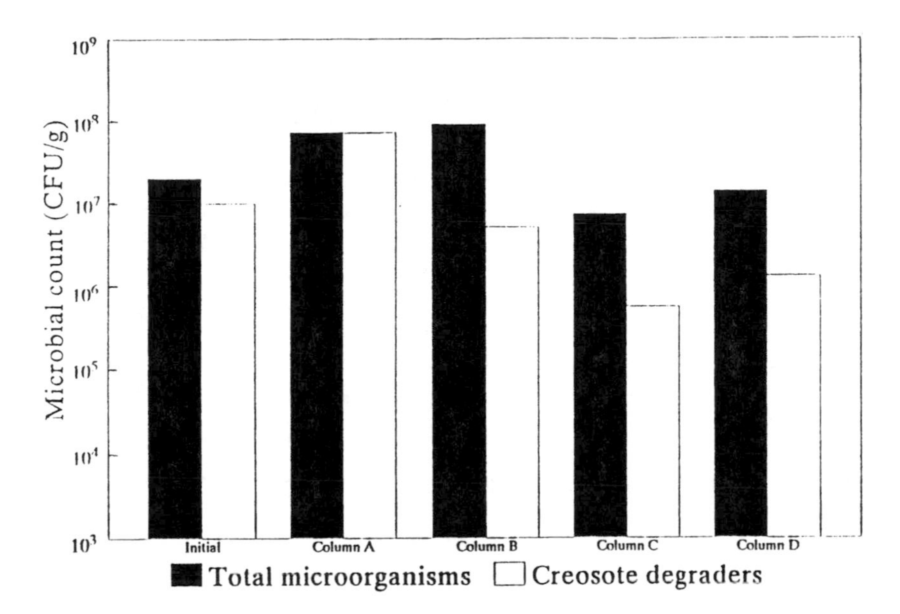

FIGURE 6. Total microbial count and number of creosote-degrading microorganisms in the soil columns after 170 days treatment.

soil. Chemical analyses show that the pollutants have been there for many decades without being degraded. Even easily degradable 2- and 3-ring PAHs are found in high concentrations. Their presence indicates that environmental conditions limit the degradation in the soil.

The column experiments show that of the three in situ bioremediation techniques used, forced air aeration combined with nutrient addition results in the highest PAH removal rates. This technique shows a 67% reduction of total PAH (16 EPA) in 170 days at 16°C. The main reduction is observed in the content of 2-, 3-, and 4-ring PAHs, while 5- and 6-ring PAHs only show minor reductions.

Forced aeration without nutrient addition shows only a 38% reduction of total PAHs, mainly 2- and 3-ring PAHs. Almost similar results are found in the reference column with free gas diffusion (34% reduction).

Circulation of air-saturated water shows only a minor reduction in PAH content (14%), mainly naphthalene and anthracene. A slightly higher decrease in PAH content is found in the lowest 18 cm of the column, where the oxygen-rich water enters the column. This might indicate that the oxygen load which can be transported in the water phase is limiting the biodegradation process.

PAH removal by leaching in the water phase contributes only to a very limited extent to the overall reduction in PAH content in the soil (estimated maximum removal 10 mg total PAH/kg).

CO_2 measurements show that the respiration rate in the combined aeration and nutrient addition column has decreased to low levels. This is surprising given that still considerable amounts of degradable 2- and 3-ring PAHs are present. The results indicate that the residual PAH content is not available for the microorganisms. The high organic carbon content is believed to contribute to the limited availability by forming asphaltic soil aggregates and chemical matrix binding of the pollutants.

The forced aeration column shows a good correlation between CO_2 production and reduction in TOC content in the soil samples. This indicates that evaporation contributes only to a limited extent to the overall removal of organic compounds in the soil column. For single compounds (e.g., naphthalene), it was not possible to quantify the exact evaporation. Based on the detection limits, the total evaporation during the experiment is estimated to be below 10 mg total hydrocarbons/kg. In the combined forced aeration and nutrient addition column, TOC removal is higher than CO_2 production, indicating a leaching of dissolved organic carbon with the water phase. However this was not quantified.

The experiments show that highly creosote-contaminated soil has a good biodegradation potential under proper environmental conditions. However a high residual concentration remains in the soil. This indicates the need to develop methods to increase the bioavailability in situ.

ACKNOWLEDGMENTS

This project has been funded by the Norwegian State Railways (NSB), Norwegian Telecom, the State Pollution Control Authorities (SFT), Norsk Impregneringskompani, AGA, Impregnor, and Norenergi. The authors would like to thank Rob Kockelkoren and Torgeir Rødsand for their excellent laboratory assistance.

REFERENCES

van Eyk, J., and C. Vreeken. 1989. "Venting-mediated removal of hydrocarbons from subsurface soil strata as a result of simulated evaporation and enhanced biodegradation." *Med. Fac. Landbouww. Rijksuniv. Gent 53 (4b)*, pp. 1873-1884.

Hinchee, R. E., and R. N. Miller. 1990. "Bioreclamation of hydrocarbons in the unsaturated zone." In W. Pillmann and K. L. Zirm (Eds.), *Haz. Waste Manag., Contaminated Sites and Ind. Risk Assess., Proc. Envirotech Vienna*, pp. 641-650. Int. Soc. for Environmental Protection, Vienna, Austria.

Lund, N.-Ch., J. Świniański, G. Gudehus, and D. Maier. 1991. "Laboratory and Field Tests for a Biological *In situ* Remediation of a Coke Oven Plant." In R. E. Hinchee and R. F. Olfenbuttel (Eds.), *In situ Bioreclamation: Applications and Investigations for Hydrocarbon and Contaminated Site Remediation*, pp. 396-412. Butterworth-Heinemann, Boston, MA.

Staps, J. J. M. 1990. *International Evaluation of In-situ Biorestoration of Contaminated Soil.* RIVM report no. 738708006, National Institute of Public Health and Environmental Protection, Bilthoven, the Netherlands.

Werner, P. 1991. "German Experiences in the Biodegradation of Creosote and Gaswork-Specific Substances." In R. E. Hinchee and R. F. Olfenbuttel (Eds.), *In Situ Bioreclamation: Applications and Investigations for Hydrocarbon and Contaminated Site Remediation*, pp. 496-517. Butterworth-Heinemann, Boston, MA.

LABORATORY EVALUATION OF IN SITU BIOREMEDIATION FOR POLYCYCLIC AROMATIC HYDROCARBON-CONTAMINATED AQUIFERS

G. R. Brubaker, W. F. Lane,
D. A. English, and D. Hargens

ABSTRACT

A soil column study was performed to compare the effects of four in situ treatment scenarios on the removal of polycyclic aromatic hydrocarbons (PAHs) from soils: (1) low dissolved oxygen (DO) (less than 0.5 mg/L) with no added nutrients, (2) moderate DO (8 mg/L) with nutrients, (3) high DO (40 mg/L) with nutrients, and (4) high DO with nutrients and inoculum. The removal rates for PAHs were identical among all four treatment scenarios. A strong correlation was observed between the rate of removal of individual PAHs and the log of their K_{ow} values. Under the conditions of the study, rapid groundwater movement and a large concentration of biodegradable constituents, in situ bioremediation provided no significant enhancement to the rate of remediation. Performance modeling of these processes, based on the oxygen demand of the soil and literature K_d values, indicates that a very small performance enhancement would result from the addition of in situ bioremediation to a full-scale pump-and-treat process in these soils.

INTRODUCTION

A majority of the sites treated with in situ bioremediation techniques have been contaminated by light hydrocarbons such as gasoline, diesel fuel, and jet fuels. However, many sites, including former manufactured gas plants (MGPs), refineries, petrochemical facilities, and wood-treating sites contain heavy, phase-separated product. These materials can create a continued source of high-molecular-weight constituents, such as PAHs, that affect the groundwater long term. Although the low solubilities and high retardation coefficients associated with these compounds are known to restrict the performance of conventional pump-and-treat processes, the importance of desorption processes on in situ bioremediation is not well understood (Brubaker 1991; Brubaker & Stroo 1992).

In principle, in situ bioremediation creates a high density of bacteria close to the sorbed contaminants, biodegrading dissolved constituents, thus allowing more solubilization and remediation than occurs with pump-and-treat methods (Rittmann et al. 1992). The overall process kinetics for in situ bioremediation processes (in saturated soils) typically are controlled by the transport of dissolved oxygen, because oxygen transport typically is very slow compared with desorption and biodegradation (Thomas & Ward 1989).

In 1991, a Record of Decision (ROD) was signed for a former MGP in Fairfield, Iowa, which required that in situ bioremediation be evaluated as a means to remediate PAHs present in soils below the depths of practical excavation. Because the rate of desorption and biodegradation of four- and five-ring PAHs from this type of coal tar matrix is often several orders of magnitude slower than the biodegradation rate of petroleum constituents, the most common target of in situ bioremediation processes (Middleton et al. 1991), it seems possible that desorption/biodegradation, rather than oxygen transport, would limit this process. Pilot and laboratory treatability/design studies were implemented to evaluate these issues, as well as to develop the data required for full-scale implementation. This paper describes the laboratory treatability studies; a description of the site and pilot study has been published elsewhere (English et al. 1992).

The laboratory treatability studies were designed to evaluate the relative performance of a pump-and-treat process combined with several options for in situ bioremediation. Although the studies also included examining nutrient interactions with soil and enumeration of microbial populations within the site, the laboratory studies focused on two questions:

1. How much oxygen would be required to meet the biological oxygen demand of this contaminant matrix under in situ bioremediation conditions?
2. Would the rate of remediation under an in situ biodegradation process be limited by the rate of desorption and dissolution of the organic constituents, as well as by oxygen transport?

Two experiments were performed to evaluate these issues. First, a column study was performed that included twelve soil columns, using three columns of different lengths under four treatment scenarios. Monitoring the rate of oxygen breakthrough in various columns provided a measure of oxygen demand and, thus, an indication of whether desorption or oxygen transport controlled biodegradation under the conditions of the study. Second, a slurry-phase respirometry study was performed to estimate the total oxygen demand of the contaminant matrix under conditions that would yield maximum degradability. Comparing the results of the two studies provides a measure of efficiency of in situ bioremediation relative to a slurry-phase treatment and provides insight into the mechanisms that control the efficiency of both the pump-and-treat and the in situ bioremediation options at this site. Each of these studies is described below.

LABORATORY STUDIES

Column Study

The soil column study compared the effect of four influent compositions on the removal of PAHs and TOC (an indicator of the general contaminant matrix) from soils: (1) low DO (less than 0.5 mg/L) with no added nutrients, (2) moderate DO (8 mg/L) with nutrients, (3) high DO (40 mg/L) with nutrients, and (4) high DO with nutrients and inoculum. Twelve soil columns were tested over a 20-week treatment period, using three column lengths for each of the four scenarios. The experimental setup and nutrient formulations are shown in Figure 1 and summarized in Table 1. The soils used for the column study contained 4,040 ± 500 mg/kg of TOC (Method SW 8310), of which 551 ± 27 mg/kg were identifiable PAHs (Method SW 9060). In the absence of experimental data regarding the total oxygen demand of residues from MGPs, the oxygen demand was estimated by assuming that 1,000 mg/kg of the TOC was natural soil organic matter that would not degrade and that 75% of the balance would be degradable under the conditions tested. (The respirometry study had not been completed when these estimates were performed.) By assuming that 2.5 g of oxygen would be required to degrade each gram of biodegradable TOC, the oxygen demand of the soil was calculated to be 5.625 g/kg. The soils used for the laboratory study were collected from sandy lenses from the site, through which most of the nutrient-amended groundwater would travel. The soils were classified as a fine to medium sand, containing about 18% fines (particles with diameters less than 0.075 mm).

Each 6-in. (15 cm) soil column holds approximately 0.5 kg of soil, and these were estimated to contain 2.8 g of oxygen demand. Flowrates through the column were selected at 0.4 mL/min (576 mL/day). Under these conditions, soil columns that received 40 mg/L of dissolved oxygen received 3.23 g of dissolved oxygen during the 20-week study.

If the rate of biodegradation were limited primarily by oxygen delivery, then oxygen would move through the column as a front and oxygen breakthrough would occur after 17 weeks of treatment. Under these conditions, the oxygen breakthrough would occur suddenly with a rapid increase in the concentration of dissolved oxygen in the effluent. If a significant portion of the potentially biodegradable TOC were not accessible to microorganisms under the conditions tested, breakthrough would occur more rapidly. If biodegradation were even slower than oxygen transport because constituents were only gradually available to biodegradation, the breakthrough of dissolved oxygen would be gradual and effluent DO concentrations would increase very slowly as the oxygen demand is satisfied (Brubaker 1991). This type of slow desorption-limited biodegradation could result in oxygen breakthrough in the 6-in., 12- and 24-in. (15, 30 and 61 cm) columns, depending upon the magnitude of the effects.

The influent and effluent water from the soil columns was sampled throughout the 20-week study and analyzed for PAHs, TOC, DO, ammonium ion, orthophosphate, and pH, as well as for total heterotrophic and PAH-degrading bacteria (using phenanthrene as a sole carbon source). Bromide tracer studies were performed

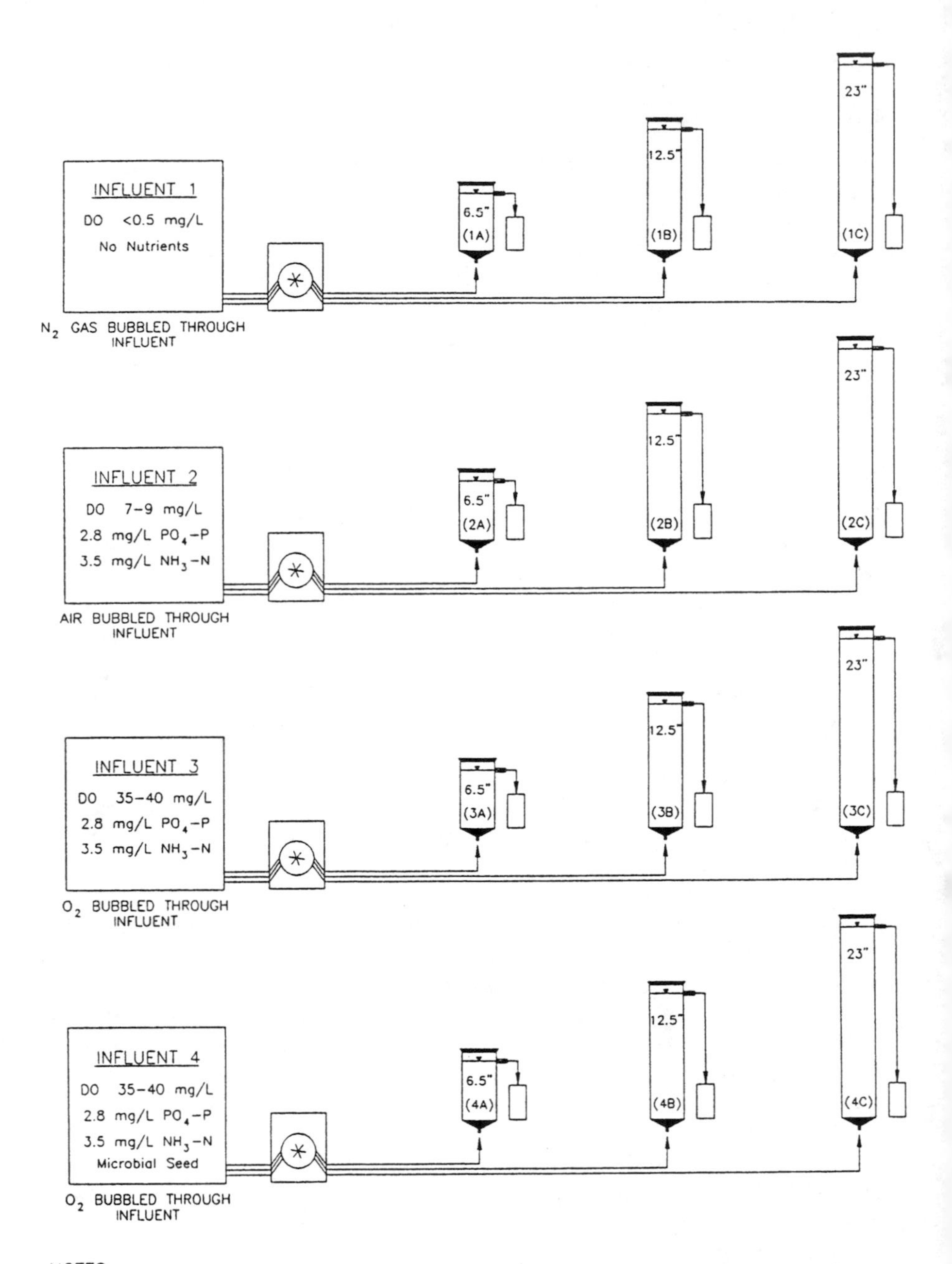

NOTES:

- Effluent storage on ice.
- Seed added was developed from a "known PAH/BTEX-degrading culture".

FIGURE 1. Schematic design of laboratory study.

TABLE 1. Mass of soil contained in study columns.

Treatment	Column	Length (inches/cm)	Mass of Soil (kg, dry weight)
Low DO (<0.5 mg/L)	1A	6.5 (16.5)	0.473
	1B	12.5 (32)	0.913
	1C	23.0 (58)	1.696
Moderate DO (8 mg/L)	2A	6.5 (16.5)	0.461
	2B	12.5 (32)	0.963
	2C	23.0 (58)	1.724
High DO (40 mg/L)	3A	6.5 (16.5)	0.478
	3B	12.5 (32)	0.928
	3C	23.0 (58)	1.734
High DO (40 mg/L) with Inoculum	4A	6.5 (16.5)	0.472
	4B	12.5 (32)	0.914
	4C	23.0 (58)	1.656
	Average A	6.5 (16.5)	0.470
	Average B	12.5 (32)	0.930
	Average C	23.0 (58)	1.702

on four of the columns to measure the average linear velocity of the water through the column and to calculate the effective porosity of the soil and the volume of water required to displace 1 pore volume of water. The results of this monitoring included the following observations:

1. The flowrate of 0.4 mL/min yielded an average linear velocity of 2.5 ± 0.26 ft/day (0.76 m/d). The effective porosity of the soil column was 40%. One pore volume of the 12-in. (30-cm) column was about 220 mL.
2. During the 20-week study, each column received 80.64 L of influent. This represented about 600 pore volumes of water for the 6-in. (15-cm) columns, 340 for the 12-in. (30-cm) columns, and 186 for the 24-in. (61-cm) columns (the actual column lengths were 6.5 in. [16.5 cm], 12.5 in. [32 cm], and 23 in. [58 cm]). Concentrations of ammonium ion and orthophosphate decreased about 90% during transport through the columns, but measurable quantities of each were detected in the effluent of all columns where nutrients were added.
3. Although the effluent was analyzed for dissolved organic constituents (TOC and PAHs), the mass detected in the effluent represented less than 20% of the mass lost from the soil, even in the control column (Treatment 1). Although some losses may have resulted due to adsorption and/or biodegradation in the effluent tubing, the cause of this discrepancy has not been fully resolved.

4. The concentrations of DO detected in the effluents of all columns were less than 0.5 mg/L throughout the entire 20-week study.
5. Microbial populations were within acceptable ranges in all of the column soils and effluents during the study. Soil samples showed between 10^5 and 10^6 CFU/mL (CFU = colony-forming units, a measure of viable bacteria) of total bacteria, of which about 10% were PAH-degraders. The concentrations of bacteria in the effluent samples were about 10 times lower than in soil. There were no consistent differences in the microbial densities among the four treatments.

Soils were analyzed at the beginning and end of the study by collecting one sample from the bottom (influent end) of each column and two samples from the top (effluent end) of each. This procedure provided the equivalent of samples from 0 to 2, 4 to 6, 10 to 12, and 22 to 24 in. (0 to 5 cm, 10 to 15 cm, 25 to 30 cm, and 56 to 61-cm) within a single column. Figure 2 compares the removal of total PAHs and TOC as a function of column length for each of the four treatments. Examination of these data indicates that no significant benefit resulted from the addition of oxygen, nutrients, and/or an inoculum of PAH-degrading bacteria under the conditions tested. This conclusion suggests that extraction of the contaminants was as effective as a combined extraction/biodegradation process under the conditions studied.

Figure 3 compares the removal rates of the various PAH constituents (using the average of all four treatments) as a function of the octanol-water partitioning coefficient, K_{ow}, of the PAH and the column length. These results show that PAHs with small K_{ow} values were more readily removed than those with higher K_{ow} values. The overall contaminant matrix (measured as TOC) behaved as if it had a log K_{ow} value near 6, suggesting that the unidentified constituents partition strongly to soil. These trends indicate that the rate of removal was controlled, at least partially, by the partitioning of constituents between the solid and aqueous phases. Both extraction and biodegradation would be expected to show this trend under some treatment conditions. The extent to which the groundwater reached equilibrium with the soils during the study can be evaluated by examining the data presented in Table 2.

Comparison of the mass of either TOC or PAHs removed from various columns shows that roughly twice as much removal occurred in the 12-in. (30-cm) columns as in the 6-in. (15-cm) columns, and almost twice as much removal occurred in the 24-in. (61-cm) columns as in the 12-in. (30-cm) columns. This trend indicates that the removal of these constituents was limited by the contact time between the soil and groundwater and that the contact time was too short to achieve equilibrium partitioning through these soil columns. If the soil and groundwater had sufficient time to fully equilibrate in the first 6 in. (15 cm) segment of the column, no net loss of constituents would have occurred in the remaining portions of the column, and the total mass of extracted constituents would be the same for each of the columns. Thus, the contact time required to equilibrate the constituents within this matrix is longer than the retention time within the 24-in. (61-cm) columns, which is approximately 17 hours, and much

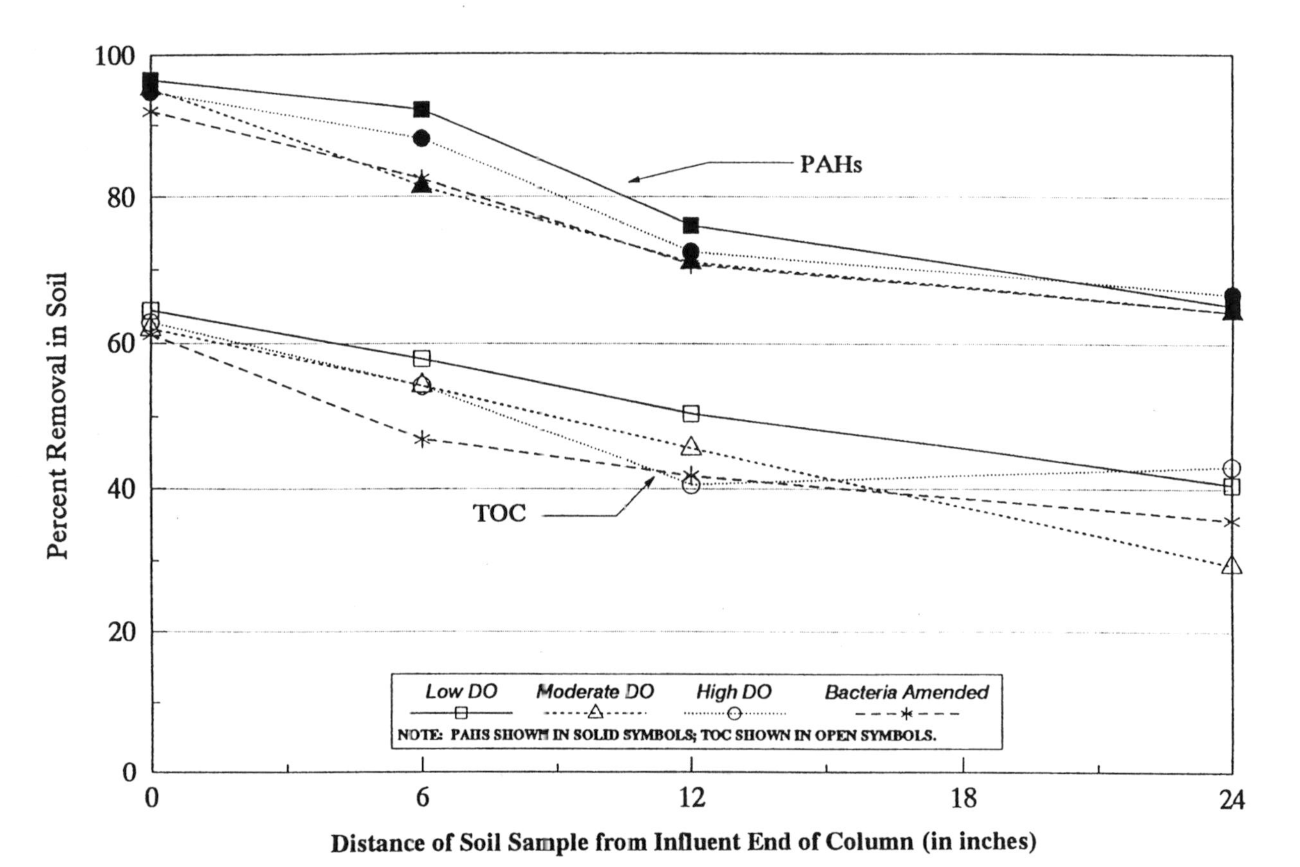

FIGURE 2. Removal of PAHs and TOC as a function of treatment.

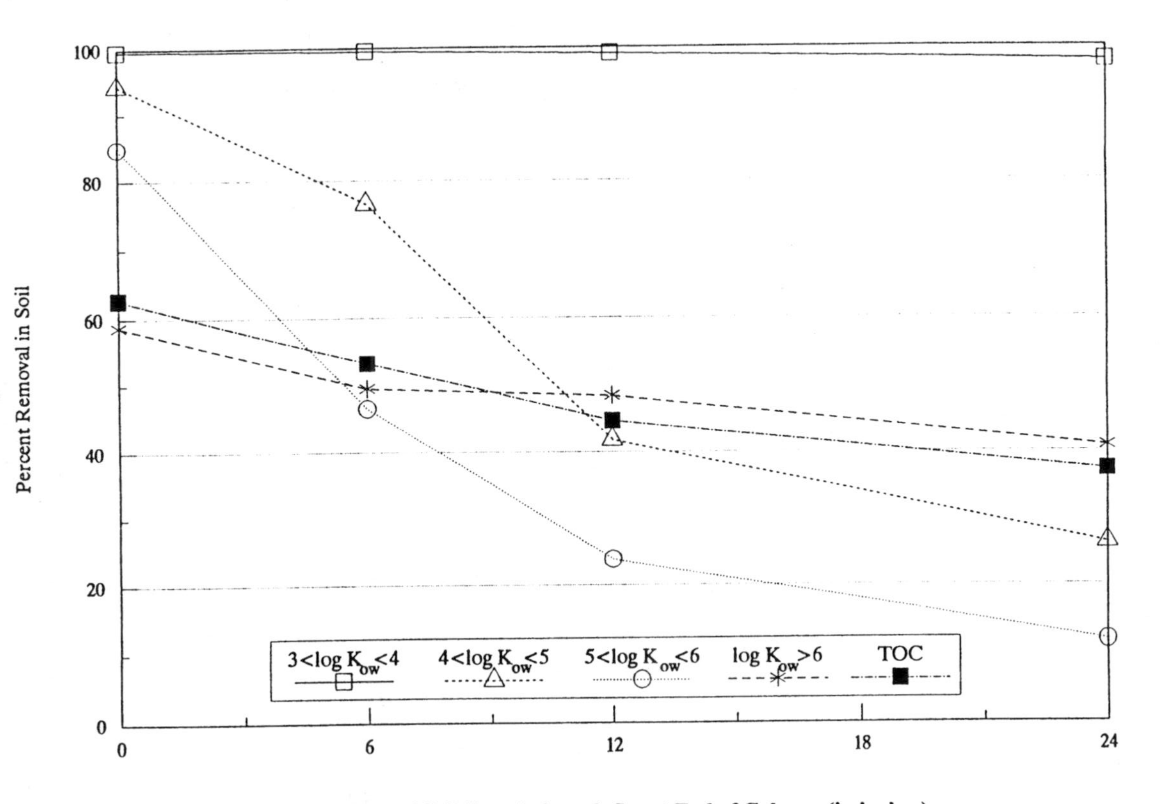

FIGURE 3. Reduction of PAHs and TOCs in soil.

TABLE 2. Comparison of initial and final soil data from column study.

	No Oxygen or Nutrients			8 mg/L DO Plus Nutrients			40 mg/L DO Plus Nutrients			40 mg/L DO, Nutrients and Inoculum		
	6-in.[a]	12-in.	24-in.	6-in.	12-in.	24-in.	6-in.	12-in.	24-in.	6-in.	12-in.	24-in.
TOC[b] - Initial Soil (mg)	1900	3700	6900	1900	3900	7000	1900	3700	7000	1900	3700	6700
TOC - Final Soil (mg)	740	1600	3300	800	1800	3800	800	1800	3300	900	1800	3400
TOC Removed (mg)	1200	2100	3600	1100	2100	3200	1100	1900	3700	1000	1900	3300
Percent TOC Removed	63%	57%	52%	58%	54%	46%	58%	51%	53%	53%	51%	49%
Total PAHs[c] - Initial Soil (mg)	260	550	940	260	530	950	260	510	960	260	510	920
Total PAHs - Final Soil (mg)	80	180	350	90	210	390	88	200	370	93	200	350
PAHs Removed (mg)	180	370	590	170	320	560	170	310	590	170	310	570
Percentage PAHs Removed	69%	67%	63%	65%	60%	59%	65%	61%	61%	65%	61%	62%

(a) See Table 1 for actual column lengths.

(b) Total organic carbon was determined using Method SW 9060 from "SW846 - Test Methods for the Evaluation of Solid Waste, Physical/Chemical Methods," 3rd ed., EPA, September 1986.

(c) PAHs were determined using Method SW 8310.

longer than the 4-hour retention time of the 6-in. (15-cm) columns. (The complete removal of PAHs with log K_{ow} values of less than 4 does not allow interpretation of the controlling mechanism for these compounds.)

This trend is further illustrated in Figure 4, which shows the change in the concentration of soil-bond PAH during the study as a function of column length and K_{ow} value. If dissolution is fast and no significant biodegradation occurs, the concentration of PAH in the effluent of columns of various lengths will be constant, the mass of dissolved phase material from each column will be constant, and the change in concentration will be inversely proportional to column length. This trend is generally illustrated by the data from those PAHs with K_{ow} values less than 6. By contract, if dissolution is slow, the concentration of dissolved PAH will increase with longer contact time (longer column length), so that the change in concentration will be independent of length. This trend is observed for the PAHs with larger K_{ow} values. These observations generally are consistent with other studies that suggests a log-log relationship between the rate of dissolution and the partitioning coefficients of hydrophobic organic compounds (Brusseau 1991).

Groundwater flowrates generally are slower in full-scale operations than those used in column studies, and the distances between the point of injection and the point of recovery are much greater. Thus, the contact time between groundwater and soil is much longer during actual site remediation and equilibration is more likely to occur. From this perspective, at least, groundwater extraction would be more efficient in the field than observed in the column studies. As a result, the performance modeling of groundwater extraction and in situ bioremediation discussed later in this paper assumes complete equilibration.

Slurry-Phase Respirometry

Two 100-g soil samples were mixed with 1.0 L of water in duplicate, sealed, stirred reactors of a BI-1000 Automated Respirometer. This unit neutralizes carbon dioxide (produced through biodegradation) and generates oxygen to replace that which is consumed, while recording the oxygen uptake of the system. Each reactor was amended with nutrients and buffered to enhance microbial activity. The slurry reactor was operated for 62 days, at which time the uptake rate of oxygen had decreased to less than 1.0 mg per kg of soil per hour. Both TOC and PAHs in soil were analyzed before and after the slurry treatment.

The average cumulative oxygen uptake at the conclusion of the study was 5,100 ± 1,500 mg/kg. The TOC content of the soil decreased from 2,300 mg/kg to 1,183 mg/kg for a 49% reduction in TOC. The total PAH concentration decreased by 69% (from 171 to 53 mg/kg) during the study, showing the usual trend of greatest biodegradation occurring for those PAHs with the smallest K_{ow} values (those that are most soluble). The observed oxygen demand was 2.2 ± 0.6 g of oxygen demand per gram of TOC detected in the original soils. Because the theoretical oxygen demand for this type of matrix is in the range of 2.5 to 3.0, this value indicates that, under ideal conditions, the contaminant matrix from this site is very biodegradable.

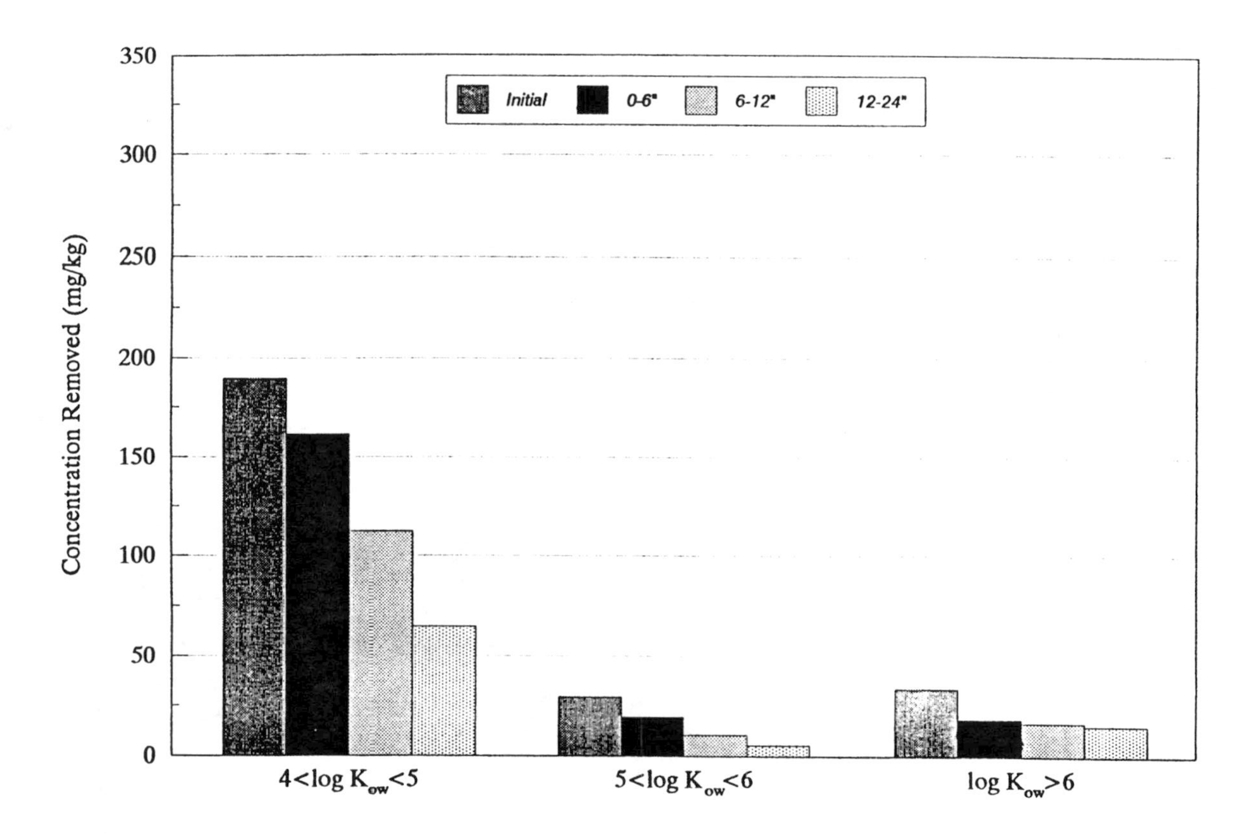

FIGURE 4. PAHs removed from soils during column studies.

SUMMARY OF LABORATORY FINDINGS

Although the lack of a total mass balance prevents a direct analysis of the fate of the removed organic constituents or a direct indication of the mechanisms controlling their fate, insight into the factors controlling contaminant removal can be obtained through careful examination of the available data. Three observations appear to be critical to an understanding of the operative mechanisms.

1. The extent of biodegradation of the contaminant matrix from this site is similar in a soil column and in a slurry-phase reactor, and a large fraction of the total contaminant matrix is biodegradable. The oxygen demand of the soil is approximately 2.2 g of oxygen per gram of initial TOC.
2. The flowrates used in the column study were too fast to allow for full equilibration of the water with the organics in the soil matrix. As a result, the water continued to accumulate dissolved contaminants as it passed through the soil column, but never reached an equilibrium concentration.
3. The overall extent of contaminant removal observed under the four scenarios tested was essentially identical, with a slight advantage for the control scenario (no nutrients or oxygen).

Collectively, these data suggest that, under the conditions of the study, the contaminants that were biologically degraded (consuming 40 mg/L of dissolved oxygen) in Treatments 3 and 4 were the same contaminants that were extracted in the absence of oxygen in Treatment 1. That is, the rate of desorption (in addition to the rate of oxygen transport) limited the rates of both removal processes. Under these conditions, biodegradation and extraction were competitive rather than additive processes.

Under conditions where the water passed through the columns more slowly, the processes would be additive rather than competitive, because the water would become fully equilibrated with contaminated soils even after it was depleted of oxygen but before it left the contaminated zone. In soils with lower oxygen demand, the results of a column study would have been easier to discern. However, the respirometry study provides strong evidence that, even under these conditions, the addition of 40 mg/L of dissolved oxygen would provide very little performance benefit (as measured by PAH reduction) at this site, because only about 10% of the dissolved oxygen added to the site would be used to biodegrade the constituents of concern.

IMPLICATIONS FOR FULL-SCALE IMPLEMENTATION

The objectives of the laboratory study were to determine (1) the mass of oxygen that would be required to achieve the desired level of biodegradation, and (2) whether the rate of desorption of PAHs would limit the rate of biodegradation. This section discusses the implications of the laboratory results for the potential implementation of full-scale in situ bioremediation at the Fairfield site.

The performance of flushing and in situ bioremediation for the site was simulated using a spreadsheet-based flushing and biodegradation model. The model estimates the concentration of various constituents in the soil and dissolved phase of a soil matrix after various amounts of flushing, either with or without biodegradation. The aquifer is simulated as a batch reactor mixed with 1 pore volume of water, drained, and then replenished. Desorption of compounds from soil into the aqueous phase is based on equilibrium-partitioning relationships. The biodegradation function assumes that 2.2 g of oxygen is required to degrade 1 g of contaminant, that the relative rate of biodegradation of different constituents is equal to their relative concentrations in the aqueous phase, and that the oxygen delivered in each pore volume of water is total consumed.

For this analysis, the composition of the matrix was approximated by 11 PAHs. The initial concentration of each PAH was set equal to the concentration determined for the soils used in the column study. Compounds with similar structure and partitioning coefficients were grouped to simplify the analysis. The nonidentified constituents, which make up 86% of the TOC, were represented by two "matrix" terms. Matrix A is assumed to readily desorb and biodegrade in a manner similar to that of two- or three-ring PAHs. Matrix B is assumed to desorb and biodegrade more slowly, similar to four-ring PAHs. Table 3 presents the initial soil-phase concentrations assumed for input to the model. Based on the

TABLE 3. Assumptions for performance modeling: Initial soil characterization (mg/kg).

Constituents	High Concentration	Low Concentration
Matrix "A"[(a)(b)]	2408	602
Matrix "B"[(a)(c)]	1032	258
Acenaphthene	16	4
Acenaphthylene	108	27
Anthracene	24	6
Benz(a)anthracene	28	7
Chrysene	28	7
Naphthalene	96	24
Fluoranthene	16	4
Fluorene	56	14
2-Methylnaphthalene/Dibenzofuran	108	27
Phenanthrene	56	14
Pyrene	24	6
TOTAL	4,000	1,000

(a) Based on the initial characterization of soil and the respirometry study, the soil used in the column study had 4,000 mg/kg of total organic constituents and 70:30 ratio for Matrix "A" and Matrix "B".
(b) Matrix "A" is similar to a PAH compound with 2 or 3 rings.
(c) Matrix "B" is similar to a PAH with 4 rings.

initial soil characterization, the soil used in the column study had a high concentration (4,000 mg/kg) of total organic constituents. The respirometry study indicated that approximately 70% of the organic matrix was degradable, so the Matrix A:Matrix B ratio was set at 70:30.

Four model analyses were performed in two sets of two calculations each. Table 4 lists the treatment methods for the two analyses. Each set of calculations compared the results of contaminant removal by flushing alone (pump-and-treat) to contaminant removal by flushing plus the biodegradation achieved after addition of 40 mg/L of DO. For each treatment method, one set of calculations assumed an initial concentration of a coal tar of 1,000 mg/kg and one assumed 4,000 mg/kg.

Figure 5 illustrates the removal of naphthalene, present at initial concentrations of 96 mg/kg and 24 mg/kg during 30 pore volumes of treatment. Based on this analysis, the addition of 40 mg/L of DO shortened the time required to achieve a 50% reduction of naphthalene by about 2 pore volumes of treatment for the 4,000-mg/kg soil and by 5 pore volumes for the 1,000-mg/kg soil. The naphthalene concentrations in the 4,000-mg/kg soil after 30 pore volumes were 33.5 mg/kg (flushing only) and 30 mg/kg (flushing and biodegradation). These values correspond to 65% and 69% removal, respectively. The final naphthalene concentrations in the 1,000 mg/kg soil were 8.4 mg/kg (flushing only) and 5.1 mg/kg (flushing and biodegradation), which indicated 65% to 79% naphthalene removal, respectively. (It should be noted that the benefits would be proportionally greater if higher concentrations of oxygen were delivered and lower if less oxygen were delivered.)

If this model and the soils used for the treatability study fairly represent this former MGP site, then very little performance benefit (as measured by PAH reduction) would be expected by adding in situ bioremediation to a traditional pump-and-treat process at this site. This trend can be generalized to other sites where low permeable soils and high concentrations of nontargeted biodegraded constituents are present, but the reader is cautioned that in situ bioremediation can still be cost-effective at sites where incremental benefit appears small, especially when the incremental cost of adding oxygen is small relative to total project costs.

TABLE 4. Performance model analyses.

	1	2	3	4
Matrix A:Matrix B	70:30	70:30	70:30	70:30
Total Concentration (mg/kg)	1,000	1,000	4,000	4,000
Treatment Method	Flushing only	Flushing and in situ bioremediation	Flushing only	Flushing and in situ bioremediation

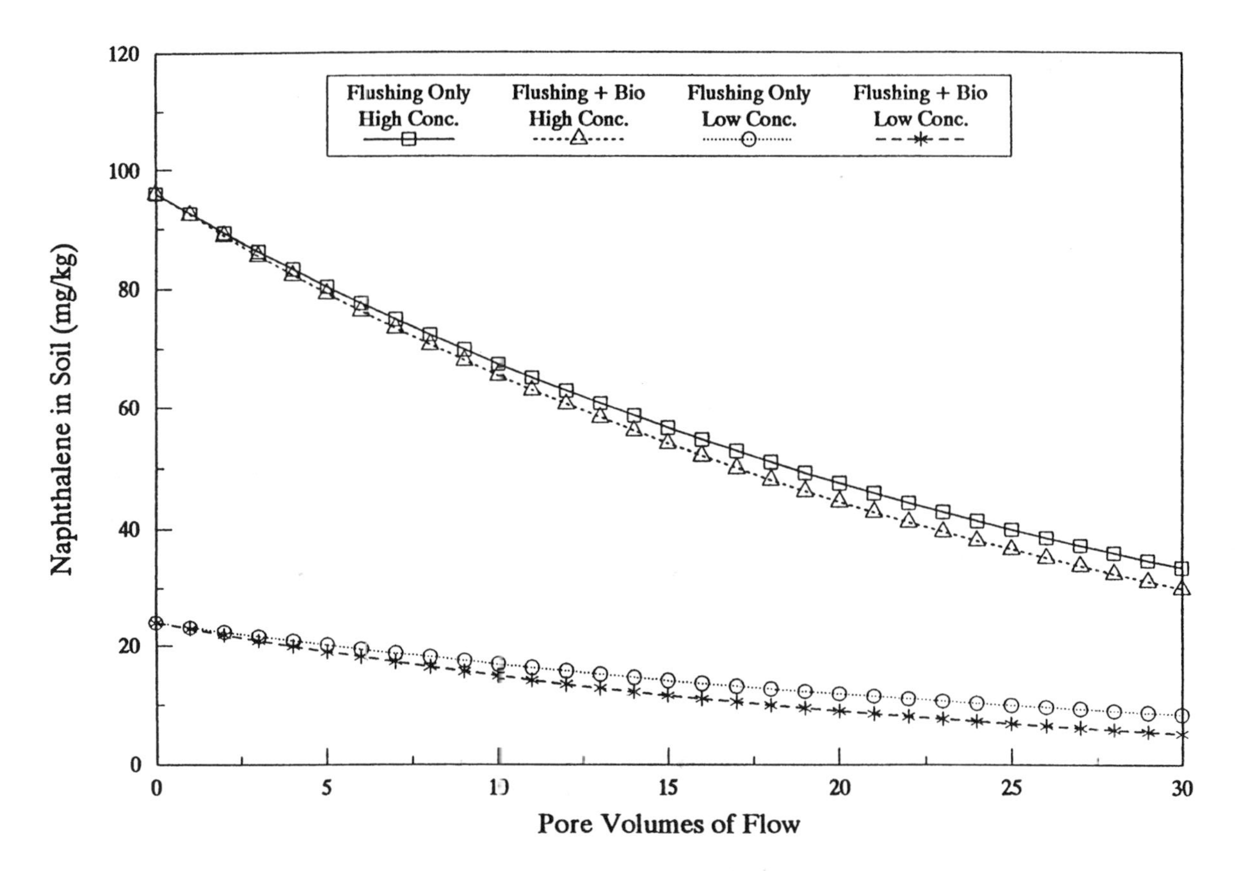

FIGURE 5. Predicted removal of naphthalene based on flushing model.

REFERENCES

Brubaker, G. R. 1991. "In Situ Bioremediation of PAH-Contaminated Aquifers." In *Petroleum Hydrocarbons and Organic Chemicals in Ground Water: Prevention, Detection and Restoration,* pp. 377-391. National Ground Water Association.

Brubaker, G. R., and H. A. Stroo. 1992. "In Situ Bioremediation of Aquifers Containing Polyaromatic Hydrocarbons." *J. Hazardous Mater.* 32: 163-177.

Brusseau, M. L. 1991. "Nonequilibrium Sorption of Organic Chemicals: Elucidation of Rate-Limiting Processes." *Environ. Sci. Technol.* 25(1): 134-142.

English, D. A., G. R. Brubaker, and D. Hargens. 1992. "Ground Water Removal and Treatment Enhanced with In Situ Bioremediation." Presented at Air & Waste Management Association, Kansas City, MO.

Middleton, A. C., D. V. Nakles and D. G. Linz. 1991. "The Influence of Soil Composition on the Bioremediation of PAH-Contaminated Soils." *Remediation* 1(4): 391-406.

Rittmann, B. E., A. J. Valocchi, E. Seagren, C. Ray, B. Wrenn, and J. R. Gallagher. 1992. *A Critical Review of In Situ Bioremediation.* Gas Research Institute Topical Report," GRI-92/0322, Chicago, IL.

Thomas, J. M., and C. H. Ward. 1989. "In Situ Biorestoration of Organic Contaminants in the Subsurface." *Environ. Sci. Technol.* 23(7): 760-766.

DEVELOPMENT OF A METHODOLOGY TO DETERMINE THE BIOAVAILABILITY AND BIODEGRADATION KINETICS OF TOXIC ORGANIC POLLUTANT COMPOUNDS IN SOIL

R. Govind, C. Gao, L. Lai, X. Yan,
S. Pfanstiel, and H. H. Tabak

ABSTRACT

The objectives of this research are to quantify the bioavailability and biodegradation kinetics of organic chemicals in surface and subsurface soil environments; to examine the effects of soil matrices and soil conditioning (drying, aging, compaction) on biodegradation rates and bioavailability of these compounds; to examine the threshold inhibition of these organics on microbiota in contaminated soils; and to develop a predictive model for biodegradation kinetics applicable to soil systems. Results have been presented on the degradation of six phenolic compounds in soil slurry reactors using electrolytic respirometry. The oxygen uptake data have been analyzed to quantitatively derive the Monod parameters for biodegradation in soil slurry systems using a mathematical model incorporating the effect of adsorption in soil. The compound adsorption/desorption isotherm parameters and biomass/soil adsorption parameter were determined experimentally using soil suspensions wherein biodegradation was not allowed to occur. These studies show that biokinetics in soil suspensions can be obtained by electrolytic respirometry using a mathematical model which incorporates the effect of compound desorption and adsorption of biomass in soil.

INTRODUCTION

Successful application of bioremediation to Superfund sites requires a fundamental understanding of the kinetics and factors that control biodegradation rates (Anderson & Domsch 1973, Brunner et al. 1985, Goring & Hamaker 1972, Tabak et al. 1992). Quantification of biodegradation kinetics can provide useful insights into the range of environmental parameters which will enhance bioremediation rates (Boethling & Alexander 1979, Tabak & Govind 1991).

Although extensive studies on biodegradation of contaminants in aqueous systems have been conducted, there is little information on bioremediation kinetics in soil environments for two reasons.

1. A large number of soil parameters can significantly impact biodegradation kinetics.
2. Changes in the bioavailability of the chemical due to irreversible binding to the soil matrix and diffusion into the pores.

This paper presents results on the biodegradation of five phenolic compounds (phenol, *p*-cresol, catechol, resorcinol, and 2,4-dimethyl phenol) in soil slurry reactors. The studies incorporate the use of soil microcosms for acclimation of soil microbiota, measurement of adsorption/desorption kinetics and equilibria, and measurement of respirometric O_2 uptake in soil slurry reactors.

BACKGROUND

Three types of parameters influence the rate of biodegradation: (1) those that determine the availability and concentration of the compound to be degraded; (2) those that affect the microbial population site and activity; and (3) those that control the reaction rate. The key factors that affect the degradation of synthetic chemicals in soil appear to be soil type; depth of soil; chemical concentration; soil microorganisms and acclimation; physical environment (including pH, temperature, oxygen availability, redox potential, and moisture content of the soil); and external carbon source (Goring & Hamaker 1972).

Several research studies in the soil bioremediation/treatability area using respirometry have recently been reported. Long-term respirometric biological oxygen demand (BOD) analysis was applied to bench-scale studies to continuously monitor bacterial respiration during growth in mixed organic wastes from contaminated water and soil (Graves et al. 1991). This information was used to make an initial determination regarding the need to further explore bioremediation as a potential remedial action technology using on-site, pilot-scale testing. A treatability study used electrolytic respirometer and biometer to determine the biodegradation potential of crude oil petroleum-based wastes (drilling mud, tarry material, and heavy hydrocarbons) as contaminants of a polluted soil site area (Anderson 1982). The treatability data provided biotreatment efficiencies of the petroleum wastes and were used to ascertain the bioremediation cleanup time.

Most biodegradation kinetics models have neglected sorption of the contaminant on soil particles. This factor has been shown to be important in contaminant transport and kinetically is manifested as a two-phase process, with an initial fast stage (< 1 hour) followed by a slower long phase (days), controlled by diffusion to internal adsorption sites (McDonald et al. 1991).

The kinetics of phenolic compound removal in soil were studied by Namkoong et al. 1988, and zero-order and first-order kinetic models were evaluated to determine their adequacy to describe the removal of 17 phenolic compounds in soil. Both models were shown to describe the removal of these phenols from soil with high-correlation coefficients. A kinetic model that incorporates microbial growth may be desirable to describe the removal of certain compounds when removal is associated with growth.

METHODOLOGY

Establishment of Soil Microcosm Reactors

The soil microcosm reactor is designed to simulate undisturbed soil systems. Its purpose is to acclimate indigenous microbiota to selected chemicals. The soil microcosm, shown in Figure 1, is an airtight rectangular reactor (50 cm × 30 cm × 30 cm) constructed of glass, and its frame is supported by stainless steel panels. The glass cover has six stainless steel liquid atomizing sprayers (Spraying Systems Inc., Chicago, Illinois) placed equidistantly to evenly spray the nutrient solution on top of the soil bed at predetermined intervals of time. The bottom panel of the reactor is equipped with ports to allow for drainage of a leachate from the soil microcosm reactor. A controlled flowrate of CO_2-free air, achieved by passing the air through KOH solution, is maintained over the soil surface. The headspace air exiting from the microcosm reactor is bubbled through a high-efficiency absorption bottle (Fisher-Milligan, Fisher Scientific) filled with KOH solution, to adsorb the CO_2 generated in the microcosm. Changes in pH of the KOH trapping solution are periodically measured to determine the amount of CO_2 absorbed.

Natural, uncontaminated forest soil was cut carefully and placed inside to fit the microcosm reactor. The soil bed was contaminated by spraying the top surface of the soil with a solution of the phenolic compounds. Nutrients, prepared following the OECD (Organization of Economic Cooperation and Development) medium guidelines (OECD 1981), was sprayed periodically to maintain the initial soil moisture content in the microcosm reactor.

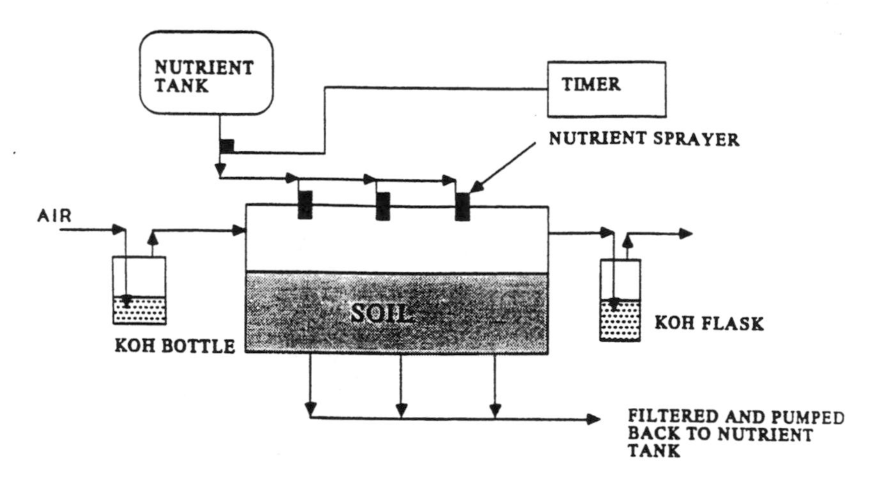

FIGURE 1. Schematic of a soil microcosm reactor system.

Each microcosm reactor represents a controlled contaminated site, which eventually selects out acclimated microorganisms for the contaminating organics. Samples of soil are then taken from the microcosm reactors and used as a source of acclimated microbial inoculum for measuring O_2 uptake respirometrically and for other studies.

Studies on Oxygen Uptake by Soil Slurries Using Electrolytic Respirometry

Studies were conducted with soil slurry reactors, wherein the O_2 uptake was monitored respirometrically (Sapromat B-12 electrolytic respirometer, Voith-Morden, Appelton, Wisconsin and Heidenheim, Germany).

The extent of biodegradation by soil microbiota and the Monod kinetic parameters for 6 phenolic compounds were determined from O_2 uptake data. Various concentrations of soil (2, 5, and 10%) and compounds (50, 100, and 150 mg/L) were mixed with a synthetic medium consisting of inorganic salts, trace elements, and either a vitamin solution or a solution of yeast extract and were stirred in the respirometric reactor flasks. The flasks were connected to the O_2 generation flask and to the pressure-indicator cells.

An OECD synthetic medium (OECD 1981) was used as the nutrients. The soil served as a source of inoculum. The concentration of forest soil in the reactor flask varied from 5 to 15% by weight, using the dry weight of soil as the basis. The total volume of the slurry in the flask was 250 mL. A more comprehensive description of the procedural steps of the respirometric tests is presented elsewhere (Tabak et al. 1992).

Kinetic Analysis of Soil Slurry Oxygen Uptake Data

The O_2 uptake data were analyzed by computer simulation techniques and curve fitting methods, using the Monod equation combined with a nonlinear adsorption isotherm, to determine the Monod equation biokinetic parameters. The model equations are summarized as follows:

$$X_s = K_b X_w \tag{1}$$

$$S_s = K_d S_w^{1/n} \tag{2}$$

$$dX_w/dt = u_w X_w S_w/(K_w + S_w) - b K_w X_t/(K_w + S_w) \tag{3}$$

$$dS_w/dt = -(1/Y)u_w X_w S_w/(K_w + S_w) \tag{4}$$

$$dS_p/dt = -Y_p(dS_t/dt) \tag{5}$$

$$X_t = X_s + X_w \tag{6}$$

$$S_t = S_s + S_t \tag{7}$$

$$O_2 = (S_{to} - S_t) - (X_t - X_{to}) - (S_p - S_{po}) \tag{8}$$

$$X_t = X_{to} \text{ when } t = 0 \tag{9}$$

$$S_t = S_{to} \text{ when } t = 0 \tag{10}$$

$$S_p = S_{po} \text{ when } t = 0 \tag{11}$$

where subscripts s, w, and p represent the soil, water, and degradation products, respectively. Subscript t represents the total concentration (soil and water) in the system. S, X, and S_p are concentrations of compound, biomass, and degradation products, respectively. The u_w, K_w, Y, and Y_p are the Monod equation maximum specific growth rate parameter, Michaelis constant, biomass yield, and product yield coefficient, respectively. The b is the biomass decay coefficient, K_d is the soil adsorption isotherm parameter, (1/n) is the soil adsorption intensity coefficient, and K_b is the biomass adsorption parameter. O_2 is the cumulative oxygen demand and t denotes time.

The model assumes that biodegradation occurs predominantly in the liquid phase and the effect of adsorption/desorption kinetics is neglected in the model. These assumptions need to be tested by further experimentation.

The soil adsorption/desorption equilibria isotherm parameters (K_d, n) are determined experimentally using abiotic soil suspensions, as described in the following sections.

The biomass adsorption parameter (K_b) is determined by incubating soil microbiota with radiolabeled phenol in a respirometric reactor until an oxygen uptake plateau was obtained indicating that all phenol had either biodegraded into $^{14}CO_2$ or ^{14}C biomass. After the soil suspension settles, 1 mL of supernatant is sampled and its ^{14}C activity is determined using liquid scintillation counting. Equilibrium amounts of the ^{14}C biomass adsorbed to the soil are determined by subtracting the ^{14}C present in the biomass in suspension and the ^{14}C present as CO_2 absorbed in the KOH solution from the total ^{14}C added initially. The ratio of the biomass adsorbed to the soil and the biomass present in the suspension gives the biomass/soil adsorption isotherm parameter, K_b.

The experimental values of the oxygen uptake are matched with the theoretically calculated values, and the best fit for the Monod equation parameters (u_w, K_w, Y, Y_p) is obtained using an adaptive random search method.

Adsorption-Desorption Equilibria and Kinetics

Adsorption Studies. Soil adsorption kinetics and equilibria are measured using batch well-stirred bottles. The soil is initially air-dried and then sieved to pass a 2.00-mm sieve. Then 10 g of soil sample is placed in each bottle and mixed with 100 mL of distilled deionized water containing various concentrations of the compound and mercuric chloride to minimize biodegradation. The soil:solution ratio is expressed as the oven-dry equivalent mass of adsorbent in grams per volume of solution.

The liquid is sampled after 2, 4, 6, 8, 10, 12, 14, 16, 18, 20, and 36 hours. After a predefined time has elapsed, the bottle contents are centrifuged and the liquid sample is taken using a syringe connected to a 0.45 μm porous silver membrane filter. The filter prevents soil particles from entering the sample. The concentration of the chemical compound in the liquid sample is analyzed using standard extraction and gas chromatography/mass spectroscopy (GC/MS) technique. Methylene chloride is used to extract the water sample. From the initial amount of compound and analysis of the liquid phase, the amount of compound adsorbed in the soil is obtained by the difference.

Equilibrium is defined when the liquid concentration reaches a stationary value, which is usually attained in 24 hours. The equilibrium data are used to obtain Freundlich isotherm parameters.

Desorption Studies. Desorption studies are conducted by first adsorbing the chemical in the soil until equilibrium is achieved. Then desorption studies are conducted by withdrawing a specified volume of the liquid and then adding deionized distilled water. The concentration in the liquid phase is measured to obtain the desorption kinetics and equilibria.

To 20 g of soil and specified concentration of chemical is mixed 100 mL of deionized distilled water. After equilibrium is attained, the sample is centrifuged and 90 mL of liquid is filtered out and replaced with an equal volume of deionized distilled water with 20 g/L of mercuric chloride to inhibit biodegradation. At 4, 8, 16, 24, 48, 72, 96, and 120 hours, a 20-mL sample is withdrawn.

Each sample is withdrawn from a separate adsorption bottle. The sample is extracted with methylene chloride and analyzed using GC/MS technique to determine the concentration of solute.

RESULTS AND DISCUSSION

With the use of these specially designed microcosm reactors, it was possible to acclimate the indigenous microbiota to the six phenolic compounds. Soil samples from the microcosm reactors can be used as a source of acclimated microbiota to determine biodegradation kinetics.

Methodology was developed to determine adsorption and desorption equilibria and kinetics for six phenolic compounds (phenol, *p*-cresol, catechol, resorcinol, hydroquinone, and 2,4-dimethyl phenol). Representative curves for phenol adsorption in soil are illustrated in Figure 2 and for desorption of phenol from soil in Figure 3.

Table 1 provides data on the adsorption/desorption Freundlich isotherm parameters for the six phenolic compounds. A very good fit ($R^2 > 0.98$) between the experimental data and the Freundlich isotherm equation was obtained for both adsorption and desorption equilibria. The isotherm parameters (K_d, $1/n$), obtained by fitting the Freundlich equation to the experimental data, were used in the analysis of the oxygen uptake data. The desorption isotherms for all six phenolic compounds are nearly identical.

The biomass/soil adsorption data was fitted to a linear isotherm (equation 1) and a good fit ($R^2 = 0.98$) was obtained. The biomass/soil adsorption parameter (K_b) was used in analyzing the oxygen uptake data.

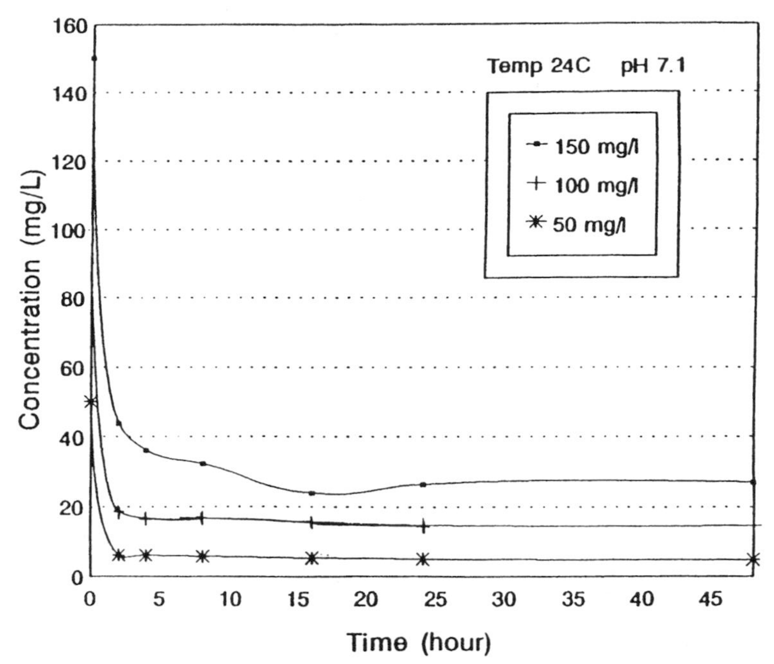

FIGURE 2. Abiotic adsorption curves for phenol in soil slurry system with initial concentrations of 50, 100, and 150 mg/L.

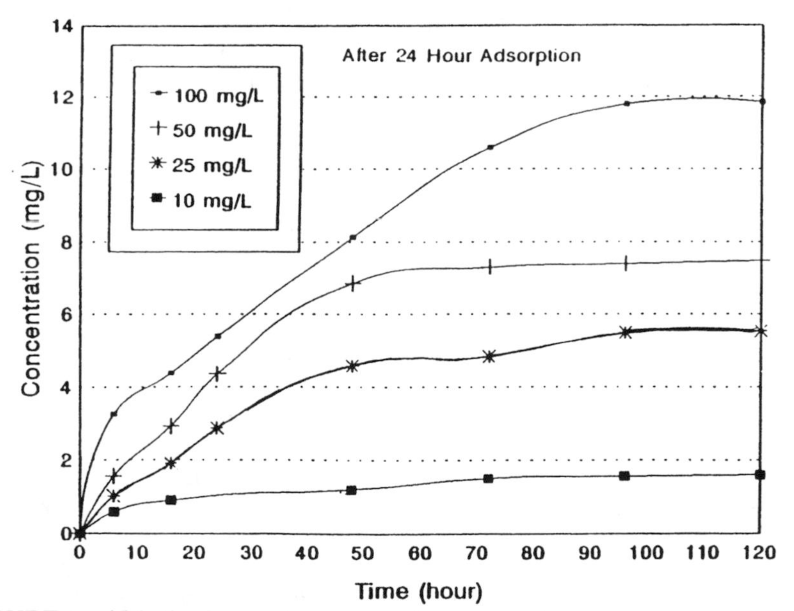

FIGURE 3. Abiotic desorption curves for phenol in soil slurry system with initial concentrations of 10, 25, 50 and 100 mg/L.

TABLE 1. Adsorption/desorption isotherm parameters for six phenolic compounds.

	Adsorption		Desorption	
Compound	K_d	1/n	K_d	1/n
Phenol	17.5	0.78	5.0	1.0
p-Cresol	24.6	0.50	5.0	1.0
2,4-Dimethylphenol	18.7	0.89	5.0	1.0
Catechol	16.3	0.84	4.8	1.0
Hydroquinone	17.7	0.92	5.0	1.0
Resorcinol	33.1	0.77	5.0	1.0

K_b = 167.0 L/Kg
Units: K_d:[(mg/Kg)(L/mg)$^{1/n}$]

Respirometric oxygen uptake data were generated for 5 phenolic compounds (phenol, *p*-cresol, catechol, resorcinol, 2,4-dimethyl phenol) using soil slurry systems. Three different compound concentrations (50, 100, and 150 mg/L) and three different soil concentrations (2, 5, and 10%) were selected in these studies. The soil samples were completely mixed with nutrients and compounds, and the oxygen uptake generated was measured respirometrically.

Figure 4 shows the representative oxygen uptake curve for phenol at an initial concentration of 100 mg/L, at soil slurry concentrations of 0, 5 and 10% by weight of dry soil. No significant oxygen uptake occurred when no soil was present due to the absence of soil microorganisms. The total oxygen uptake for 10% soil was higher than for 5% soil, and the lag time for 10% soil also was lower than that for 5% soil. Similar results were obtained for the other compounds, except hydroquinone which did not biodegrade at the experimental concentrations. Clearly, either hydroquinone was toxic to soil microbiota or acclimation prior to respirometric studies will have to be conducted.

The oxygen uptake curves for 5% soil concentration at various initial concentrations of phenol (0, 50, 100, and 150 mg/L) are shown in Figure 5. At 0% phenol, the oxygen uptake occurs mainly due to the organic matter present in the soil. As the phenol concentration increases, the oxygen uptake increases also, due to degradation of phenol.

It should be noted that the theoretical oxygen demand for 150 mg/L of phenol in the aqueous phase is 357.4 mg of oxygen. However, when soil is present, the actual final oxygen demand at 150 mg/L was 300 mg/L, which indicated that some of the phenol may have diffused into the soil particle micropores and hence was inaccessible to the bacteria (Roy et al. 1987).

The oxygen uptake data for the selected phenolic compounds were analyzed and fitted using an adaptive random search technique to a nonlinear mathematical model comprised of the Monod equation and a nonlinear adsorption isotherm. Experimentally determined values for the soil/compound adsorption isotherm

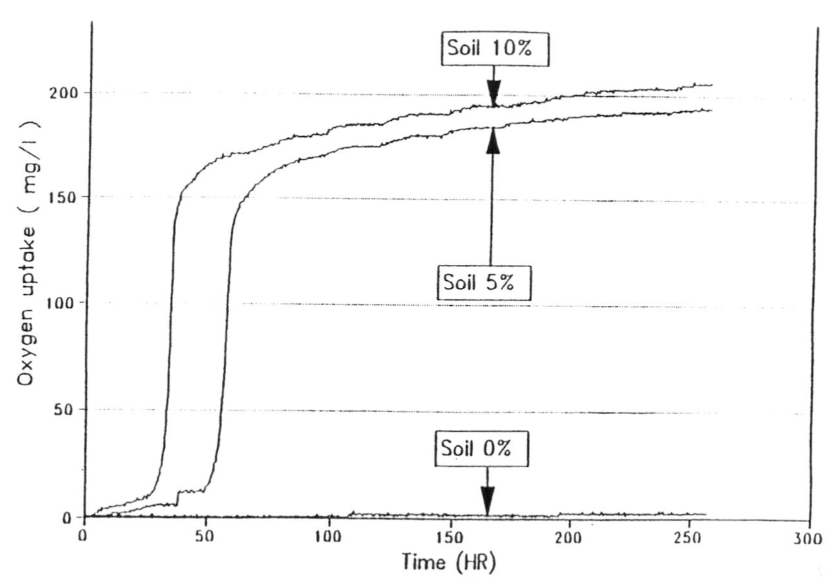

FIGURE 4. Oxygen uptake curves for phenol in soil slurry reactor at an initial concentration of 100 mg/L and soil slurry concentrations of 0, 5, and 10% by weight of dry soil.

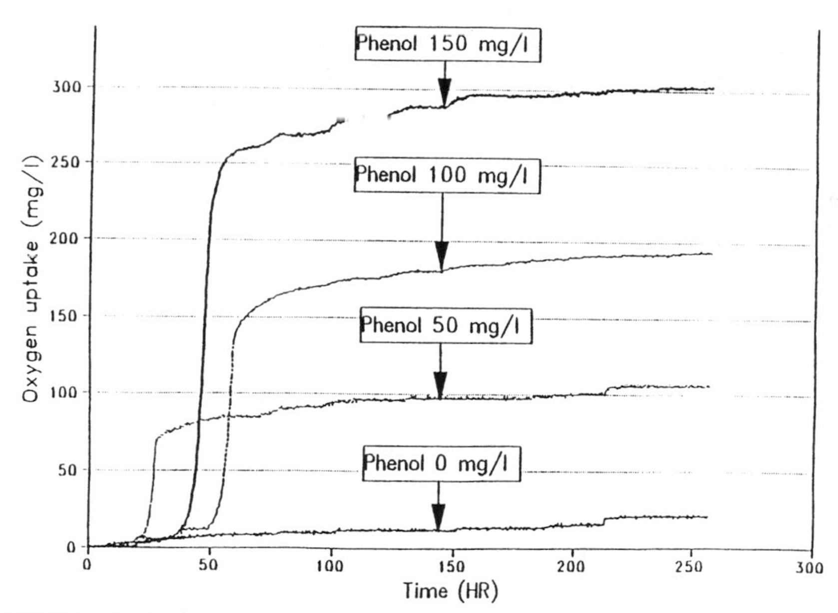

FIGURE 5. Oxygen uptake curves for phenol in soil slurry reactors at various initial concentrations of 0, 50, 100 and 150 mg/L and 5% soil slurry concentration.

parameters and the soil/biomass isotherm parameter were used in the analysis. The goodness of fit, defined as a mean relative residue error (RREM), between the model and the experimental oxygen uptake data, obtained for various soil and compound concentrations, is defined as follows:

$$\mathrm{RREM} = \left(\sum_i^n \left(o_2 - o_2^{exp} \right)_i^2 / (n-1) / o_{2,max} \right)^{1/2} \tag{12}$$

where o_2 is the theoretical oxygen uptake calculated from the model, o_2^{exp} is the experimental oxygen uptake value, $o_{2,max}$ is the maximum value of oxygen uptake (plateau value), and n is the number of experimental points.

A mean relative residue error of less than 5% was used to determine the best fit Monod kinetic parameters.

The Monod constants for five phenolic compounds are summarized in Table 2. Except for K_s, the other biokinetic parameters did not vary significantly with changes in compound or soil concentration. The specific growth rate parameter, u_w, the biomass yield, Y, and bioproduct yield coefficient, Y_p, did not vary significantly with soil concentration.

Adsorption for all the compounds studied was significantly faster than desorption and biodegradation (refer to Figures 3, 4). Thus, it was found that using desorption isotherm parameters produced a significantly better fit of the model to the experimental oxygen uptake data. Hence it was concluded that compounds initially adsorbed before significant biodegradation had occurred and only the desorption step had a significant impact on the biodegradation rates. In all the compounds studied it was found that desorption from soil controlled the rate of compound mineralization.

CONCLUSIONS

Respirometric studies with soil slurry reactors provide valuable insight into the biodegradation kinetics of compounds in the presence of soil. It has been

TABLE 2. Monod biokinetic parameters for five phenolic compounds obtained from experimental oxygen uptake data.

Compound	u_w (hr^{-1})	K_w (g/L)	Y (g/g)	$Y_p \times 10^4$ (g/g)	$b \times 10^4$ (hr^{-1})
Phenol	0.415	3.47	0.428	6.93	4.38
p-Cresol	0.279	8.76	0.476	8.33	7.95
Resorcinol	0.360	5.02	0.226	10.10	2.07
Catechol	0.560	6.11	0.483	74.0	61.60
2,4-Dimethylphenol	0.492	1.23	0.343	5.90	9.53

shown that a Monod kinetic equation in conjunction with a nonlinear desorption isotherm can provide reliable estimates of the Monod kinetic parameters.

Further studies are planned for continuing the respirometric O_2 uptake for other compounds. Findings from these studies would enable the generation of a database of Monod kinetic parameters and soil adsorption coefficients for various compounds. Eventually, this database will be used to develop predictive models using structure-activity relationships, so that the biokinetics of a wide variety of compounds in soil systems can be estimated.

ACKNOWLEDGMENT

This project has been funded by the U.S. Environmental Protection Agency Risk Reduction Engineering Laboratory, Cincinnati, Ohio.

REFERENCES

Anderson, J. P. E. 1982. "Soil Respiration, Methods of Soil Analysis," Part 2, *Chemical and Microbiological Properties. Agronomy Monograph,* No. 9, 2nd ed., pp. 831-871.

Anderson, J. P. E., and K. H. Domsch. 1973. "Quantification of Bacterial and Fungal Contributions to Soil Respiration." *Arch. Mikrobiol. 93*: 113-127.

Boethling, R. S., and M. Alexander. 1979. "Microbial Degradation of Organic Compounds at Trace Levels." *Environ. Sci. Technol. 13*: 989-991.

Brunner, W., F. H. Sutherland, and D. D. Focht. 1985. "Enhanced Biodegradation of Polychlorinated Biphenyls in Soil by Analog Enrichment and Bacterial Inoculation." *J. Environ. Qual. 14*(3): 324-328.

Goring, C. A. I., and J. W. Hamaker. 1972. *Organic Chemicals in the Soil Environment.* Marcel Dekker, New York, NY.

Graves, D. A., C. A. Lang, and M. E. Leavitt. 1991. "Respirometric Analysis of the Biodegradation of Organic Contaminants in Soil and Water." *Applied Biochem. Biotechnol. 28/29*: 813-826.

McDonald, J. P., C. Baldwin, and L. E. Erickson. 1991. "Rate Limiting Factors for In-Situ Bioremediation of Soils Contaminated with Hydrocarbons." Paper presented at the Fourth International IGT Symposium on Gas, Oil, and Environmental Biotechnology.

Namkoong, W., R. C. Loehr, and J. F. Malina. 1988. "Kinetics of Phenolic Compounds Removal in Soil." *Hazardous Waste and Hazardous Materials 5*(4): 321-328.

OECD. 1981. *OECD Guidelines for Testing of Chemicals,* Section 3, Degradation and Accumulation, Method 301C, Ready Biodegradability: Modified MITI Test (I) adopted May 12, 1981 and Method 302C Inherent Biodegradability: Modified MITI Test (II) adopted May 12, 1981, Director of Information, Org. for Economic Cooperation and Development, Paris, France.

Roy, W. R., I. G. Krapac, S. F. J. Chou, and R. J. Griffin. 1987. *Batch-Type Procedures for Estimating Soil Adsorption of Chemicals.* U.S. Environmental Protection Agency Technical Report, EPA/530-SW-87-006-F, Risk Reduction Engineering Laboratory, Cincinnati, OH.

Tabak, H. H., and R. Govind. 1991. "Determining Biodegradation Kinetics with use of Respirometry for Development of Predictive Structure-Biodegradation Relationship Models." Presented at the 4th International IGT Symposium, Colorado Springs, CO.

Tabak, H. H., and R. Govind, C. Gao, I.S. Kim, L. Lai, and X. Yan. 1992. "Development of Methodology for Determining Bioavailability and Biodegradation Kinetics of Toxic Organic Compounds in Soil." Presented at the 85th Annual Meeting of the Air and Waste Management Association, Kansas City, MO.

BIOREMEDIAL PROGRESS AT THE LIBBY, MONTANA, SUPERFUND SITE

M. R. Piotrowski, J. R. Doyle, D. Cosgriff, and M. C. Parsons

ABSTRACT

An integrated, biological treatment program has been implemented to remediate soils, groundwater, and aquifer sediments contaminated by wood preservatives (creosote and pentachlorophenol) at the Libby Superfund Site in northwest Montana. Key features of the site's contamination include 45,000 yds^3 (~34,405 m^3) of contaminated soil, 27,000 yds^3 (~20,645 m^3) of contaminated rock debris, and a groundwater contaminant plume approximately 1 mi (1.6 km) in length. Progress toward the development of the bioremedial design implemented at the site began with conducting key pilot-scale and prototype demonstrations of the capabilities of the selected bioremediation technologies under site conditions during the preparation of the Feasibility Study (completed September 1988) and the Remedial Action Plan. The bioremedial design includes the operation of land treatment units (LTUs) for soil treatment, the operation of aqueous-phase bioreactors for aboveground treatment of heavily contaminated groundwater, spray irrigation using bioreactor effluent for treatment of contaminated rock debris, and the injection of hydrogen peroxide and oxygen for in situ aquifer treatment at several on-site locations. Portions of the remedial program have been in operation since 1989, and system installation is being completed. Operation of the complete remedial system will be under way shortly. This paper summarizes progress made during implementation of the remedial design.

INTRODUCTION

This paper provides a status report on the accomplishments and activities to implement an integrated bioremediation program to treat contamination in soil and groundwater derived from past wood-treating operations at a site located in Libby, Montana. The site was among the first listed as a Superfund site in 1983, and the U.S. Environmental Protection Agency's (EPA's) Record of Decision for bioremediation of the contaminated soil and groundwater was issued in December 1988.

The remedial design consists of a land treatment program to remediate the contaminated soils, a groundwater extraction and aboveground treatment system to treat the heavily contaminated groundwater, and two oxygen injection systems for in situ biological treatment of the contaminated aquifer (Figure 1). Remediation of the Libby site was initiated in the spring and summer of 1989.

Construction activities for the soil treatment program were conducted from 1989 through 1991. Lifts of contaminated soil have been treated to reduce the contaminant concentrations to established cleanup levels since 1989 and through 1992, nine lifts totaling 5,686 yds^3 (4,347 m^3) have been treated.

Construction of the groundwater extraction and treatment system began in 1989. The aboveground treatment system consists of aerobic, heated bioreactors operated in series to remove contamination from groundwater extracted from the most highly contaminated aquifer region around the major contaminant source area.

The in situ bioremediation portion of the remedial program has been under way at the site since 1989 when the pilot-scale system that was successfully used to evaluate the technological approach was expanded. More recently, a second water-conditioning and injection system has been installed along the downgradient boundary of the site to treat the contaminant plume in off-site areas.

These systems have been operated since that time with reasonable success. Adjustments have been made as required to better address certain technical limitations in the original design and to meet overall site remediation requirements. Principal among the technical issues is the presence of nonaqueous-phase liquids (NAPLs) in the groundwater.

This project represents one of the first comprehensive applications of bioremediation technology to remediate a complex, multimedia, contaminated site. Champion International, Inc. and Woodward-Clyde Consultants, the technical consultant throughout this project, believe the efforts and results to date indicate that this approach can be as effective and less costly than other approaches involving more traditional remedial technologies.

SITE DESCRIPTION

The site is an active, lumber-production facility located in an alluvial valley in northwest Montana where wood-treating operations were conducted from 1946 through 1969. Surface soil grain sizes range from clays to gravels where deposits of fine-grained materials often occur adjacent to deposits of large-grained materials. Uncontrolled releases of two types of wood preservatives, creosote and pentachlorophenol (PCP), to the surface soils occurred in a waste pit and associated areas. The primary contaminants of concern are the polycyclic aromatic hydrocarbons (PAHs) derived from the creosote and PCP. These soil areas became the focus of the soil remediation effort.

In addition to the contaminated soils, the groundwater beneath the site is also contaminated with PCP and PAHs. Remedial efforts have been focused on the shallowest groundwater-bearing unit (the Upper Aquifer) due to its potential

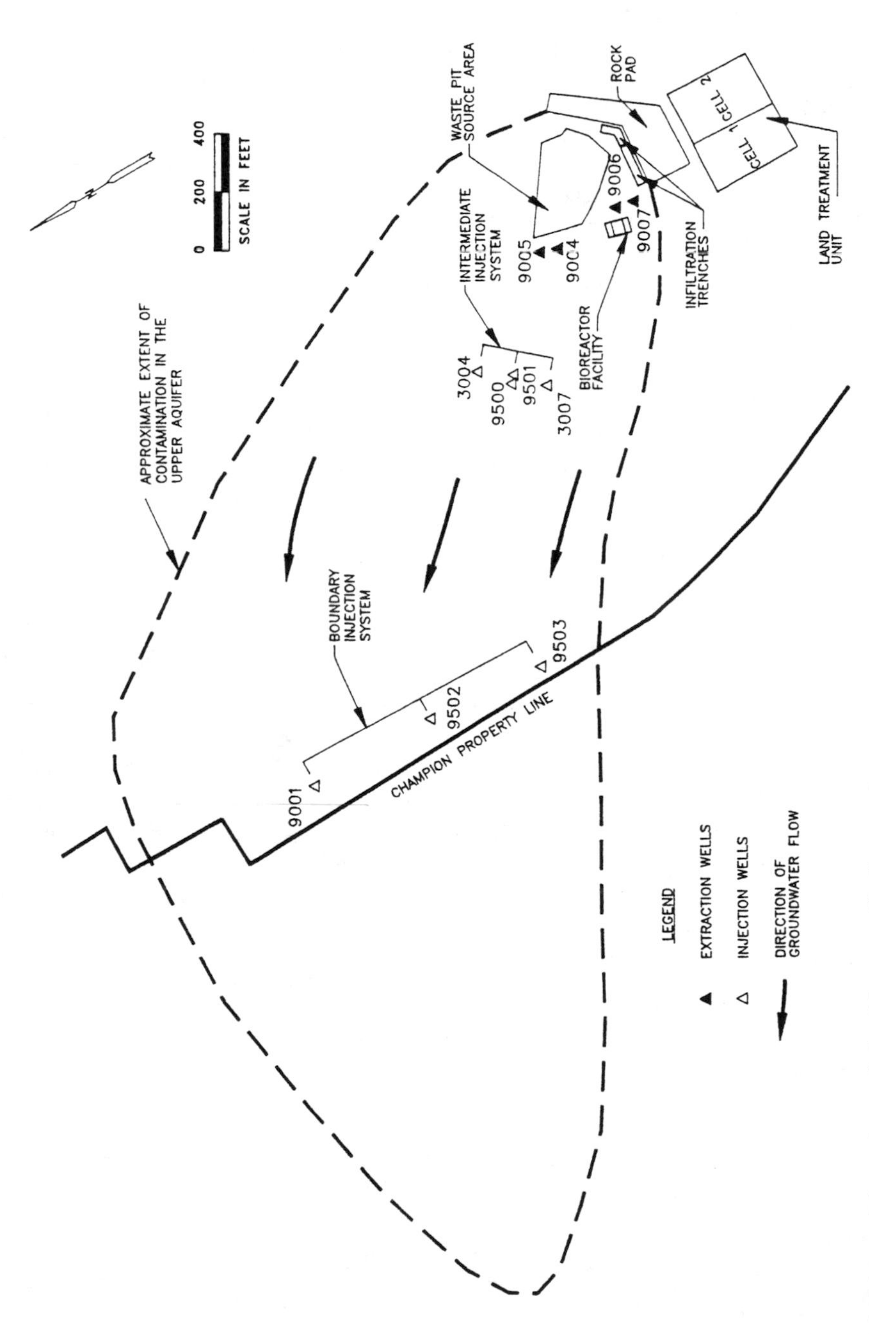

FIGURE 1. Plan view of the bioremediation systems.

use as a domestic water supply. The Upper Aquifer is located at depths of ~18 to ~70 ft (~5.5 to ~21 m) below ground surface (bgs) and is underlain by a low hydraulic conductivity zone that contains silt and clay. The contaminant plume in the Upper Aquifer is nearly 1 mi (1.6 km) in length.

Because the subsurface geology of the site is complex (with clay lenses often occurring adjacent to gravel lenses), the hydrogeology of the site is also complex and local groundwater flow patterns are poorly predictable. However, the large-scale direction of groundwater flow in the Upper Aquifer is to the northwest.

REMEDIAL PERFORMANCE SUMMARIES

Soil Treatment Program

The objective of the Soil Treatment Program at the Libby site is to reduce the concentrations of the PCP and PAHs in the affected soil materials to below the remedial goals established for each contaminant (Table 1). The program consists of several treatment actions: in situ land treatment of de-rocked, contaminated soils stored in the former waste pit; ex situ treatment of de-rocked, contaminated soils in two 1-acre (0.4 hectare) land treatment units (LTUs); and irrigation treatment of contaminated rock piles using effluent from the above-ground treatment system for groundwater.

Contaminated soils were excavated down to the water table in the source areas of the site in the spring of 1990. The excavated materials were passed through an automated de-rocking device to separate the contaminated soils from rocks greater that 1-in (2.54 cm) in diameter. Approximately 72,000 yds^3 (~55,050 m^3)

TABLE 1. Summary of remedial action goals established by EPA for soil and groundwater at the Libby, Montana site.[a]

Medium	Contaminant	Remedial Goal
Soil	Carcinogenic PAH[b]	88 mg/kg
	Naphthalene[c]	8.0 mg/kg
	Phenanthrene[c]	8.0 mg/kg
	Pyrene[c]	7.3 mg/kg
	Pentachlorophenol[c]	37.00 mg/kg
Groundwater	Total PAH[d]	400 ng/L
	Carcinogenic PAH	40 ng/L
	Pentachlorophenol	1.05 mg/L

(a) Source: U.S. EPA 1988.

(b) Carcinogenic PAH = (suspected) carcinogenic priority pollutant polycyclic aromatic hydrocarbon (PAH) compounds.

(c) Goal based on proposed treatment requirement for land disposal of K001 RCRA waste.

(d) Total PAH = Total concentrations of 16 Priority Pollutant PAH compounds.

of contaminated materials were excavated from the source areas, consisting of 45,000 yds^3 (~34,405 m^3) of soils and 27,000 yds^3 (~20,645 m^3) of rocks.

Construction of the first LTU (Cell 1) was completed in May 1989, and it has been used to treat soil lifts since that time. The second LTU (Cell 2), completed in the spring of 1991, received its first lift of contaminated soil in late July of that year.

Soil Pretreatment. The de-rocked, contaminated soils were placed in the excavated waste pit for storage and pretreatment before delivery to the LTUs for final treatment. Pretreatment consists of weekly tilling and periodic irrigation of the surface soils to initiate and maintain a degree of biodegradation of the organic contaminants in the soils during the warmer seasons of the year. No formal monitoring program has been conducted to evaluate the effect of the pretreatment process on contaminant biodegradation rates in the surface soils in the former waste pit.

Soil Treatment in the LTUs. Typically at the beginning of each summer, two lifts of the pretreated surface soil (~600 to 700 yds^3 [~460 to 535 m^3] of soil per lift) are excavated from the former waste pit and placed on the LTUs. Each lift is spread over the LTU to an average depth of approximately 8 to 10 in (20 to 25 cm).

The concentrations of total organic carbon, total Kjeldahl nitrogen, and total phosphorus are assessed in the soil after lift placement. A stoichiometric evaluation of the ratios of carbon:nitrogen:phosphorus in the soil lift is then performed, and if nitrogen or phosphorus are found to be at concentrations that could potentially limit microbial activities, nutrients are added to the soil lift.

Lift treatment involves weekly tilling and periodic irrigation with water from an on-site pond to maintain soil moisture levels in the lift. These actions stimulate appreciable rates of biodegradation of the contaminants of concern which are evaluated using an EPA-approved monitoring program. Once the remedial goals for the contaminants of concern have been attained in a soil lift (Table 1), a second lift of contaminated soils is placed on the treated lift, and treatment of the new lift is initiated.

Performance Characteristics of Soil Treatment in the LTUs. The performance of the land-treatment process is evaluated through an EPA-approved monitoring program. Sampling intervals may vary from weekly to monthly during treatment of a specific lift. During each sampling period, four composited soil samples are collected, one from each of four designated quadrants in each treatment cell. The soil samples are analyzed for the concentrations of PCP (EPA Method 8040) and PAHs (EPA Methods 8100 or 8310).

The time requirements for lift treatments have varied from 32 to 163 days of active treatment and the time requirement for an individual lift depends on the initial concentrations of the contaminants in the lift, the time of year lift treatment was begun, the rates of biodegradation achievable for the organic contaminants of concern, and climate characteristics (temperature and precipitation) during lift treatment. Figures 2 through 4 summarize the concentrations

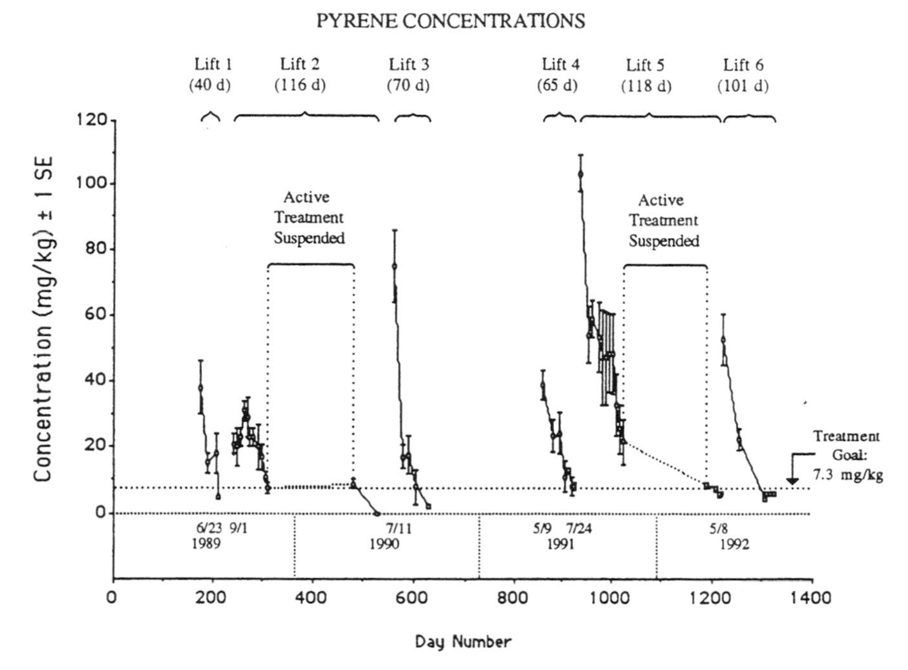

FIGURE 2. Time line of mean (±1 standard error [SE]) pyrene concentrations (in gm/kg) in soil lifts treated in the Cell 1 land treatment unit at the Libby, Montana site, 1989-1992. Beneath each lift number is the number of days of active treatment required for successful treatment of the lift.

pyrene, PCP, and carcinogenic PAHs (CPAHs) in the lifts treated in Cell 1 from 1989 through 1992.

To date, the foremost determinant of the time period required for successful lift treatment is the biodegradation rate achieved for pyrene (Figure 2). This compound is the most recalcitrant of the contaminants of concern with respect to its remedial goal (7.3 mg/kg), and attainment of the goal has been difficult at times. As a result of the inability of land treatment to readily reduce pyrene concentrations below the remedial goal, three lifts (one in 1989 and one in 1991 in Cell 1, and one in 1991 in Cell 2) were not successfully treated before the onset of winter in the respective treatment years. Their treatment continued during the following springs until satisfactory pyrene reductions were attained (Figure 2).

These ranges in time requirements for effective lift treatment represent periods of active treatment (i.e., tilling and irrigation). Three of the time periods do not include periods when treatment was suspended (i.e., during the winters of 1989-90 and 1991-92).

Biodegradation rates of the other contaminants of concern (e.g., PCP and CPAHs) characteristically have been much higher (Figures 3 and 4). If treatment time requirements for lifts depended only on reductions in concentrations of PCP or CPAHs, the times required would be much lower (Table 2), and the overall soil processing rate would increase.

Nevertheless, by the end of 1992, approximately 4,232 yds^3 (3,236 m^3) of contaminated soil (seven lifts) had been successfully treated in Cell 1 and approximately 1,454 yds^3 (1,111 m^3) of contaminated soil (two lifts) had been treated in Cell 2. Therefore, approximately 5,686 yds^3 (4,347 m^3) of contaminated soil have been effectively treated in the LTUs over a 4-year period with active treatment each year confined to late spring, summer, and fall.

Toxicity Reduction Evaluation. Most assessments of the effectiveness of bioremediation programs for treating contaminated soils or waters focus on reducing concentrations of the priority pollutant organic compounds of concern at a particular site. This approach does not address the potential for the biodegradative process to produce intermediate breakdown products that could impart greater toxicity than the parent organic compounds.

Recognition of this potential has resulted in uncertainty in the overall evaluation of bioremediation as an effective tool in treating contaminated soil or water. This uncertainty is further accentuated when multiple organic contaminants are present in the medium being treated, as is the case with the contaminated soils and waters at the Libby site.

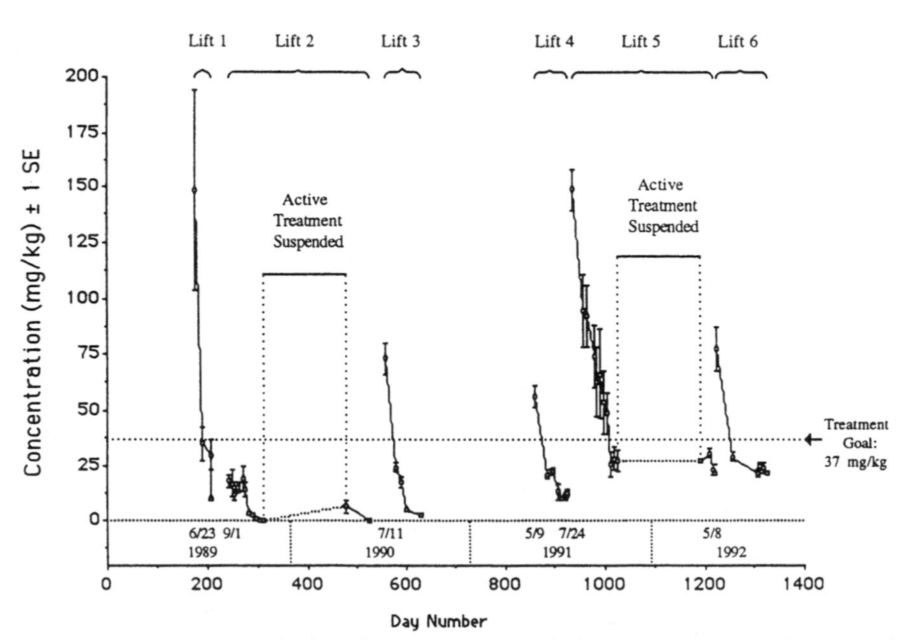

FIGURE 3. Time line of mean (±1 standard error) PCP concentrations (in mg/kg) in soil lifts treated in the Cell 1 land treatment at the Libby, Montana site, 1989-1992.

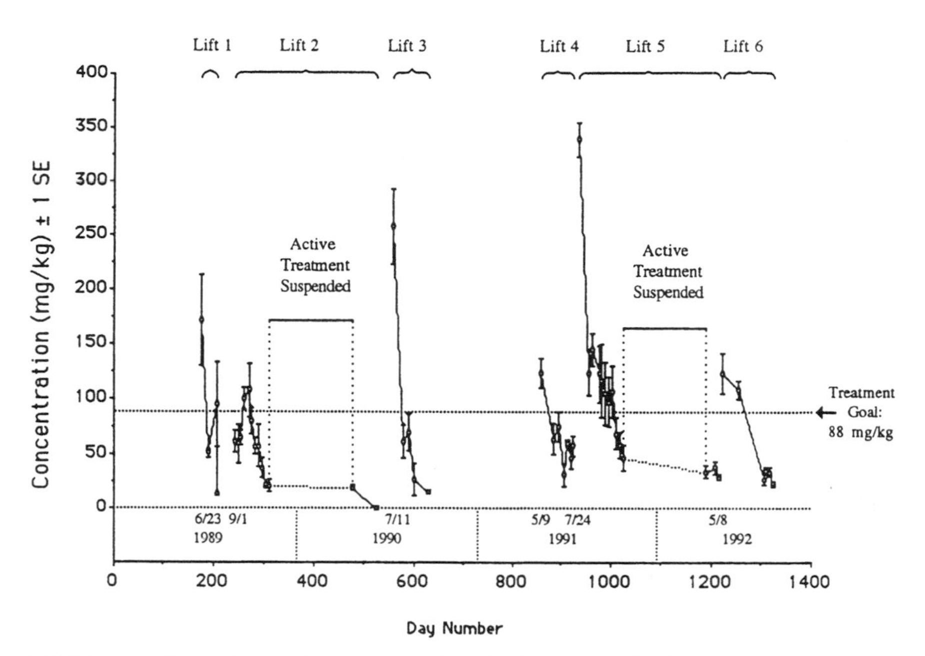

FIGURE 4. Time line of mean (±1 standard error) CPAH concentrations (in mg/kg in soil lifts treated in the Cell 1 land treatment unit at the Libby, Montana site, 1989-1992.

Therefore, although the soil treatment program at the Libby site demonstrated reduced concentrations of the priority pollutant contaminants of concern to below the remedial goals, questions remained regarding the potential development of toxic intermediates. A study conducted by researchers from Texas A&M University during the 1989 Land Treatment Demonstration, however, has provided independent evidence that this soil treatment program is indeed appreciably reducing the toxicity of the treated soils (Donnelly et al. 1992).

The Texas A&M study involved the collection and toxic screening of multiple soil samples from two lifts in Cell 1 and background soil samples from the Libby surroundings. One lift (Lift 1) had undergone land treatment for 3 months, but the second lift (Lift 2) had been treated for only 5 weeks.

Treatment of Lift 1 had been completed approximately 10 weeks prior to the Texas A&M sampling event as the concentrations of contaminants of concern had been reduced below the remedial goals. At the end of Lift 1 treatment, the mean (±1 Standard Error) concentration of PCP was 10 ± 0.3 mg/kg, the mean pyrene concentration was 5 ± 0.5 mg/kg, and the mean CPAH concentration was 13 ± 2 mg/kg. In addition, the mean concentration of total PAHs (TPAHs) was 20 ± 2.5 mg/kg.

TABLE 2. Summary of time period requirements for successful treatment of soil lifts[a] in land treatment units at the Libby, Montana site, 1989-1992, based on individual contaminants or contaminant categories.

		Treatment Time Period Requirements (Days) Based on:		
Cell Number	Lift Number	PCP	CPAHs	Pyrene
1	1	32	34	34
1	2	NA[b]	41	116[c]
1	3	21	21	70
1	4	22	22	65
1	5	75	75	117[c]
1	6	33	82	94
2	1	73	73	163[c]
Range in Time Periods (days):		21-75	21-82	34-163
Mean (±1 SE) Time Period (days):		43 ± 10	50 ± 10	94 ± 16

(a) Successful lift treatment defined as attaining the remedial goal for the organic contaminant(s) of concern. The remedial goals of concern for this table are as follows: PCP, 37 mg/kg; CPAHs, 88 mg/kg; and pyrene, 7.3 mg/kg.
(b) NA = Not Applicable. The concentrations of PCP at the beginning of lift treatment were below the remedial goal of 37 mg/kg.
(c) Time period requirement does not include period when active treatment of a lift was suspended for the winter.

Treatment of Lift 2 was under way during the Texas A&M sampling event, and the mean concentrations of the contaminants in the lift at that time were PCP, 14 ± 3.4 mg/kg; pyrene, 23 ± 3 mg/kg; and CPAHs, 79.8 ± 12.2 mg/kg. Furthermore, the mean concentration of TPAHs was 204 ± 61 mg/kg.

The Texas A&M researchers assessed the toxicity of extracts from the soils samples using the *Salmonella*/microsome bioassay (the Ames test). Their results indicated that the toxicity levels of the fully treated (Lift 1) samples were within the range of the values for the Libby background samples and the toxicity levels of the partially treated (Lift 2) samples were twice as high as the background toxicity levels. Furthermore, the toxicity levels of the untreated waste pit samples were 1-to-2 orders-of-magnitude higher that the toxicity levels observed in the treated soils (Table 3).

The results of this study, combined with the results of the soil monitoring program, indicate that reductions in the concentrations of the PAHs and PCP correlated with reductions in the toxicity levels of the soil samples as indicated by the Ames bioassay. These findings indicate that biological treatment of soil contaminated by multiple organic contaminants did not result in the formation of persistent, intermediate breakdown products that were more toxic than the parent organic contaminants. Therefore, the combined results of these studies provide evidence that bioremediation can decontaminate and detoxify contaminated soils, even those contaminated with a variety of hazardous compounds.

Aboveground Groundwater Treatment Program

The objective of the aboveground treatment program for groundwater is to remove heavily contaminated groundwater from the Upper Aquifer and biologically treat the extracted groundwater to reduce the concentrations of PCP and PAHs. Removal of the heavily contaminated groundwater is expected to assist in the in situ biological treatment of the Upper Aquifer.

The aboveground treatment system for groundwater consists of the following components: four extraction wells; an equalization tank and nutrient amendment system; two fixed-film, upflow, heated, aerobic bioreactors operated in series;

TABLE 3. Summary of contaminant concentrations and toxicity levels in soil lifts and related samples during the 1989 Land Treatment Demonstration at the Libby, Montana site[(a)].

Parameter	Background[(b)]	Lift 1[(c)]	Lift 2[(d)]	Waste
Contaminants[(e)]				
PCP	NT[(f)]	10 ± 0.3	14 ± 3.4	NT
Pyrene	NT	5 ± 0.5	23 ± 3	NT
CPAH	NT	13 ± 2	79.8 ± 12.2	NT
TPAH	NT	20 ± 2.5	204 ± 61	NT
Toxicity Data[(g)]				
CH_2CL_2 Fraction				
−S9	NT	0-7	0-25	0-"Toxic"
+S9 (High)	66	71-114	59-174	873-8,873
CH_3OH Fraction				
−S9	0	0	0-8	81-"Toxic"
+S9 (High)	0	1-102	5-331	879-5,911

(a) Contaminant data for Lifts 1 and 2 from Woodward-Clyde Consultants (1990); toxicity data summarized from Donnelly et al. (1992).
(b) One background sample from site tested.
(c) Lift 1 treatment completed approximately 10 weeks prior to toxicity sampling.
(d) Lift 2 treatment not completed at time of toxicity sampling.
(e) Contaminants: PCP, pentachlorophenol; CPAH, carcinogenic polycyclic aromatic hydrocarbons; TPAH, total polycyclic aromatic hydrocarbons.
(f) NT = Not tested by Donnelly et al. (1992).
(g) Weighted activities expressed as ranges in the numbers of net *Salmonella* tester strain TA98 His^+ revertants per g of material; CH_2CL_2 Fraction = methylene chloride fraction extracted from samples; CH_3OH Fraction = methanol fraction extracted from samples; −S9 = strain not metabolically activated; and +S9 (High) = strain metabolically activated. "Toxic" descriptors from Donnelly et al. (1992).

and an effluent discharge system to the contaminated rock piles or infiltration trenches. The extraction wells consist of two sets of well pairs located immediately downgradient from the former waste pit area. Each well pair contains a well with a screened interval located in the upper portion of the Upper Aquifer and a second well with a screened interval located in the lower portion of the aquifer. This well design allows flexibility in selecting groundwater extraction rates for specific depths in the aquifer. Extracted groundwater is delivered to the aboveground treatment system where it is treated and then discharged to the rock piles or infiltration trenches.

Operation of the Treatment System. The groundwater treatment system was constructed in late 1989 and began operating in the spring of 1990. The nutrient amendment system delivers aqueous solutions of nitrogen and phosphorus to the system flow upstream of the bioreactors to maintain microbial activity in the reactors.

Each bioreactor has a tank capacity of 10,000 gal (37,853 L), and the paired bioreactors have a combined hydraulic retention time (HRT) of 8 hours at the maximum design flowrate of 40 gpm (2.5 L/s). The average flowrate for the system from 1990 through 1992 was 8 gpm (~0.5 L/s). The influent to the first bioreactor is heated by the steam system of the wood-production facility to stimulate an elevated degree of microbial activity in the bioreactors, and the heat carries over to the second bioreactor.

Inoculation of the bioreactors was accomplished using indigenous microorganisms contained in the extracted groundwater. The operation of the bioreactors in series resulted in the development of different microbial consortia in the two reactors. The consortium in the first bioreactor has developed the capability to degrade the PAHs, and biodegradative performance of the consortium is not sensitive to temperature fluctuations over a range of 18 to 30°C (Cosgriff, unpublished data, 1992). PAH biodegradation began shortly after system start-up.

The second bioreactor has developed a consortium with the capability to degrade PCP. However, appreciable PCP biodegradation did not take place until several weeks after system startup. In addition, the PCP-biodegradative capability of this consortium is sensitive to temperature fluctuations and exhibits a marked reduction in the ability to degrade the contaminant once the water temperature falls below 22°C (Cosgriff, unpublished data, 1992). As a result, the heat supplied to the bioreactor system has been adjusted to maintain a water temperature of 22° C in the second bioreactor.

Performance Characteristics of the Groundwater Treatment System. At the average flowrate for the 1990 to 1992 period of 8 gpm (~0.5 L/s), approximately 11.7 million gal (~44.3 million L) have been extracted and treated by the system. Figure 5 summarizes the 1992 monthly performance of the bioreactor system in reducing aqueous concentrations of TPAHs and PCP in the extracted groundwater.

System upsets have occurred periodically and have included fouling of the aerators in the bioreactors, extraction pump malfunctions, and fluctuations in the contaminant concentrations in the extracted groundwater. The latter has been

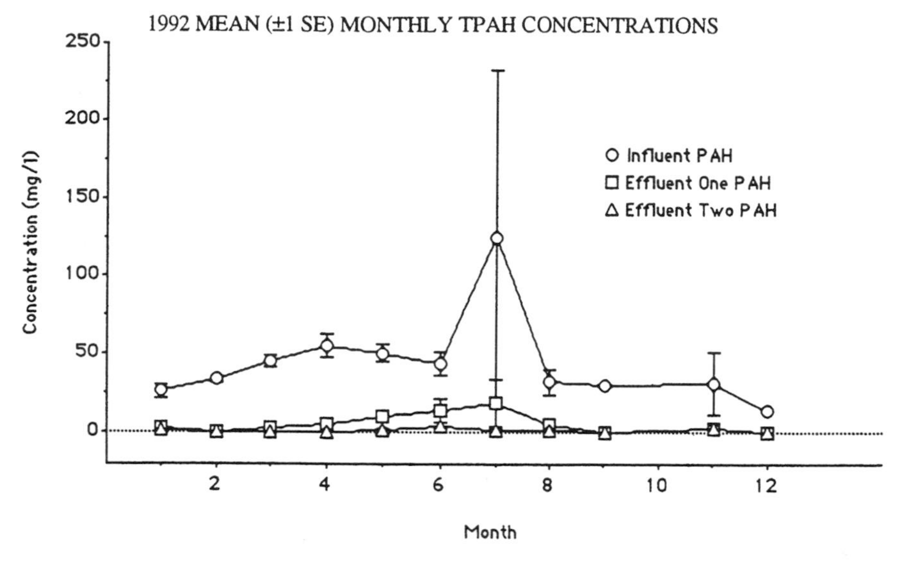

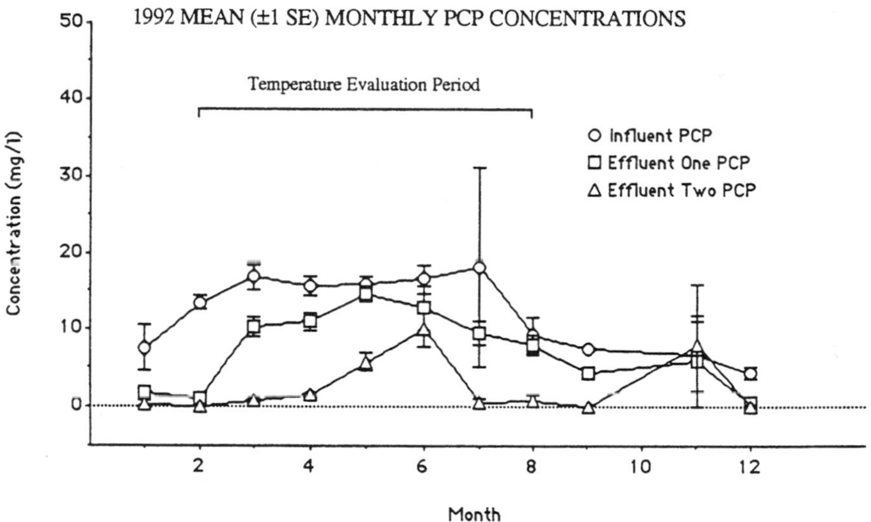

FIGURE 5. Mean (±1 standard error) 1992 monthly concentrations of TPAH (top graph) and PCP (bottom graph) in water samples collected from sampling locations along the bioreactor treatment system. "Influent" refers to the influent to Bioreactor 1; "Effluent One" refers to the effluent from Bioreactor 1; and "Effluent Two" refers to the effluent from Bioreactor 2.

the most common problem (see below), and has resulted in periods when the nutrient amendment system has not delivered the appropriate amount of nutrients to stoichiometrically balance the organic contaminants in the influent to the bioreactors.

Nevertheless, the treatment system typically has been able to produce an average reduction in concentration of TPAHs of approximately 97% and a reduction of PCP of 60% throughout this period. Recent performance of the bioreactor system indicates that PCP removal has exceeded 88%.

A second operational problem involves the coextraction of NAPLs with the groundwater. The NAPLs, with a specific gravity of approximately 1.02 g/mL, characteristically occur as dispersed droplets in the extracted groundwater due to the shearing action of the centrifugal pumps in the extraction wells. High densities of NAPL droplets in the extracted groundwater can result in excessive organic loading to the bioreactors, which reduces the capability of the treatment system to produce appreciable contaminant reductions in the extracted groundwater. Several attempts at NAPL separation, including gravitation, dissolved air flotation, induced air flotation, and coalescence, have not produced satisfactory results.

In response, a program has been initiated to expand the capacity of the aboveground treatment system by installing two 50,000-gal (189,265-L) suspended-growth bioreactors to be operated in series. The expanded capacity is expected to absorb the organic loading induced by the coextraction of NAPLs, and the upgraded system was placed on line for testing in September 1992.

In Situ Groundwater Treatment Program

The last major form of remedial action at the Libby site involves injection of oxygenated water at strategic locations along the groundwater contaminant plume of the Upper Aquifer. The objective of the injection program is to stimulate in situ biodegradation of the dissolved and adsorbed contaminants in the aquifer. This approach was adopted after a pilot-scale study conducted from 1987 to 1988 indicated it could reduce dissolved contaminant concentrations in the aquifer (Piotrowski 1989, 1991; Woodward-Clyde Consultants 1988). To date, two areas of the contaminant plume are receiving the oxygenated water: an intermediate area and an area along the boundary of the facility (Figure 1).

Intermediate Injection System. The Intermediate Injection System was installed approximately 400 ft (~121 m) downgradient from the former waste pit area. The system consists of three injection well sites: 3004, 3007, and 9500/9501 (Figure 1). Each well site consists of one well screened in the upper portion of the aquifer and one well screened in the lower portion. The total injection rate for the system is approximately 100 gpm (6.3 L/s).

The system currently injects water amended with hydrogen peroxide at a delivered concentration of approximately 100 mg/L. The concentration decomposes in the subsurface to produce a dissolved oxygen (DO) concentration of approximately 50 mg/L. The use of an alternative, potentially cost-effective oxygen source for this injection system is currently being evaluated. Nutrients (nitrogen and phosphorus) are added to the infiltration water to produce nutrient concentrations that are stoichiometrically balanced with the concentration of injected oxygen.

Boundary Injection System. Construction of the Boundary Injection System was completed in the fall of 1992 in a site approximately 1200 ft (~364 m) downgradient from the former waste pit area (Figure 1). The system consists of three injection well sites (9001, 9502, and 9503) that will deliver oxygenated water across much of the vertical thickness of the Upper Aquifer. Based on preliminary testing, the system is expected to operate at a total, combined injection rate of 265 gpm (16.7 L/s) and the higher injection rate (relative to the Intermediate Injection System) is attributed to the higher permeability of the Upper Aquifer in this area of the site.

Two oxygen delivery systems will be tested for use at the Boundary Injection System during 1993. The first is a deep-well oxygen exchange system that uses pressure to enhance the penetration of the oxygen gas into the water. This system has produced injected DO concentrations ranging up to 70 mg/L. The second is a prototype system that uses multiple, semipermeable filaments that contain oxygen gas under pressure and delivers oxygen to the water without off-gassing. This system, currently being tested, creates DO concentrations that approach 40 mg/L. Both systems will be operated concurrently to evaluate operational differences.

Operational Characteristics of the In Situ Groundwater Treatment Program. A performance monitoring program has been adopted to track the progress of the injection program and this monitoring program focuses on DO, PCP, and PAH concentrations in the monitoring wells downgradient from the injection systems. Observations of elevated DO concentrations in the wells characteristically have correlated with reduced concentrations of dissolved contaminants (Figure 6).

The Intermediate Injection System has been under full-scale operation since January 1991. Three monitoring wells located downgradient from the injection sites currently exhibit elevated DO concentrations (~20 mg/L) and reduced dissolved contaminant concentrations (< the laboratory detection limit of approximately 3 µg/L for the each contaminant). Notable among these wells is a monitoring well located 200 ft (61 m) downgradient from the injection system that has exhibited elevated DO and reduced contaminant concentrations for more than 5 years.

A second well, located approximately 150 ft (~45 m) downgradient from the injection system with a screened interval approximately 66 ft (20 m) bgs, has developed this pattern of oxygen and contaminant concentrations more recently. Oxygen breakthrough (>15 mg/L) and contaminant reduction (from 73 µg/L to <3 µg/L for TPAH and 420 µg/L to <3 µg/L for PCP) occurred in November 1991.

Despite these positive indications, the remedial goals for a number of the contaminants of concern in the groundwater of the site (Table 1) currently are below the detectable limits for commercial analytical laboratories. As such, final evaluation of the performance of the injection program awaits improvements in analytical technology.

In addition, a number of other wells in the downgradient vicinity of the injection system have not exhibited the pattern of elevated DO and reduced contaminant concentrations. These observations indicate that the complex hydrogeology of the site, combined with variable concentrations of the organic contaminants in the aquifer (including NAPLs) and variable delivery of the oxygenated water throughout this region of the aquifer, result in variable performance of the injection system.

SUMMARY AND CONCLUSIONS

Results of more than 4 years of bioremedial activities at the Libby site show progress in treating contaminated soil in land treatment units; heavily

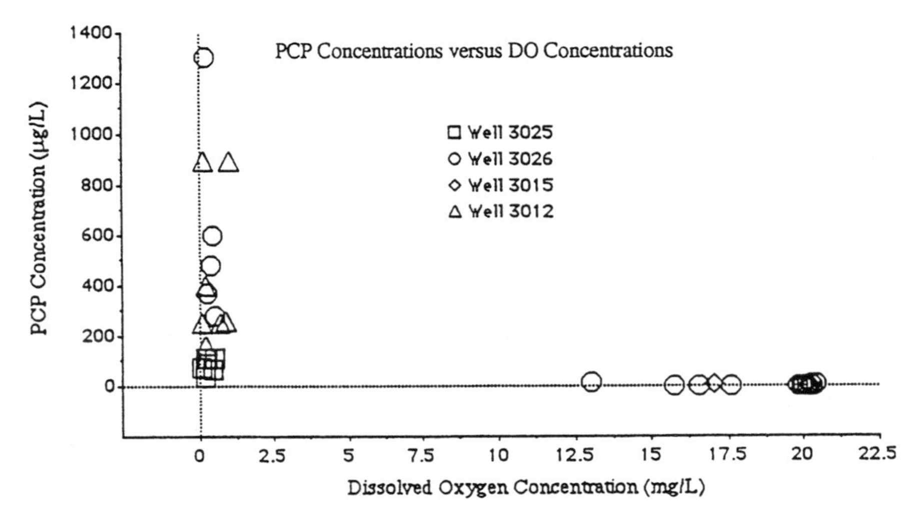

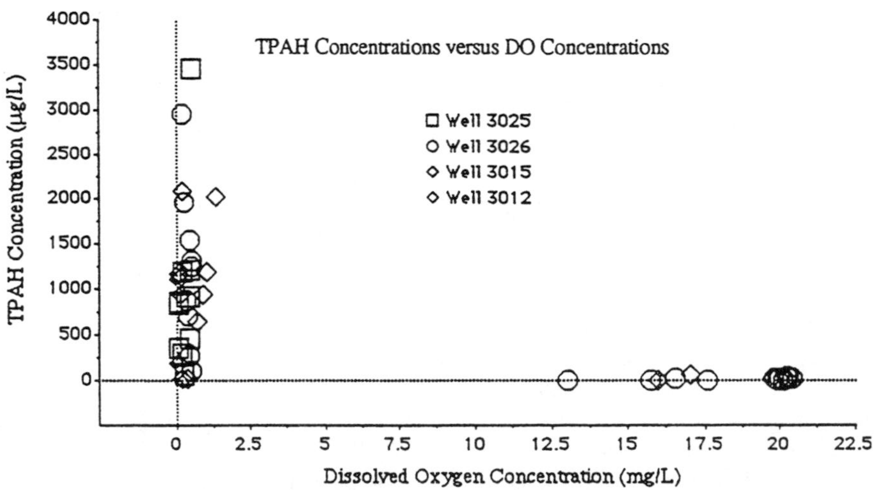

FIGURE 6. Scatter plots of PCP (top graph) and TPAH (bottom graph) concentrations versus DO concentrations in selected monitoring wells from the Libby, Montana site, 1987-1992.

contaminated groundwater in an aboveground, biologically based treatment system; and and directly within the contaminated regions of the Upper Aquifer by means of an in situ bioremediation program. An independent study of toxicity reduction in the biologically treated soils provided evidence that biological treatment has reduced the concentrations of the priority pollutant contaminants of concern without producing intermediate breakdown products of elevated toxicity. These results indicate that biological treatment is an effective treatment approach for soils contaminated by multiple, hazardous organic compounds.

Finally, in overview, the rapid remedial progress achieved at this site is due to the constructive interactions between the client, the regulatory agencies, the public, and the consultants' multidisciplinary technical experts. The approach used to move forward with the remedial program at this site is an excellent model for the remediation industry.

ACKNOWLEDGMENTS

Champion personnel instrumental in the success of this project include James Carraway, Ralph Heinert, James Davidson, Gerald Cosgriff, Virginia Hesselink, and Michael Funk. We also acknowledge the support and assistance of the following former and present EPA personnel: Kenneth Wallace, Henry Elsen, Julie Dalsoglio, James Harris, Scott Huling, Bert Bledsoe, and John Matthews.

REFERENCES

Donnelly, K. C., C. S. Anderson, J. C. Thomas, K. W. Brown, D. J. Manek, and S. H. Safe. 1992. "Bacterial Mutagenicity of Soil Extracts from a Bioremediation Facility Treating Wood-Preserving Waste." *J. Hazardous Materials 30*: 71-81.

Piotrowski, M. R. 1989. "In Situ Biogeochemical Reduction of Hydrocarbon Contamination of Groundwater by Injecting Hydrogen Peroxide: A Case Study in a Montana Aquifer Contaminated by Wood Preservatives." Ph.D. dissertation, Boston University, Boston, MA. UMI No. 8913768.

Piotrowski, M. R. 1991. "U.S. EPA-Approved, Full-Scale Biological Treatment for Remediation of a Superfund Site in Montana." In E. J. Calabrese and P. T. Kostecki (Eds.), *Hydrocarbon Contaminated Soils*, Vol. 1, pp. 433-457. Lewis Publishers, Chelsea, MI.

U.S. Environmental Protection Agency. 1988. *Record of Decision: Libby Ground Water Superfund Site, Lincoln County, Montana.* U.S. EPA Region VIII, Montana Operations Office. December 1988.

Woodward-Clyde Consultants. 1988. *Feasibility Study for Site Remediation, Libby Montana.* Report prepared by Woodward-Clyde Consultants, Denver, CO, for Champion International, Inc. November 1988.

Woodward-Clyde Consultants. 1990. *No Migration Petition Report, Land Treatment Units, Libby, Montana.* Prepared by Woodward-Clyde Consultants, Denver, CO. February 1990.

OBTAINING REGULATORY APPROVAL AND PUBLIC ACCEPTANCE FOR BIOREMEDIATION PROJECTS WITH ENGINEERED ORGANISMS IN THE UNITED STATES

D. J. Glass

ABSTRACT

The first approved uses of genetically engineered microorganisms (GEMs) in bioremediation may soon be in the offing. Until recently, government biotechnology regulation and negative public perceptions have led many in the biotreatment industry to shy away from the use of GEMs, even if technically feasible or desirable. But as the federal government relaxes its regulations on biotechnology, barriers to the use of GEMs in bioremediation are falling. This paper describes these changes in federal regulation and presents case studies for obtaining U.S. Environmental Protection Agency (EPA) approval for environmental uses of GEMs. One such project is Envirogen, Inc.'s planned use of a GEM bioreactor to remediate trichloroethylene (TCE) from contaminated groundwater. It is possible to obtain EPA approval for GEM biotreatment projects, both in reactors and in situ, in an efficient, straightforward manner, within acceptable time frames. Similarly, public concerns about open-environment use of GEMs has subsided, so that local support for such projects can be obtained. Government biotechnology regulation and public perceptions should no longer be considered barriers to field use of GEMs in bioremediation.

THE PROMISE OF GENETIC ENGINEERING FOR BIOTREATMENT

Commercial bioremediation to date has made use only of natural or indigenous microorganisms. However, there are several legitimate reasons for using engineered organisms: to engineer a naturally occurring degradative pathway so that it is constitutively expressed in the absence of molecules otherwise needed to activate the pathway (e.g., Winter et al. 1989); to artificially create hybrid degradative pathways for xenobiotics (e.g., Rojo et al. 1987); or to introduce a novel enzymatic activity created by in vitro protein engineering (e.g., Ornstein 1991).

Although there has been considerable research activity by many industrial and academic groups, numerous factors have deterred the commercial use of GEMs in waste remediation. These include the lack of a compelling technical need to use genetic engineering and concerns over the added regulatory burdens that have been placed on the use of engineered microorganisms in the environment.

FEDERAL BIOTECHNOLOGY REGULATION

Commercial biotechnology is regulated in the United States on a product-by-product basis under a "Coordinated Framework" (U.S. Office of Science and Technology Policy 1986), using statutory authority of the Food and Drug Administration (FDA), the EPA, and the U.S. Department of Agriculture (USDA). Under existing programs, the FDA regulates most biotechnology products (such as drugs and diagnostic tests), the EPA oversees biological pesticides and new chemicals produced by biotechnology, and the USDA regulates animal biologics (e.g., vaccines) and plant and animal pathogens.

Beyond this existing authority, new rules or procedures have been adopted to regulate the environmental uses of GEMs. The USDA issued new regulations to cover transgenic plants and some agricultural microorganisms, and the EPA oversees environmental releases of microorganisms under two laws, the Federal Insecticide, Fungicide and Rodenticide Act (FIFRA) and the Toxic Substances Control Act (TSCA). These agencies in general have required risk assessments and regulatory review for all open-field uses of engineered organisms, regardless of scale.

The earliest proposed tests often generated controversy, particularly in the communities where the tests would occur (Krimsky & Plough 1988, Piller 1991), and sometimes were approved only after long regulatory reviews or legal battles. However this has changed dramatically; from 1987 to 1991 approximately 100 field tests of genetically modified microorganisms took place in the United States, about 40 of which involved GEMs (Figure 1). About 10 other tests of GEMs took place in 1992. Through November 1992, the USDA issued 335 permits for field tests of genetically engineered plants. Numerous trials of transgenic plants and GEMs have occurred outside the U.S. as well (GBF 1990). All of these tests have been conducted safely, with no unexpected incidents and little negative public reaction (Casper & Landsmann 1992, Drahos 1991, McKenzie & Henry 1991).

As a result of the cumulative safety record, U.S. government policies have begun to relax. In February 1992, the Bush Administration introduced a new policy, known as the "scope policy," that was intended to remove the special oversight currently given recombinant organisms (U.S. Office of Science and Technology Policy 1992). This policy is meant to institute an approach that is based solely on the risks of the organism under review, and should make it easier to release GEMs for testing or deployment in the field.

This watershed change in U.S. policy indicates that the federal government recognizes the relative risks of biotechnology and its importance to the global competitiveness of the United States. In response to the policy, the USDA has

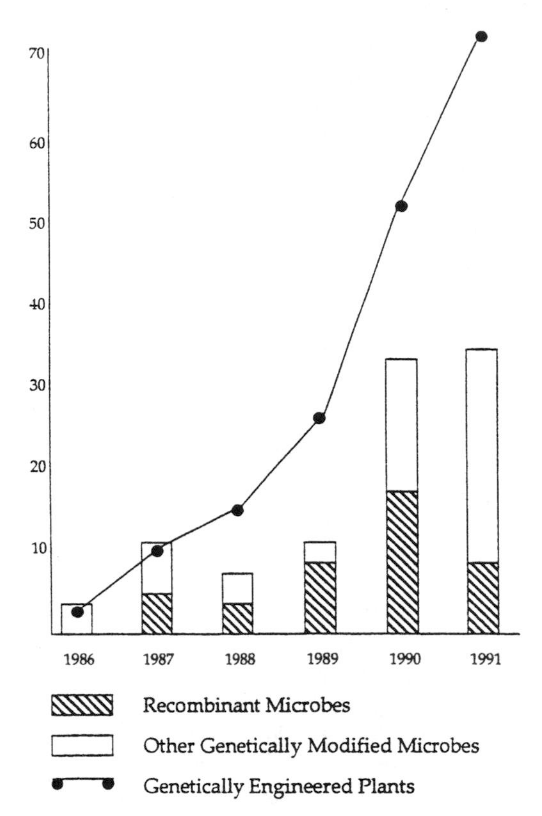

FIGURE 1. U.S. field tests of genetically engineered organisms, 1986-1991.

proposed new regulations that would allow many field tests of transgenic plants simply upon USDA notification (U.S. Department of Agriculture 1992). Although EPA policies have not yet been changed, adoption of the scope policy should facilitate the use of engineered microbes in waste treatment by reducing the number of small-scale environmental uses (i.e., less than 10 acres) requiring EPA oversight. EPA has, in fact suggested such a change for biopesticides in recently published proposed regulations under FIFRA (U.S. Environmental Protection Agency 1993).

EPA BIOTECHNOLOGY REGULATION

Most GEMs field tested in the United States have been microbial pesticides. Although early biopesticide field tests required permits, most tests now are approved with only a brief EPA review. Several engineered microbial pesticides have undergone expanded field testing, and three insecticides containing killed recombinant microorganisms have been approved for commercial sale.

Environmental uses of non pesticidal GEMs are regulated by EPA under TSCA. This program also covers certain contained manufacturing activities involving new microorganisms (i.e., industrial enzyme manufacture), which pose fewer questions of environmental risk than do open-field releases.

Under TSCA, EPA regulates the entry into commerce of all new chemicals not regulated under other federal laws, including the Food, Drug and Cosmetic Act and FIFRA. Manufacturers of new chemicals must file a premanufacture notice (PMN) with EPA at least 90 days before commencing manufacture. The PMN must describe the chemical and its process of manufacture, and must include data regarding its effects on human health and the environment.

The EPA is applying TSCA to biotechnology by requiring PMNs for uses of "new" microorganisms for commercial purposes in specialty chemical production, enzyme manufacture, bioremediation, and other environmental uses. The EPA currently defines "new" microorganisms as those containing genetic material from more than one taxonomic genus ("intergeneric" organisms), regardless of how derived. However, by treating vectors according to the genus from which they were first isolated, the EPA considers virtually every recombinant microorganism to be intergeneric. Within TSCA's statutory reach, all commercial uses of new microbes, even for contained manufacturing, require PMN, while indoor research is exempt from reporting. However, the EPA wants to review all environmental uses of intergeneric microbes, with voluntary PMN reporting requested for research field tests. Academic research is presently exempted, because EPA restricts TSCA regulation to commercial activities.

In contrast, uses of intrageneric, classically mutated and naturally occurring organisms are exempted from notification. Thus, most ongoing biotreatment activities (even bioaugmentation with enriched cultures) are not regulated. A 1988 EPA proposal to require notification for naturally occurring microbes attracted widespread opposition, so that EPA does not now intend to regulate nonengineered organisms under TSCA. By way of comparison, under the Canadian equivalent of TSCA, that country may require manufacturers of microbial cultures to notify Environment Canada regardless of whether the microbes are genetically altered.

Procedures and data requirements for biotechnology PMNs are slightly different than those for chemicals, and EPA generally conducts more thorough risk assessments than for most chemical PMNs. Biotechnology PMNs for contained manufacturing have routinely been approved within 90 days; PMNs for environmental releases may require up to four months. There have been approvals for 10 agricultural biotechnology field tests under TSCA since 1987 (Table 1), but as of March 1993, there has not been a PMN for a waste treatment microorganism.

EPA REGULATION OF BIOTREATMENT USING ENGINEERED MICROBES

The EPA is attempting to adopt new rules under TSCA to relax its current policies. In June 1991, EPA released a draft proposal that would create a number

TABLE 1. Microbial PMNs approved under the Toxic Substances Control Act (as of March 1993).

Company	PMNs Approved
Agricultural Field Tests	
BioTechnica	19 PMNs for 5 field tests
Research Seeds	5 PMNs for 4 field tests
Monsanto	1 PMN for 1 field test
Contained Enzyme Manufacture	
Novo	4
Gist-Brocades	3
Genecor	1
Enzyme Biosystems	1
IMC	1
Pesticide Intermediates	
Mycogen	16
Biotreatment Organisms	none to date

of exemptions for commercial activities and research field tests that now involve PMN reporting (U.S. Environmental Protection Agency 1991). The proposed regulations would clarify EPA's definition of contained activities, and would also make it easier to conduct research field tests of GEMs for in situ bioremediation applications.

In the past, clear lines have been drawn between contained PMNs (e.g., for enzyme manufacture) and environmental release PMNs. Waste treatment bioreactors fall somewhere in between because most could not be operated in compliance with the guidelines on which EPA's 1986 definition of containment was based. These guidelines, the National Institutes of Health (NIH) guidelines for large-scale fermentations of recombinant bacteria, initially required extensive design features and procedural controls to minimize or prevent escape of microbes from the fermentation vessel, such as sterilization of all live microorganisms before disposal. The NIH Guidelines were later relaxed (U.S. Department of Health and Human Services 1991) to grant institutions greater leeway in choosing appropriate containment conditions and procedures, based on risks posed by the microorganism. For example, specific inactivation procedures for discharges of organisms are no longer required.

In recent years, EPA has begun to clarify that it interprets its containment definition more flexibly. For example, EPA's 1991 draft proposed TSCA rule specified that research conducted in any sort of structure would be considered contained, and would qualify for TSCA's research and development exemption if certain conditions were met. The research must be conducted under the supervision of a technically qualified individual, who would be given fairly broad discretion to adopt appropriate procedures to ensure limited access to the facility, inactivation of waste microbes, and controls on emissions of microbes.

The ongoing interagency deliberations over the scope policy have prevented the 1991 proposal from being formally proposed as a rule. If this proposal is adopted, many biotreatment reactors would be judged to be contained, even without strict controls on discharges. In the summer of 1992, the EPA issued its first ruling that a bioreactor for waste treatment is contained. This sets a precedent so that many GEM bioreactors will not be regulated as deliberate releases, but instead as contained activities. This ruling is discussed in more detail below.

EPA's June 1991 document also proposed a streamlined process for agency review of small-scale field tests of new microorganisms, for which full PMNs are now requested under the voluntary policy. This procedure, the TSCA Experimental Release Application (TERA), would reduce the data requirements and shorten the review time for field experiments. Furthermore, as a result of the scope policy, EPA has been considering several options whereby some field tests could be reviewed and approved by local oversight committees.

APPROVAL FOR OPEN-FIELD USE OF ENGINEERED MICROBES

Under the present system, obtaining EPA approval for outdoor field tests of GEMs has become routine. The following case study describes several of the earliest agricultural field tests to be approved under TSCA. These projects are described in more detail elsewhere (Glass 1989a, Glass 1989b).

Five of the TSCA-approved field tests have been conducted by BioTechnica International (BTI). These involved the nitrogen-fixing bacteria *Rhizobium* and *Bradyrhizobium*, which convert atmospheric nitrogen to ammonia in the roots of legume plants. Currently, legume farmers use commercial products containing these bacteria to provide nitrogen for their crops. BTI was developing improved products for soybeans and alfalfa by enhancing the bacterial genes controlling nitrogen fixation.

BioTechnica began prenotice consultations with EPA staff in the summer of 1986 to keep EPA informed of planned field tests, and to negotiate agreement on testing protocols and data requirements. Each test involved the filing of PMNs describing the strains to be tested, the steps taken to genetically construct these strains, and the results of laboratory and greenhouse environmental fate studies. The proposals also included descriptions of the test, including features designed to restrict dissemination of the microorganisms beyond the test plot, and procedures to monitor the environmental behavior of the organisms.

EPA's risk assessments featured review by internal and external scientists, were coordinated with the USDA and state regulators, and often included a visit to the field test site. These assessments focused on the potential environmental effects of the proposed field test, weighed against its benefits, and included evaluation of the potential survivability of the microorganisms, their possible effects on target and nontarget plants, and on dispersal of the microbes or transfer of the introduced genetic material to other organisms (e.g., as discussed in Alexander 1985).

As a result of its assessments, EPA concluded that each proposed test presented low risks. To allow a proposed field test under TSCA without approving unlimited manufacture of the microorganisms, EPA and the company entered into consent orders, limiting allowed activities to the specific field test and monitoring protocols approved by the agency. The terms of these orders could be changed by mutual consent. Although approval of BTI's first proposal in 1987 was delayed several months because of the need to refine the monitoring plan, subsequent proposals were approved in approximately 90-100 days.

BioTechnica's field tests were also reviewed by state governments. Today, many states have specific biotechnology laws (Glass 1991). Regulation by a state agency may involve a separate scientific review, occasionally including consultation with experts from local universities. State review may raise local concerns that escape the attention of federal regulators, but the states almost always have concurred with federal approvals.

Because of the community concerns that arose over earlier field tests, each of BTI's proposals was accompanied by community relations programs to inform local citizens of the company's plans and to involve the public in the decision-making process (Glass 1989a, Glass 1989b). The most extensive program was conducted in rural Pepin County, Wisconsin, in 1987, because at that time no institution had yet performed an approved outdoor GEM field test.

The cornerstones of company policy were openness with the public about all aspects of the test and availability to provide information and answer questions. The community, through its elected officials and other respected citizens, learned of all developments from the company, rather than from the press or the rumor mill. The company notified key community leaders before submitting the PMNs, and used press releases and brochures to describe the project in nontechnical terms to the broader community. BTI sponsored a public meeting in the county, to answer questions from local citizens, and attended a number of other public forums. Finally, citizens and members of the press were invited to tour the field site at various times in the project, including the day of planting.

Although this first field test attracted considerable community interest and drew the opposition of a small local group, community support was obtained. The test was begun in April 1988 and attracted no demonstrations or protests, despite heavy media interest. Two subsequent field test proposals in 1989 near Madison, Wisconsin attracted no controversy, in spite of the proximity of the site to the homes of two prominent local politicians. Most of the microbial field tests conducted around the United States to date have encountered no opposition, and these projects have ceased being controversial. Although well-publicized lawsuits hindered early field tests, there has not been an attempt by any public interest group to use the courts to block the environmental release of an altered organism in the United States since 1987.

APPROVAL FOR BIOREACTOR USE OF ENGINEERED MICROBES

Most of the first proposals for use of GEMs in waste treatment are expected to involve contained bioreactors. One company conducting research that may lead in this direction is Envirogen, Inc. The company is focusing on several categories of toxic waste including polychlorinated biphenyls and TCE. Its approach includes isolating natural organisms capable of degrading these wastes and, in some cases, enhancing their performance through genetic modification, for use in special bioreactor systems.

One of Envirogen's projects involves the use of microorganisms in contained bioreactors to biodegrade vapor-phase TCE that has been volatilized by air-stripping or vacuum extraction from contaminated groundwater. Envirogen's technology is based on a naturally occurring bacterial strain that can degrade TCE through cometabolism with toluene (Winter et al. 1989). In initial field trials in a bioreactor in New York State, this strain showed an average 90% destruction of TCE in air stripped from contaminated groundwater. In the long term, the company also is contemplating commercial use of a genetically engineered version of this strain, which is expected to utilize a less expensive, less toxic organic inducer molecule.

In the expectation of beginning field use of a developmental GEM bioreactor, the company initiated discussions with EPA in 1992 regarding the possible regulation of this project under TSCA. The engineered strains likely will be subject to TSCA jurisdiction because they are intergeneric (i.e., new) and are not used as a pesticide, food or drug. Although the bioreactor was designed to adequately contain microorganisms, the company wanted to ascertain that it met EPA's containment criteria. If it did not, PMN reporting of proposed research uses would be requested under the voluntary policy, and proposed commercial uses would trigger substantive questions about environmental risk. The company also wished to begin a dialogue with EPA about the data requirements and testing protocols that would eventually be needed for PMNs for its genetically modified strains.

Envirogen held a preliminary meeting with EPA's biotechnology staff early in 1992, to describe the makeup of the GEMs and the design of the reactor. Envirogen later submitted this information in a letter, requesting EPA's opinion on the adequacy of containment and the possible data requirements for commercial use of the reactor.

In July 1992, EPA responded with its determination that, although Envirogen's GEMs would be considered new under the TSCA policy, the company's developmental bioreactor would be considered a contained structure, so that use of the GEMs in the reactor would qualify for a research and development exemption under TSCA. EPA reserved comment on specific data requirements until notified by the company of specific conditions of use.

EPA's letter to Envirogen was the first time that the agency formally stated that a bioreactor for waste treatment would be considered contained. This decision sets a valuable precedent and is indicative of EPA's progressive attitude towards biotreatment, and growing comfort with the use of GEMs in the environment.

EPA is likely to consider other well-designed reactors to be contained, and is planning to develop guidance documents for bioreactor users.

During pilot use of the reactor with naturally occurring microbes, Envirogen will collect data on the number of microorganisms released in off-gases and liquid effluents, and on the effectiveness of its waste-inactivation process. Subsequent to these field trials, Envirogen expects to submit its first PMNs for biotreatment using GEMs. Other data that might be needed for PMNs would be similar to those required for agricultural GEM field tests, including characterization of the engineered strains and assessment of potential environmental effects. Envirogen plans on conducting public education and information campaigns prior to field use of its GEM reactors to allay any potential concern over their use by public interest groups.

STRATEGY FOR OBTAINING APPROVALS FOR USE OF GEMs IN BIOREACTORS OR IN SITU BIOREMEDIATION

In today's regulatory climate, there should be no obstacles to obtaining EPA approval to use GEMs in commercial bioremediation, particularly in a bioreactor or a well-designed field release. As outlined in the flowchart in Figure 2, the process should require only six to eight months, with half that time for EPA and state biotechnology review and half for data generation and preparation. Approval under the biotechnology policy may come more quickly than authorization for field use under applicable hazardous waste laws, particularly the Resource Conservation and Recovery Act or the Superfund program.

As shown in Figure 2, a key component of a company's regulatory plan must be presubmission consultations with the EPA. This step is essential to begin the dialogue with EPA on data requirements and the testing protocols needed to acquire these data. Although potential biotechnology data requirements are summarized in EPA's "Points to Consider" document (U.S. Environmental Protection Agency 1990), the agency has a fair amount of leeway in determining which data are needed for each GEM based on its characteristics and intended use. More important, it is essential to know that any required testing will be carried out using protocols that meet EPA approval and address crucial scientific concerns. Four types of data are the most important.

1. *Strain-Specific Data* — EPA expects a complete description of each engineered strain and how it was constructed. Laboratory data must be submitted assessing the biological characteristics of the strain, comparing the GEM to the natural organism from which it was derived. Most of this is information that is routinely collected during early phases of research.
2. *Site-Specific Data* — Use of GEMs in bioreactors will not require extensive environmental effects testing, but the few data that may be needed must reflect the site(s) at which the reactor will be used. The most likely

test will be a 4- to 6-week experiment comparing the ability of the GEM and its parent to survive in representative soil or groundwater from the site. Such testing certainly will be needed for any proposed field release.

3. *Validation of Reactor Containment* — Although EPA will consider many reactors to be contained, relevant data may be required. The number of microbes in liquid effluents and possibly off-gases and the efficacy of the chosen inactivation procedures will probably require documentation, often in a trial run with the parental microorganisms.
4. *Monitoring Plan for Field Releases* — Proposals to EPA for field experimentation with GEMs must be accompanied by detailed monitoring protocols, that indicate how the behavior of the GEM in the environment will be studied. These must include appropriate methods for distinguishing the introduced GEM from the background microbiota. Monitoring data from early research field tests are used to support requests for expanded testing or commercial use.

As shown in Figure 2, all of these activities can go on in parallel, culminating in the preparation and submission of a PMN document. It may even be possible for initial commercial uses of GEM bioreactors to be approved under test marketing exemptions, which would cover only a limited number of sites and clients, but which might be approved more easily.

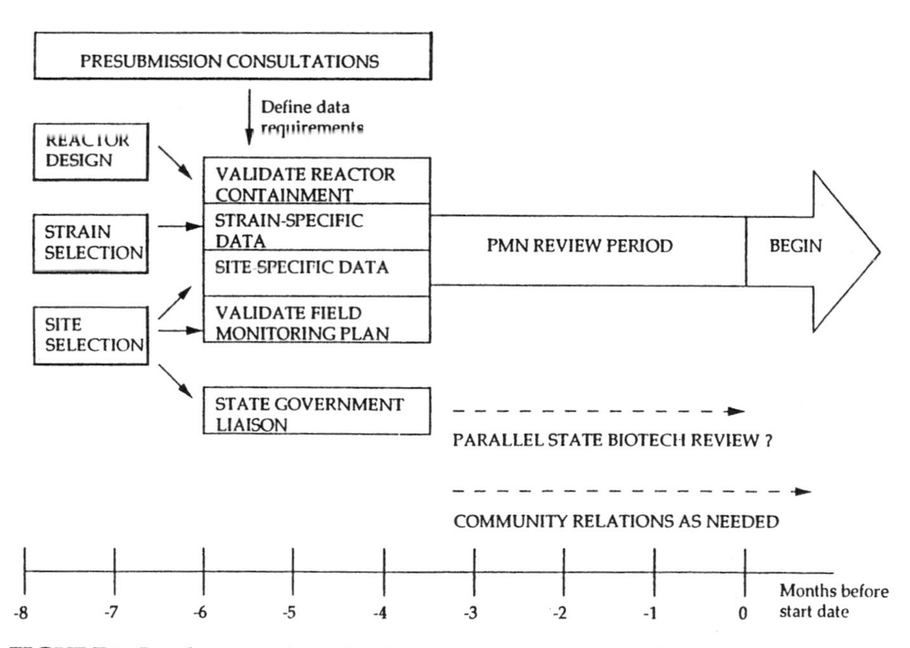

FIGURE 2. Implementation plan for regulatory approval for GEM bioremediation projects.

Finally, by conducting programs to ensure community outreach, companies can expect to avert any possible public opposition, and in fact gain community support for the company's plans. Such successes have been demonstrated for agricultural biotechnology; for biotreatment activities that entail the cleanup of hazardous wastes, community support should be easier to obtain.

CONCLUSIONS

As EPA gains more experience with uses of genetically modified organisms in bioreactors and the environment, many existing barriers are beginning to disappear. The track record of environmental introductions has been uniformly positive, supporting the predictability and safety of the technology. Growing public acceptance, coupled with changes in federal biotechnology policy, should make it easier to bring environmental biotechnologies to the marketplace. Within the next several years, these combined trends will lead to the first commercial uses of advanced biotechnology in waste treatment in both reactors and in situ use.

REFERENCES

Alexander, M. 1985. "Genetic Engineering: Ecological Consequences." *Issues in Science and Technology 1(3)*: 57-68.

Casper, R., and J. Landsmann (Eds.). 1992. *The Biosafety Results of Field Tests of Genetically Modified Plants and Microorganisms*, Braunschweig, Germany.

Drahos, D. J. 1991. "Field Testing of Genetically Engineered Microorganisms." *Biotech. Adv. 9*: 157-171.

GBF (Gesselschaft für Biotechnologische Forschung mbH). 1990. *GENTEC Update*.

Glass, D. J. 1989a. "Managing Government Regulation and Public Perception of Biotechnology Field Tests: A Case Study." In *Biotreatment '89*, pp. 93-97. Hazardous Materials Control Research Institute, Silver Spring, MD.

Glass, D. J. 1989b. "Regulating Biotech: A Case Study." *Forum for Applied Research and Public Policy 4(3)*: 92-95.

Glass, D. J. 1991. "Emerging Issues in State Biotechnology Regulation." In T. J. Mabry, S. C. Price, and M. D. Dibner (Eds.), *Commercializing Biotechnology in the Global Economy*, pp. 155-165. IC^2 Institute, Austin, TX.

Krimsky, S., and A. Plough. 1988. *Environmental Hazards: Communicating Risks as a Social Process*, pp. 75-129. Auburn House Publishing Company, Dover, MA.

McKenzie, D. R., and S. C. Henry. 1991. *Biological Monitoring of Genetically Engineered Plants and Microbes*. Agricultural Research Institute, Bethesda, MD.

Ornstein, R. L. 1991. "Rational Redesign of Biodegradative Enzymes for Enhanced Bioremediation: Overview and Status Report for Cytochrome P450." In *Proceedings of National Research and Development Conference on the Control of Hazardous Materials '91*, pp. 314-319, Hazardous Materials Control Research Institute, Greenbelt, MD.

Piller, C. 1991. *The Fail Safe Society: Community Defiance and the End of American Technological Optimism*, pp. 78-117. Basic Books.

Rojo, F., D. H. Pieper, K. H. Engesser, H. J. Knackmuss, and K. N. Timmis. 1987. "Assemblage of Ortho-Cleavage Route for Simultaneous Degradation of Chloro- and Methylaromatics." *Science 238*: 1395-1398.

U.S. Department of Agriculture. 1992. "Genetically Engineered Organisms and Products: Notification Procedures," *Federal Register 57*: 53036-53043.

U.S. Department of Health and Human Services. 1991. "Recombinant DNA Research: Actions Under the Guidelines." *Federal Register 56*: 33174-33183.

U.S. Environmental Protection Agency. 1990. "Points to Consider in the Preparation and Submission of TSCA Premanufacture Notices for Microorganisms."

U.S. Environmental Protection Agency. 1991. "Microbial Products of Biotechnology: Proposed Regulation under the Toxic Substances Control Act". Draft, June 21, 1991.

U.S. Environmental Protection Agency. 1993. "Microbial Pesticides: Experimental Use Permits and Notifications." *Federal Register 58*: 5878-5902.

U.S. Office of Science and Technology Policy. 1986. "Coordinated Framework for Regulation of Biotechnology." *Federal Register 51*: 23302-23393.

U.S. Office of Science and Technology Policy. 1992. "Exercise of Federal Oversight Within Scope of Statutory Authority: Planned Introductions of Biotechnology Products into the Environment." *Federal Register 57*: 6753-6762.

Winter, R. B., K. M. Yen, and B. Ensley. 1989. "Efficient Degradation of Trichloroethylene by a Recombinant *Escherichia coli*." *Bio/Technology 7*: 282-285.

SUBSURFACE APPLICATION OF SLIME-FORMING BACTERIA IN SOIL MATRICES

I. C.-Y. Yang, Y. Li,
J. K. Park, and T. F. Yen

ABSTRACT

Experiments showed that the strength of soil can be enhanced by either applying slime-forming bacteria directly to the soil matrix to produce a biopolymer inside it, or by applying biopolymer from slime-forming bacteria or commercially available products such as poly-3-hydroxybutyrate (PHB), xanthan gum, and sodium alginates to the soil matrix. Cross-linked support networks can be obtained for the interaction of biopolymer with the soil matrix. Zonal bioremediation and environmental encapsulation are potential applications. Improvement of soil erosion problems and enhancement of soil strength on the foundations of hydraulic systems also may have good prospects in geotechnical and agricultural concerns.

INTRODUCTION

Slime-forming bacteria may be derived from either *Alcaligenes faecalis* or *A. viscolactis*. The chemical composition of the slimes consists mostly of polysaccharides and minor amounts of lipids and proteins (Magee & Colmer 1960b, Punch 1966). Viscosity up to 50,000 cps can be achieved in the proper medium (Magee & Colmer 1960a, Stamer 1963).

The organic soil fraction, soil organic matter (SOM), existing in dynamic equilibrium, is continuously subjected to polymerization, condensation, and degradation processes. SOM is more susceptible to ecological changes than is the inorganic soil fraction. Consequently, the strength of the soil matrix will be reduced by the large losses of SOM under exploitative cultivation practices or under erosion.

Soil colloids are roughly classified generally as inorganic colloids (such as hydrated sesquioxides, allophanes, etc.) and organic colloids (such as humic acids, polysaccharides, and other organic components of humus). Many microbially derived biopolymers, such as PHB, the slimes from slime-forming bacteria, or the gellan gum of polysaccharides, etc., behave as humus which acts as a hydrophilic colloid with numerous functional groups as donor atoms. Humus is closely associated with positively charged colloids such as sesquioxides, allophanes, and other clay constituents. The interfacial chemical interactions within the domains and microaggregates of the soil matrix are either adhesion or binding. As a result, the binders and the newly introduced biopolymers are expected to increase the strength of the soil.

It is possible to chemically alter the structure of biopolymers by a cross-link through coordination of transition metals. The interaction between the soil matrix and biopolymers can be enhanced by curing. For example, heating a soil matrix that contains a biopolymer can open the functional groups in the biopolymer and cross-link the soil matrix and biopolymer by free radicals and unsaturated backbones. If the strength of the soil matrix actually can be enhanced by applying a biopolymer to it, this technology can be applied to locations that are vulnerable to contamination (such as the periphery of landfills and underground storage tanks) to control leachate migration from hazardous wastes less expensively. Biopolymer has been used to seal the pores of sand, grain, and soil to reduce the permeability in porous media, so it can act as a barrier of pollutants. The reduction of permeability to a certain extent by some slime-forming bacteria has been reported (Jack et al. 1989, Li 1993).

With this objective in mind, experiments were designed to test the feasibility and possibility of the effect of strengthening a soil matrix by either growing bacteria inside the soil matrix or applying a biopolymer to it. The strength of the soil matrix was tested by either the unconfined compression test or the torvane test.

EXPERIMENTAL

Growth of Slime-Forming Bacteria

Two kinds of slime-forming bacteria were used in this investigation. *A. faecalis* (ATCC # 49677) and *A. viscolactis* (ATCC # 21698). These two bacteria can produce very hydrophilic and viscous slimes under the proper environment. Both were grown in an appropriate medium. The simpler medium consisted of nutrient broth only. The other recommended formula consists of many ingredients that we felt were difficult to simulate in the natural environment (Magee & Colmer 1960b, Stamer 1963, Wilson & Knight 1952). But evidently, the yield of biopolymer is larger in multiple-ingredient media. Also, for reference, commercially available polymers such as PHB (0.1% solution), xanthan gums (3% solution), and sodium alginate (2% solution) were studied.

After 4 days of incubation, the bacteria growth was examined by using a microscope at 200X magnitude. The *A. viscolactis* in the whey medium, however, showed poor growth.

Our first attempt is to apply the bacteria culture to a sterilized soil matrix and wait for the production of biopolymer inside the soil matrix to see if there is any enhancement in the strength of the soil. But practically, it is very difficult to keep the soil matrix in a shape rigid enough to test the soil strength during sterilization. So the second attempt is to grow the bacteria in the medium first, isolate the biopolymer, and then apply the biopolymer to the soil matrix instead of growing bacteria inside the soil matrix. In this way, a better soil matrix for the test can be obtained.

Preparation of Biopolymer

To make the gum solution from the two medium with the slime-forming bacteria, the medium was centrifuged at 2,000 rpm for 40 minutes. After centrifugation, the viscous precipitates were removed from the bottom of centrifugation tube, redissolved in distilled water, and stored for further use. To make the gum solution from PHB, xanthan gums, and sodium alginates, the biopolymer powder was just dissolved into distilled water with mechanical shaking.

Test of Strength of Soil Matrix

Soil is composed of three different types of soil matrix, including sand, clay, and silt. In this study, various types of soil matrix — sand (washed and dried), silica (SiO_2, about 240 mesh), and clay (gray modeling clay) — were used to determine how biopolymers are effective in increasing the strength of different types of soil matrix. Depending on the type of matrix, one of the methods was used to measure the strength. The unconfined compression test was used for cohesive materials such as clay and silica, and the torvane test was used for a very soft to stiff soil matrix such as sand (Holtz & Kovacs 1981).

RESULTS

The test result are listed in Table 1 for the three different types of soil matrix with different sample treatment. A detailed description of the results follows.

Sand

The torvane test compared the shear strength of sand. Higher shear strength means that the sand can withstand more shear stress under allowable strain or deformation. Many factors may affect the shear strength of the sand, such as the void ratio or relative density, particle shape and size, grain-size distribution, particle surface roughness, water, and overcondensation or prestress. We used same type of sand (Mallinkrodt 7062, washed and dried) to avoid error due to various types of sands, and the shear stress was measured under different moisture content in order to figure out how moisture content would affect the strength.

PHB is known to be capable of plugging the porous medium and forming a barrier around it. We found that the soil matrix strength with a sand sample mixed with PHB solution had a higher torvane stress (0.19 kg/cm^2) than one without PHB solution (0.15 kg/cm^2) under the same moisture content, showing that PHB can enhance the strength of the sand.

Humic material such as that from lignite is quite abundant in natural soils and in landfill sites. The effect of humic material on the soil matrix strength also

TABLE 1. Results of strength test for different types of soil matrix.

Soil Matrix	Sample treatment	Test method	Ratio of sample stress[c] to control[d] stress	Remarks
Sand	PHB added	T.T[a]	1.27	PHB can enhance the strength of sand
	Humic material added (pH = 4)	T.T.	1.96	Around the PZC (point of zero cross) of humic material
	Humic material added (pH = 13)	T.T.	1.96	Attraction force between the organics
	Humic material added (pH = 1)	T.T.	0.61	Repulsion force between the organics
	Distilled water and *A. faecalis* added	T.T.	1.19	Nutrient supply is important for bacteria to grow well
	Nutrient broth added	T.T.	1.21	Composition of nutrient broth may enhance strength
	Nutrient broth and *A. faecalis* added	T.T.	1.26	Proper nutrient supply with bacteria can get more biopolymer and function well
Silica	PHB added	U.C.T.[b]	2.86	PHB can enhance the strength of silica
	Distilled water and *A. faecalis* added	T.T.	1.17	Tentative data only because T.T. may not good for measuring the strength of silica
Clay	Nutrient broth added	T.T.	1.04	Same remark as above
	Nutrient broth and *A. faecalis* added	T.T.	1.08	Same remark as above
	Biopolymer from *A. faecalis* in recommended medium added	U.C.T.	1.62	Good yield of biopolymer and appropriate functionality of the biopolymer
	Xanthan gums added	U.C.T.	1.58	Based on its special structure and properties
	PHB added	U.C.T.	1.27	Smaller enhancement of strength can be observed
	Biopolymer from *A. faecalis* in nutrient broth medium added	U.C.T.	1.20	Smaller enhancement of strength can be observed
	Biopolymer from *A. viscolactis* in nutrient broth medium added	U.C.T.	1.09	Smaller enhancement of strength can be observed
	Sodium alginates	U.C.T.	1.02	Smaller enhancement of strength can be observed

(a) T.T.: Torvane test.
(b) U.C.T.: Unconfined compression test.
(c) Stress: Torvane stress in torvane test and axial stress in unconfined compression test.
(d) Control: Soil matrix mixed with distilled water only.

was studied. The sand sample mixed with humic material from lignite gave high torvane stress at pH = 4, which is around the PZC (point of zero cross) of humic material, and at pH = 13, because of the attraction force between the organics. The opposite trend for low pH such as pH = 1 is due to the repulsion force within the organics.

If nutrient is well provided, bacteria can grow in the soil matrix and form biopolymers to bind with it. The result shows that a sand sample mixed with nutrient broth and *A. faecalis* can achieve the highest torvane stress compared to others without nutrient supply or without bacteria. Therefore the growth of slime-forming bacteria inside the sand with proper nutrient can enhance strength more effectively because more biopolymers that function well with the sand can be produced.

Silica

The silica used here is about 240 mesh, which is very fine, so the saturated silica is somewhat like cemented soils that retain their intrinsic strength after removal of confining pressure. The unconfined compression test is more suitable to measure the strength of silica than the torvane test. The result indicates that the sample containing distilled water mixed with *A. faecalis* obtained the highest torvane stress compared to others. This set of data is only tentative, because the torvane test may not be a good measurement for the strength of silica.

Using the unconfined compression test to measure the strength of silica, we found that a silica specimen mixed with PHB solution yielded almost 3 times higher compressive stress than a silica specimen mixed with water only (control condition). Thus, the strength of silica is highly enhanced by PHB.

Clay

Different concentrations of biolymer solutions were prepared to see how these solutions perform in clay samples. Biopolymer from *A. faecalis* growing in the recommended medium can most enhance the strength of clay, probably due to the good yield and appropriate functionality of the biopolymers. Xanthan gum also worked well, based on its special structures and properties. The polymer backbone of xanthan is made up of β-1,4-linked *D*-glucose residues and is therefore identical to the cellucose molecule. It provides high solution viscosity at low concentration. As for PHB, biopolymers from *A. faecalis* growing in nutrient broth, biopolymers from *A. viscolactis* growing in nutrient broth, and sodium alginates show different smaller enhancements. Thus, all these biopolymers can enhance the strength of clay.

DISCUSSION

Different concentrations of biopolymer solutions were used in this study, so it is difficult to conclude which biopolymer would most enhance the strength of the soil matrix because they have different properties and structures. Evidently, regardless of concentration, soil matrix strength is enhanced. In this study, we concentrated on the potential and feasibility to apply this technology, but a number of factors may affect the capability of biopolymers to enhance soil matrix strength. For example, different concentrations and viscosities may cause different enhancements; and different nutrient supplies and different temperatures or pH also may have different effects. The ultimate concentration of a biopolymer solution with the best nutrient supply, and correct temperature and pH can maximize the increase of strength while saving on cost and giving the best result.

Soil Erosion

Soil erosion is certainly a worldwide problem, and is now more severe than ever. It may cause unstable foundations and the loss of soil in geotechnical, civil engineering, or agricultural endeavors. Improving soil strength to prevent erosion will be the solution. The study showed that the soil strength can be enhanced by applying biopolymers because of the cross-link interaction of biopolymers and the soil matrix. We anticipate that this technology can be applied to different types of soil.

Liquefaction

Liquefaction of sand also may present a problem. For example, the failure of Ft. Peck Dam in Montana is said to have been due to this phenomenon (Terzaghi & Peck 1948). When loose saturated sands are subjected to strains or shocks, sands tend to decrease in volume. This tendency may cause a positive increase in pore pressure, which results in a decrease in effective stress within the soil mass. Once the pore pressure becomes equal to the effective stress, the sand will lose all its strength, and it is said to be in a state of liquefaction. The strength of sand also can be enhanced via this application. The bases of large hydraulic complexes such as dams would be able to bear a heavier loading under conditions of large statically induced strains like earthquakes to avoid a collapse, if this technology is developed.

Bioremediation

In situ bioremediation to clean hazardous sites has become increasingly popular. As the study showed, microbially derived polymers can be used efficiently as binding agents to help the soil matrix become stronger and less permeable and thus stop the migration of hazardous leachates or aggressive liquids. Also, biopolymers can prevent fluid migration by controlling the generation of biopolymer-filled soil layers to generate a capsule around the spill such as a slurry wall. This technology can be extended to minimize risks associated with land subsidence due to fossil energy exploitation. It is expected that biopolymers will encapsulate the oil-contaminated soil in a dynamic natural environment and seal off pollutant plumes so that naturally occurring bioremediation can take place preferentially in the remaining regions (Yen & Chen 1990). The creation of a shield also may eventually prevent further migration in conjunction with other techniques such as air stripping, vacuum venting, etc.

ACKNOWLEDGMENTS

The authors would like to express their appreciation to Dr. R. E. Goodman, Director of Laboratories, Biological Science, and Dr. J. P. Bardet, Associate Professor, Civil Engineering, University of Southern California for technical assistance. We would also like thank Kelco Co. for providing the gellan gum samples as well as information for this study. This work has been financially supported by the National Science Foundation under MSS-9118234.

REFERENCES

Holtz, R. D., and W. D. Kovacs. 1981. *An Introduction to Geotechnical Engineering*. Prentice-Hall Inc., Englewood Cliffs, N.J.

Jack, T. R., J. Shaw, N. Wardlaw, and J. W. Costerton. 1989. "Microbial Plugging in Enhanced Oil Recovery." In E. C. Donaldson, G. V. Chilingarian, and T. F. Yen (Eds.), *Microbial Enhanced Oil Recovery*, Chapter 7, pp. 125-149. Elsevier Scientific Publishers, Amsterdam, The Netherlands.

Li, Y., I. C.-Y. Yang, K. Lee, and T. F. Yen. 1993. "Subsurface Application of *Alcaligenes eutrophus for plugging of porous media*." In E. T. Premuzic and A. Woodhead (Eds.), *Proceedings of the 4th International Conference of Microbial Enhanced Oil Recovery*, Elsevier Scientific Publishers, Amsterdam, The Netherlands.

Magee, L. A., and A. R. Colmer. 1960a. "Factors Affecting the Formation of Gum by *Alcaligenes faecalis*." *J. of Bacteriology 80*: 477-483.

Magee, L. A., and A. R. Colmer. 1960b. "Properties of Gums Formed by *Alcaligenes faecalis*." *J. of Bacteriology 81*: 800-802.

Punch, J. D. 1966. "The Production and Composition of the Slime of *Alcaligenes viscolactis*." Ph.D. Dissertation, University of Minneapolis, MN.

Stamer, J. R. 1963. "A Study of Slime Formation by *Alcaligenes viscolactis*." Ph.D. Dissertation, Cornell University, Ithaca, NY.

Terzaghi K., and R. B. Peck. 1948. *Soil Mechanics in Engineering Practice*. John Wiley & Sons, Inc., New York, NY.

Wilson, P. W., and S. G. Knight. 1952. *Experiments in Bacterial Physiology*. Burgess Publishing Co., Minneapolis, MN.

Yen, T. F., and J. Chen. 1990. "Transport of Microorganisms to Enhance Soil and Groundwater Bioremediation." In T. Burszyski (Ed.), *Proceedings, Hazmacon*, Vol. II, pp. 95-100. Association of Bay Area Government, Anaheim, CA.

IN SITU BIOLOGICAL ENCAPSULATION: BIOPOLYMER SHIELDS

Y. Li, I. C.-Y. Yang, K.-I. Lee, and T. F. Yen

ABSTRACT

Alcaligenes eutrophus, which produces a massive amount of the intracellular polyester-poly-3-hydroxybutyrate (PHB), as high as 70% of the cell weight, was selected for porous media plugging studies. To simulate the subsurface environment, both static drainage and pressurized pumping flow systems of *A. eutrophus* living cells and PHB suspensions through laboratory sand packs were investigated. In the static drainage flow system, the effluent rate was reduced about 280-fold. For the pressurized pumping flow system, the relative permeability (K_i/K_o) was reduced by a factor of 1 million. Both dead cells and living cells of *A. eutrophus* plugged the sand pack columns in the static drainage flow system. The dead cell suspension reduced the effluent rate by 4.3 times within 5 hr, while the living cell suspension made a 280-fold reduction. A PHB water solution, a commercial product in powder form which disperses well but is not totally dissolved in water, showed the plugging effects to be solely dependent on the concentration of PHB. These findings signify that *A. eutrophus* and its microbial product, PHB, are both efficient plugging agents. They have potential applications in filling the soil and forming a shield because of their relative nonagglomerating cell size; their rod shape of 0.7 µm in diameter and 1.8 to 2.6 µm in length; and their lack of any exopolymer in culture solutions, especially the beneficial biopolymer (PHB) produced internally in cells.

INTRODUCTION

The direct and indirect disposal of hazardous waste, with minor nonpoint sources such as urban infiltration, agriculture activities, or mining operations are main sources of subsurface contaminants. Major direct sources come from landfill, underground injection leaks, or spills from storage, treatment, and process facilities. Current contaminant migration controls, e.g., sheet piling cutoff walls, grout curtains, and slurry walls, are inefficient and often cause leakage.

In landfill sites, leachates may result from draining at the sides of a fill or from breaks in liners meant to remain intact beyond the life-span of the most incalcitrant compounds in the fill. The currently available technology, i.e., the excavation of contaminated soils and subsequent groundwater treatment, has proven to be very costly and ineffective. A new technology is needed to better mitigate or treat leachates entering the surrounding soils in situ to inhibit their migration into the aquifer.

Microbial polyesters recently have attracted a lot of industrial attention because degradable thermoplastics have a wide range of applications. A number of biopolymers can be found in soil or in contaminated regions. It is possible that the migration of water fluid could be prevented by biopolymers produced by microbial activity. Both biological industrial waste treatment processes and the recovery of oil from reservoirs in the petroleum industry indicate that microbial biopolymers can be used to prevent fluid migration by controlling the generation of biopolymer-filled soil layers to generate a capsule around the spill. It is expected that the application of such biopolymers will encapsulate the oil-contaminated soil in a dynamic natural environment and seal off fuel plumes so that naturally occurring biodegradation can occur preferentially in the remaining less-contaminated regions. The creation of a shield may also, by preventing further migration, work well in combination with other techniques such as in situ soil vacuum extraction process.

Bacteria plugging in an oil field was observed in an oil production process called microbial enhanced oil recovery (MEOR), in which bacteria are used to improve the recovery of oil (Yen 1990). It has been reported that some slime-producing bacteria can reduce permeability to a certain extent but cannot reach zero permeability (Jack et al. 1989).

PHB is an intracellular storage polymer with the function of providing a reserve of carbon and energy for the cells in which it is produced. It is a highly crystalline thermoplastic with a melting temperature around 180°C and a glass-transition temperature around 4°C (Doi 1990). It is known in polymer science that the glass-transition temperature is the temperature that material becomes glassy and material with a low glass-transition temperature is good for ambient applications. It was discovered that a wide variety of prokaryotic organisms accumulate PHB as a reserve material (Table 1). *A. eutrophus* was selected to produce PHB for the plugging test by considering factors such as product, capital cost, and the ease or difficulty of product separation.

This marks at the first time that *A. eutrophus* (ATCC 17699) and its bioproduct, PHB, have been studied in the application of plugging. More promising applications have been expected for intracellular polymer-producing bacteria such as *A. eutrophus*. In particular, the bioproduct of *A. eutrophus*, PHB, has many unique characteristics that have drawn our attention.

The preliminary investigation has not been done by any people before in geotechnical application. We are interested in investigating the application by applying viable bacteria to the intended sites for the purpose of plugging. Some publications have reported the application of certain particles such as dead bacteria on plugging. Since viable bacteria should have more advantages than dead bacteria because viable bacteria are motile other than Brownian motion; they can migrate, transport, and follow the nutrient vector; and they can produce and

TABLE 1. Poly(3-hydroxybutyrate)-accumulating microorganisms.

Acinetobacter	Clostridium	Micrococcus	Sphaerotilus
Actinomycetes	Derxia	Microcoleus	Spirillum
Alcaligenes	Ectothiorphodospira	Microcystis	Spirulina
Aphanothece	Escherichia	Moraxella	Streptomyces
Aquaspirillum	Ferrobacillus	Mycoplana	Syntrophomonas
Azospirillum	Gamphosphaeria	Nitrobacter	Thiobacillus
Azotobacter	Haemophilus	Nitrococcus	Thiocapsa
Bacillus	Halobacterium	Nocardia	Thiocystis
Beggiatoa	Hyphomicrobium	Oceanospirillum	Thiodictyon
Beijerinckia	Lamprocystis	Paracoccus	Thiopedia
Caulobacter	Lampropedia	Photobacterium	Thiosphaera
Chlorofrexeus	Leptothrix	Pseudomonas	Vibrio
Chlorogloea	Methylobacterium	Rhizobium	Xanthobacter
Chromatium	Methylocystis	Rhodobacter	Zoogloea
Chromobacterium	Methylosinus	Rhodospirillum	

Modified after Dawes et al. 1973.

excrete PHB. Thus, the objective of this investigation is to use viable *A. eutrophus* to produce PHB and both *A. eutrophus* and PHB can act as plugging agents.

EXPERIMENTAL

PHB Polymer Stability Test

Solubility is considered as an indicator of stability. The solubility of PHB in water under different pHs (at room temperature, 20 to 30°C) and temperatures (at normal pH value of distilled water, pH 6.5 to 7.0) was tested in this research (Figures 1 and 2). The suspension was allowed to dissolve in 1 hour.

Static Drainage System

A simple device was used to differentiate the plugging effects of test fluids (Li et al. 1993). This system consists of a suspension container, a sand column, and an effluent collector. The sand columns were made of 20-g 32 to 100 mesh standard Ottawa sand held in glass tubes with an average porosity of 46.4%. Distilled water with pH 6.5 to 7.0 was used in the experiments and the experiments were run at room temperature (20 to 30°C).

Three types of fluids were run in this static drainage system: (1) a living cell suspension of *A. eutrophus* with 2.14×10^7 cells/mL determined by the growth curve (viable method) in nutrient broth (Becton Dickinson BBL nutrient broth, prepared at pH 6.5 to 7.0), (2) a dead cell suspension of *A. eutrophus* also with 2.14×10^7 cells/mL determined by the growth curve (viable method) in nutrient broth because fluids (1) and (2) should be identical cell suspension and the number

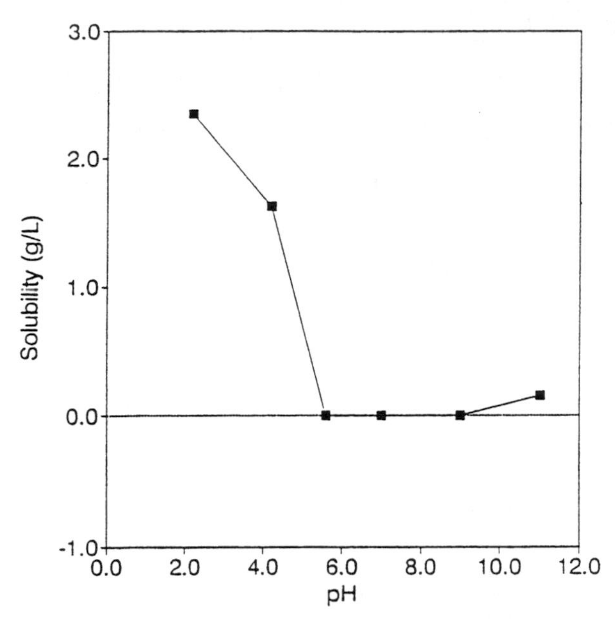

FIGURE 1. PHB (commercial products) stability vs. pH (in water) at room temperature 20 to 30°C.

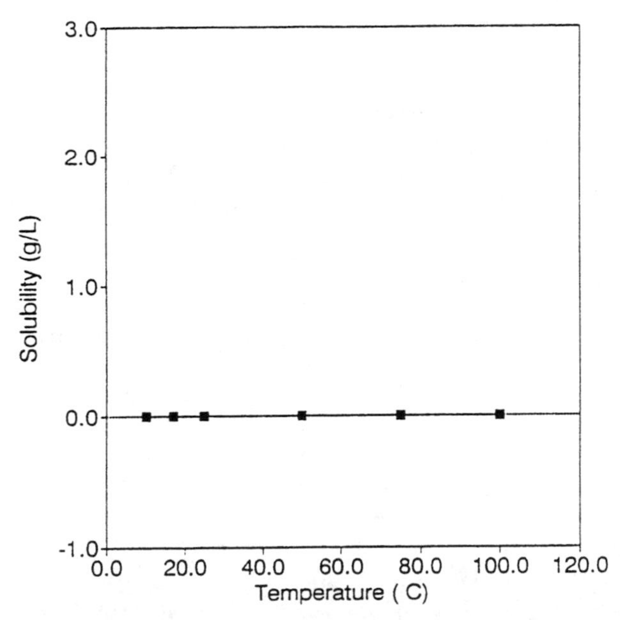

FIGURE 2. PHB (commercial products) stability vs. temperature (in water) at pH = 6.5 to 7.0.

of bacteria should not change in the suspension after being autoclaved, and (3) commercial PHB water solutions. *A. eutrophus* in nutrient broth has a 1-day lag phase and a 2-day exponential phase, the living cell suspension was a culture of 2-day growth of *A. eutrophus* in nutrient broth. The dead cell suspension was gained by sterilizing the living cell suspension in autoclave for 20 minutes. Three PHB water solutions, labeled A, B, and C, were made from one original solution. The original solution was 0.913 g of PHB powder in 1 L of distilled water, sonicated for 16 hr at 250 rpm setting to disperse the PHB well in the water, because PHB has very little solubility in water. After allowing to sit overnight, 250 mL of the solution was filtered out through filter paper. The filtrate was collected as solution A and had a concentration of 0.380 g/L. Solution B was 250 mL of the original solution without filtration, containing 0.630 g in 1 L of water. The remaining original solution without filtration became solution C in which 1.193 g/L was determined, and the reason why the concentration in solution C was higher than the original solution was because solution C was the bottom portion from the original solution.

The draining time for the test fluids from the upper mark down to the bottom mark was recorded at regular intervals. Because the distance between the two marks was fixed for all the columns, the ratios of the drainage rates can be expressed as $r_i/r_o = t_o/t_i$, where t_o is the initial time of each run, t_i is the drainage time for the fluids passing down the two fixed marks on the column, r_i is the drainage rate at time t_i, and r_o is the drainage rate at initial time t_o.

Pressurized Pumping Flow System

Further experiments were conducted with a carefully designed experimental apparatus so that the actual hydrodynamic conditions for bacteria transport in porous media could be simulated (Yen et al. 1991). An acrylic cylinder column of 10.0 cm in length and 2.5 cm in diameter is the key part of this system. Standard Ottawa sand, SX 75, 30 to 50 mesh size, obtained from Matherson Coleman & Bell, was packed inside the column for tests. Before packing the column, the sand was thoroughly washed with distilled water to remove fines and dried at 80°C for at least 48 hr, until totally dried. To prepare the bacterial suspension, *A. eutrophus* was grown in a nutrient broth for 2 days. Bacterial suspensions were replaced by fresh 2-day suspensions after being recycled in the test system for 3 days to maintain the live cell solution state being used in the plugging tests. According to the *A. eutrophus* growth curve made in this laboratory by the viable counts method, the doubling phase of *A. eutrophus* starts after 1 day of growth in nutrient broth and lasts 2 days until the third growth day. A number of cells were still alive after 5 days.

In this system, the column was horizontally mounted, and the pressure-recording device was connected across the column. Before pumping the bacterial suspension in each run, sterilized distilled water was pumped through the flow system at the same rate as the bacterial suspension to clean the system. The pressure difference of complete plugging in the column was 30 psi, determined by closing the outlet control valve. A peristaltic pump provided this pressure difference constantly to force the suspension through the system. Prior to plugging, the flowrate (determined by collecting the effluent at the outlet of the system) was measured at around 3.65 mL/min, and the pressure difference was zero. As the pressure difference increased, the measured flowrate decreased. The pressure difference across the sand pack column was continuously monitored and recorded while a bacterial suspension was injected into the sand pack column, the rate of the effluent being measured at regular intervals. The expression of the plugging effect can be derived from permeability changes during the experiments. The permeability values at each interval of injection were calculated based on Darcy's equation,

$$K = Q\,L\,\mu\,/(A\,\Delta p) \tag{1}$$

where K is the permeability in Darcy, Q is the flowrate in cm^3/sec, L is the column length in cm, μ is the liquid viscosity in cp, A is the column cross sectional area in cm^2, and Δp is the pressure difference in atmosphere. The ratios of permeability values at different intervals to the initial permeability value, K_i/K_o, are used to indicate the plugging effect. In equation (1), L and A are apparently constants. When similar bacterial suspensions are injected, and the effect of the temperature change from day to night in the laboratory on the μ value is negligible, μ also can be considered as a constant. In this case, equation (1) can be simplified as follows:

$$K_i/K_o = (\Delta p_o\,Q_i)\,/\,(Q_o\,\Delta p_i) \tag{2}$$

where i refers to any interval when the pressure difference and effluent rate are recorded.

RESULTS

The stability test of PHB in water under different pHs shows that PHB is fairly stable at the range of pH 5.5 to 9.0, as the solubility is almost zero (Figure 1). Both increasing and decreasing outside this range, the solubility of PHB increases to 0.12 g/L at pH = 11.0 and 2.35 g/L at pH = 2.2. Temperature shows no significant effect on the stability of PHB, because its solubility from 10 to 100°C was indistinguishable (Figure 2).

The differences in plugging effects of the test fluids were screened out by draining through small sand pack columns. Figure 3 shows the results between living cells and dead dells of *A. eutrophus* indicated by the change in the ratio of drainage rate (r_i/r_o) with the test time. The nutrient solution, which is the medium without inoculum, had no change in drainage rate while passed down the sand columns. Within 5 hr, the dead cell suspension of *A. eutrophus* made a 4.3-fold reduction in effluent rate. Its r_i/r_o value is 0.232 in Figure 3. The effluent rate was reduced about 90-fold by the living cell suspension in 5 hours drainage time. Unlike the dead cell suspension, this decreasing trend was kept for living ones

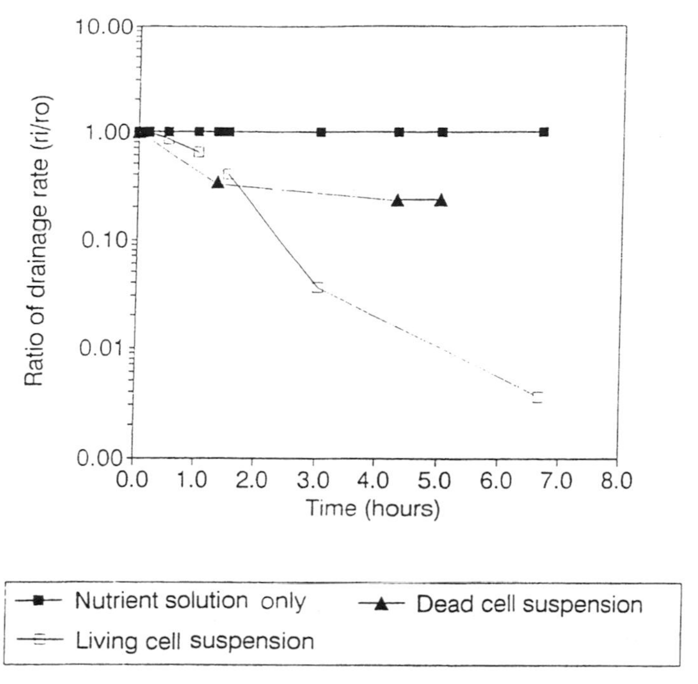

FIGURE 3. Plugging effect of *A. eutrophus* cell suspension in sand pack columns determined by static drainage flow system. (1) nutrient solution only; (2) dead cell suspension; (3) living cell suspension. The cell suspensions were 2-day growth cultures. The dead cell suspension was prepared under autoclave operation.

after 5 hours, showing a 2.65 cycles drop of logarithmic scale (in logarithmic scale, one log cycle refers to 10-times change in quantity) in the r_i/r_o axis that is about 280-fold reduction at 7 hours.

It was realized that the inaccuracy of those results, due to uncontrolled parameters (e.g. pressure variation with liquid level in the test column), constrained the detailed conclusions that could have been drawn from this system. However, certain indications of a complex system with multiple factors were indeed gained from such data both in this research and in other previous works. McCalla (1950) reported that the plugging effects of adding a sucrose solution resulted from its percolation through columns packed with organic-deficient Peorian loess. Similar studies have been carried out using sand-packed columns rather than soil to observe the permeability reduction by Gruesbeck & Collins (1982). Furthermore, the results of the present experiments agreed with published results of other researchers (Jack et al. 1989). Their data also showed a higher plugging effect of living cells than that of dead cells, despite the fact that they used a *Pseudomonas* species.

The plugging effect of commercial PHB powder in water was revealed by draining three different concentration solutions through the sand pack columns. Figure 4 shows that concentration is the sole factor that affects the plugging results. A low PHB concentration of 0.380 g/L is no different from distilled water in the plugging of the columns. The slight plugging of sand columns showed up as the PHB concentration increased to 0.630 g/L (indicated by a small drop of r_i/r_o). More significant plugging occurred, however, for the solution with 1.193 g PHB in 1 L of water, the sharply dropping value of r_i/r_o being calculated as a 100-fold reduction in the drainage rate within 3 hours. Those results express that PHB powder water solution, though not a soluble solution, can be used for porous media plugging. It is possible to control the extent of plugging by adjusting the concentration of PHB injected. Certain treatments, such as sonication, may be needed to help disperse the PHB in water, because PHB powder has a strong tendency to clump.

In a pressurized pumping flow system, the *A. eutrophus* living cell suspension greatly reduced the K_i/K_o value (indicating the plugging effect) down to 4 log cycles after 16 days and 6 log cycles after 11 days for two repeat experiments (Figure 5).

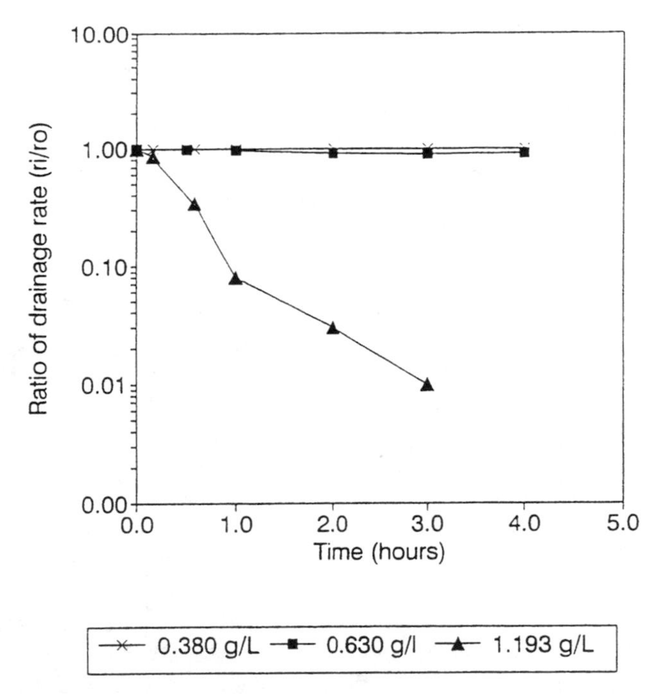

FIGURE 4. Plugging effect of commercial poly-3-hydroxybutyrate (PHB) powder dispersed in distilled water in sand pack columns determined by static drainage flow system. Solutions were sonicated at 250 setting for 16 hours. (A) distilled water and 0.380 g/L; (B) 0.630 g/L; (C) 1.193 g/L.

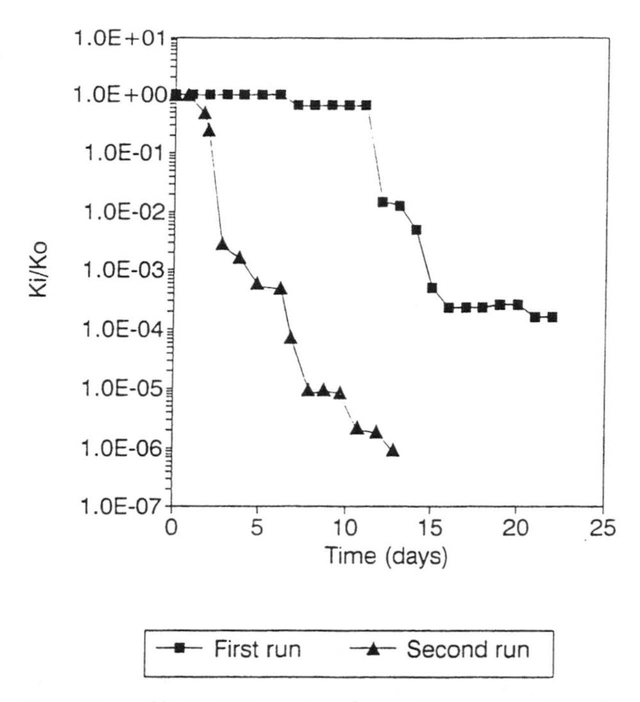

FIGURE 5. Plugging effect of *A. eutrophus* cells suspension in sand pack columns determined in pressurized flow system. (1) first run; (2) second run. The suspensions were 2-day growth cultures and were replaced after 3 days of recycle in the system.

K_i/K_o values are calculated from the changes in both pressure difference across the column and flowrate (Equation 2). The pressure difference is more sensitive than the flowrate because its change was observed prior to the difference in flowrate. Even though the pressure difference reached the ultimate value of plugging (30 psi), effluent still trickled down from the column. This effluent is probably due to the nonuniform pore sizes caused by the irregular shapes of the sand used. That means that most of the pores were plugged, but there were a few through which the effluent could escape. The nonuniform pore sizes of the packed sand were observed under a light microscope with a magnification of 5.

DISCUSSION

The unique properties of PHB make its applications useful. Even though the positive effect of polymer production on plugging has been proved by previous workers (Jack et al. 1989), PHB could be the best selection among polymers for three main reasons. First, the low glass-transition temperature (4°C) is important to plug the pores of soil. Low glass-transition temperature enables PHB to reach a glassy state at 4°C, which may easily be filled into the soil pores without other special operations. After the soil pores are filled with PHB polymer, a barrier is formed that would be able to prohibit the movement of fluids because of its stable characteristics of high temperature and water with pH values of 5.5 to 9.0. On the other hand, not all the insoluble polymers are suitable for plugging, only insoluble polymers with a low glass-transition temperature can be used because polymers with a high glass-transition temperature are easily cracked at filed temperature and can not act as plugging agents. Second, because PHB is produced by a variety of microorganisms, it is possible to find indigenous microorganisms, that can produce PHB at the sites where barriers are to be built. Also, the carbon sources that produce PHB are common small acids that perhaps can be obtained without cost from a landfill site where they are considered hazardous wastes. In special cases, those small acids may be contained in the fluid water that is intended to be isolated. The combination of isolation and treatment of the hazardous water is naturally achieved in those cases. Last, another cost-effective option is that naturally occurring PHB is commercially available even if the above advantages do not exist.

Two general mechanisms of microbial plugging likely to occur in a porous media were suggested by Jack et al. (1989), particulate plugging and plugging due to biofilm formation. In the former, bacterial cells alone cause permeability reduction or drainage rate decrease by plugging and filter cake formation. Both dead and living cells are capable of this type of plugging. The latter resulted from the buildup of biofilms in the porous media by viable microorganisms. Obviously, this happens with living cell suspension plugging. Away from these conclusions, plugging by many kinds of extracellular products has been well documented.

In the *A. eutrophus* culture, plugging the porous media is not assisted by any slimes or polysaccharides because it is not an extracellular product producing bacterium. Other possible interfering factors existing in the culture can also be neglected for the results in the drainage flow system because the dead cell suspension was a half portion of the living cell suspension, sterilized in autoclave. It is implied that those two cell suspensions had the same chemical components in their cultures. Therefore, the drainage rate reduction by dead cell suspension was the result of cell aggregation in the porous media. The living cell suspension plugged the sand pack column with an aggregation of cells and biofilm formation on the sand surface. If this is true for the results obtained, the difference in drainage rate reduction between dead and living cell suspensions presents the effect of *A. eutrophus* cell growth inside the sand column. This difference is very significant, being 90-fold when compared to 4.3-fold of r_i/r_o reduction. In terms of its batch growth curve in the same medium, the *A. eutrophus* population increases only about 4 times within 5 hours, for its doubling time is 3 hours. Other factors may play a role in creating that difference (e.g., the change of cell surface characteristics due to autoclave treatment may reduce the ability of cell aggregation).

PHB has been reported to accumulate in a wide variety of bacterial species (Dawes et al. 1973). The level of PHB in cells can be drastically increased from a low percentage to more than 80% of the dry cell weight when bacterial growth is limited by the depletion of essential nutrients such as nitrogen, oxygen, phosphorus, sulfur, or magnesium. Schlegel et al. (1961) demonstrated that *A. eutrophus* accumulated PHB up to 70% of the total dry cell weight when chemolithotrophic growth was limited by ammonium consumption in a batch culture. PHB powder is commercially available at the present time and provides choices of uses ranging from in situ production by bacteria to on-site preparation in bulk.

As discussed, viable microorganism cultures have a higher plugging effect in porous media than dead bacteria. That is proven by the fact that microbial activity clogs the pores with the products of growth, cells, slimes, or polysaccharides. Only few a publications have reported on plugging with intracellular polymers. The current research is trying to fill this gap. *A. eutrophus* does not produce extracellular products at all, but instead produces the biopolymer PHB which can be detected in the cultures used in this research by ultraviolet spectrophotometer. The plugging of the sand pack column by the *A. eutrophus* living cell suspension is an accumulative effect as shown by the steps at the beginning of the two curves in Figure 5. In this process, the entrainment of the plugging particles occurred because the drops of Δp were recorded on the 9th day in the first run and after 7 days in the second run. Thus, the phenomenon can be explained by the mechanisms of particulate plugging of cell aggregates and plugging due to biofilm formation. The buildup of biofilm (i.e., the growth of *A. eutrophus*) on the sand surface is the main reason for the great degree of plugging, because these pluggings occurred without the presence of any other agents such as exopolymers. During the plugging tests, on the other hand, 2 days' growth of *A. eutrophus* nutrient broth culture was used in the exponential phase of its growth curve, and those nutrients needed for growth also were available. This is only an initial trial for subsurface plugging. More studies should be done on the in situ bacteria growth kinetics, seeding protocol, and injection strategy.

CONCLUSIONS

Four conclusions can be drawn from this study:

First, because PHB is insoluble in water after being separated, even when heated to 100°C, it is a good material to prevent water solution from migrating when used to build up a barrier. This function can be applied to isolate toxic underground water where the temperature is higher than that in the atmosphere. Its solubility at pH values from 5.5 to 9.0 is nondetectable, so it may also build a good barrier for holding landfill leachates because they are water fluid with pH values mostly from 5.3 to 8.5 (Lu 1985, Ehrig 1989).

Next, with the *A. eutrophus* suspension, especially the PHB polymer inclusions, the plugging results obtained in this laboratory are better than any others ever reported. This means that PHB functions better to plug porous media. The results indicate that it is sufficient to stop any aggressive liquid movement.

Third, the plugging of porous media by *A. eutrophus* culture may last longer than with other plugging bacteria that do not contain any polyester. Based on this research, three types of plugging have occurred in the *A. eutrophus* culture system: (1) living cell suspension, (2) dead cell suspension, (3) intracellular product, i.e., a PHB solution. Plugging by living cell suspension of *A. eutrophus* can possibly last for long period of time even though some changes in this system may occur. If cells were to die, a new plugging agent, PHB, would be released.

Fourth, both a static drainage flow system and a pressurized pumping flow system have plugging results which match the plugging mechanisms of particulate plugging by microbial cells and viable bacterial plugging through biofilm formation.

ACKNOWLEDGMENTS

The authors would like to thank Dr. R. E. Goodman, Microbiological Laboratory, University of Southern California, for his technical assistance. They also wish to acknowledge part of the initial sand column plugging test done by Gary K. Wong. Partial financial support from Naval Civil Engineering Laboratory N-47408-90-C-1170 and full support from the National Science Foundation MSS-9118234 are appreciated.

REFERENCES

Dawes, E. A., and P. J. Senior (Eds.). 1973. "The Role and Regulation of Energy Reserve Polymers in Microorganisms." *Adv. Microb. Physiol. 10*: 135-266.

Doi, Y. 1990. *Microbial Polyesters.* VCH Publishers, Inc., New York, NY.

Ehrig, H.-J. 1989. "Water and Element Balances of Landfills." In P. Baccini (Eds.), The Landfill: Reactor and Final Storage, pp. 83-116. Springer-Verlag, Berlin.

Gruesbeck, C., and R. E. Collins. 1982. "Entrainment and Deposition of Fine Particles in Porous Media." *Soc. Pet. Eng. J. 22*: 847-856.

Jack, T. R., J. Shaw, N. Wardlaw, and J. W. Costerton. 1989. "Microbial Plugging in Enhanced Oil Recovery." In E. C. Donaldson, G. V. Chilingarian, and T. F. Yen (Eds.), *Microbial Enhanced Oil Recovery*," Chapter 7, pp. 125-149. Elsevier Scientific Publishers, Amsterdam, The Netherlands.

Li, Y., I. C.-Y. Yang, K. Lee, and T. F. Yen. 1993. "Subsurface Application of *Alcaligenes eutrophus* for Plugging of Porous Media." In E.T. Premuzic and A. Woodhead (Eds.), *Proceeding of the 4th International Conference of Microbial Enhanced Oil Recovery*, Elsevier Scientific Publishers, Amsterdam, The Netherlands.

Lu, J. C. S., B. Eichenberger, and R. Stearns. 1985. *Leachate from Municipal Landfills: Production and Management.* Appendix C, Noyes Publications, Park Ridge, NJ.

McCalla, T. M. 1950. "Studies of the Effect of Microorganisms on Rate of Percolation of Water Through Soil." *Soil Sci. Proc., 14*: 182-186.

Schlegel, H. G., G. Gottschalk, and R. Von Bartha. 1961. "Formation and Utilization of Poly-β-hydroxybutyric Acid by Knallgas Bacteria (*Hydrogenomonas*)." *Nature 191*: 464-465 (London).

Yen, T. F. 1990. *Microbial Enhanced Oil Recovery: Principle and Practice.* CRC Press, Buca Raton, FL.

Yen, T. F., J. K. Park, K. I. Lee, and Y. Li. 1991. "Fate of Surfactant Vescicles Surviving from Thermophilic, Halotolerant, Spore Forming, Clostridium *Thermohydrosulfuricum*." In E.C. Donaldson (Ed.), *Microbial Enhancement of Oil Recovery — Recent Advances, 31*, pp. 297-309. Elsevier Scientific Publishers, New York, NY.

GROUNDWATER QUALITY IN SEVERAL LANDFILL AREAS OF JAKARTA, INDONESIA

W. S. Winanti and I. Mawardi

ABSTRACT

The condition of groundwater in Jakarta, Indonesia, has become very bad because an effective waste management system has not yet been instituted. It has already been decided by the government that industries should treat their own waste. However, the problem of domestic waste is still becoming more and more serious. Jakarta, a city of 7 million people, produces about 23,000 m^3/day, as well as a large amount of domestic wastewater. The existence of waste landfill and wastewater creates the potential to pollute groundwater. Observation of groundwater quality conducted in Srengseng, Cempaka Putih, and Cakung-cilincing showed that the groundwater in the surrounding areas is contaminated. The amounts of chemical oxygen demand (COD), biological oxygen demand (BOD), chloride, and iron in Srengseng are 349.1 mg/L, 220 mg/L, 2.675 mg/L, and 180 mg/L, respectively. Waste distribution from the landfill area goes to the north or in the same direction as the groundwater as it flows to the sea. In Cakung-cilincing, the amounts of COD, BOD, chloride, and iron reach 1,378 mg/L, 650 mg/L, 520 mg/L, and 236 mg/L, respectively. The direction of waste distribution is not toward the sea, but rather to the south (land). The plume developed because Cakung-cilincing is located near the seashore and there is seawater intrusion. The same conditions exist in the Cempaka Putih area. The values of COD, BOD, chloride, and iron there reach 369.8 mg/L, 210 mg/L, 3,692 mg/L, and 12.2 mg/L, respectively. Other parameters showing high values are nitrates, nitrites, ammonia, phenol, and coliforms.

INTRODUCTION

Disposal of organic wastes is a continuing problem, particularly in cities such as Jakarta, Bandung, and Surabaya. Municipal waste production in Jakarta reached about 6,800 metric tonnes/day (Wibowo & Shocchib 1988). Until recently, these wastes were placed in open landfills. Several suitable sanitary landfill areas are still open or have already been closed such as Srengseng (West Jakarta) and Cakung-cilincing (North Jakarta).

The heaps of waste in the open landfill systems influence the water quality all around the disposal site. The large, established populations in these areas use the groundwater for their daily needs, making it imperative to establish accurate information concerning the groundwater quality in the area.

The people who live near the landfills generally use the groundwater as the standard water for drinking and bathing. Therefore, the water should fulfill the water requirements of the B groups as stipulated in Government Regulation Number 20 (1990).

People in the surrounding areas have registered complaints about diseases, such as skin diseases/dermatitis, influenza, throat infections, and stomach ulcers, which they believe to be caused by contaminants in the sanitary landfills (Erowati 1989).

Observations of the water quality showed that the groundwater in several locations in Jakarta contained impurity levels far above those allowed by the B group water quality standards.

GROUNDWATER QUALITY IN THE SRENGSENG AND CAKUNG-CILINCING LANDFILL AREAS

The Srengseng landfill area in West Jakarta comprises 3.5 Ha and has operated since 1972. This location is relatively flat, with an elevation of 25 m above sea level. It is situated between two hills back toward the north-south and is parallel to Pesanggrahan stream. Around this location (within the radius of 50 m from the landfill area center), there are legal and illegal settlements.

Groundwater quality and surrounding inhabitants' wells were tested at 23 points around the landfill locations up to the radius of 200 m, as illustrated in Figure 1. The results of these water sample analyses around the Srengseng landfill area show several parameters whose values exceed the limits of the B group water quality standard (Government Regulation No. 20 [1990]). The parameters are chlorides, ammonia, phenol, iron, COD, BOD, and a high amount of *Escherichia coli* bacteria.

Chloride

Water with chloride content exceeding 100 mg/L has a brackish/salty taste (Salvato 1982). If the chloride contamination is in the form of sodium chloride and the water is used as drinking water, excessive sodium consumption may cause problems in sensitive individuals, particularly those with hypertension or heart diseases. Chloride in the groundwater may be caused by disposal of household waste such as drink and food, and also from domestic wastewater.

Monitoring of the Srengseng landfill area showed the chloride content at the center point of the landfill area to be 185 mg/L. At a radius of 50 m, the chloride content rose to 880 mg/L toward the southwest and to 157 mg/L toward the northwest; however, east of the landfill, the chloride content was still below

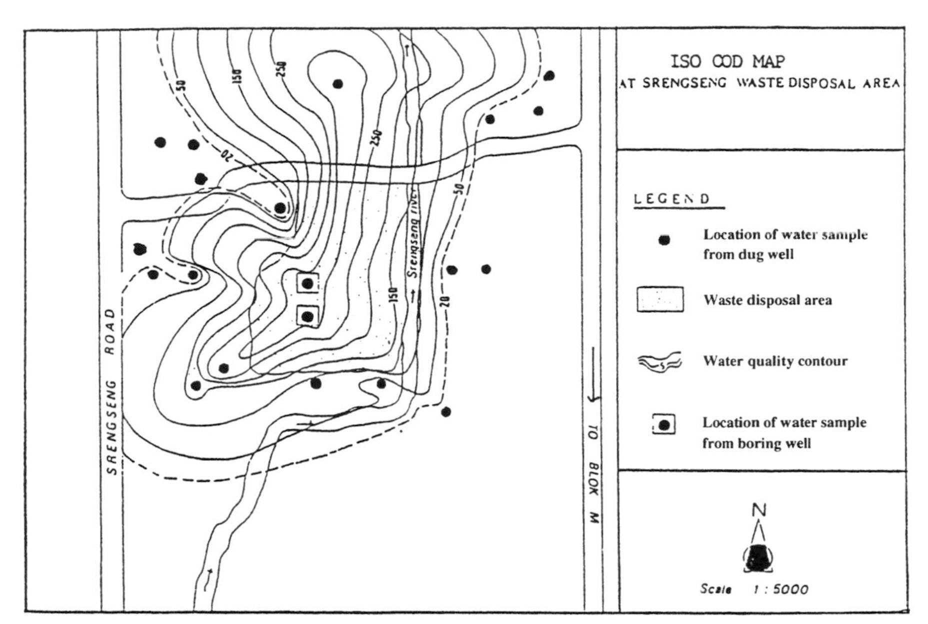

FIGURE 1. Iso chemical oxygen demand concentration of groundwater in Srengseng waste disposal area.

100 mg/L. These levels are in accord with the direction of groundwater flow toward the southwest and northwest.

For the Cakung-cilincing landfill area, the chloride content ranged from 269 to 613 mg/L at a radius of 50 m, and from 92 to 439 mg/L up to the 100 m radius. At a radius of 150 m the chloride concentration was even higher, especially toward the southwest, where it reached 730 mg/L.

Ammonia (NH_3)

The ammonia within groundwater can originate from the anaerobic decomposition of organic materials that form nitrite and nitrate. In water, ammonia can be toxic to several fish species when the concentration reaches 0.2 to 2.0 mg/L (Salvato 1982).

The result of monitoring in the Srengseng landfill area showed that, up to a radius of 200 m toward the north, the concentration was below 3.08 mg/L, whereas the ammonia concentration at the center point was 6.41 mg/L (see Figure 2). In the Cakung-cilincing landfill area, the ammonia content is still beneath the allowable quality standard limit.

Phenol

While chlorinated phenols are more active, all the phenols are pretty nasty, and phenol or substituted phenols in the water will impart a medicinal taste.

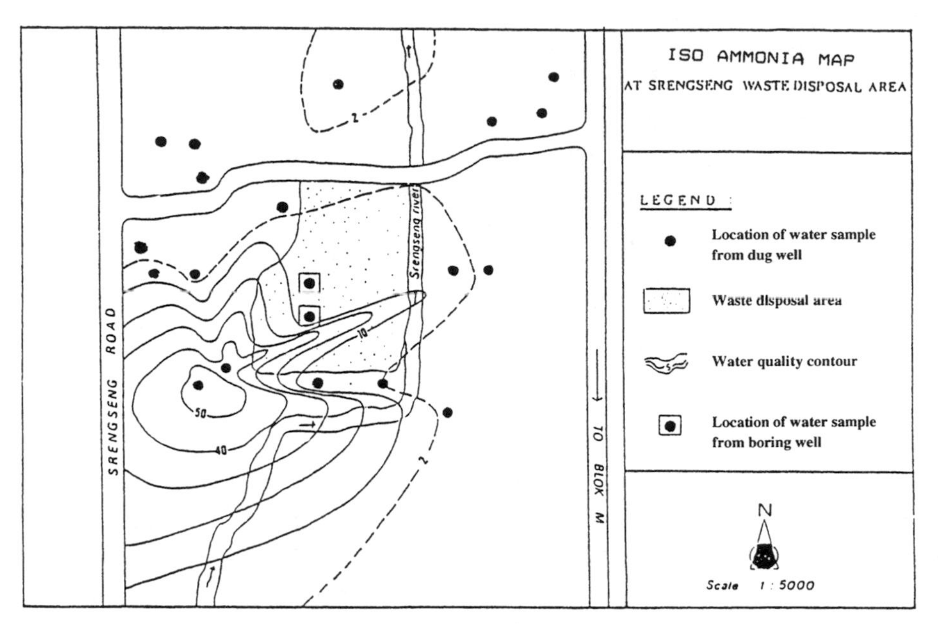

FIGURE 2. Iso ammonia concentration of groundwater in Srengseng waste disposal area.

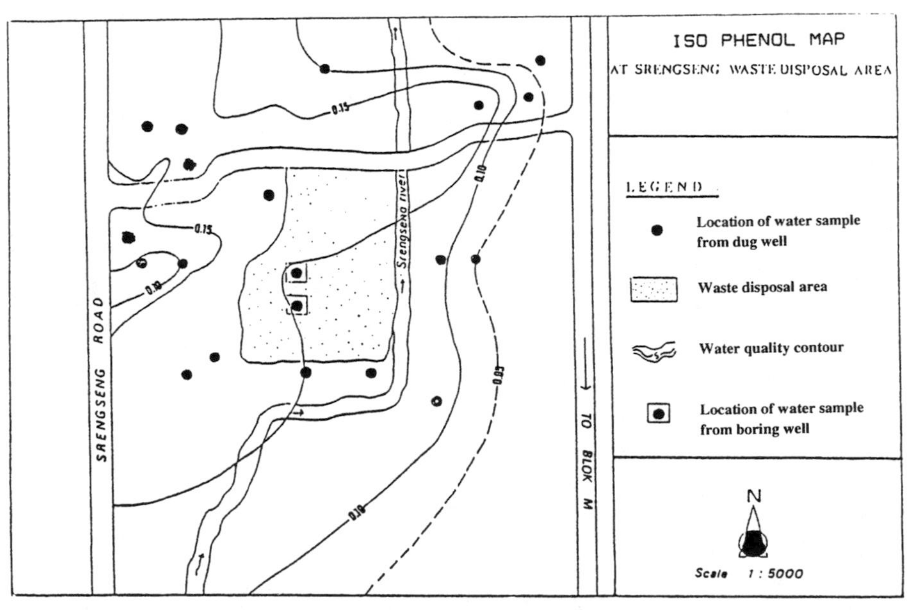

FIGURE 3. Iso phenol concentration of groundwater in Srengseng waste disposal area.

In the B group water quality standard (PP No. 20/1990), the phenol concentration limit is 0.002 mg/L.

The phenol content in the Srengseng landfill area, from the center point out to a radius of 150 m, shows a value higher than the quality standard limit (ranging from 0.08 to 0.16 mg/L). The highest concentration was found within the radius of 50 to 100 m (see Figure 3).

Phenol concentrations are higher in the Cakung-cilincing landfill area. The concentration reached 1.24 mg/L at the center. At a radius of 150 m toward the southwest, the concentration level was 0.17 mg/L.

Iron (Fe)

Iron is a common natural component in groundwater. Excessive iron concentration in water can discolor household utensils, porcelain, and textiles (Salvato 1982). The B group water quality standard specifies that the iron concentration should not exceed 5 mg/L.

The monitoring result at the center point of the Srengseng landfill area showed an iron concentration of 180.2 mg/L. Up to a radius of 50 m toward the southwest and south, the content still reach 44 to 54 mg/L. Toward the north up to a radius of 200 m, iron showed a high concentration. The highest iron content in the Cakung-cilincing landfill area was 256 mg/L, concentrated only at the center. At a radius of 150 m toward the southwest, the concentration declined to 8.2 mg/L (see Figure 4).

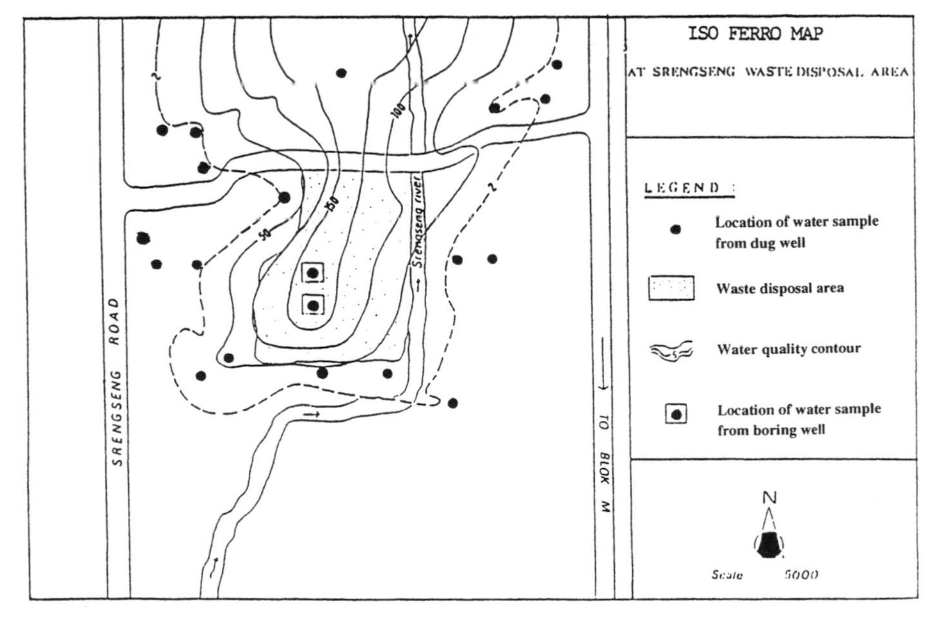

FIGURE 4. Iso ferro concentration of groundwater in Srengseng waste disposal area.

Chemical Oxygen Demand (COD)

COD is a good general indicator of the quality of domestic or industrial water. Based on the 1988 stipulation of the Governor of Jakarta district no. 1608, the COD content limit for the A group of the water quality standard, for water that can be processed for drinking water, is 10 mg/L. (The A group water quality standard from the Jakarta Governor's stipulation is the same as that of the B group in Government Regulation No. 20 [1990].)

In the Srengseng landfill area, the COD content at the center area is 345 mg/L. At a radius of 200 m toward the north, the concentration was 344 mg/L; toward the other directions, it was below 10 mg/L (see Figure 1). In the Cakung-cilincing landfill area the COD content at the center point reached 1,378 mg/L, but only concentrated within a radius of 50 m (see Figure 5).

Biological Oxygen Demand (BOD)

BOD also provides a good general measurement of pollution. The limit for B group water according to Government Regulation No. 20 (1990) is 10 mg/L.

Measurements in the Srengseng landfill area showed a BOD at the center point of 220 mg/L. Toward the north, up to a radius 200 m, a concentration of 120 mg/L was found. In the other direction, at a radius of 100 m, the BOD level was below 10 mg/L and thus within the allowable limit.

In the Cakung-cilincing landfill area, the BOD level at the center point was 650 mg/L but, just as with the COD, this water pollution was concentrated only

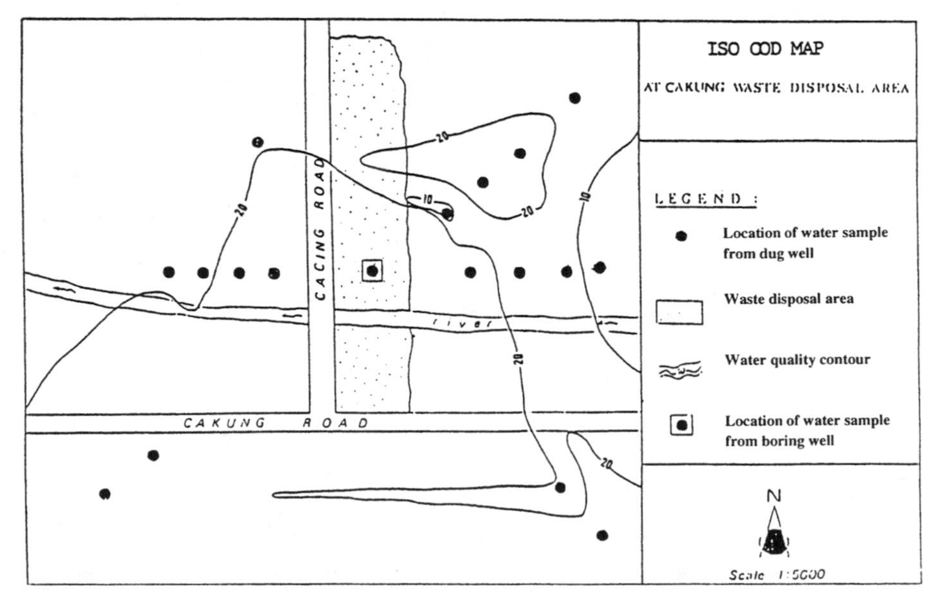

FIGURE 5. Iso chemical oxygen demand concentration of groundwater in Cakung-cilincing waste disposal area.

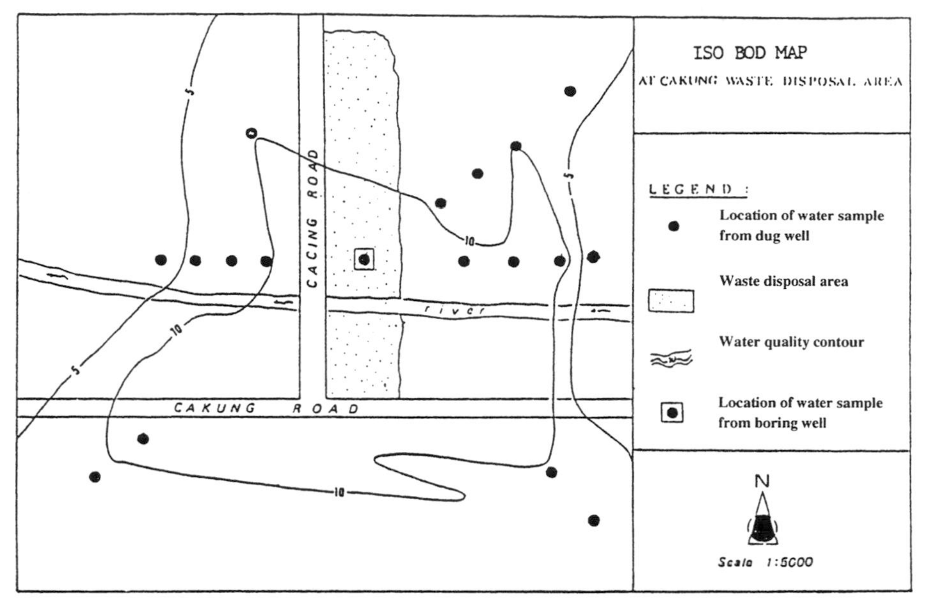

FIGURE 6. Iso biological oxygen demand concentration of groundwater in Cakung-cilincing waste disposal area.

within a radius of 50 m. Outside this distance, its value decreased until below 10 mg/L (see Figure 6).

E. Coli Bacteria

The presence of *E. coli* bacteria in water generally indicates pollution by manure and excrement. According to B group water standards, its value should not exceed 20.10^2 MPN/100 mg (Maximum Probable Number/mg). *E. coli* bacteria in the water indicates the presence of other bacteria such as salmonella and streptococci (Salvato 1982).

The *E. coli* concentration at the center point of the Srengseng landfill area is 150,000 MPN/100 mg, and thus exceeds the water quality limit. The effect of *E. coli* concentration showed in all directions except to the north. Its presence is caused by household waste, especially unprocessed excrement. The center point of the Cakung-cilincing landfill area reaches 240,000 MPN/100 mg, and its impact is less dominant in all other directions.

DISCUSSION

The observations made at the two landfill areas in Srengseng and Cakung-cilincing showed that the groundwater has become contaminated by landfill activities. The decrease in groundwater quality results from the increased concentrations of BOD, COD, phenol, ammonium, iron, and *E. coli*. The increase of pollutant concentrations is consistent with the characteristics and compositions of the waste in the vicinity of the two landfill areas.

Movement of groundwater contamination depends on the direction and flow of groundwater at the two landfills. At Srengseng, groundwater flows toward the sea (northwest and northeast). On the other hand, Cakung-cilincing groundwater does not flow to the sea because of seawater intrusion (Sampara 1991). Groundwater contamination is more concentrated at the center of landfill.

Observation at other landfills in Jakarta, such as Cempaka Putih (center of Jakarta) and Cawang (east Jakarta), showed contamination similar to that of Srengseng and Cakung-cilincing.

Erowati (1989) reported that there is a relationship between the existence of landfill and the diseases suffered by surrounding communities. The dominant diseases with the highest percentages are diarrhea, throat infections, and ulcers.

CONCLUSION

As a result of this study, the following conclusions can be drawn:

1. The groundwater located in a radius 200 m from the center of the landfill in Jakarta has already been contaminated.
2. Pollutant types that have increased concentrations are chloride, ammonia, phenol, iron, COD, BOD, and *E. coli* bacteria.

REFERENCES

Erowati, D. A. 1989. *Pengaruh Lokasi Pembuangan Akhir (LPA) Sampah Terhadap Kesehatan Penduduk* [The Influence of Landfill to the Health of the Surrounding Community], Direktorat Teknologi Pemukiman dan Lingkungan Hidup, Badan Pengkajian dan Penerapan Teknologi, Jakarta, Indonesia.

Local Government of Central District of Jakarta. 1988. *The Decree of the Governor of the Central District of Jakarta No. 1608*, Jakarta Central District, Indonesia.

Salvato, J. A. 1982. *Environmental Engineering and Sanitation.* A Wiley-Interscience Publication, John Wiley and Sons, New York, NY.

Sampara, J. T. 1991. *Laporan Sementara Penelitian Arah Aliran Air pada Sekitar Lokasi Pembuangan Sampah Akhir Srengseng dan Cakung-cilincing.* [The report of groundwater direction observation surrounding the Srengseng and Cakung-cilincing landfills.] Proyek Perlindungan dan Pengembangan Lingkungan, Direktorat Teknologi Pemukiman dan Lingkungan Hidup, Bandan Pengkajian dan Penerapan Teknologi, Jakarta, Indonesia.

State Minister of Population and Environment. 1990. *Peraturan Pemerintah.* [The Government Regulation] No. 20, Tahun 1990, Republik Indonesia.

Wibowo, S., and Shocchib, R. 1988. *Pemanfaatan Barang Bekas dari Sampah dan Pengaruhnya terhadap Pengelolaan Sampah di DKI Jakarta.* [The effect of recycling of domestic solid waste on the waste management in Jakarta.] Kelompok Sanitasi Lingkungan, Direktorat Riset Operasi dan Managemen, Badan Pengkajian dan Penerapan Teknologi, Jakarta, Indonesia.

ELECTROOSMOTIC BIOREMEDIATION OF HYDROCARBON-CONTAMINATED SOILS IN SITU

E. A. Nowatzki, R. J. Lang, M. C. Medellin, and S. M. Sellers

PURPOSE OF INVESTIGATION AND DESCRIPTION OF EXPERIMENT

The purpose of this investigation was to assess the feasibility of using electroosmosis in conjunction with bioremediation to treat hydrocarbon-contaminated soils of low permeability in situ. The concept that is investigated is the use of electroosmosis to move both bacteria and chemical nutrients through soil to achieve bioremediation.

To accomplish this, an electrochemical cell was built by filling a specially constructed container with nontoxic, hydrocarbon-contaminated, saturated soil and inserting two hollow, perforated electrodes or "wells" into the soil. An electric potential was applied between the two electrodes after a solution of bacteria and nutrients was introduced at the anode. As electroosmotic flow of pore fluid occurred from anode to cathode, a steady-state flowrate of fluid collecting in the cathode "well" was removed and circulated back to the anode. Biological and chemical tests were performed to determine if the bacteria withstood the applied potential and to verify that bacteria and nutrients were indeed being transported through the soil.

EXPERIMENTAL SETUP AND PROCEDURES

Soil Bins. Two soil bins having interior dimensions of 19.5 × 19.5 × 47.5 in. (0.5 × 0.5 × 1.2 m) were constructed from plywood (see Figure 1). The containers were then lined with tailored sheets of 10-mil polyvinyl chloride (PVC) to make them watertight and electrically nonconducting. One container was used for the actual test, and the other was used for control.

Placement of Soil. Nontoxic, hydrocarbon-contaminated soil, obtained from a local oil storage facility was placed in each of the containers in two 6-in (0.15 m) lifts under controlled compaction conditions to obtain a uniform total unit weight of approximately 125 lb/ft^3 (2000 kg/m^3) at the natural moisture content of 49.2%.

Electrochemical Cell. The electrochemical cell used to induce electroosmotic flow consisted of a 20-V (nominal) DC power source connected to two electrodes embedded vertically into the soil approximately 30 in (0.75 m) apart along the longitudinal centerline of the box. The electrode "wells" were made from 18-in (0.45-m) lengths of conventional ½-in (12.7-mm) copper tubing by perforating them on ½-in (12.7-mm) intervals along their embedded lengths. Transported fluids were recirculated to keep electrical currents in the range of 370 to 400 mA at an applied electrical potential of 35 V (DC), and to keep the soil from drying out.

Bacteria and Nutrient. The bacteria used in this research are proprietary and were donated by Solomar Corp. of Orange, California, who identified them only as anaerobic bacillus of the family subtillus (Advanced Bio Cultures Formulation L-104). The nutrient consisted of a commercially available 20-20-20 soluble plant food that was diluted in water and introduced into the system at the anode (along with the bacteria) as a solution containing 400 mg/L nitrogen and phosphate salts.

RESULTS OF TESTS

Tests were performed to determine the soil's pertinent physical and electrical properties, the final relative concentration of bacteria in the cell, the phosphorus concentration in the cell as a measure of nutrient transport, and the electroosmotic flowrate.

Soil Characteristics. Pertinent physical and electrical properties of the soil were determined by using standard testing procedures. The results of those tests are summarized in Table 1.

Microbiological Testing. A series of microbiological tests was conducted to determine the extent of bacteria transport and to obtain a qualitative estimate of their spatial concentrations with respect to the anode and cathode. A standard plate count procedure (Greenberg 1975) was used for all microbiological tests. Samples of soil were taken from 12 locations within the electrochemical cell 74 hours after the bacteria had been introduced at the anode. The sampling pattern is shown in Figure 1. Bacteria concentrations at each sampling location were visually compared to a baseline concentration corresponding to the natural soil prior to its exposure to the bacteria. Comparisons of such concentrations revealed the presence of bacteria at all of the sampled locations. In general, concentrations decreased in the longitudinal direction from anode to cathode.

Chemical Tests. Studies of nutrient transport through a clay by electroosmosis conducted by Segall and Bruell (1992) indicated that phosphates tend to precipitate at the anode when either iron or graphite electrodes are used. To investigate the general applicability of Segall and Bruell's observations, pilot tests were performed in the electrochemical cell using copper electrodes. The test results, summarized in Table 1, suggest that although it was possible to move phosphorus through the clay, the movement occurred at a much slower rate than the dispersement

TABLE 1. Summary of test results.

Test	Value of Pertinent Parameters
Fall Head Permeability (Bowles 1992)	$k_h = 5.84 \times 10^{-8}$ cm/sec $k_h = 6.45 \times 10^{-8}$ cm/sec
Consolidation (ASTM D-2435)	cv = 1×10^{-4} cm^2/sec
Harvard Miniature Compaction (Bowles 1992)	Max. dry unit weight = 1600 kg/m^3 (99.3 lb/ft^3)
Grain Size Analysis by Mechanical Sieving (ASTM D-421) Hydrometer Method (ASTM D-422)	Fine-grained soil – more than 50% passing the #200 sieve, i.e., smaller than 0.074 mm.
Liquid Limit (LL), Plastic Limit (PL), Plasticity Index (PI) and Natural Moisture Content (w_n) (ASTM D-4318)	LL = 58% PL = 32% PI = 26 w_n = 49.2%
Unified Soil Classification	CH = high plasticity silty clay
Electrical Resistivity Range (CalTrans 1978)	890 ohm-cm at 17.5 V to 2100 ohm-cm at 3.9 V

Phosphate concentration determined by vanadomolybdophosphoric yellow method:

Specimen Location (Refer to Figure 1)	Phosphate Concentration (mg/L dry weight)
E	6
F	8
G	10
C	15
K	17
H	45

of the bacteria. Visual inspection of the anode after the test revealed that phosphate precipitation also had occurred.

Electroosmotic Flowrate. The electroosmotic flowrate was determined by collecting the overflow at the cathode "well" during the application of electric potential and recording the volume collected over specific periods of time. The applied voltage was held constant during a collection period, but because of slight changes in moisture content (resistivity) induced by the electric potential, currents ranged from 370 to 400 mA. Under these conditions, the average electroosmotic

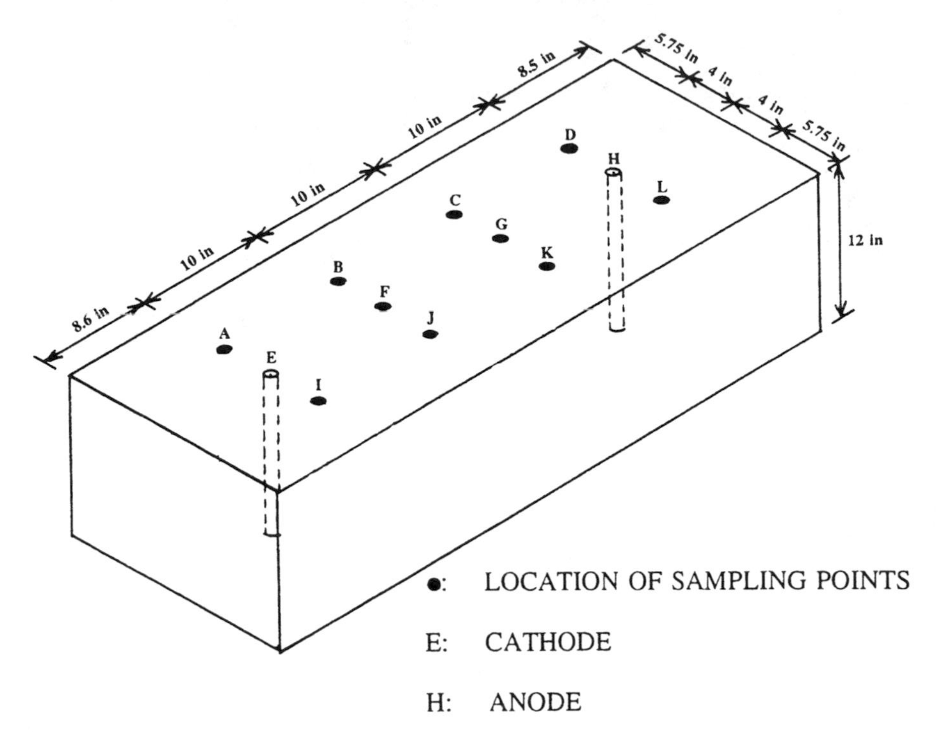

FIGURE 1. Sketch of electrochemical cell.

flowrate was determined to be 3.67×10^{-4} cm^3/sec at an average electrical gradient of 0.46 V/cm along the longitudinal centerline.

DISCUSSION

This experiment was performed to determine the feasibility of transporting bacteria and nutrients through a clay soil under steady-state conditions. Although no empirical verification of the achievement of steady-state electroosmotic transport was made, a time-scale analysis indicates that steady-state was likely to have been achieved. Many factors could subsequently affect the ultimate performance of a bioremediation project which uses this technique for distributing the bacteria and nutrients to the contamination sites. The movement of nutrients could be affected by either precipitation or ion exchange within the soil matrix, limiting the growth potential of the bacteria. Electrical potential could similarly have a significant effect on the growth rate of the bacteria. Transport of bacteria could also be affected by an electrophoretic mechanism, which would have to be assessed in an independent experiment.

CONCLUSIONS AND RECOMMENDATIONS

This experiment has demonstrated that anaerobic, hydrocarbon-reducing bacteria can be transported by electroosmosis through a low permeability clay without harm at acceptable flowrates. Voltages (and currents) required are in the range of those typically used in field applications of electroosmosis for dewatering soils. The results of preliminary tests performed to demonstrate the feasibility of transporting phosphate-based nutrients by electroosmosis were inconclusive. Additional research is needed to identify the factors affecting the transport of phosphates by electroosmosis and to measure the soil hydrocarbon concentrations as a function of time to demonstrate the potential of this procedure for bioremediation.

REFERENCES

ASTM. 1992. *Annual Book of ASTM Standards,* Vol. 04.08, *Soil and Rock; Building Stones.* American Society for Testing and Materials, Philadelphia, PA.

Bowles, J. E. 1992. *Engineering Properties of Soils and Their Measurement,* 4th ed. McGraw-Hill Book Company, New York, NY.

Caltrans. 1978. "Method for Estimating the Service Life of Steel Culverts: Part 4 - Laboratory Method for Determining Minimum Resistivity." In *California Standard Test 643.* Department of Transportation, Division of Construction, Sacramento, CA.

Greenberg, A. E. 1975. "Test 907 - Standard Plate Count." In *Standard Methods for the Examination of Water and Wastewater,* 14th ed., p. 908. American Public Health Association, American Water Works Association, and Water Pollution Control Federation.

Mitchell, J. K. 1976. *Fundamentals of Soil Behavior.* John Wiley and Sons, New York, NY.

Segall, B. A., and C. J. Bruell. 1992. "Electroosmotic Contaminant Removal Processes." *ASCE Journal of Environmental Engineering* 118(1):84-100.

SCALE-UP IMPLICATIONS OF RESPIROMETRICALLY DETERMINED MICROBIAL KINETIC PARAMETERS

P. J. Sturman, R. R. Sharp, J. B. DeBar, P. S. Stewart, A. B. Cunningham, and J. H. Wolfram

INTRODUCTION

Successful scale-up of remediation processes requires an understanding of the extent to which they are influenced by scale-dependent phenomena such as mass transport and interfacial transfer limitations. Such effects may introduce rate limitations in field scale bioremediation which were not present in the laboratory. Thus, what appears to be a viable bioremediation strategy in the lab may be unsuccessful in the field (Goldstein et al. 1985). A better understanding of the requirements of the scale-up process is necessary to successfully predict which information translates across scales.

The goal of this research was to assess the adequacy of respirometrically determined kinetic parameters to predict field-scale biotransformation via bioprocess modeling. Electrolytic respirometry was used to determine kinetic parameters for an indigenous microbial consortium at a site contaminated with dissolved alkylbenzenes at concentrations up to 10 mg/L. Mean values and 95% confidence intervals (CIs) were determined for kinetic parameters. These parameters were used in a bioprocess model to estimate the active zone of biotransformation for a field remediation system. To assess the effects of kinetic parameter variation on the contaminant mass biotransformed predicted by the model, the values of two important parameters (μ_{max} and K_s) were varied through their CIs in the input to the bioprocess model. The μ_{max} (maximum specific growth rate) is the consortia growth rate (μ) under optimal conditions for a given substrate (S), whereas K_s (half-saturation coefficient) is the substrate concentration at which μ is one-half μ_{max}. Where $K_s >> S$, μ responds in first-order fashion to changes in S. Where $S >> K_s$, μ responds in zero-order fashion to changes in S.

METHODS

The microbial consortium used in the respirometry experiments was collected from a site that has been exposed to dissolved benzene, toluene, ethylbenzene, and xylenes (BTEX) continuously for the last 15 years. Hydrocarbon degraders were isolated by plating on selective (BTEX-rich) media. Biosciences electrolytic respirometers (500-mL capacity) were filled with 400 mL of mineral salts solution,

1 mL of concentrated cell inoculum, and 4.4 to 22.8 mg total BTEX (equal quantities of each compound). The final contaminant concentration in the reaction vessel varied from 11 to 57 mg/L. Total oxygen demand from each respirometry run was automatically recorded. A typical accumulated O_2 demand curve is shown in Figure 1.

The O_2 demand curve generated in each respirometry experiment was fitted with a 5-parameter Monod model with cell decay and least sum squares regression techniques. The model equation used for the change in accumulated O_2 demand with time was:

$$\frac{dO_2}{d_t} = Y_{o/x}\left(\frac{\mu_{max}S}{K_s + S}\right)X + 1.42bX \qquad (1)$$

where $Y_{o/x}$ = mass O_2 utilized per mass biomass produced. Other parameters as defined in Table 1. Kinetic parameters (μ_{max}, K_s, decay rate, and biomass yield) and initial active biomass concentration (X_0) were estimated by continuously varying the input parameters until a curve of best fit was generated. Parameter means and 95% CIs were thus obtained (Table 1).

Consortia kinetic parameters obtained from respirometry and physical data obtained from the actual contaminated site (Table 1) were entered as inputs into a two-dimensional porous medium bioprocess model (Ewing et al. 1984). This deterministic model considers advection, diffusion, dispersion, and biological reaction kinetics. Bulk fluid flow is described by Darcy's law and bioreactions by double Monod kinetics (both O_2 and substrate potentially limit biodegradation

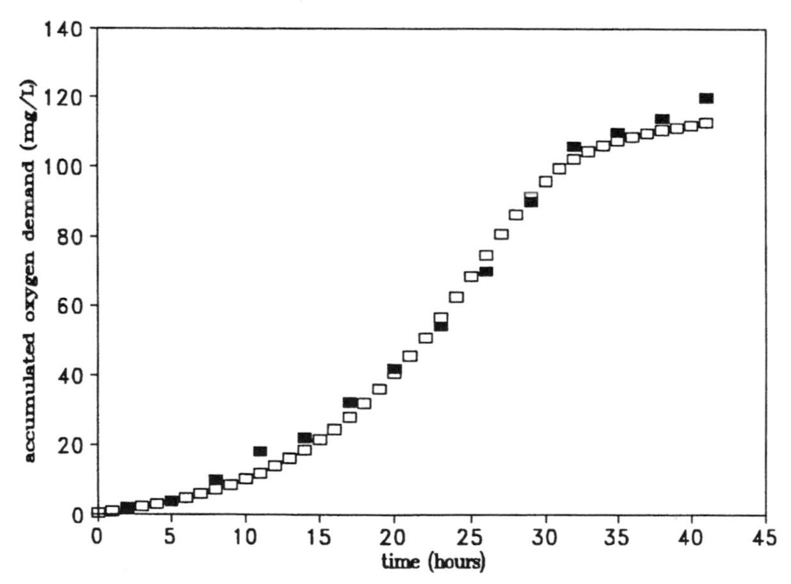

FIGURE 1. Typical oxygen demand curve from respirometry (■), and model-generated curve (□).

TABLE 1. Kinetic parameters from respirometry and bioprocess model inputs.

		Mean	Std.Dev.
Microbial Parameters			
μ_{max} [a]	maximum specific growth rate	0.23 hr^{-1}	0.01 hr^{-1}
K_S [a]	substrate half-saturation coefficient	16.9 mg L^{-1}	1.9 mg L^{-1}
$Y_{x/s}$ [a]	biomass (X) yield from substrate	0.67 gX gS^{-1}	0.31 gX gS^{-1}
$Y_{x/e}$	biomass (X) yield from e^- acceptor	0.32 gX gE^{-1}	
b [a]	endogenous respiration rate (decay)	0.02 hr^{-1}	0.004 hr^{-1}
K_E	e^- acceptor half-saturation coefficient	0.1 mg L^{-1}	
Site Parameters			
ϕ	porosity	0.3	
k	permeability	0.32 Darcy	
d_l	longitudinal dispersivity	4.5 m	
d_t	transverse dispersivity	0.45 m	
C_S	initial substrate concentration	10 mg L^{-1}	
C_X	initial biomass concentration	1.24 mg L^{-1}	
X	formation length	152 m	
Y	formation width	152 m	
Transport Parameters			
Q	injection rate	2.04 m^3 d^{-1} m^{-1} aquifer thickness	
C_E	background e^- acceptor concentration	8 mg L^{-1}	
μ	viscosity	1.31 centipoise	

(a) Respirometrically determined parameters.

rate). The model assumes that the contaminant is fixed in place within the aquifer, while the electron acceptor is transported with the bulk fluid flow. This mimics a field situation where a sorbed contaminant desorbed at a rate similar to that at which biodegradation occurs (i.e., desorption continuously replenishes the aqueous phase with contaminant until no sorbed contaminant remains). Dissolved O_2 at 8 mg/L is transported into the formation through a single injection well (see Figure 2), whereas BTEX is transformed in situ. Numerical simulation of the model uses mixed finite elements, the Modified Method of Characteristics, and methods for solving stiff ordinary differential equations. Model runs were performed with nine combinations of the kinetic coefficients μ_{max} and K_s. Minimum, maximum, and mean values for μ_{max} and K_s were used in combination (minimums and maximums determined by CIs).

RESULTS AND DISCUSSION

Respirometry experiments were performed at initial total BTEX concentrations of 11, 33, 47, and 57 mg/L. Each experiment generated an oxygen demand curve,

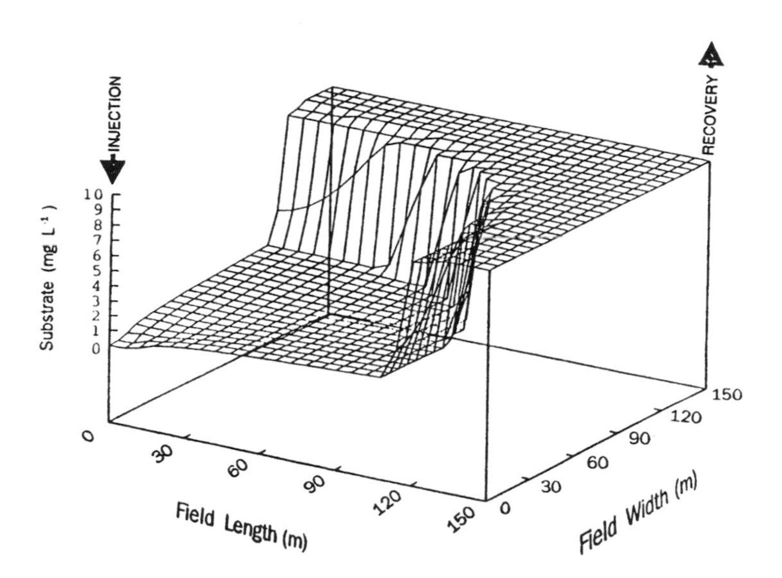

FIGURE 2. Substrate (contaminant) concentration within flow field of bioprocess model. Run time 4,800 days.

to which a curve of best fit was applied via least sum squares error analysis. A typical curve fit is indicated in Figure 1. Each of the four respirometry-generated O_2 demand curves resulted in an estimate of four kinetic parameters. The means and 95% CIS are shown in Table 1. Although only four data points are considered in this initial experimentation, the CIs for μ_{max}, K_s, and b are relatively small. The calculated $Y_{x/s}$ has a considerably larger CI (0.039 to 1.3), although this did not cause excessive variation in the other kinetic coefficients. The average μ_{max} and K_s predicted were 0.23 hr^{-1} and 19.9 mg/L, respectively. Few literature values for alkylbenzene compounds exist, but the above values fall within the ranges of 0.18 to 0.78 hr^{-1} for μ_{max} and 6 to 500 mg/L for K_s reported by Grady and Lim (1980) for mixed organic wastes.

Each bioprocess model run resulted in predictions of the percent of total contaminant mass biotransformed. The volume biotransformed is shown in Figure 2 for a typical model run. O_2-rich water is injected at the left side of the figure and removed via a recovery well at the right. Similar plots were generated for nine combinations of μ_{max} and K_s through the respirometrically determined CIs (Figure 3, inset). As expected, the highest μ_{max} and lowest K_s resulted in the greatest mass biotransformed in 4,800 days. As μ_{max} decreased and K_s increased, the contaminant mass biotransformed decreased by only 3% from 47 to 45% of the total mass present (Figure 3).

This result is somewhat surprising in light of the dramatic impact the maximum specific growth rate has on microbial utilization of substrate in batch situations, even where small variations in μ_{max} are used. Several factors may account

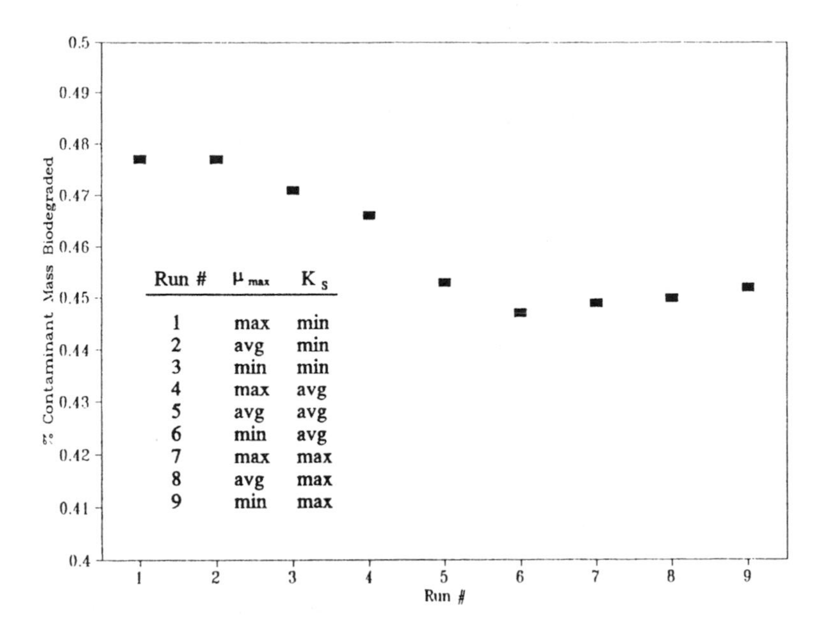

FIGURE 3. Comparison of the percent of the total contaminant mass biodegraded in 4,800 days with various kinetic parameter inputs. Maximum and minimum μ_{max} and K_s are defined by the 95% confidence interval determined from respirometry.

for the observed lack of sensitivity to variation in kinetic parameters. The influence of kinetics may be overshadowed by the importance of advective transport of electron acceptor to the reaction site. The steep substrate front in Figure 2 indicates a very narrow zone of reaction, which supports this idea. Substrate degradation appears to be limited by the presence of oxygen, rather than by the intrinsic kinetics of the consortium present. The implications of this result to scale-up are twofold: (1) efforts to maximize bioremediation should focus on increasing O_2 transport to the reaction site, and (2) small changes in the intrinsic kinetics of the site consortium may have little impact on the rate of biodegradation.

REFERENCES

Ewing, R. E., T. F. Russell, and M. F. Wheeler. 1984. *Computer Methods in Applied Mechanics and Engineering*. North-Holland, NY.

Goldstein, R. M., L. M. Mallory, and M. Alexander. 1985. "Reasons for possible failure of inoculation to enhance biodegradation." *Appl. Environ. Micro. 50*(4):977-983.

Grady, C.P.L., and H. C. Lim. 1980. *Biological Wastewater Treatment*. Marcel Dekker, New York, NY.

MINERALIZATION OF NAPHTHALENE, PHENANTHRENE, CHRYSENE, AND HEXADECANE WITH A CONSTRUCTED SILAGE MICROBIAL MAT

P. Phillips, J. Bender, J. Word, D. Niyogi, and B. Denovan

INTRODUCTION

A constructed freshwater consortium of bacteria and cyanobacteria (blue-green algae), or microbial mat, degraded four petroleum hydrocarbon compounds.

Cyanobacteria, although photosynthetic, are known to use exogenous organic substrates under both lighted and dark conditions (heterotrophy) as a portion of the total carbon requirement for growth (Fogg et al. 1973). Additionally, cyanobacteria are well known for their ability to exist in the most inhospitable and caustic environments (Des Marais 1990). Below the cyanobacteria photozone, facultative bacteria colonize. These cyanobacteria/bacteria biofilms form multilayered laminated mats in the sediment region of shallow water. With these characteristics, microbial mats may represent ideal organisms for bioremediation.

In this study, we examined the potential capabilities of our mats for mineralizing petroleum hydrocarbon compounds: naphthalene, a two-ring polycyclic aromatic hydrocarbon (PAH); phenanthrene, a three-ring PAH; chrysene, a four-ring PAH; and hexadecane, a paraffin. Experimental treatments determined the amount of $^{14}CO_2$ produced using ^{14}C-hydrocarbons as the sole carbon source for the microbial mats.

METHODS

Mats were developed in laboratory trays by constructing an artificial ecosystem consisting of a soil base, a charcoal-filtered tap-water column, a floating layer of ensiled grass clippings (silage), and cyanobacteria inocula, principally *Oscillatoria* spp. (Bender et al. 1989). Ensiled grass clippings added organic acids, principally lactic and acetic acids, as well as a microbial consortium of fermentative anaerobes to the system. Within days a spontaneous succession of bacteria species, migrating from the soil base, colonized the floating silage. The grass clippings additionally served to stabilize the floating mat. The final product was a thick, gelatinous green mat.

The experimental design included a lighted and a dark series for each petroleum hydrocarbon treatment. The purpose of this was to detect mineralization rate differences between all the microbial mat's autotrophs and heterotrophs (lighted series) and only the mat's heterotrophs (dark series). Uniform mat "plugs" (1.00 to 1.69 g; s.d.=0.15) were individually placed in 250-mL acid-washed glass flasks containing 100 mL of Allen/Arnon Modified Medium (Allen & Arnon 1955). These were sealed with a Teflon™-coated stopper to eliminate a potential reaction between the test hydrocarbon and the stopper material (Bauer & Capone 1985).

The ^{14}C-naphthalene, phenanthrene, chrysene (Amersham/Searle Corp., Arlington Heights, Illinois), and hexadecane (Sigma Chemical Co., St. Louis, Missouri) were spiked at greater than 4,800 disintegrations per minute (dpm)/mL/flask (naphthalene, 9,324; phenanthrene, 5,195; chrysene, 9,639; hexadecane, 9,030). Each petroleum hydrocarbon experimental treatment was conducted in triplicate for both lighted and dark series. Only one hydrocarbon was added to each flask. Mercuric chloride was added (1 mL of 0.125 M $HgCl_2$) to the control flasks to kill all microorganisms.

A 7-mL scintillation vial containing 1 mL of 0.3 M KOH was hung inside each flask above the surface of the medium. Sealed flasks were placed in a 33°C incubator. Lighted flasks were held under continuous fluorescent lighting. At 11, 24, 40, 60, and 90 days, the KOH trap from each flask was removed and scintillation-counted. New KOH traps were replaced. Mineralization rates were calculated for days 11, 24, 40, 60, and 90 as a percentage of the initial amount of hydrocarbon spiked into each flask (measured as dpm/flask).

RESULTS

In experimental lighted flasks, the mat plug was rapidly covered with a new green growth of cyanobacteria, in the form of biofilms coating the inside of the flask. By day 11, new growth represented approximately double the original surface area. From days 11 to 24, mats in the experimental dark flasks showed a white filamentous growth. After this time, these plugs became brown and much of the physical integrity of the mat was lost. At the end of the experiment (day 90), these mats were thoroughly decomposed and only detritus remained. Control mats were brownish colored and slightly deteriorated.

The general trends in $^{14}CO_2$ as dpm/mL in KOH were as follows for all four hydrocarbons. In lighted experimental flasks, after day 11, $^{14}CO_2$ amounts in KOH traps dropped to near zero. Dark experimental flasks produced KOH trap counts much higher than their counterpart lighted flasks. Naphthalene, phenanthrene, and hexadecane treatments all showed a steady decrease in the amount of $^{14}CO_2$ assimilated in the KOH trap during 90 days. The greatest chrysene value occurred at day 60.

After termination of the experiment, an unweighed and washed sample of microbial mat representing each hydrocarbon from the lighted series had the following total dpms: naphthalene, 1,242; phenanthrene, 8,510; chrysene, 40,800; and hexadecane, 1,496. These values indicated that the labeled compounds were attached to or incorporated into the mat cellular contents.

Table 1 shows the cumulative percent mineralization trends. Percent mineralization was calculated based on initial dpm spiking. The near-zero increase in slope of experimental lighted flasks curves after day 11 is due to the extremely low KOH trap dpm counts. Mineralization in the dark experimental flasks increased through day 90, reaching values of 19.0% for naphthalene, 24.1% for phenanthrene, 20.5% for chrysene, and 9.3% for hexadecane. All control flasks (lighted and dark) had mineralization rates less than 2% after 90 days, except for dark hexadecane (2.6%).

DISCUSSION

The fate of radiolabeled hydrocarbons was likely due to (1) adsorption to the glass flask, (2) adsorption to the microbial mat, (3) volatilization and perhaps later adsorption to the dry glass, (4) chemical degradation to $^{14}CO_2$ and water, or (5) mineralization to $^{14}CO_2$ by the microbial mat. The $^{14}CO_2$ formed during mineralization may then be reincorporated into living tissue by either photosynthesizing cyanobacteria or synthesizing bacteria.

The control flasks, with all biota killed by mercuric chloride, indicate the fate of the hydrocarbons was controlled by all the above processes except mineralization. The higher mineralization values for the lighted naphthalene, phenanthrene, and chrysene control flasks (compared to the dark controls) indicate that photochemical oxidation of these compounds may occur to some degree. Mineralization rates for lighted, and especially the dark, hexadecane controls remain unexplained.

In our more recent research, *Oscillatoria* sp. alone was found to degrade chrysene and hexadecane, although at a slower rate than our complete microbial mat (data not shown). *Oscillatoria* sp. and other cyanobacteria have been shown to degrade naphthalene (Cerniglia et al. 1980a, Cerniglia et al. 1980b). Low $^{14}CO_2$ values for lighted experimental flasks suggest that rapidly growing cyanobacteria mats may be incorporating all available $^{14}CO_2$ produced by mineralization before it can reach the trap. Thus, carbon dioxide may have been limiting to the photosynthesizing cyanobacteria. As a result of this reincorporation, the calculated mineralization rates are likely underestimated and are not comparable to the values for dark flasks. In dark flask chrysene treatments, this high-molecular-weight PAH initially was mineralized more slowly than lower-molecular-weight naphthalene and phenanthrene. However, by day 90, the overall mineralization of chrysene was similar to that of naphthalene and phenanthrene. Hexadecane, a paraffin, was least degraded (less than 6% mineralized after correcting for control degradation) after 90 days of incubation. Unlike for photosynthesizing cyanobacteria, dark series mineralization would most likely be due to bacteria and heterotrophic cyanobacteria.

Our results clearly suggest that microbial mats dominated by cyanobacteria have the capability to bioremediate petroleum-contaminated sites. Our findings are substantiated by recent observations of the extensive appearance of cyanobacteria mats along the Saudi Arabian coastline associated with oil deposited

TABLE 1. Cumulative percent mineralization of the original spiked ^{14}C-labeled compound versus days of incubation. All experimental flask calculations are the mean of a triplicate series. The control values are calculated from one flask. N = Naphthalene, P = Phenanthrene, C = Chrysene, H = Hexadecane.

	Day 11	Day 24	Day 40	Day 60	Day 90
N-Dark	6.06	9.92	13.26	16.27	19.00
Control	0.02	0.06	0.08	0.12	0.16
N-Ligh	10.0	10.0	10.0	10.11	10.13
Control	0.136	0.21	0.28	0.35	0.41
P-Dark	6.09	11.39	16.23	20.5	24.06
Control	0.10	0.23	0.32	0.43	0.54
P-Light	0.85	0.87	0.88	0.88	0.90
Control	0.64	1.06	1.34	1.61	1.88
C-Dark	1.15	4.41	8.21	15.27	20.46
Control	0.02	0.07	0.09	0.13	0.17
C-Light	0.14	0.14	0.14	0.15	0.15
Control	0.21	0.32	0.38	0.45	0.55
H-Dark	3.26	5.67	7.38	8.55	9.32
Control	0.10	0.72	1.39	1.97	2.57
H-Light	2.89	2.91	2.92	2.93	2.95
Control	0.01	0.02	0.02	0.11	0.33

during the Gulf War (Sorkhoh et al. 1992). We have found that our microbial mats tolerate hypersaline conditions (up to 100 parts per thousand) with no pre-adaptation. Our current research focuses on the hydrocarbon mineralization rates of individual microbial mat constituents, as well determining the cellular location of ^{14}C-compounds.

ACKNOWLEDGMENT

Infrastructure support was provided by U.S. Environmental Protection Agency Grant No. CR81868901.

REFERENCES

Allen, M. B., and D. I. Arnon. 1955. "Studies on Nitrogen-Fixing Blue-Green Algae. I. Growth and Nitrogen Fixation by *Anabaena cylindrica* Lemm." *Pl. Physiol. Lancaster 30*: 366-372.

Bauer, J. E., and D. G. Capone. 1985. "Degradation and Mineralization of the Polycyclic Aromatic Hydrocarbons Anthracene and Naphthalene in Intertidal Marine Sediments." *Applied and Environmental Microbiology 50*(1): 81-90.

Bender, J., Y. Vatcharapijarn, and A. Russell. 1989. "Fish Feeds from Grass Clippings." *Aquacultural Engineering 8*: 407-419.

Cerniglia, C. E., C. van Baalen, and D. T. Gibson. 1980a. "Metabolism of Naphthalene by the Cyanobacterium *Oscillatoria* sp., Strain JCM." *J. Gen. Microbiol. 116*: 483-494.

Cerniglia, C. E., D. T. Gibson, and C. van Baalen. 1980b. "Oxidation of Naphthalene by Cyanobacteria and Microalgae." *J. Gen. Microbiol. 116*: 495-500.

Des Marais, D. J. 1990. "Microbial Mats and the Early Evolution of Life." *Trends in Ecology and Evolution 5*(5): 140-144.

Fogg, G. E., W. D. P. Stewart, P. Fay, and A. E. Walsby. 1973. *The Blue-Green Algae*. Academic Press, New York, NY.

Sorkhoh, N., R. Al-Hasan, S. Radwan, and T. Hopner. 1992. "Self-Cleaning of the Gulf." *Nature 359*: 109.

KINETICS OF PHENANTHRENE DEGRADATION BY SOIL ISOLATES

W. T. Stringfellow and M. D. Aitken

INTRODUCTION

Polycyclic aromatic hydrocarbons (PAHs) are hydrophobic chemicals characterized by extremely low water solubilities. PAHs are naturally occurring constituents of coal and crude oil, and are commonly found in soil contaminated with petroleum, coal tar, and related products (Srivastava et al. 1990; Thomas et al. 1991). PAHs are also major constituents in contaminated soil at sites associated with wood-preserving operations (Rosenfeld & Plumb 1991; U.S. Environmental Protection Agency 1990).

PAHs are biodegradable, and the pathways for the degradation of naphthalene and the three-ring PAHs anthracene and phenanthrene have been described (Cerniglia & Heitkamp 1989). However, little quantitative information exists on rates of PAH degradation, even in the aqueous phase. This study focused on aqueous-phase degradation kinetics as an initial step in the quantification of biodegradation rates in heterogeneous systems. Phenanthrene was chosen as a model PAH because it is biodegradable and sufficiently soluble to permit conventional biodegradation kinetics assays. Like other PAHs, phenanthrene has a high octanol-water partition coefficient (Mackay et al. 1992), and therefore can be expected to exhibit typical hydrophobic partitioning behavior in soils and in other heterogeneous systems.

METHODS

Media for the Cultivation of Phenanthrene Degraders. Phenanthrene-degrading cultures were grown either in peptone (PEP; 5 g/L) or in tap water buffer (TWB) containing phenanthrene as the sole carbon source (PAT/TWB). For PAT/TWB medium, a 2 g/L solution of phenanthrene in hexane was filter-sterilized and added to presterilized culture flasks, and the hexane evaporated. In this way, the phenanthrene was deposited on the bottom of the flask in a thin film and TWB could be added to give a final concentration of 0.5 g/L phenanthrene. TWB was prepared by combining Na_2HPO_4-$7H_2O$ (1.5 g), KH_2PO_4 (1.0 g), NH_4Cl (2.0 g), and 2 $Na_2S_2O_3$-$5H_2O$ (0.02 g) in 1 L of tap water (final pH 6.5).

Measurement of Phenanthrene Oxidation Kinetics. The oxidation of dissolved phenanthrene by phenanthrene-degrading isolates was measured with a Clark oxygen electrode (Yellow Springs Instruments, Yellow Springs, Ohio) at 25°C. Cultures grown on either PEP or PAT/TWB medium were harvested in late-growth stage (before the onset of stationary phase), centrifuged, washed in dechlorinated tap water, centrifuged again, and resuspended in TWB. The cell suspension was placed in a stirred, water-jacketed cell (1.7 mL volume) fitted with the Clark electrode. An initial endogenous oxygen uptake rate was measured; then the desired amount of phenanthrene was delivered into the cell in 10 mL of methanol. The oxygen uptake rate was determined immediately after injection, and the endogenous rate was subtracted from the postinjection rate. After each rate measurement, the cell suspension was removed and placed in a spectrophotometer for determination of optical absorbance at 420 nm. Each absorbance unit was determined to correspond to 500 mg/L biomass. Oxygen uptake results are expressed in terms of specific oxygen uptake rate (SOUR) and reported as milligrams of oxygen consumed per minute per gram of bacteria. Preliminary experiments confirmed that injection of methanol alone had no effect on oxygen uptake.

Kinetic coefficients were determined by nonlinear regression of data according to the Michaelis-Menten equation. Parameter values and 95% confidence intervals (CI) were obtained using SYSTAT (Evanston, Illinois).

RESULTS AND DISCUSSION

Enrichment of Phenanthrene Degraders from a Wood Treatment Site Soil. A soil sample from a wood treatment site was enriched directly on PAT/TWB. The initial enrichment turned turbid and became brown. Eight isolates were recovered from this enrichment, five of which degraded phenanthrene. The five degrading strains were identified by standard biochemical tests as belonging to two species. Two organisms (PAT-15 and PAT-16) were selected as representative of the two groups and used for further study. Isolate PAT-15 was identified as a *Pseudomonas vesicularis*; and PAT-16 was a *Pseudomonas stutzeri*.

Measurement of Phenanthrene Oxidation Kinetics by Oxygen Electrode. Phenanthrene oxidation was measured by respirometry using resting cells, and oxidation rate was determined as a function of soluble phenanthrene concentration. Comparisons were made between *Ps. stutzeri* PAT-16 and *Ps. vesicularis* PAT-15. Oxidation rates by each of the microorganisms after growth in two different media (PAT/TWB, peptone) also were examined.

When PAT-16 and PAT-15 were grown in PAT/TWB, the kinetic profiles of the oxidation rates were very similar (Figure 1). Both cultures exhibited Michaelis-Menten saturation kinetics in relation to phenanthrene concentration:

$$\mathrm{SOUR} = \frac{(\mathrm{SOUR}_m)\,S}{K_s + S}$$

where SOUR = specific oxygen uptake rate, SOUR_m = maximum specific rate, and K_s = half-saturation coefficient.

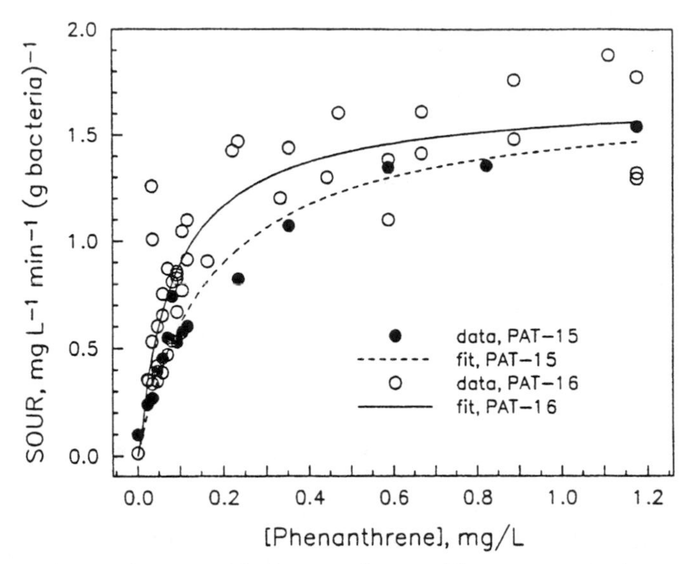

FIGURE 1. Phenanthrene oxidation rate (as specific oxygen uptake rate, SOUR) as a function of phenanthrene concentration by resting cells of isolates PAT-15 and PAT-16 after growth on phenanthrene. Lines shown are best fits to Michaelis-Menten model.

The $SOUR_m$ and K_s values for *Ps. stutzeri* PAT-16 and *Ps. vesicularis* PAT-15 were not significantly different (Table 1). It should be noted that the K_s values for these organisms are well below the solubility limit of phenanthrene (1.2 mg/L), and that maximum oxidation rates also were achieved at concentrations below the solubility limit. Oxidation of phenanthrene by each culture after growth on peptone is shown in Figure 2. PAT-16 oxidized phenanthrene even when grown on peptone in the absence of phenanthrene, suggesting that phenanthrene oxidation is constitutive for this organism. PAT-15 had a very low SOUR over the

TABLE 1. Oxygen uptake kinetics for phenanthrene isolates.

Organism	Media[a]	K_s[b]	95% CI	$SOUR_m$[c]	95% CI
PAT-16	PEP	0.20	(0.11, 0.29)	2.40	(1.90, 2.90)
PAT-16	PAT/TWB	0.10	(0.06, 0.15)	1.86	(1.55, 2.17)
PAT-15	PAT/TWB	0.18	(0.12, 0.24)	1.69	(1.49, 1.89)

(a) See text for media descriptions.
(b) K_s = half-saturation constant (mg/L).
(c) $SOUR_m$ = maximum specific oxygen uptake rate (mg O_2/L/min/g biomass).

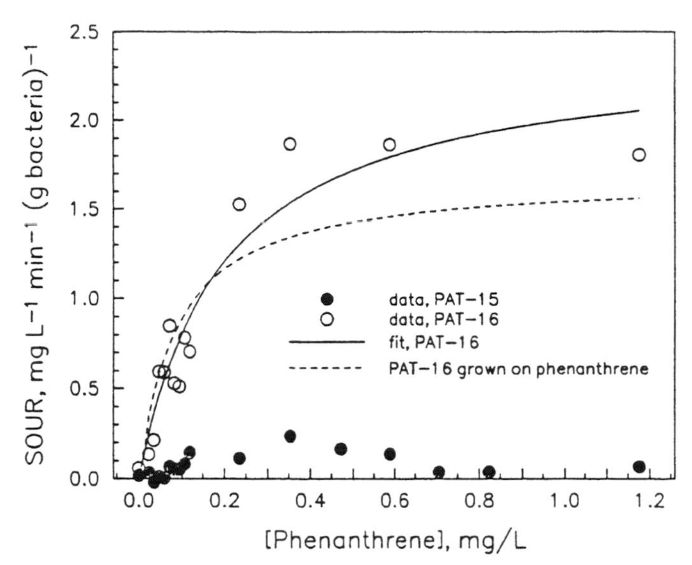

FIGURE 2. Phenanthrene oxidation rate as a function of phenanthrene concentration by resting cells of isolates PAT-15 and PAT-16 after growth on peptone. The curve obtained for PAT-16 after growth on phenanthrene is shown for comparison.

same range of phenanthrene concentrations, and actually appeared to be inhibited at higher concentrations of phenanthrene. Phenanthrene oxidation in this organism therefore appears to be induced.

ACKNOWLEDGMENTS

This work was funded jointly by the United States Geological Survey (grant number 14-08-0001-G2103) and The University of North Carolina Water Resources Research Institute (grant number 20162). We also thank Paul Flathman of OHM Remediation Services, Inc., for providing the soil sample.

REFERENCES

Cerniglia, C. E., and M. A. Heitkamp. 1989. "Microbial degradation of polycyclic aromatic hydrocarbons (PAH) in the aquatic environment." In U. Varanasi (Ed.), *Metabolism of Polycyclic Aromatic Hydrocarbons in the Aquatic Environment*, pp. 41-68. CRC Press, Boca Raton, FL.

Mackay, D., W. Y. Shiu, and K. C. Ma. 1992. *Illustrated Handbook of Physical-Chemical Properties and Environmental Fate for Organic Chemicals*, Vol. 2: *Polynuclear Aromatic Hydrocarbons, Polychlorinated Dioxins and Dibenzofurans*. Lewis Publishers, Ann Arbor, MI.

Rosenfeld, J. K., and R. H. Plumb, Jr. 1991. "Ground water contamination at wood treatment facilities." *Ground Water Monitor. Rev. 11*(1):133-140.

Srivastava, V. J., J. J. Kilbane, R. L. Kelley, W. K. Gauger, C. Akin, T. D. Hayes, and D. G. Linz. 1990. "Bioremediation of former manufactured gas plant sites." In *44th Purdue Industrial Waste Conference Proceedings*, pp. 49-60. Lewis Publishers, Chelsea, MI.

Thomas, A. O., P. M. Johnston, and J. N. Lester. 1991. "The characterization of the subsurface at former gasworks sites in respect of in situ microbiology, chemistry and physical structure." *Haz. Waste Haz. Mater. 8*:341-365.

U.S. Environmental Protection Agency. 1990. *Approaches for Remediation of Uncontrolled Wood Preserving Sites.* EPA/625/7-90/011, Cincinnati, OH.

GROUNDWATER BIOREMEDIATION SYSTEM DESIGN: BACTERIAL EVALUATION PHASE

G. J. Boettcher, R. E. Moon, and E. K. Nyer

INTRODUCTION

An investigation was completed to evaluate the feasibility of groundwater bioremediation at a chemical manufacturing facility. Approximately 1.2 hectares (3 acres) of the 8.1-hectare (20-acre) site were impacted for the last 15 years with production chemicals accidently released to a porous limestone aquifer. The most abundant organic compounds detected in groundwater samples were benzene, toluene, sulfolane, and diethyl ether.

Hydrogen sulfide vapors also were detected at the site. Hydrogen sulfide contributed to frequent odor complaints and metal corrosion. The presence of elevated hydrogen sulfide concentrations (up to 7,000 mg/m^3) and low oxygen concentrations (less than 5%) in the vadose zone indicated that bacterial degradation of organic compounds was occurring under anoxic conditions. It was believed that hydrogen sulfide production was produced by both obligate anaerobes and facultative bacteria.

A bacterial evaluation was completed to evaluate the feasibility of implementing groundwater bioremediation. The task was designed to quantify aerobic (and/or facultative) heterotrophic bacteria, evaluate if the aerobic bacteria can degrade organic constituents, and determine if a nutrient solution would interact with samples of the limestone aquifer. An evaluation of anaerobic bacteria was not completed because the bioremediation system was designed to inactivate obligate anaerobes by inducing aerobic conditions, and to replace the community with aerobic (and/or facultative) bacteria that would degrade organic compounds.

PROGRAM OBJECTIVES AND METHODS

The bacterial evaluation objectives were to (1) collect and prepare limestone and groundwater samples, (2) enumerate aerobic heterotrophic bacteria, (3) evaluate the capability of aerobic heterotrophic bacteria to metabolize selected carbon sources, and (4) determine if inorganic nutrients interact with samples of the limestone aquifer.

To meet the objectives, six limestone cores and three groundwater samples were collected throughout the manufacturing facility. These samples represented three general facility locations (background, production, and solvent handling).

Previous investigations showed that the production area had high organic chemical concentrations, high hydrogen sulfide, and low oxygen concentrations, and the solvent-handling areas showed high organic chemical concentrations, moderate hydrogen sulfide, and reduced oxygen concentrations.

The limestone cores were collected using an air-rotary drill rig equipped with a (disinfected) 10.2-cm (4-in.) coring device. Core sections selected for analyses were approximately 30.5 cm long (12 in. long) and collected at a depth between 6.7 and 7.0 m (22 and 23 ft) below land surface. All core samples consisted of white limestone and contained small to large solution channels. The groundwater samples were collected from monitoring wells using (disinfected) stainless steel bailers.

The bacterial evaluation methods were designed to enumerate aerobic heterotrophic bacteria from aquifer material (limestone and groundwater) and to determine whether the bacteria (extracted from limestone) can degrade compounds detected in the groundwater. The first procedure was designed to quantify heterotrophic bacteria using standard plating techniques (Clesceri et al. 1989). A portion of each crushed limestone sample was extracted with a sterile basal mineral buffer solution, and serial dilutions were prepared (10^{-1} to 10^{-7}). Aliquots from the serial dilutions were plated (in duplicate) on standard plate count agar and incubated at ambient temperature (28°C). Groundwater samples also were diluted and plated (in duplicate) on standard plate count agar. The plates were enumerated within 7 days.

The second procedure was designed to determine the extent of aerobic bacterial acclimation in the subsurface. The procedure was modified from standard plate count techniques where bacteria were then incubated on a medium that does not contain a source of carbon (purified agar). A specific carbon source (sulfolane, benzene, and diethyl ether) was introduced either directly during preparation of the bacterial medium (sulfolane at 1,000 mg/L) or introduced by vapor diffusion (benzene, diethyl ether) via volatilization. The bacterial extracts were plated as previously described, then incubated at ambient temperature (28°C) and enumerated after 14 days.

The third procedure was designed to indicate qualitatively the capability of indigenous aerobic bacteria to degrade selected organic compounds and to estimate the relative inhibitory concentration for each constituent. Three bacterial cultures were prepared from locations described above. An aliquot of each extract was inoculated into sterile nutrient broth and incubated at 32°C until turbid. The bacterial cultures were transferred to purified agar plates by removing a 200-μL aliquot and spreading the bacterial suspension onto compound specific purified agar plates.

Sterile aqueous solutions containing toluene, diethyl ether, and sulfolane were prepared at 0 mg/L, 0.5 mg/L, 5 mg/L, 10 mg/L, 50 mg/L, and 100 mg/L concentrations. Adsorbent disks were prepared by aseptically saturating the disks in each solution and evenly distributing the disks around each inoculated purified agar plate. One set of disks was placed on each plate. The plates were incubated for 7 days and evaluated for growth and zones of inhibition.

In addition to each bacterial evaluation, these environmental condition analyses were completed from crushed limestone samples: (1) pH (U.S. EPA SW-846 Method 9045); (2) soil moisture (Standard Method 2540 G); (3) total organic carbon (TOC; Standard Method 5310 B), soluble *o*-phosphate (Standard Method 4500-P), and soluble ammonia (Standard Method 4500-NH_3 F) (Clesceri et al. 1989).

A column study was completed to determine if nutrients introduced to the aquifer during bioremediation would interact with limestone and limit availability. Two columns of 3.8 cm inner diameter, 15.2 cm long (1.5 in. inner diameter, 6 in. long) were packed with crushed limestone. End caps were placed on each column and a neutral nutrient solution (108 mg/L ammonia and 58 mg/L *o*-phosphate) was introduced to the head of one column. Simultaneously, distilled water was introduced onto the second column (control).

A flow (1 mL per min) was maintained, and the effluent from each column was collected. Ten pore volumes were collected and analyzed for pH, *o*-phosphate, ammonia-nitrogen, and TOC.

EVALUATION RESULTS

The heterotrophic bacterial enumeration analyses results showed that indigenous bacteria were present in limestone core samples and ranged from 2.2×10^4 to 2.1×10^7 colony-forming units (CFUs) per gram (Table 1). The highest concentrations of bacteria were in samples collected from the solvent-handling area and background (S-01, S-02, and B-0). The bacterial concentrations in samples collected from the production area were lowest (M-01, M-02, and M-03).

The results of the bacterial enumeration from the groundwater samples showed that heterotrophic bacteria were present. They ranged from 2.1×10^4 to 3.5×10^4 CFU/g (GW-01, GW-02, GW-03) (Table 1).

The specific carbon degrader plate count enumerations were low. Bacterial counts ranged from no growth to an estimated quantity of 2.4×10^3 CFU/g (Table 1).

The results of the disk-agar diffusion evaluation showed that bacterial growth occurred around each disk and was not inhibited by carbon sources (diethyl ether, toluene, and sulfolane). Control disks (distilled water) showed no growth.

The results of environmental condition analyses completed on limestone samples (Table 1) showed that TOC ranged from 22 to 110 mg/kg, ammonia-nitrogen ranged from <0.02 mg/kg to 3 mg/kg, and *o*-phosphate concentrations ranged from <2 mg/kg to 156 mg/kg. The limestone moisture content ranged from 14 to 23%. The pH of the six core samples was between 7.2 and 7.4 standard units.

The results of the limestone column evaluation showed that ammonia and *o*-phosphate were not retained by the limestone, and the application of an ammonia (108 mg/L) and *o*-phosphate (58 mg/L) solution at pH 7.6 had no affinity for limestone. This lack of affinity was shown by no significant change in concentrations analyzed in each pore volume. The control column results indicated that neither ammonia nor *o*-phosphate was present or liberated from the limestone column.

TABLE 1. Results of the bacterial and environmental conditions analyses.

Sample ID	Sample Location	Medium	Bacterial Data: Plating Medium Carbon Source	Bacterial Data: Total Bacteria CFU/g[(a)]	Environmental Condition Data: Total Organic Carbon mg/kg[(b)]	Environmental Condition Data: Available Ammonia Nitrogen mg/kg[(b)]	Environmental Condition Data: Available o-Phosphate mg/kg[(b)]	Environmental Condition Data: Moisture Content %	Environmental Condition Data: pH s.u.[(c)]
B-0	Background	Limestone	Complex	1.8e+07	104	<0.02	35	14	7.3
			Diethyl Ether	4.4e+02					
			Benzene	4.7e+01					
			Sulfolane	1.5e+03					
S-01	Solvent Handling	Limestone	Complex	1.7e+07	110	<0.02	<2	19	7.2
			Diethyl Ether	1.4e+03					
			Benzene	3.7e+02					
			Sulfolane	2.4e+03					
S-02	Solvent Handling	Limestone	Complex	2.1e+07	93	<0.02	<2	23	7.3
			Diethyl Ether	3.7e+02					
			Benzene	(d)					
			Sulfolane	(d)					
M-01	Production Area	Limestone	Complex	2.2e+04	25	3	84	17	7.4
			Diethyl Ether	6.5e+02					
			Benzene	(d)					
			Sulfolane	(d)					
M-02	Production Area	Limestone	Complex	2.7e+04	63	<0.02	<2	15	7.3
			Diethyl Ether	(d)					
			Benzene	(d)					
			Sulfolane	(d)					
M-03	Production Area	Limestone	Complex	7.8e+05	22	<0.02	156	16	7.3
			Diethyl Ether	5.7e+02					
			Benzene	(d)					
			Sulfolane	7.8e+02					
GW-01	Background	Groundwater	Complex	2.4e+04	(e)	(e)	(e)	(e)	(e)
GW-02	Solvent Handling	Groundwater	Complex	2.1e+04	(e)	(e)	(e)	(e)	(e)
GW-03	Production Area	Groundwater	Complex	3.5e+04	(e)	(e)	(e)	(e)	(e)

(a) Colony-forming unit, dry weight; values less than 300 are estimates. (b) Dry weight. (c) Standard unit. (d) No growth; quantitative detection limit was approximately 300 CFU/g.
(e) Analyses not completed.

BACTERIAL EVALUATION CONCLUSIONS

The results of the bacterial analyses showed that limestone and groundwater contained viable heterotrophic bacteria indicating the ability to thrive in a potentially toxic environment. The bacterial concentrations in the limestone were lowest near the production area compared to a sample collected from background and solvent-handling locations. These data correlate with site data showing high organic chemical concentrations, high hydrogen sulfide, and low oxygen concentrations in the production area (Moon et al. 1991). These data indicate that the high concentration of organics and the lack of oxygen may have inhibited the growth of aerobic bacteria near the production area.

The environmental conditions analyses showed that inorganic nutrients necessary to support enhanced aerobic bacterial growth may be low, and the limestone column study confirmed that if nutrients were added to the aquifer, sorption would be minimal.

In summary, the bacterial evaluation showed that the limestone aquifer contained viable heterotrophic aerobic bacteria capable of degrading selected organic compounds. Environmental condition analyses indicate that oxygen and inorganic nutrients should be added to the aquifer to enhance aerobic bacterial growth and to minimize hydrogen sulfide production. The limestone column study confirmed that inorganic nutrient sorption would be negligible.

REFERENCES

Clesceri, L. S., A. E. Greenberg, and R. R. Trussell (Eds.). 1989. *Standard Methods for the Examination of Water and Wastewater*, 17th ed. American Public Health Association, American Water Works Association, Water Pollution Control Federation, Washington, DC.

Moon, R. E., R. Ackart, J. A. Whitehead, and K. Chellman. 1991. "Soil-Gas Remediation of Reduced Sulfur Compounds in a Limestone Aquifer." *Proceedings of the First Annual Hazardous Materials and Environmental Management Conference/South*, pp. 431-442. Tower Conference Management Company, Atlanta, GA.

Moon, R. E., E. Nyer, and P. Gurney-Read. 1992. "In-Situ Groundwater Bioremediation in a Karst Aquifer." *Proceedings from the Superfund 92*, Hazardous Material Control Resources Institute, Washington, DC.

POLYCYCLIC AROMATIC HYDROCARBON REMOVAL RATES IN OILED SEDIMENTS TREATED WITH UREA, UREA-FISH PROTEIN, OR AMMONIUM NITRATE

R. F. Lee and M. Silva

INTRODUCTION

Hydrocarbon degradation rates increase in marine sediments after the addition of petroleum or petroleum-based products. One explanation for this degradation increase is that the numbers of hydrocarbon-utilizing microbes increase in sediments after petroleum exposure (Leahy & Colwell 1990). In addition to the concentration of hydrocarbon-degrading microbes, factors that can influence hydrocarbon degradation rates in sediments include nutrients, oxygen, and temperature.

A number of studies have shown that hydrocarbon biodegradation rates in oiled sediments can be enhanced by the addition of nutrients (Atlas 1991, Leahy & Colwell 1990, Lindstrom et al. 1991, Pritchard & Costa 1991). The focus of our study was to determine how the addition of ammonium nitrate, urea, or a urea-fish meal mixture to oiled sediments influenced hydrocarbon removal rates. Hydrocarbon removal rates were determined by decreases in concentrations of polycyclic aromatic hydrocarbons (PAHs) and degradation rates by mineralization of ^{14}C-labeled PAHs.

MATERIALS AND METHODS

Sediments collected from the marsh near the Skidaway Institute of Oceanography had a composition of 61% sand, 25% silt, and 14% clay. Fiberglass tanks, each containing 10 kg of this sediment, were used for the various treatments. Treatments included (1) 20 g of Prudhoe crude oil mixed with the sediment; (2) 20 g of Prudhoe crude oil mixed with the sediment followed by the addition of 100 mL of 6M urea; (3) 20 g of Prudhoe crude oil mixed with the sediment followed by the addition of 100 mL of 6M urea containing 1 g of fish meal (American Protein); (4) 20 g of Prudhoe crude oil mixed with the sediment followed by the addition of 100 mL of 6M ammonium nitrate. Each day, 2 L of seawater were added to each tank to keep the sediment moist. There were drains to the outside in each tank. Three sediment cores were taken at each sampling period for PAH analysis. Sampling periods were 0, 20, 40, and 60 days after the addition of oil. The experiments were done in April-May when temperatures of water and sediment ranged from 15 to 19°C.

The analytical procedures used were based on procedures described in MacLeod et al. (1985). Approximately 20 g of freeze-dried sediment were extracted sequentially with methanol, 1:1 methanol:methylene chloride, and methylene chloride. The organic phase was concentrated and fractionated on columns of silica gel over alumina packed over activated copper to remove elemental sulfur. The aromatic hydrocarbon fraction was further purified by Sephadex LH-20 chromatography (Ramos & Prohaska 1981). PAHs were quantified and identified by capillary gas chromatography-mass spectrometry using full-scan and selected ion monitoring modes. A DB-5 column mounted in a Hewlett Packard 5890 gas chromatography interfaced to a Finnigan Incos 50 mass spectrometer/data system was used. The PAHs selected for analysis included phenanthrene, methylphenanthrenes, fluoranthene, fluorene, methylfluorenes, pyrene, chrysene, acenaphthene, naphthalene, benz(a)anthracene, and acenaphthylene.

For testing of ^{14}C-labeled hydrocarbons, sediments were collected from both oiled and oil plus urea tanks at t = 40 days. Control sediments had no treatment. Microbial degradation of either 9-^{14}C-fluorene (95.1 MBq/mmol) or [12-^{14}C]benz(a)-anthracene (777 MBq/mmol) was determined by adding one of the compounds to sediment-seawater mixtures (1 g of sediment and 5 mL of seawater in a 125-mL bottle capped with silicon stopper). The final concentration of each radiolabeled compound was 2.5 μg/g sediment. Sterile controls were bottles containing mercuric chloride at a concentration of 1 mg/g sediment. At the end of each of the various incubation periods, the respired $^{14}CO_2$ was collected by trapping it on filter paper soaked with phenethylamine (details of this procedure are given in Lee & Ryan 1983). All samples were in triplicate for each time interval.

RESULTS

The concentration of PAHs in the sediments after the addition of crude oil (time = zero) ranged from 52 to 67 μg/g dry sediment (Figure 1). Before the addition of oil, PAH concentrations in the sediments ranged from 1.2 to 0.9 μg/g. The addition of urea, urea-fish meal or ammonium nitrate to oiled sediments greatly accelerated the removal of PAHs and at rates significantly different from the PAH removal rates in oiled sediment with no treatment ($p < 0.01$). Removal rates were significantly higher in oiled sediments treated with urea and urea-fish meal compared with ammonium nitrate treatment ($p < 0.05$). There were no significant differences in the PAH removal rates between urea and urea-fish meal treatments. In oiled sediment treated with urea or urea-fish meal, the PAH concentrations after 60 days ranged from 1.1 to 10 μg/g. Half-lives for PAHs in oiled, ammonium nitrate treated, urea-fish meal treated, and urea treated were 80, 50, 32, and 30 days, respectively.

The addition of ^{14}C-labeled hydrocarbons to sediment resulted in slow mineralization to $^{14}CO_2$ in untreated control sediments (Figures 2 and 3). No mineralization occurred in sterilized sediment. Oiled sediments showed degradation of

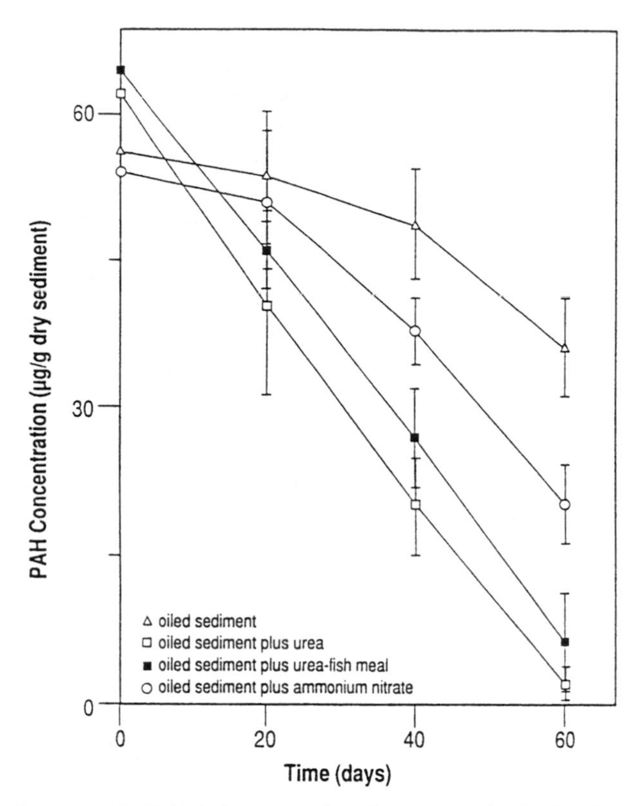

FIGURE 1. Decrease in PAHs in treated and untreated oiled sediments. Error bars are standard deviation (n = 3). The concentrations of urea, ammonium nitrate, and fish meal added to sediments are given in Materials and Methods.

both fluorene and benz(a)anthracene which were accelerated by the addition of urea. Calculated half-lives for fluorene in oiled sediments plus urea, oiled sediments, and control sediments were 10, 14, and 115 days, respectively. Calculated half-lives for benz(a)anthracene in oiled sediments plus urea, oiled sediments, and control sediments were 8, 37, and 242 days, respectively.

DISCUSSION

A number of studies have shown that the addition of nutrients, including nitrogen and phosphorus, to oiled soils or sediments resulted in increases in hydrocarbon degradation rates (Atlas 1991, Leahy & Colwell 1990, Wang et al. 1990). The results of our studies showed that addition of organic nitrogen (urea) or inorganic nitrogen (ammonium nitrate) accelerated removal rates of PAHs relative to untreated oiled sediments. The significant differences in PAH removal rates between urea- and ammonium nitrate-treated sediments suggests that urea is a better source of nitrogen for petroleum degraders than is inorganic nitrogen. Fish meal was added to the urea to act as a source of slow-release nitrogen and a readily biodegradable carbon source that could act to initiate biodegradation of PAHs. Concentrated urea solutions readily dissolve many insoluble proteins, including fish meal. In the experiments reported here we found no significant differences in PAH removal rates between urea and urea-fish meal treated oiled sediments. We speculate that a urea-fish meal mix may be useful with large volumes of oil in sediment where a slow-release form of nitrogen, such as fish meal, may prove effective in bioremediation.

We are assuming that most of the PAHs removed from the sediments were due to degradation. Because seawater was added to sediments, losses by washout cannot be excluded. The experiments using ^{14}C-labeled hydrocarbons clearly demonstrated that rapid degradation of hydrocarbon does occur in oiled sediments

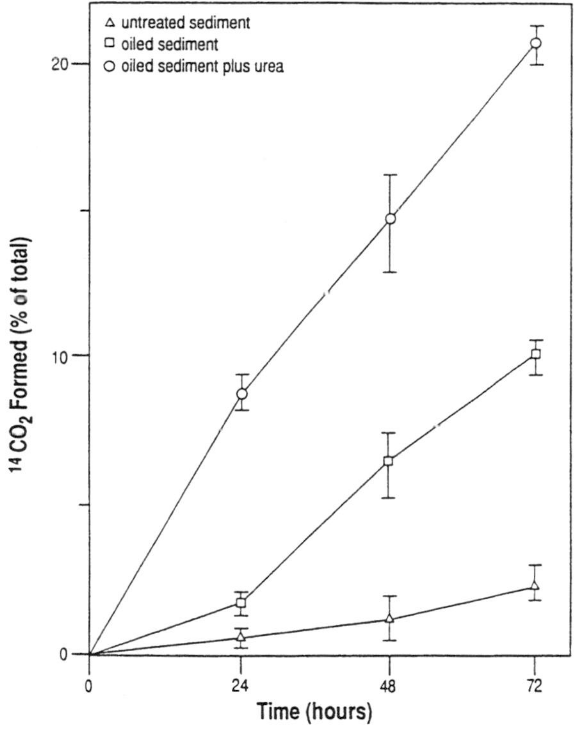

FIGURE 2. ^{14}C-fluorene (2.5 µg/g sediment) was added to either previously oiled sediments or to untreated control sediments. Error bars are standard deviation (n = 3). The calculated half-lives for fluorene in oiled sediments plus urea (25 mg), oiled sediments, and control sediments were 10, 14, and 115 days, respectively.

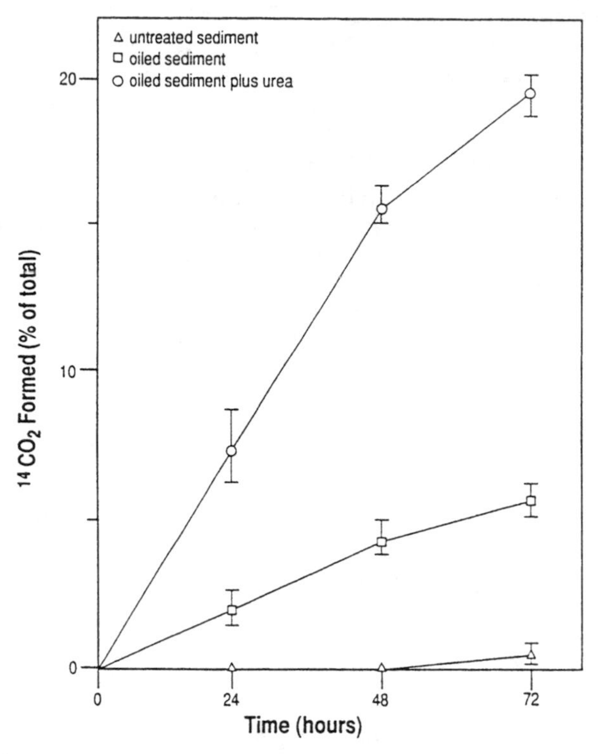

FIGURE 3. **^{14}C-benz(a)anthracene (2.5 µg/g sediment) was added to either previously oiled sediments or to untreated control sediments. Error bars are standard deviation (n = 3). The calculated half-lives for benz(a)anthracene in oiled sediments plus urea (25 mg), oiled sediments, and control sediments were 8, 37, and 242 days, respectively.**

compared with control sediments. This increase in potential hydrocarbon degradation rates was apparently due to microbial adaptation to oil (Leahy & Colwell 1990, Lee & Ryan 1983, Lee & Hoeppel 1991). Microbial adaptation likely resulted from an increase in the population of hydrocarbon-degrading microbes. The addition of urea to oiled sediment along with ^{14}C-hydrocarbons resulted in increases in the rate of degradation to $^{14}CO_2$. Increases in the population of hydrocarbon degraders in sediment after an oil spill is well documented (Leahy & Colwell 1990, Lindstrom et al. 1991). The addition of nutrients also could increase the production of microbial surfactants that would facilitate the removal of PAHs from sediment without necessarily degrading these compounds (Schulz et al. 1990). We hypothesize that the addition of nutrients to sediments accelerates the removal of PAHs from sediments by increases in the population of petroleum degraders and possibly by production of bacterial surfactants.

Our experiments were with fresh oil. Recent experiments suggested that PAHs in weathered crude oil (1 year in the sediments) were removed at significantly lower rates after the addition of nutrients compared with removal rates of fresh oil (Lee, unpublished data).

In summary, our results showed that the addition of either organic or inorganic nitrogen to oiled sediments resulted in an increase in PAH removal rates, and in the mineralization rate of added ^{14}C-hydrocarbons. More studies are necessary to demonstrate the possible usefulness of slow-release nitrogen mixtures, such as fish protein. It may be that when high concentrations of highly weathered oil are present in sediments, it would be useful to have slow-release forms of nitrogen and natural surfactants present in the bioremediation mixtures.

REFERENCES

Atlas, R. A. 1991. "Bioremediation of Fossil Fuel Contaminated Soils." In R. E. Hinchee and R. F. Olfenbuttel (Eds.), *In Situ Bioreclamation*, pp. 14-32. Butterworth-Heinemann, Stoneham, MA.

Leahy, J. G., and R. R. Colwell. 1990. "Microbial Degradation of Hydrocarbons in the Environment." *Microbial Revs. 54*: 305-315.

Lee, R. F., and C. Ryan. 1983. "Microbial and Photochemical Degradation of Polycyclic Aromatic Hydrocarbons in Estuarine Waters and Sediments." *Can. J. Fish. Aquat. Sci. 40(2)*: 86-94.

Lee, R. F., and R. Hoeppel. 1991. "Hydrocarbon Degradation Potential in Reference Soils and Soils Contaminated with Jet Fuel." In R. E. Hinchee and R. F. Olfenbuttel (Eds.), *In Situ Bioreclamation*, pp. 570-580. Butterworth-Heinemann, Stoneham, MA.

Lindstrom, J. E., R. C. Prince, J. C. Clark, M. J. Grossman, T. R. Yeager, J. F. Braddock, and E. J. Brown. 1991. "Microbial Populations and Hydrocarbon Biodegradation Potentials in Fertilized Shoreline Sediments Affected by the T/V Exxon Valdez Oil Spill." *Appl. Environ. Microbiol. 57*: 2514-2522.

MacLeod, W. D., D. W. Brown, A. J. Friedman, D. G. Burrows, O. Maynes, R. W. Pearce, C. A. Wigren, and R. G. Bogar. 1985. "Standard Analytical Procedures of the NOAA National Analytical Facility 1985-1986." *Extractable Toxic Organic Compounds.* U.S. Department of Commerce, NOAA/NMFS. NOAA Tech. Memo. NMFS F/NWC-92.

Pritchard, P. H., and C. F. Costa. 1991. "EPA's Alaska Oil Spill Bioremediation Project." *Environ. Sci. Technol. 25*: 372-379.

Ramos, L., and P. G. Prohaska. 1981. "Sephadex LH-20 Chromatography of Extracts of Marine Sediments and Biological Samples for the Isolation of Polynuclear Aromatic Hydrocarbons." *J. Chromatogr. 211*: 284-289.

Schulz, D., A. Passeri, M. Schmidt, S. Lang, F. Wagner, V. Wray, and W. Gunkel. 1990. "Marine Biosurfactants, I. Screening for Biosurfactants among Crude Oil Degrading Marine Microorganisms from the North Sea." *Naturforsch. 46C*: 197-203.

Wang, X., X. Yu, and R. Bartha. 1990. "Effect of Bioremediation on Polycyclic Aromatic Hydrocarbon Residues in Soil." *Environ. Sci. Technol. 24*: 1086-1089.

USE OF LABORATORY SOIL COLUMNS TO OPTIMIZE IN SITU BIOTRANSFORMATION OF TETRACHLOROETHENE

C. M. Morrissey, S. E. Herbes, O. M. West,
A. V. Palumbo, T. J. Phelps, and T. C. Hazen

INTRODUCTION

Tetrachloroethene (perchloroethylene, PCE) is a major groundwater contaminant at many U.S. Department of Energy (DOE) sites, often at concentrations of several milligrams per liter (Davis et al. 1988). At the Savannah River Integrated Demonstration (SRID) site, PCE, together with trichloroethene (TCE), has been detected in both soil and groundwater (Eddy et al. 1991). Biological degradation of the combination of PCE and TCE may prove difficult in the subsurface gas injection scheme at the SRID. The injection of methane into the subsurface at SRID will provide aerobic, methanotrophic conditions that are favorable for TCE degradation. However, PCE is degraded by a reductive dechlorination process under anaerobic conditions (Freedman & Gossett 1989; DiStefano et al. 1991) but has proven recalcitrant under aerobic, methanotrophic conditions (Fogel et al. 1986). The research reported herein includes three main objectives: (1) to evaluate the potential for anaerobic dechlorination of PCE at high mg/L concentrations; (2) to evaluate the potential for PCE removal by coupling anaerobic PCE dechlorination with aerobic, methanotrophic degradation of the anaerobic products; and (3) to simulate the subsurface environment of the SRID gas injection test site in the laboratory and evaluate the fate of PCE in this laboratory simulation.

To accomplish these objectives, two types of experiments were conducted: (1) saturated sand column experiments inoculated with either anaerobic digester sludge or soil from local waste disposal sites; and (2) soil column experiments packed with soil from the SRID site and purged with either 3% CH_4 in air or N_2. This technical note describes the experimental methods used in and the results obtained from these ongoing experiments.

METHODS

Saturated Sand Column. A 70-cm by 5-cm-diameter glass column was packed with coarse sand and equipped with five intermediate sampling ports. For the first set of column experiments, the sand column was inoculated with anaerobic digester sludge obtained from the primary digester at the local sewage treatment plant. The sludge was diluted to 10% in the revised anaerobic mineral medium (RAMM) of Shelton and Tiedje (1984) supplemented with 1 g/L sodium acetate and 50 mg/L yeast extract. The diluted sludge was applied to the column using a peristaltic pump in an upward flow mode. The supplemented RAMM was fed to the column for three weeks in order to establish biological activity.

After biological activity was established in the column (based on a steady production of CH_4 containing off-gas), a flow of PCE in RAMM was initiated. Influent PCE concentrations ranged from approximately 30 to 90 mg/L. A flowrate of 2 mL/hr (hydraulic retention time of 10 days) was maintained with a syringe pump. Effluent gas was collected over acidified water in a 40-mL vial.

The same column was used for experiments that attempted to combine anaerobic PCE dechlorination with methanotrophic degradation of the dechlorination products. These experiments involved the addition of methanotrophs of groundwater origin to the upper half of the sand column and the continuous injection of aerobic mineral medium at the column midpoint.

Soil Columns. Subsurface soil collected during drilling at the SRID site was packed into two 100-cm by 5-cm-diameter glass columns. The packing arrangement involved the addition of soil collected from the saturated zone at the SRID site to the lower one-fourth of the column and soil from the unsaturated zone to the remainder of the column. One column was purged with 3% CH_4 in air, and the other with N_2. PCE was applied to the soil columns as a vapor (approximately 2 mg/L) in the purge gas stream by allowing the purge gas to equilibrate with an aqueous solution of PCE to simulate the subsurface gas injection methods employed at the SRID site.

Analytical Methodology. Volatile chlorinated compounds in aqueous samples from the sand column were analyzed by headspace gas chromatography. In the SRID soil column experiments, PCE in the vapor phase was analyzed directly using an electron capture detector. Samples of sand column off-gas and SRID soil column inlet and outlet gas were analyzed for CH_4, CO_2, N_2, and O_2 using a gas chromatograph (GC) equipped with thermal conductivity and flame ionization detectors. Acetate concentrations were determined using a high-performance liquid chromatography system equipped with an ion-exclusion column and conductivity detector.

RESULTS

Saturated Sand Column. Initially, the sand column was inoculated with anaerobic digester sludge and remained anaerobic along its entire length. At an influent PCE concentration of 75 mg/L, PCE was reduced below detection within the first 15 cm of the column (Figure 1) after a two-month adaptation period. The dechlorination products were TCE and both the *cis-* and *trans-*isomers of 1,2-dichloroethene (DCE). No further dechlorination was observed beyond

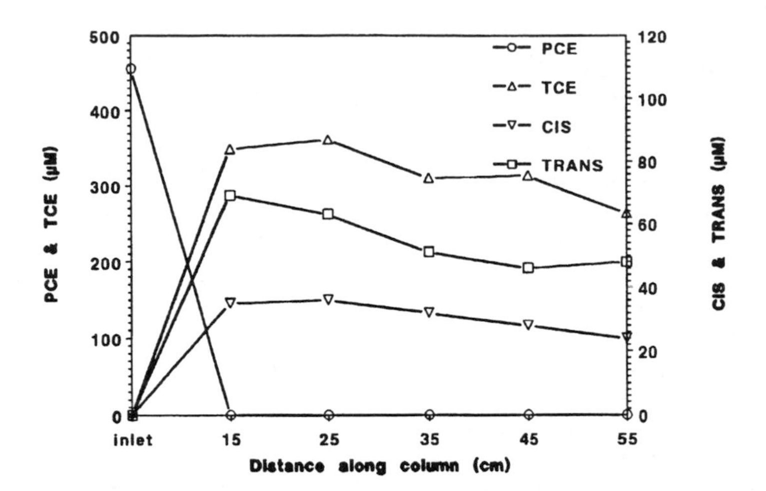

FIGURE 1. Dechlorination of 75 mg/L influent PCE in the saturated, anaerobic sand column inoculated with anaerobic digester sludge.

the first sampling port, and no vinyl chloride (VC) was detected. The reduction in concentration of *cis*- and *trans*-1,2-DCE was mainly due to the expected approximately 50% dilution. However, cometabolic degradation was confirmed by the detection of both *cis*- and *trans*-1,2-DCE epoxide.

The anaerobic column remained actively methanogenic whether or not PCE was present. Acetate was used rapidly, and an average of 15 mL of off-gas was collected each day, with a typical composition of 40% CH_4, 50% CO_2, and 10% N_2.

When the influent PCE concentration to the anaerobic column was reduced to 35 mg/L, PCE was again dechlorinated within the first 15 cm of the column. This reduced influent concentration resulted in the first appearance of VC in the column (Figure 2). Although not quantified, VC was estimated from the GC response to be the major volatile compound present.

When methanotrophs were added to the column and oxygenated mineral medium (Wittenbury et al. 1970) was supplied at the 35-cm port at approximately the same flowrate as the influent anaerobic medium, the concentrations of all four dechlorination products were reduced (Figure 3). The reduction in concentration of *cis*- and *trans*-1,2-DCE was mainly due to the expected approximately 50% dilution. However, cometabolic degradation was confirmed by the detection of both *cis*- and *trans*-1,2-DCE epoxide (Figure 4) which were identified by mass spectral analysis (Arvin 1991). Further experiments in which the anaerobic and aerobic processes are separated in different columns are presently in progress.

Soil Columns. Soil columns designed to simulate the SRID subsurface gas injection process at the water table have been in operation for several months. Both the CH_4/air and the N_2 columns are being purged at a rate of approximately

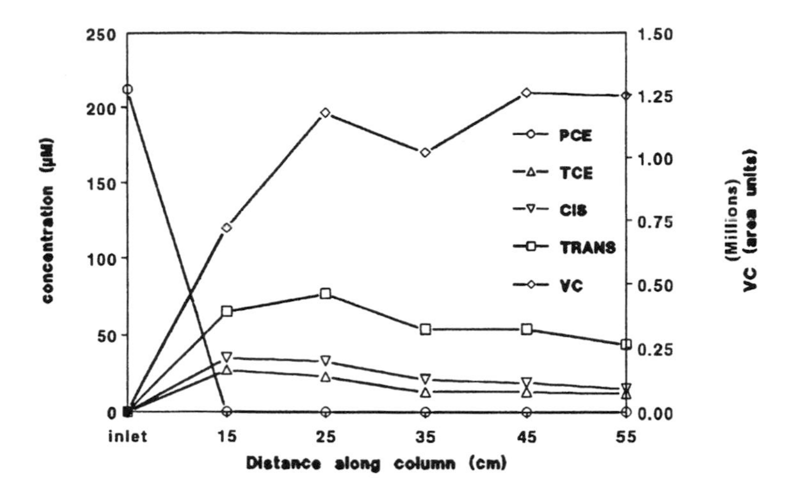

FIGURE 2. Dechlorination of 35 mg/L influent PCE in the saturated, anaerobic sand column inoculated with anaerobic digester sludge.

5 mL/min, which is the slowest gas flowrate that can be maintained in the columns. The result is an unsteady flow with an average gas retention time of about 2.5 hours (discounting the obvious channeling that occurs as the gas flows upward through the columns).

Treatments of the columns to date have included replacement of the liquid phase in the saturated zone of each column with (1) distilled water, (2) SRID

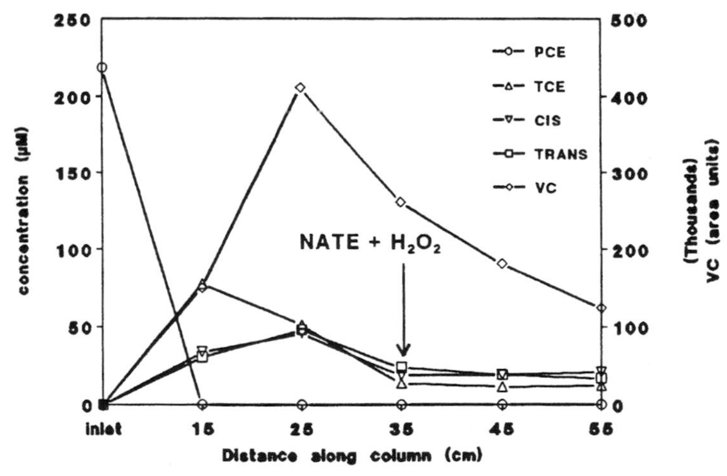

FIGURE 3. Changes in PCE dechlorination products in the aerobic zone of the saturated sand column after the addition of methanotrophs, NATE medium, and H_2O_2 to the upper half of the column.

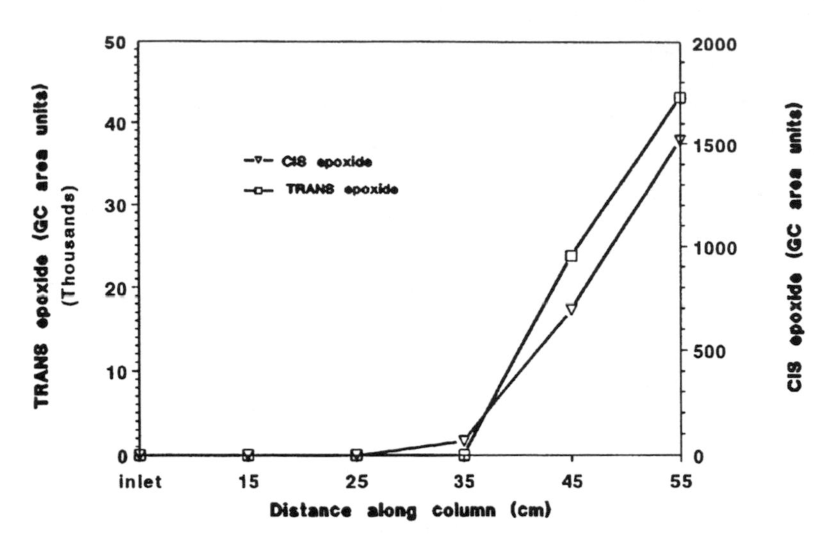

FIGURE 4. Formation of *cis*- and *trans*-1,2-DCE epoxides by methanotrophs in the aerobic zone of the saturated sand column.

groundwater, and (3) SRID groundwater supplemented with NATE medium. Each treatment regime has been monitored for 2 to 4 weeks. Monitoring of PCE in the influent and effluent gas to date has shown no indication of PCE degradation in either column under any of the treatment regimes. After groundwater from the SRID site was added to the columns, the use of CH_4 through the CH_4/air column was observed. Most probable number analyses of soil from the CH_4/air purged column confirm that the growth of methanotrophs has been stimulated.

These experiments are still in progress. Studies with DCE, known to be degradable by methanotrophs, and with the sequential addition of an electron donor (acetate) and a known PCE-degrading anaerobic inoculum to the columns are currently being conducted. Also, larger columns with online analytical capabilities are being designed to permit investigation of the effect of substantially longer column residence times on PCE degradation.

DISCUSSION

The saturated sand column experimental results indicate that high concentrations of PCE that may be encountered during an in situ remediation of groundwater can be rapidly dechlorinated anaerobically, in the presence of necessary nutrients and an electron donor, to a mixture of TCE, *cis*- and/or *trans*-DCE, and, under some conditions, VC. We have shown that the potential exists for combining an aerobic biological process in situ with anaerobic dechlorination to degrade the anaerobic dechlorination products. Further work is needed to demonstrate the maximum removal rates that can be expected.

The column tests designed to simulate the SRID in situ process conditions have thus far shown that no biodegradation of PCE occurs under the treatment regimes employed. Additional treatment regimes will be evaluated to identify conditions under which PCE may be degradable within the operational bounds of the SRID gas injection process. Treatment methods to produce these conditions might then be applied to the SRID project to maximize PCE removal.

ACKNOWLEDGMENTS

Support for this work has been provided by the U.S. Department of Energy, Environmental Restoration and Waste Management Program, Office of Technology Development. Oak Ridge National Laboratory is managed by Martin Marietta Energy Systems, Inc. for the U.S. Department of Energy under Contract DE-AC05-84OR21400.

REFERENCES

Arvin, E. 1991. "Biodegradation Kinetics of Chlorinated Aliphatic Hydrocarbons with Methane Oxidizing Bacteria in an Aerobic Fixed Biofilm Reactor." *Water Research 25*: 873-881.

Davis, H. A., D. K. Martin, and J. L. Todd. 1988. *Savannah River Site Environmental Report for 1988*, WSRC-RP-89-59-1, Vol. II, Westinghouse Savannah River Company, Savannah River Site, Aiken, S.C.

DiStefano, T. D., J. M. Gossett, and S. H. Zinder. 1991. "Reductive Dechlorination of High Concentrations of Tetrachloroethene to Ethene by an Anaerobic Enrichment Culture in the Absence of Methanogenesis." *Applied and Environmental Microbiology 57*(8): 2287-2292.

Eddy, C. A., B. B. Looney, J. M. Dougherty, T. C. Hazen, and D. S. Kaback. 1991. *Characterization of the Geology, Geochemistry, Hydrology and Microbiology of the In-situ Air Stripping Demonstration Site at the Savannah River Site.* WSRC-RD-91-21, Westinghouse Savannah River Company, Savannah River Site, Aiken, S.C.

Fogel, M. M., A. R. Taddeo, and S. Fogel. 1986. "Biodegradation of Chlorinated Ethenes by a Methane-Utilizing Mixed Culture." *Applied and Environmental Microbiology 51*: 720-724.

Freedman, D. L. and J. M. Gossett. 1989. "Biological Reductive Dechlorination of Tetrachloroethylene and Trichloroethylene to Ethylene under Methanogenic Conditions." *Applied and Environmental Microbiology 55*: 2144-2151.

Shelton, D. R. and J. M. Tiedje. 1984. "General Method for Determining Anaerobic Biodegradation Potential." *Applied and Environmental Microbiology 47*: 850-857.

Wittenbury, R., K. C. Phillips, and J. F. Wilkson. 1970. "Enrichment, Isolation, and Some Properties of Methane-Utilizing Bacteria." *Journal of General Microbiology 61*: 205-218.

SIMULTANEOUS DEGRADATION OF CHLOROBENZENE, TOLUENE, XYLENE, AND ETHANOL BY PURE AND MIXED *PSEUDOMONAS* CULTURES

S. Keuning and D. Jager

INTRODUCTION

The European Space Agency (ESA-ESTEC) has initiated with support from NIVR the development of a biological air filtration (BAF) system to remove of gaseous and airborne contaminants from space cabins. Within the framework of this project, biodegradation studies have been performed to investigate the feasibility of the microbial degradation concept.

The BAF is based on an advanced ecological concept for removing contaminants of both biological and nonbiological origin such as material off-gassing and gas leaks (Binot & Paul 1989). The system is based on the principle of combining of a support/sorbant material colonized by selected microorganisms in a near-resting state, and catabolizing the contaminants to harmless compounds, mainly carbon dioxide, water, and salts. Several advantages are expected over pure physicochemical systems, namely adaptability to unexpected contaminants and ability to recover normal efficiency after accidental poisoning.

To prove the feasibility of the BAF concept, it must be demonstrated that mixtures of contaminants can be degraded effectively by microbial populations composed of different specialized bacteria.

KINETIC STUDIES

Although many synthetic organic compounds have been shown to be biodegradable by microorganisms, relatively few studies have evaluated the basic parameters describing the kinetics of contaminant use. Because the efficient removal of contaminants in a biological treatment system depends largely on the kinetic properties of the microbial population, determination of these parameters is essential to develop and properly dimension such systems. Microbial growth usually is described by Monod kinetics, i.e., the specific growth rate (μ) of an organism is determined by the concentration of the available substrate (S), the maximal specific growth rate (μ_{max}), and the Monod constant (K_s): $\mu = \mu_{max} S/(K_s+S)$. The Monod constant is an important parameter for biodegradation processes, for it determines if low (growth-limiting) substrate concentrations can be degraded efficiently by the organisms.

The aromatic hydrocarbons toluene, chlorobenzene, and *m*- and *p*-xylene were chosen as model substrates for kinetic studies. Aromatic hydrocarbons generally are degraded via the formation of catechols, followed by ring cleavage of the catechols and further degradation. Ring cleavage can proceed by two different routes: *meta*-cleavage (intradiol cleavage) or *ortho*-cleavage (extradiol cleavage). In the case of chlorinated aromatics, only *ortho*-cleavage leads to complete mineralization.

The microbial strains *Pseudomonas* GJ40 and *Pseudomonas* GJ31 can both use toluene as a sole growth substrate (Oldenhuis et al. 1989). Strain GJ31 also can use chlorobenzene as a sole source of carbon and energy. Strain GJ40 lacks this ability, because chlorobenzene is toxic for this strain. Cultures of strain GJ40 that are supplied with chlorobenzene turn black due to the formation and accumulation of toxic intermediates that poison the cells of GJ40. Presumably strain GJ40 can use only the *meta*-route for catechol cleavage. *Pseudomonas* GJ8 can use toluene, *p*-xylene, or *m*-xylene as the sole carbon source.

The *Pseudomonas* strains GJ40, GJ31, and GJ8 were grown in pure and mixed cultures in a continuous-flow bioreactor that was operated as a chemostat (working volume 0.75 L; minimal salts medium pH 7; 30°C; dilution rate 0.1 to 0.5 h^{-1}; air flowrate 1.5 $L.h^{-1}$) with single and mixed volatile aromatic hydrocarbons as the carbon source, which were dosed via the gas phase. The Monod constants (K_s) for the microbial degradation of the single aromatic compounds were determined by measuring residual substrate concentrations in the culture medium after steady-state growth was reached at several fixed dilution rates. Concentrations of aromatic compounds in the culture fluid and in the effluent gas were measured by gas chromatography (GC).

RESULTS

Kinetic parameters for the biodegradation of toluene, chlorobenzene, and *p*-xylene by pure cultures of, respectively, *Pseudomonas* GJ40, *Pseudomonas* GJ31, and *Pseudomonas* GJ8 have been determined from experiments with continuous cultures in a chemostat (Table 1). In pure cultures of strain GJ31 and strain GJ40, low K_s values (0.05 µM and 0.07 µM, respectively) were measured for the biodegradation of toluene. A continuous culture of strain GJ31 growing on chlorobenzene as the sole carbon source showed a K_s value for chlorobenzene of 0.36 µM.

Pseudomonas GJ8 demonstrated rather unstable growth behavior when grown as a pure chemostat culture on *p*-xylene as carbon source. The culture medium frequently turned yellow, probably due to the accumulation of (toxic) intermediates, and washout of the cells occurred. An actual steady state with stable growth and removal of *p*-xylene could not be established, so no reliable kinetic parameters could be determined. Strain GJ8 appeared to be very sensitive to pH changes in the culture medium. At pH > 7.5 growth was severely hindered, and pH values of 8 and higher were lethal for the cells. When the dilution rate of the chemostat

TABLE 1. Kinetic parameters for pure cultures of *Pseudomonas* GJ40, *Pseudomonas* GJ31, and *Pseudomonas* GJ8 for growth on toluene, chlorobenzene, and *p*-xylene.

Strain	Compound	μ_{max} (h^{-1})	K_s (μ M)
GJ40	toluene	0.6	0.07
GJ31	toluene	0.6	0.05
GJ31	chlorobenzene	0.4	0.36
GJ8	*p*-xylene	0.2[a]	200-400[a]

(a) No stable growth.

was increased, the removal of *p*-xylene improved. At a dilution rate of 0.2 h^{-1} the cells did not wash out, as predicted from the apparent μ_{max} value determined earlier in the culture liquid, perhaps because of the lower concentrations of toxic intermediates in the chemostat due to the higher dilution rate and resulting higher growth rate. Addition of a mixture of *p*- and *m*-xylene as the carbon source also resulted in repeated washout of the culture. Addition of toluene as an extra growth substrate improve neither the growth of GJ8 nor the removal of the xylenes.

In the case of *Pseudomonas* strain GJ31, a significant positive influence of the cultivation on mixed substrates (50% toluene and 50% chlorobenzene) was observed. The removal capacity for chlorobenzene was enhanced in the presence of toluene as an additional substrate, resulting in a very low steady-state concentration of chlorobenzene in the culture medium and an apparent K_s value of < 0.02 µM. When the substrate feed was changed to 80% chlorobenzene and 20% toluene, the steady-state residual concentrations of toluene in the culture medium were not affected. The chlorobenzene concentration in the chemostat showed a slight increase to 0.1 µM at this ratio, that, nevertheless, was still lower than in the absence of toluene. Probably toluene is a better inducer of the catabolic enzymes than chlorobenzene. This hypothesis requires further research.

Mixed Cultures. In mixed cultures of strain GJ31 and strain GJ40 fed with toluene, the toluene was degraded to the same low level as in separate cultures of these strains. Under steady-state conditions, a stable cell ratio of 4:1 for GJ31 and GJ40 was observed in the chemostat during growth on toluene as determined by plating out samples of the culture fluid on agar plates.

Strains GJ40 and GJ31 were grown together on a 1:1 mixture of toluene and chlorobenzene. Toluene and chlorobenzene were degraded to the same low levels as in a pure culture of GJ31. In contrast to a pure culture of strain GJ40, a mixed culture of strain GJ40 and GJ31 growing on toluene did not turn black when chlorobenzene was added to the growth medium, nor did GJ40 wash out from the continuous culture. Efficient degradation of chlorobenzene by strain GJ31 probably prevented the buildup of toxic transformation products of chlorobenzene by strain GJ40 and poisoning of the culture.

Subsequently, ethanol was supplied as a third and easily biodegradable substrate via the incoming minimal salts medium to a chemostat culture of GJ31 and GJ40 that was growing on a mixture of toluene and chlorobenzene. Both *Pseudomonas* strains are capable of fast growth on ethanol. The residual steady-state concentrations of toluene and chlorobenzene in the culture fluid did not change significantly when ethanol was added. Both aromatic compounds were removed (under substrate-limiting conditions) as effectively in the presence of ethanol as in the absence of this easily degraded compound. Addition of ethanol in concentrations ranging from one-fifth to five times the total amount of chlorobenzene supplied resulted only in a higher biomass content in the chemostat that simultaneously degraded ethanol, toluene, and chlorobenzene. A mixed culture of strain GJ31 and GJ40 that was grown on ethanol as the sole growth substrate for some time, and then suddenly spiked with chlorobenzene and toluene, required less than 2 hours to reestablish the earlier low steady-state values of toluene and chlorobenzene (Figure 1).

When strain GJ8 was cultivated together with the microbial strains GJ31 and GJ40 on a mixture of toluene, *p*-xylene, and *m*-xylene, the growth behavior and

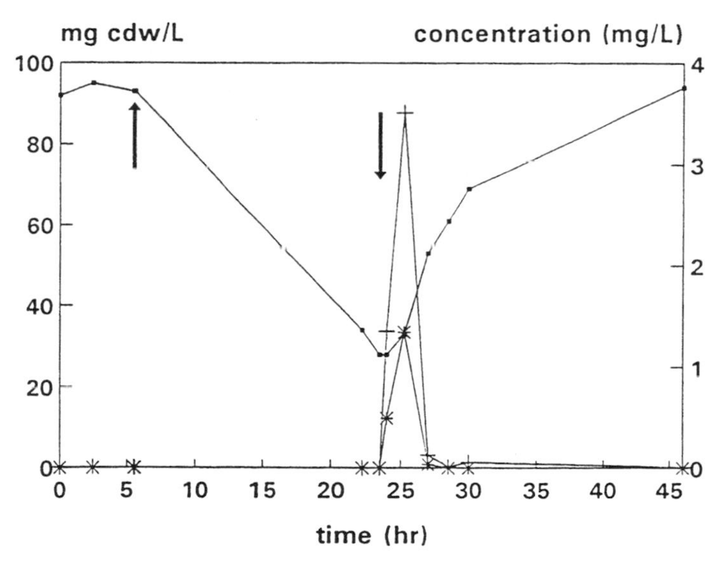

FIGURE 1. Chemostat cultivation of GJ31 and GJ40 fed with toluene, chlorobenzene, and ethanol. The cell density is shown as mg cell dry weight per liter (mg cdw/L). The toluene and chlorobenzene concentrations in the culture medium are also shown. The first arrow indicates when the addition of toluene and chlorobenzene was temporarily stopped. After 18 hours of growth on only ethanol (five volume changes of the chemostat), toluene and chlorobenzene were added again (second arrow).

stability of GJ8 dramatically improved. Furthermore the degradation of *p*- and *m*-xylene was remarkably enhanced resulting in xylene concentrations that dropped below the detection limits (<17 µg/L). Probably GJ31 and GJ40 catabolize intermediate degradation products of *m*- and *p*-xylene formed by strain GJ8 that otherwise accumulate and poison the cells of strain GJ8. This catabolizing process currently is under investigation. A mixed culture of GJ31, GJ40, and GJ8 could also be grown in the bioreactor on a mixture of toluene, *p*- and *m*-xylene, and chlorobenzene, resulting in an elimination of more than 99% of each individual compound.

CONCLUSIONS

The *Pseudomonas* strains GJ31 and GJ40 show very good kinetic properties for the removal of the selected contaminants toluene and chlorobenzene. Very low K_s values were measured for the removal of chlorobenzene and toluene, indicating that these contaminants can be effectively degraded even at very low concentrations.

Mixtures of contaminants may increase the degradation performance of microorganisms compared to pure contaminants, as shown for growth of strain GJ31 on mixtures of chlorobenzene and toluene. Increased degradation efficiency of chlorobenzene by strain GJ31, as a result of simultaneous growth on a mixture of toluene and chlorobenzene, has been demonstrated.

Under growth-limiting substrate conditions the presence of very easily degradable substrates as ethanol in substantial amounts does not necessarily impose a negative effect on the removal of more difficult to degrade compounds such as chlorinated hydrocarbons.

The stability and performance of specific microbial strains can be improved by culturing them in a mixed microbial community, as shown for *Pseudomonas* GJ8 strain. In a mixed culture with strain GJ40 and GJ31, strain GJ8 demonstrated more stable growth behavior and *p*- and *m*-xylene were degraded to much lower concentrations than in pure culture of GJ8. Apparently the additional strains stabilize strain GJ8, probably by consuming intermediates produced by GJ8 that otherwise accumulate and poison the cells of strain GJ8. These findings need further research. Furthermore it was demonstrated that a mixed culture of the *Pseudomonas* strains GJ8, GJ31, and GJ40 can efficiently and simultaneously degrade a complex mixture of toluene, *p*- and *m*-xylene, and chlorobenzene.

REFERENCES

Binot, R. A., and P. G. Paul. 1989. *BAF – An Advanced Ecological Concept for Air Quality Control.* SAE Technical paper series no. 891535.

Oldenhuis, R. L., L. Kuijk, A. Lammers, D. B. Janssen, and B. Witholt. 1989. "Degradation of Chlorinated and Non-chlorinated Aromatic Solvents in Soil Suspensions by Pure Bacterial Cultures." *Appl. Microbiol. Biotechnol. 30*: 211-217.

OPTIMIZATION OF AN ANAEROBIC BIOREMEDIATION PROCESS FOR SOIL CONTAMINATED WITH THE NITROAROMATIC HERBICIDE DINOSEB (2-*SEC*-BUTYL-4,6-DINITROPHENOL)

R. H. Kaake, D. L. Crawford, and R. L. Crawford

INTRODUCTION

For many organic contaminants, bioremediation is a cost-effective alternative to previous disposal methods such as incineration. A technology that has been successful at the bench scale, however, may have limited success when applied at the pilot scale or full scale in the field (MacRae & Alexander 1965; Spiker et al. 1992). The principal factors that often determine whether a bioremediation process can be practical (i.e., cost-effective) for field use are the ease with which the technology can be scaled up and the total throughput of the system. Physical parameters can greatly affect incubation periods required for detoxification, and hence, the system throughput. Therefore, it is important to establish optimal parameters for a process at both the pilot and bench scales so that cost-effectiveness can be assessed.

Dinoseb (2-*sec*-butyl-4,6-dinitrophenol, DNBP) is a nitroaromatic herbicide that was used on many different crops prior to its recall in 1986 by the U.S. Environmental Protection Agency (EPA 1986a,b). Dinoseb is still used on crops outside of the United States (Szeto & Price 1991). It is a highly recalcitrant molecule that persists as a soil contaminant at many rural airstrips where crop-dusting activities took place (Kaake et al. 1992). Dinoseb probably is not degraded under aerobic or microaerophilic conditions (Stevens et al. 1991). Recently, an anaerobic bioremediation process has been developed at the bench (Kaake et al. 1992) and pilot (Roberts et al. 1992, 1993) scales. This process removes all aromatic products from contaminated soil and water. Here we describe optimizations of the process that have reduced operational costs and expedited treatment periods so that the process may be used commercially.

MATERIALS AND METHODS

The soil described in Kaake et al. (1992) is a silt loam containing approximately 250 mg/kg dinoseb. This soil was contaminated by leakage of several dinoseb storage barrels at the site of an abandoned rural airstrip near Hagerman, Idaho.

The soil was collected and passed through a 2-mm (#10) sieve, placed in a sealed container, and stored at 5°C until used.

Prior experiments with the contaminated soil indicated that it did not contain a microflora capable of quickly degrading dinoseb or dinoseb biotransformation intermediates (Kaake et al. 1992). Therefore, the addition of an already treated soil previously contaminated with dinoseb and containing dinoseb-degrading microorganisms was necessary. This soil came from a successful biotreatment of a site in Washington State and is described elsewhere (Kaake et al. 1992).

The carbon source also described in Kaake et al. (1992) is a waste product from a potato processing plant. It contained (per gram) 215 mg available starch and 8×10^3 culturable amylolytic, heterotrophic bacteria. The potato waste was autoclaved 20 min to reduce the number of viable bacteria. Previous experiments (Kaake et al. 1992) showed that the microorganisms in the potato waste were incapable of degrading dinoseb, although they may modify the molecule.

Dinoseb concentrations in soil and aqueous samples were determined by high pressure liquid chromatography (HPLC) using a method previously described (Kaake et al. 1992). Redox potentials of cultures were measured at the soil/aqueous-phase interface with a platinum electrode (Orion model 96-78).

All experiments were performed in triplicate. The 200-g sample of soil, potato waste, and dinoseb-acclimated soil inoculum were placed in sterile, 500-mL, wide-mouth Erlenmeyer flasks and flooded with 200 mL of sterile 50-mM phosphate buffer. The amount of potato waste and dinoseb-degrading soil inoculum, the incubation temperature, and the pH of the buffer were varied depending on the culture parameter being studied.

RESULTS

Previous experiments (Kaake et al. 1992) had shown that anaerobic conditions could not be established without the addition of supplemental carbon such as the starchy potato waste. An experiment was designed to determine the minimum amount of potato waste necessary to generate anaerobic conditions and to complete the degradative process. As Figure 1 illustrates, a 0.5% addition (1 g potato waste per 200 g contaminated soil) of potato waste was insufficient to facilitate the degradation of dinoseb. Redox potential was monitored continuously in one of each set of flasks. All three cultures attained anaerobic conditions ($E_h < 0$ mV) in 3 days; however, the redox potential of the 0.5% culture began to increase after 4 days (data not shown). A 1% addition of potato waste allowed complete dinoseb removal by 6 days, only slightly longer than in cultures receiving a 2% addition (5.5 days). However, 2% cultures still contained an unidentified dinoseb biotransformation intermediate after the 1% cultures had removed all detectable aromatic products from the system (data not shown). The apparent increase in dinoseb concentration during the first several days is due to dinoseb desorption from the soil into the aqueous phase.

Previous experiments (Kaake et al. 1992) had shown that this soil required an inoculum of acclimated dinoseb-degrading microorganisms for efficient removal

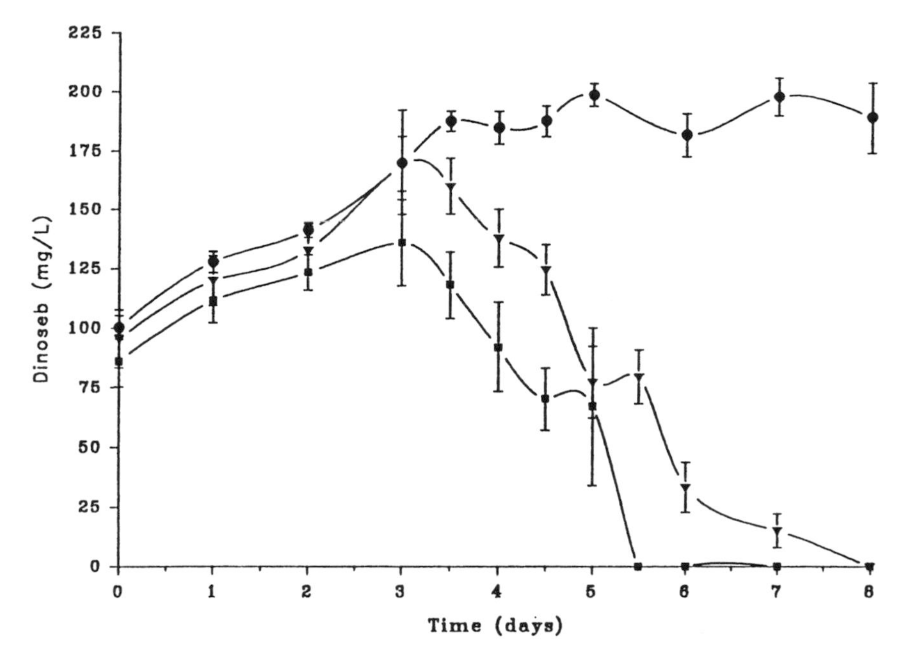

FIGURE 1. Dinoseb concentration in culture supernatants receiving 0.5% (•), 1.0% (▾), and 2.0% (■) by soil weight of starchy centrifuge cake.

of dinoseb. Identical cultures were inoculated with 0, 2, 5, 10, and 25% (w/w) of the acclimated soil inoculum. Cultures receiving no soil inoculum transformed dinoseb; however no biotransformation products were detected by HPLC, and an insoluble precipitate, consistent with a polymerized dinoseb product, was seen to form on the aqueous surface. The remaining cultures had removed dinoseb from the aqueous phase by day 6 and showed no significant difference in degradative capacities.

Incubation temperatures had a dramatic effect on dinoseb degradation rates. Cultures were incubated at 15, 22, 25, 30, 37, and 40°C. Incubation at 37°C allowed dinoseb removal from the aqueous phase in 4 days, whereas temperatures of 30 and 40°C removed dinoseb in 4.75 and 5 days, respectively. Temperatures below 30°C significantly slowed degradation rates. Dinoseb was removed by day 11 and 17 at 25 and 22°C, but dinoseb was still present after 20 days when treated at 15°C.

Buffer pH values were maintained by adjusting the pH 7 buffer to the desired pH at time zero and adjusting it daily (if necessary) using 6 M HCl or 10 M NaOH. A pH value of 8.0 facilitated the most rapid degradation, in which dinoseb was removed in 7 days (Figure 2). Dinoseb was removed from cultures after 9 days when the buffer was maintained at pH 7.5 or when the pH was uncontrolled and

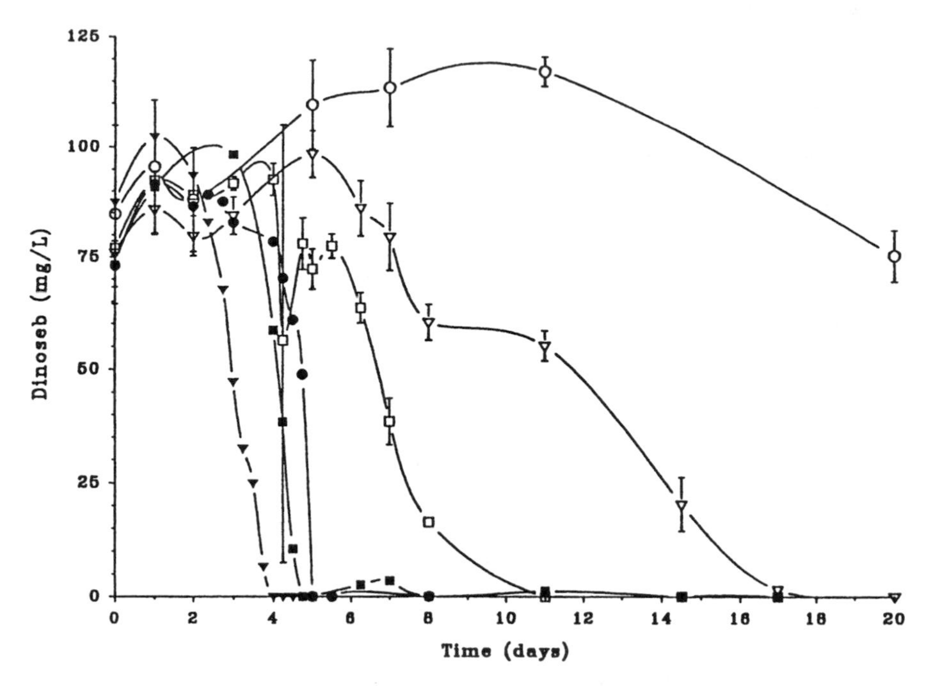

FIGURE 2. Dinoseb concentration in culture supernatants that were maintained at a certain pH. Cultures were adjusted daily to within 0.1 pH units with 6 M HCl or 10 M NaOH as needed. The pH values tested were 6.0 (○), 6.5 (▽), 7.0 (□), 7.5 (•), 8.0 (▾), and uncontrolled (■).

allowed to fluctuate naturally. Lower pH values slowed degradation further. At pH 6.0 and 6.5, little degradation had occurred by day 14 (Figure 2).

DISCUSSION

The optimum conditions for the remediation of dinoseb-contaminated soil were as follows: potato waste addition; 1% (by weight) acclimated soil inoculum; 2% (by weight) phosphate buffer; pH 7.5 to 8.0; and incubation temperature 35 to 37°C. In addition, the redox potential must be sufficiently anaerobic (<100 mV) for the dinoseb-degrading organisms to be active (Kaake et al. 1992). Oxygen diffusion may occur more rapidly in laboratory experiments with smaller volumes of soil and buffer than in large-scale reactors. Therefore, the estimate of the minimum amount of potato waste needed to establish anaerobiosis may be conservative. Under the conditions in the laboratory, dinoseb and its biotransformation intermediates were removed from soil in 31 days. This process now has been implemented at a pilot scale of 9,000 kg (Roberts et al. 1993), showing that scaleup to commercial size should be possible.

In considering the cost-effectiveness of any remediation process, one must consider both the environmental and engineering parameters associated with it. This may involve developing several different implementation strategies for a process. Although many parameters can be easily determined at the bench scale, some are media and site specific. In some cases the most cost-effective implementation may involve running the system under nonoptimal conditions. For example, if a large volume of soil were to be treated on site using an anaerobic lagoon, the costs of controlling the temperature of the system might outweigh the benefits of a shortened incubation period. Alternatively, if a smaller volume of soil were to be treated with a large mobile reactor, the incorporation and use of a thermostatic system on the reactor might be a more cost-effective solution.

Some parameters, however, must be tightly controlled in order to be assured of success. In performing a pilot-scale treatment on 9,000 kg of the soil described above (Roberts et al. 1992), hindsight showed that thorough homogenization of the soil to a consistent particle size was necessary for complete remediation of the soil. Extremely compacted soil clods measuring 10 to 20 cm in diameter were found in the reactors at the end of the treatment process. These had not been wetted and still contained dinoseb. Use of a hopper and crusher to obtain a consistent particle size would have avoided this problem. This example illustrates the importance of performing initial bench- and pilot-scale studies prior to conducting a full-scale remediation at a site. These studies allow assessments of both microbiological and engineering problems that might be encountered when proceeding to full-scale operations.

REFERENCES

Kaake, R. H., D. J. Roberts, T. O. Stevens, R. L. Crawford, and D. L. Crawford. 1992. "Bioremediation of Soils Contaminated with 2-*sec*-Butyl-4,6-Dinitrophenol (Dinoseb)." *Appl. Environ. Microbiol. 58*(5): 1683-1689.

MacRae, I. C., and M. Alexander. 1965. "Microbial Degradation of Selected Herbicides in Soil." *J. Agric. Food Chem. 13*(1): 72-75.

Roberts, D. J., R. H. Kaake, S. B. Funk, D. L. Crawford, and R. L. Crawford. 1992. "Anaerobic Remediation of Dinoseb from Contaminated Soil: An On-site Demonstration." *Appl. Biochem. Biotechnol.* In press.

Roberts, D. J., R. H. Kaake, S. B. Funk, D. L. Crawford, and R. L. Crawford. 1993. "Field Scale Anaerobic Bioremediation of Dinoseb-Contaminated Soils." In M. Gealt (Ed.), *Biotreatment of Industrial and Hazardous Wastes*, pp. 219-243. McGraw-Hill, New York, NY.

Spiker, J. K., D. L. Crawford, and R. L. Crawford. 1992. "Degradation of 2,4,6-Trinitrotoluene (TNT) in Explosives-Contaminated Soils by the White-Rot Fungus *Phanerochaete chrysosporium*: Influence of TNT Concentration." *Appl. Environ. Microbiol. 58*(9): 3199-3202.

Stevens, T. O., R. L. Crawford, and D. L. Crawford. 1991. Selection and Isolation of Bacteria Capable of Degrading Dinoseb (2-*sec*-Butyl-4,6-Dinitrophenol)." *Biodegradation* 2: 1-13.

Szeto, S. Y., and P. M. Price. 1991. "High Performance Liquid Chromatography Method for the Determination of Dinoseb: Application to the Analysis of Residues in Raspberries." *J. Agric. Food Chem. 39*: 1614-1617.

U.S. Environmental Protection Agency. 1986a. *Federal Register 51*(198): 36634-36650.

U.S. Environmental Protection Agency. 1986b. *Federal Register 51*(198): 36650-36661.

REMOVAL OF 2,4,6-TRINITROTOLUENE FROM CONTAMINATED WATER WITH MICROBIAL MATS

M. Mondecar, J. Bender, J. Ross, W. George, and J. Preslan

INTRODUCTION

There are currently thousands of toxic military sites, about 40% of which are contaminated with 2,4,6-trinitrotoluene (TNT) (U.S. Congress, Office of Technology Assessment 1991). Many of these disposal sites are on the U.S. Environmental Protection Agency's (EPA) list of national priorities for urgent cleanup. TNT has been shown to be toxic to a number of organisms (Kaplan & Kaplan 1982a), and has been classified into EPA Group C, as a possible human carcinogen (Gordon & Hartley 1989).

Many treatments have been used to remove TNT from the environment, such as incineration and detonation. Newer technologies, such as biotransformation and biodegradation, may be used more successfully on contaminated soils. However, the biotransformation products also may be toxic (Kaplan & Kaplan 1982a). Composting of explosives-contaminated soils and use of other microbial consortia to degrade TNT are currently showing some success (Kulpa & Wilson 1991, Unkefer et al. 1990, Williams 1989). Because microbial consortia contain a wide variety of microbial groups, they may represent the preferred biological treatment system for complete degradation of TNT. In mixed microbial ecosystems, the metabolic end product of one group often becomes the substrate for another member of the consortia.

Biosorption may be another cost-effective treatment for munitions cleanup. TNT and its biodegradation products are highly sorptive, binding to humus fractions in soil (Kaplan & Kaplan 1982b, 1983). The mixed consortium used in this research (silage microbial mats) has been shown to degrade petroleum distillates (Phillips et al. 1994) and chlordane (Bender et al. 1994) and to sequester heavy metals from contaminated water and sediments (Bender et al. 1989).

METHODS

Bacterial consortia (BC) were isolated from TNT-contaminated soil collected from Bangor Naval Submarine Base, Washington (Mondecar et al. 1992). These consortia were developed for tolerance to the target substrate by incubating with increasing concentrations of TNT, added as an acetone solution (TNT does not dissolve readily in aqueous media). Similarly, TNT-tolerant *Oscillatoria* spp. (TNT-OS), isolated from microbial mats obtained from J. Bender, were developed by incubation with TNT. New mats (TNT-mats) were constructed from the BC and TNT-OS by coculturing all tolerant microbes together with ensiled grass clippings. The ensiled grass added fermentative bacteria to the consortium.

Two types of microbial materials were tested for TNT degradation: the BC group and the constructed mats, described above. Additionally, cometabolism studies were performed with the BC, using a 0.1% benzoic acid solution along with the TNT.

Bacteria and mats were supplemented with minerals. Except for cometabolism experiments (and having TNT added as an acetone solution for bacteria), no carbohydrate additions were made. Because the mats are photosynthetic, they produced an internal supply of energy molecules and did not depend on the TNT nutrient value during light periods. All mat cultures were maintained in a 14:10 light:dark cycle.

Degradation of TNT was examined by an NaOH plate assay (Osmon & Klausmeier 1972) and by high-performance liquid chromatography (HPLC). The HPLC analyses were performed on a Perkin-Elmer C-18 reverse-phase column with methanol/water (1:1) as the mobile phase and a flowrate of 1 mL/min. Metabolites were derivatized with trifluoroacetic acid anhydride. Compounds were detected with an LC85B Perkin-Elmer Ultraviolet Detector at 254 nm.

RESULTS AND DISCUSSION

HPLC analyses of TNT (175 mg/L) treated with BC showed a 75 to 97% degradation in 12 days. When grown in 100 mg/L TNT, supplemented with benzoic

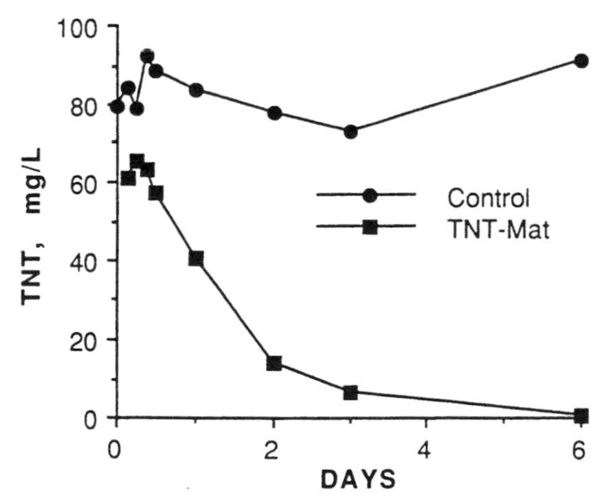

FIGURE 1. Degradation of TNT by a TNT-tolerant microbial mat.

acid, an 80% degradation was achieved in 11 days. Therefore, cometabolism did not improve the degradative efficiency of the BC.

TNT-mats exposed to 100 mg/L TNT showed >99% degradation in 6 days (Figure 1). Up to 30 minutes elapsed between pouring the TNT medium into dishes and taking the first measurement in any experiment. During this time period, adsorption to the mat could have occurred and perhaps all of the TNT was not in solution at that point, thus the initial concentration would be lower than the expected 100 mg/L. In this experiment, more than 90% of the TNT was degraded by day 3. In separate experiments with lowered concentrations of TNT (50 mg/L), the TNT-mats showed >99% degradation in less than 1 day.

Four expected metabolites increased with time, but their concentrations remained low. Combined concentrations of all detected metabolites never exceeded 10 mg/L, suggesting that further metabolism of these materials was occurring. The metabolites detected during the degradation process were 2-amino-4,6-dinitrotoluene (2-amino-DNT); 4-amino-2,6-dinitrotoluene (4-amino-DNT); 2,4-dinitrotoluene (2,4-DNT); and 2,6-dinitrotoluene (2,6-DNT). Due to the HPLC detection limits, the occurrence of trinitrobenzene (TNB) and *m*-dinitrobenzene (*m*-DNB) is uncertain. If present, the concentrations were <1 mg/L.

The NaOH plate assay gave qualitative indications of the degradation of 100 mg/L TNT by the BC group and the TNT-mats. Preliminary mineralization experiments with ^{14}C-labeled TNT showed no mineralization of TNT in a 23-day period. Expanded mineralization experiments are planned.

CONCLUSIONS

The BC group degraded TNT at rates that compared favorably with those found in similar studies thus far. TNT-mats showed much higher rates of degradation than did the BC group. Although both BC and TNT-mats were consortial groups of microbes, the TNT-mats probably contained a greater number of microbial species. In addition, the laminated structure of the mat creates discrete regions of unique chemistry characterized by oxic and anoxic regions in close proximity (Bender et al. 1989). These zones harbor anaerobic and aerobic bacteria, producing a multifunctional unit that likely degrades recalcitrant organics, such as TNT, more effectively.

Although the detected metabolites indicate a low-level persistence of aromatic compounds, those metabolites were never detected in high concentrations. The total mass of the metabolites was far less than the initial TNT. Additionally, HPLC analysis did not indicate any other metabolites. This suggests that the TNT was degraded into products not detectable by HPLC methods used in this research.

ACKNOWLEDGMENT

This project was supported by U.S. Department of Defense Grant #2, Order #89-116, 88-150 to the senior author while at Xavier University of Louisiana.

REFERENCES

Bender, J., R. Murray, and P. Phillips. 1994. "Microbial Mat Degradation of Chlordane." In J. L. Means and R. E. Hinchee (Eds.), *Emerging Technology for Bioremediation of Metals.* Lewis Publishers, Ann Arbor, MI.

Bender, J., E. R. Archibold, V. Ibeanusi, and J. P. Gould. 1989. "Lead Removal from Contaminated Water by a Mixed Microbial Ecosystem." *Water Science and Technology* 21:1661-1665.

Gordon, L., and W. R. Hartley. 1989. *Health Advisory on 2,4,6-Trinitrotoluene.* U.S. Environmental Protection Agency, Washington, D.C.

Kaplan, D. L., and A. M. Kaplan. 1982a. "2,4,6-Trinitrotoluene-Surfactant Complexes: Decomposition, Mutagenicity, and Soil Leaching Studies. *Environ. Sci. Technol. 16*:566-571.

Kaplan, D. L., and A. M. Kaplan. 1982b. "Thermophilic Biotransformations of 2,4,6-Trinitrotoluene under Simulated Composting Conditions." *Appl. Environ. Microbiol. 44*:757-760.

Kaplan, D. L., and A. M. Kaplan. 1983. *Reactivity of TNT & TNT-Microbial Reduction Products with Soil Components.* Technical Report Natick/TR-83/041, Natick, MA.

Kulpa, C., and M. Wilson. 1991. "Degradation of Trinitrotoluene by a Mixed Microbial Culture Isolated from Soil." Abstract, 91st Annual Meeting Am. Soc. Microbiol., Dallas, TX.

Mondecar, M., J. Bender, J. Ross, W. George, and K. Dummons. 1992. "Bioremediation of 2,4,6-Trinitrotoluene (TNT) by a Mixed Microbial Ecosystem." Abstract, 92nd Annual Meeting Am. Soc. Microbiol., New Orleans, LA.

Osmon, J. L., and R. E. Klausmeier. 1972. "The Microbial Degradation of Explosives." *Rev. Ind. Microbiol. 14*:247-252.

Phillips, P., J. Bender, J. Word, D. Niyogi, and B. Denovan. 1994. "Mineralization of Naphthalene, Phenanthrene, Chrysene, and Hexadecane With a Constructed Silage Microbial Mat." In R. E. Hinchee, D. B. Anderson, F. B. Metting, Jr., and G. D. Sayles (Eds.), *Applied Biotechnology for Site Remediation.* Lewis Publishers, Ann Arbor, MI.

Unkefer, P. J., M. A. Alvarez, J. L. Hanners, C. J. Unkefer, M. Stenger, and E. A. Margiotta. 1990. "Bioremediation of Explosives." In *JANNAF Safety and Environmental Protection Subcommittee Workshop: Alternatives to Open Burning/Open Detonation of Propellants and Explosives*, pp. 307-326. Chemical Propulsion Information Agency, Laurel, MD.

U.S. Congress, Office of Technology Assessment. 1991. *Complex Cleanup: The Environmental Legacy of Nuclear Weapons Production.* OTA-0-484, Washington, D.C.

Williams, R. T. 1989. "Composting of Explosives Contaminated Sediments." Abstract, Proceedings for the Workshop on Composting of Explosives Contaminated Soils. Proceedings Report No. CETHA-TS-SR-89276, New Orleans, LA.

BIOTREATABILITY EVALUATION OF SEDIMENT CONTAMINATED WITH THE EXPLOSIVE PENTAERYTHRITOL TETRANITRATE (PETN)

C. M. Swindoll, G. M. Seganti, and G. O. Reid

INTRODUCTION

Pentaerythritol tetranitrate (PETN), an aliphatic nitrate-ester explosive used principally in primers, the explosive core of detonating cord, and the base charge of blasting caps, was manufactured from 1967 to 1989 at an industrial facility near Denver, Colorado. PETN can be characterized as a sensitive, powerful high explosive with a low water solubility (0.01 g/100 mL water at 50°C) and a specific gravity of 1.77. As part of the manufacturing process, effluent wastewater contaminated with PETN and a surfactant was discharged into two surface impoundments (settling ponds).

The two settling ponds have been designated as Resource Conservation and Recovery Act (RCRA) solid waste management units (SWMUs) and will be decontaminated and closed as part of a larger remediation effort at the facility. The ponds, each 119 by 119 m, consist of a high-density polyvinyl chloride liner covered with approximately 22 to 36 cm of clayey soil and 1 to 1.5 m of water. A sampling and analysis program showed a highly heterogeneous distribution of PETN within the pond sediment, with concentrations ranging from 1 to over 200,000 mg/kg. Several remediation technologies were evaluated for the ponds, including thermal oxidation, chemical reaction, and bioremediation. This report summarizes the status of the bioremediation evaluation for the ponds.

METHODS

A biotreatability investigation consisting of laboratory and field studies was initiated in 1990. As part of the laboratory evaluation, a bioreactor was used to simulate full-scale operation of a bioremediation system. The bioreactor consisted of a 200-L drum lined with heavy plastic, a mixer, and an air pump with diffuser. As designed, the mixer suspended PETN-contaminated sediment, and the sparger supplied atmospheric oxygen for aerobic biodegradation. Approximately 45 kg of PETN-contaminated pond sediment was placed in the bioreactor with 150 L of well water. The volume of water was maintained with additions of well water.

Initially, 1 g/L disodium phosphate was added as a source of phosphorus, and a total of 500 g disodium phosphate was added over the course of the study; there were no other nutrient amendments. Calcium carbonate was added as needed to maintain the pH above 6.0.

Sediment and water samples were collected from the bioreactor for analyses of PETN, surfactant, phosphate, and ammonium and for bacteria enumeration. Prior to collecting the samples, the mixer and sparger were stopped, and the suspended material was allowed to settle for approximately 4 hours. A liquid extraction procedure using acetonitrile, coupled with a high-pressure liquid chromatograph (HPLC, Hewlett-Packard 1050), was used to analyze PETN. The surfactant was analyzed using the Chemetrics, Inc., sewage water test for detergents. Phosphate and ammonium concentrations were determined using Hach analytical kits. Temperature and pH were measured with a thermometer and laboratory pH meter. The total heterotrophic bacteria were enumerated on Difco nutrient agar using pour plate and dilution series techniques. PETN-utilizing bacteria were enumerated by spread-plating dilutions of aqueous samples on a mineral salts medium with PETN as the sole carbon source. Microbial enumeration plates were replicated in triplicate and incubated in the dark at room temperature.

RESULTS AND DISCUSSION

At the onset of this investigation, it was expected that PETN could be biodegraded under aerobic conditions due to the biodegradability of similar nitroexplosives. However, in the first bioreactor study, PETN degradation was not conclusively demonstrated over a 90-day period. A significant increase in the number of heterotrophic bacteria during the study, from an initial 10^3 colony-forming units per gram of sediment (CFU/g) to 10^7 CFU/g, indicated that carbon sources other than PETN were being consumed by the bacteria. Based on these results, the hypothesis that preferential biodegradation may be occurring was formulated, with the surfactant or other readily biodegradable compounds being consumed prior to the more recalcitrant PETN. A second bioreactor study was conducted to test this hypothesis.

In the second bioreactor study, both PETN and surfactant concentrations were significantly reduced over a 180-day study period (see Figure 1). The surfactant decreased from 13.2 mg/L at day 1 to below the 1 mg/L detection limit at day 70. The steady decrease in the surfactant was attributed to biodegradation. Although surfactant volatilization may have contributed to surfactant reduction, the persistence of the surfactant for several years in the settling ponds indicates that volatilization is not a significant fate process for the surfactant.

PETN at the beginning of the study averaged 16,533 mg/kg and did not show significant reductions over the first 70 days of the study. However, once surfactant concentrations were depleted to nondetectable levels, PETN concentrations began to decrease. From day 70 to day 165, PETN decreased from approximately 16,500 to 356 mg/kg, a 98% reduction. The chromatographs used to quantify PETN indicated that intermediate breakdown compounds were produced prior to the

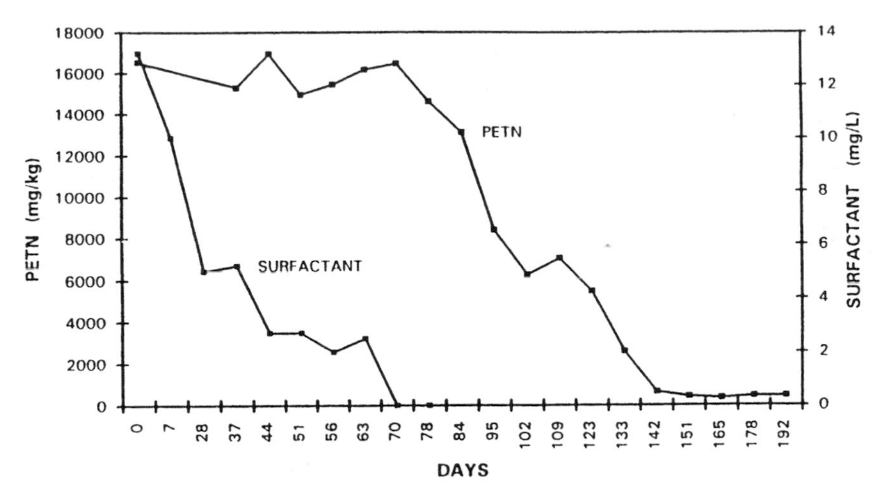

FIGURE 1. Mean PETN and surfactant concentrations over the course of the bioreactor study.

full mineralization of PETN. The concentrations of the breakdown products were highly variable and showed no clear correlation with PETN concentrations. As PETN concentrations decreased, the pH steadily decreased from greater than 6.5 to less than 5.8. The rate of PETN degradation was significantly reduced at a lower pH; therefore, sodium carbonate was used to maintain the pH at a near neutral level.

Once the PETN concentration reached approximately 350 mg/kg (day 165), no further reduction in PETN occurred during the remainder of the 180-day study; additional amendments of phosphate and sodium carbonate did not result in further PETN degradation. An unidentified condition, such as a nutrient deficiency or inhibiting by-product, may have limited further PETN degradation.

Based on the laboratory studies, it was concluded that bioremediation could be effectively implemented to decontaminate the two ponds. To test this conclusion, disodium phosphate was added and maintained in the ponds at 5 mg/L phosphorus. Following the addition of phosphate, surfactant concentrations in the ponds decreased from a mean of 71.5 to 16.5 mg/L within 6 weeks, and to 5.6 mg/L after 6 months.

To enhance the biodegradation of PETN, it was concluded that a bioremediation system would need to periodically suspend the contaminated sediment in an aerated water column, provide sufficient phosphate, and maintain a nearly neutral pH. A pilot-scale test using a 1000-m^2 test cell constructed in one of the ponds was initiated to obtain the parameters needed for the full-scale design and operation of the bioremediation system. Phosphate was maintained at 5 mg/L, and two 2-HP aeromixers were used to suspend the sediment and aerate the water column. Initial pilot cell results indicated that complete removal of the surfactant is not a prerequisite for PETN degradation. After 2 months of operation, PETN concentrations were reduced from an initial level of 3,639 mg/kg to less than 1,900 mg/kg, while surfactant concentrations remained less than 5 mg/L. Based on the bench- and pilot-scale studies, bioremediation appears a viable alternative for reducing PETN concentrations in the ponds. The full-scale implementation of bioremediation for the ponds is pending regulatory agency approval.

METHANOTROPHIC TREATMENT TECHNOLOGY FOR GROUNDWATER CONTAINING CHLORINATED SOLVENTS

D. R. Jackson and T. D. Hayes

INTRODUCTION

A program is being undertaken to conduct research and development for demonstration of a new groundwater technology based on the ability of methanotrophic microorganisms to degrade trichloroethylene. This paper describes the market for, the basic research for, the bench-scale development of, and the pilot-scale demonstration of a granular activated carbon fluidized-bed reactor (GAC-FBR) based on methanotrophic treatment technology (MTT).

MARKET DEFINITION

Groundwater contamination by chlorinated hydrocarbons, particularly trichloroethylene (TCE), is a nationwide problem. Currently, more than 1,200 final and proposed U.S. Environmental Protection Agency (U.S. EPA) Superfund sites exist in the United States (U.S. EPA 1990). Almost 40% of these sites are contaminated with chlorinated hydrocarbons. Superfund sites contaminated with these compounds are located in almost every major population center in the United States (Jackson et al. 1991). Most of these sites are within reach of the national pipeline grid, thus ensuring the availability of natural gas required for MTT.

We surveyed the waste disposal sites in the United States to estimate the total number of sites and the number of sites contaminated with chlorinated hydrocarbons (Jackson et al. 1991). From this survey, we estimate that 4.7 million tons of chlorinated hydrocarbons are present in current and future Superfund sites, U.S. Department of Defense (DOD) and U.S. Department of Energy (DOE) sites, and Resource Conservation and Recovery Act (RCRA) sites. The U.S. EPA Office of Technology Assessment has estimated that $500 billion will be spent on cleanup of hazardous waste sites in the next 50 years within the United States. Clearly, innovative and economical strategies are needed to assist in the massive remediation efforts that lie ahead. MTT is designed to meet this need.

METHANOTROPHIC TREATMENT TECHNOLOGY DESCRIPTION

MTT is based on the stimulation of natural methane-oxidizing methanotrophic bacteria to produce methane monooxygenase which is capable of degrading recalcitrant chlorinated solvents for ex situ and in situ application. Since 1985, the Gas Research Institute (GRI) has supported Radian Corporation and researchers from Cornell University, Stanford University, and Michigan Biotechnology Institute (MBI) to develop an ex situ methanotrophic process for destruction of chlorinated solvents, such as trichloroethene (TCE), in contaminated groundwater. In addition, Envirex Ltd. has constructed a 30-gpm methanotrophic fluidized-bed reactor for a pilot-scale demonstration. An in situ application of MTT was demonstrated at the Westinghouse Savannah River Laboratory (WSRL) in Aiken, South Carolina.

Because methanotrophic treatment of contaminated groundwater was an unproven technology, the development effort began with a fatal flaw analysis and a technical-economic evaluation. Both of these studies were favorable to development of MTT, providing that critical research goals be accomplished within 5 years. These research goals defined the conversion efficiency, decomposition kinetics, and cleanup levels that were necessary to compete with conventional air stripping with granular activated carbon (AS/GAC) for emission control. A preliminary economic analysis indicated that MTT had the potential of being economically competitive with AS/GAC for aboveground treatment of groundwater contaminated with TCE (Radian Corporation 1988).

METHANOTROPHIC TREATMENT TECHNOLOGY DEVELOPMENT

Research efforts have focused primarily on development of an innovative high rate methanotrophic attached film expanded bed (MAFEB) reactor system to treat TCE in groundwater. The general concept is illustrated in Figure 1. Professor W. J. Jewell at Cornell University performed feasibility studies to define the growth characteristics of methanotrophic attached film in an expanded bed reactor with and without exposure to TCE. Results of these studies showed that the MAFEB reactor was highly stable while operating at single-pass hydraulic retention times as low as 15 seconds. The methanotrophs formed thin, dense films, exceeding 100 g volatile solids per liter of attached film with microbial concentrations exceeding 40 g of volatile solids per liter of expanded bed reactor. Maximum TCE degradation rates were up to 800 mg TCE/L-d (Jewell et al. 1990). These pioneering studies formed the basis for continuing development of MTT.

A continuous flow bench-scale GAC-FBR was constructed for TCE degradation studies at MBI (Wu et al. 1992). The 12-L, 3.6-m-deep glass reactor was constructed with evenly spaced sampling ports along the side of the reactor tube. Fluid samples can be withdrawn from these ports to determine dissolved methane, oxygen, and TCE concentrations. Methane-, oxygen-, and nitrogen-saturated water

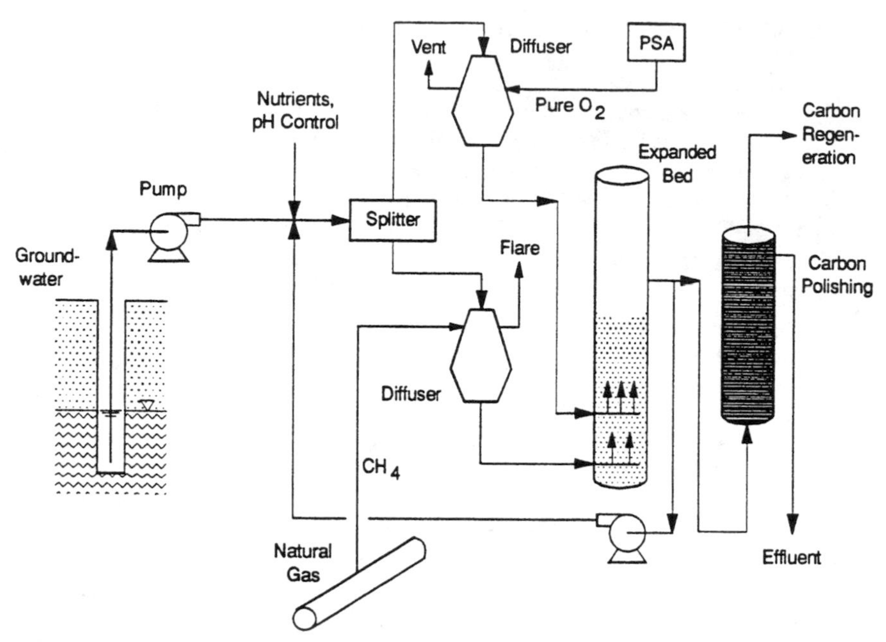

FIGURE 1. Concept diagram of the methanotrophic fluidized-bed reactor.

streams are prepared by circulating influent water through downflow bubble contact aerators. No effluent was recirculated. TCE feedwater was prepared by mixing neat TCE in a reservoir bottle to provide for complete dissolution. The feedwater was delivered into the reactor through a metering pump. Several runs were conducted using fine-grained GAC (12 × 40 mesh, 0.4 to 1.4 mm) and TCE concentrations varying between 250 and 3,300 µg TCE/L. The upflow velocity limited the flux of dissolved methane and oxygen supplied, so that only the lower ⅓ of the bed showed biological activity. After a change to coarser media (10 × 25 mesh, 0.7 to 1.7 mm) and higher flowrates, methane metabolism and TCE cometabolism were observed in the lower ⅔ of the bed (see Figure 2). The influent concentration of 430 µg TCE/L was reduced by 50% in an empty bed contact time of 3.3 minutes. The actual residence time of water in the pores of the bed was significantly lower.

METHANOTROPHIC TREATMENT TECHNOLOGY DEMONSTRATION

A pilot-scale, 30-gpm, GAC-FBR reactor was designed by a team of engineers in 1991. The process design includes features of the bench-scale unit. The height and bed depth of this reactor were 11.6 and 3.6 m, respectively, with a diameter of 0.6 m. This reactor, constructed by Envirex Ltd., was shipped to MBI where

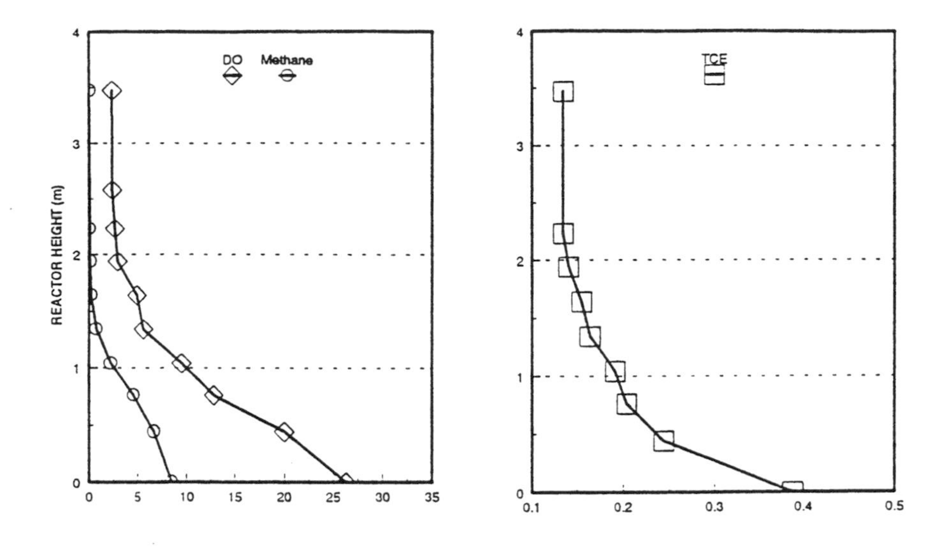

FIGURE 2. Dissolved oxygen, methane, and TCE profiles in the GAC-FBR at 42 days after startup.

it was hydraulically tested and is scheduled for initial startup in 1993. After completion of the testing phase, the reactor skid will be transported to the WSRL where it will be demonstrated using on-site groundwater as part of the DOE Integrated Demonstration Program.

REFERENCES

Jackson, D. R., K. M. Adams, B. M. Eklund, and G. E. Schrab. 1991. *Estimation of the Groundwater Remediation Market: Annual Topical Report.* Radian Corporation, Austin, TX.

Jewell, W. J. et al. 1990. *Methanotrophs for Biological Pollution Control: Feasibility of Developing an Attached Microbial Film Reactor and Kinetics of TCE Removal.* Final Report, Phase I and Phase II. Cornell University, Ithaca, NY.

Radian Corporation. 1988. *Phase I - Technical Feasibility of Methane Use for Water Treatment: Task 2. Preliminary Economic Analysis.* Final Report. Radian Corporation, Austin, TX.

U.S. Environmental Protection Agency. 1990. *Superfund Database.* Washington, DC.

Wu, W., J. Krzewinski, D. Wagner, and B. Hickey. 1992. *Pilot Test of the Granular Activated Carbon Fluidized Bed Reactor (GAC-FBR) for Methanotrophic TCE Degradation.* Quarterly Report. Michigan Biotechnology Institute, Lansing, MI.

EFFECTS OF HEAVY METALS ON THE BIODEGRADATION OF ORGANIC COMPOUNDS

P. Majumdar, S. Bandyopadhyay, and S. K. Bhattacharya

INTRODUCTION

Heavy metals such as zinc, copper, cadmium, chromium, lead, etc. commonly are found as copollutants with organic compounds in many industrial wastestreams. The inhibitory effects of metals on microbial transformation of organic compounds are of great importance in biodegradation of both solid and liquid wastes. Information on anaerobic degradation of specific compounds such as nitrobenzene and pentachlorophenol is available from literature (Haghighi-Podeh 1991, Yuan 1991). However, only a limited amount of research has been conducted to assess the biotransformation of organic wastes in the presence of heavy metals. The objective of this part of the project was to study the inhibitory effects of zinc on anaerobic biodegradation of nitrobenzene and pentachlorophenol using an acetate enrichment methanogenic culture. Other heavy metals that will be looked into in future include copper, cadmium, lead, and chromium.

MATERIALS AND METHODS

Serum bottle (50-mL culture) studies were performed with a stock acetate enrichment anaerobic culture following the method of Bhattacharya et al. (1988). A daily feed of 50 mL (1000 mg/L-d) of glacial acetic acid (99.7%; Fisher Scientific, Houston, Texas) was maintained for all the bottles until steady state was reached. The theoretical CO_2 and CH_4 production from 50 mL of acetic acid is 44 mL. Daily methane generations were determined by subtracting the theoretical CO_2 volume (22 mL) from the total gas measured daily. Cumulative volume of this methane generations over the period of the experiment were plotted in the figures. After the bottles reached steady state, calculated amounts of zinc chloride, nitrobenzene, or pentachlorophenol were added to the bottles to achieve the desired concentrations of the toxicants in the serum bottles. Table 1 shows the design of the experiments. Triplicate serum bottles were maintained at 35°C and the gas production was monitored daily using a gas measuring device following Bhattacharya et al. (1988).

TABLE 1. Experimental design and summary of serum bottle results.

Organic compound	Spiked Zinc (mg/L)	Total soluble Zn measured (mg/L)	Comments
Nitrobenzene (mg/L)			
00	00	0.00	control
10	00	0.00	no effects[(a)]
20	00	0.00	no effects
30	00	0.00	failed (no methane)
10	10	0.35	no effects
10	20	0.30	recovered[(b)]
10	30	–	failed (no methane)
20	10	0.33	recovered
20	20	0.35	failed (no methane)
20	30	0.35	failed (no methane)
30	10	0.20	recovered
30	20	0.10	recovered
30	30	–	failed (no methane)
Pentachlorophenol (mg/L)			
0.00	0.0	0.00	control
1.10	0.0	0.00	no effects
1.40	0.0	0.00	no effects
2.10	0.0	0.00	no effects
1.10	8.6	0.26	no effects
1.10	17.2	0.22	partial effect
1.10	25.8	0.29	failed (no methane)
1.40	8.6	0.18	no effects
1.40	17.2	0.19	failed (no methane)
1.40	25.8	0.32	failed (no methane)
2.10	8.6	0.22	partial effect
2.10	17.2	0.25	failed (no methane)
2.10	25.8	0.20	failed (no methane)

(a) Daily methane production equals theoretical value (22 mL).
(b) Initial drop in daily methane generation followed by a later increase to the theoretical value.

RESULTS AND DISCUSSION

Figures 1 through 3 show the anaerobic degradation of 10, 20, and 30 mg/L of nitrobenzene (NB) in the presence of 10, 20, and 30 mg/L of zinc, respectively. Toxicity and recovery were measured in terms of methane formation. Up to 20 mg/L of NB alone was not toxic, and up to 20 mg/L of zinc caused either no inhibition or only reversible inhibition of NB degradation. Interestingly, even though 30 mg/L of NB alone caused toxicity, addition of 10 and 20 mg/L of

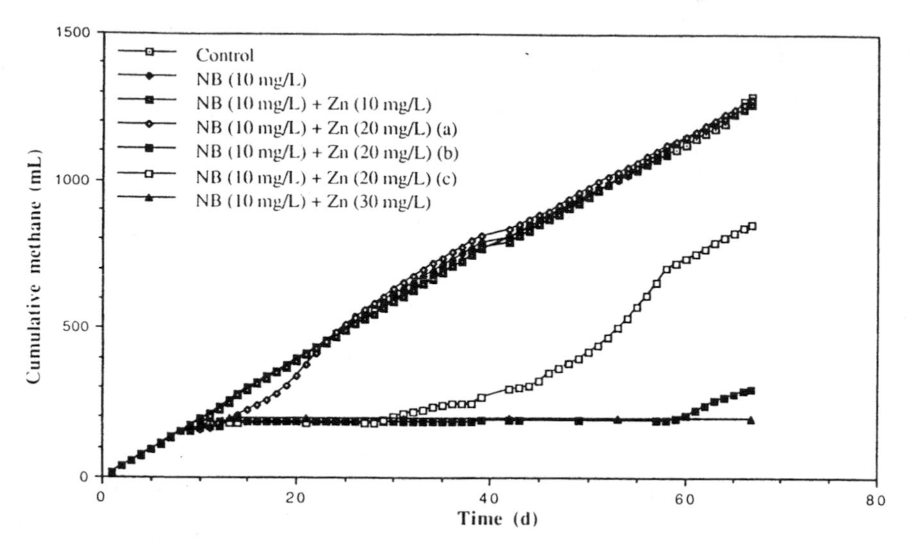

FIGURE 1. Effect of zinc on degradation of 10 mg/L of nitrobenzene. Note: (a), (b), and (c) denote triplicate bottles.

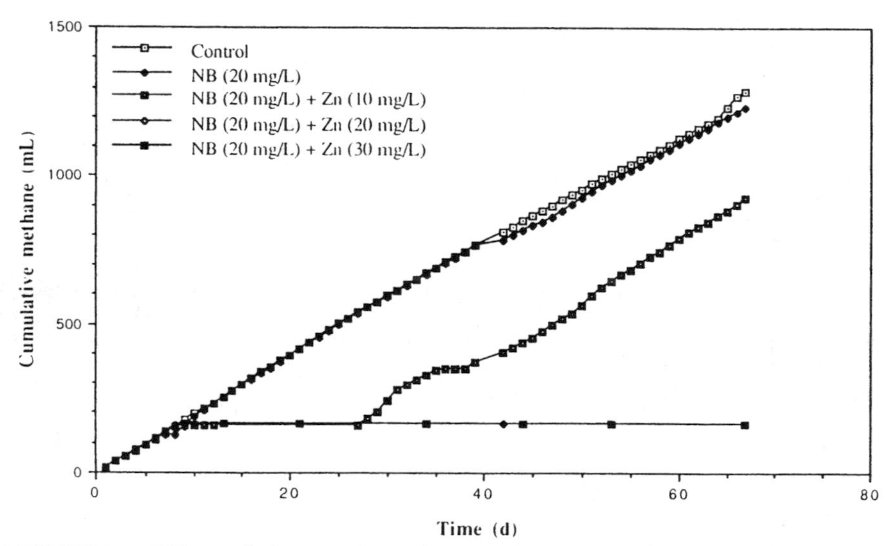

FIGURE 2. Effect of zinc on degradation of 20 mg/L of nitrobenzene.

zinc showed recovery of the process (Figure 3); 30 mg/L of zinc was toxic in all systems. It appears that zinc can behave as either a stimulant or as a toxicant in anaerobic NB degradation. With 10 mg/L of NB and 20 mg/L of zinc, the triplicate bottles showed widely varying gas production (more than 50% variation, see

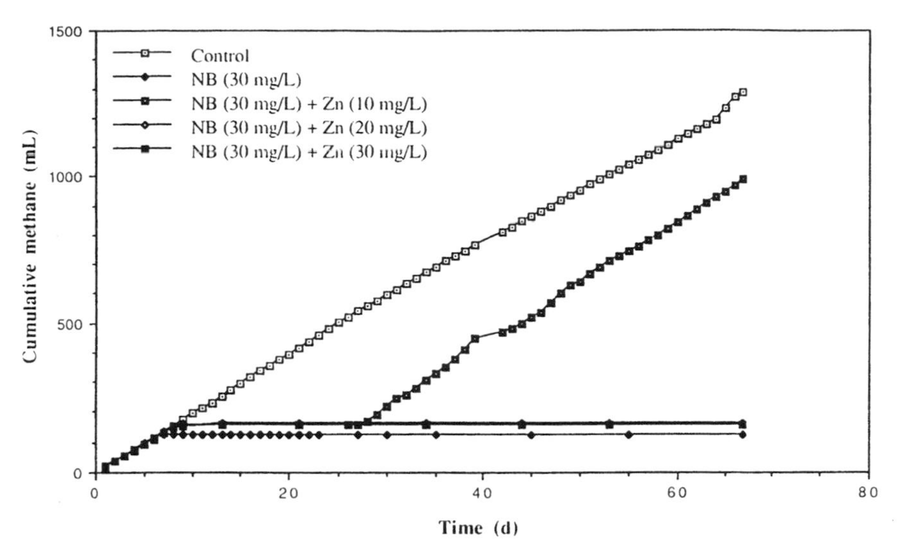

FIGURE 3. Effect of zinc on degradation of 30 mg/L of nitrobenzene.

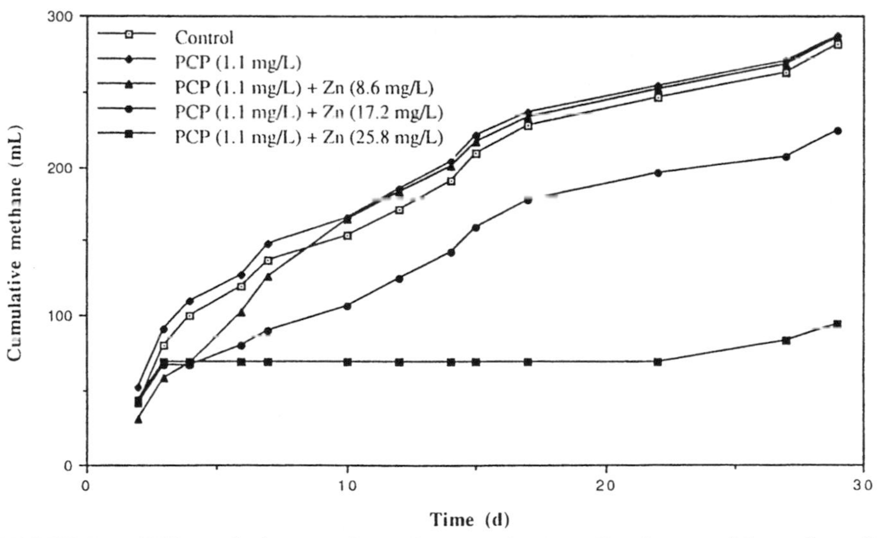

FIGURE 4. Effect of zinc on degradation of 8.6 mg/L of pentachlorophenol.

Figure 1). Because the gas production in all the other sets of triplicates varied within 10%, only the average values have been shown in the figures. High-performance liquid chromatograph (HPLC) data showed no detectable residual NB in the systems with no inhibition.

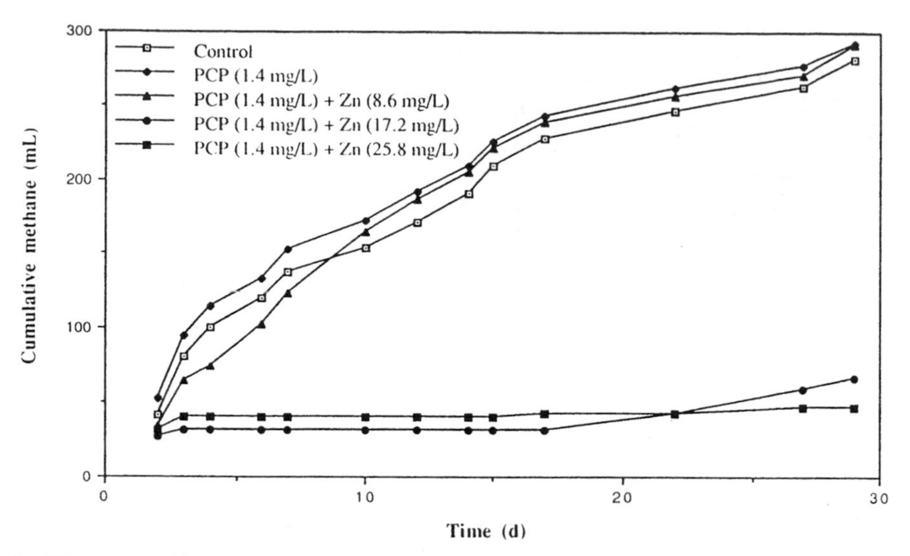

FIGURE 5. Effect of zinc on degradation of 17.2 mg/L of pentachlorophenol.

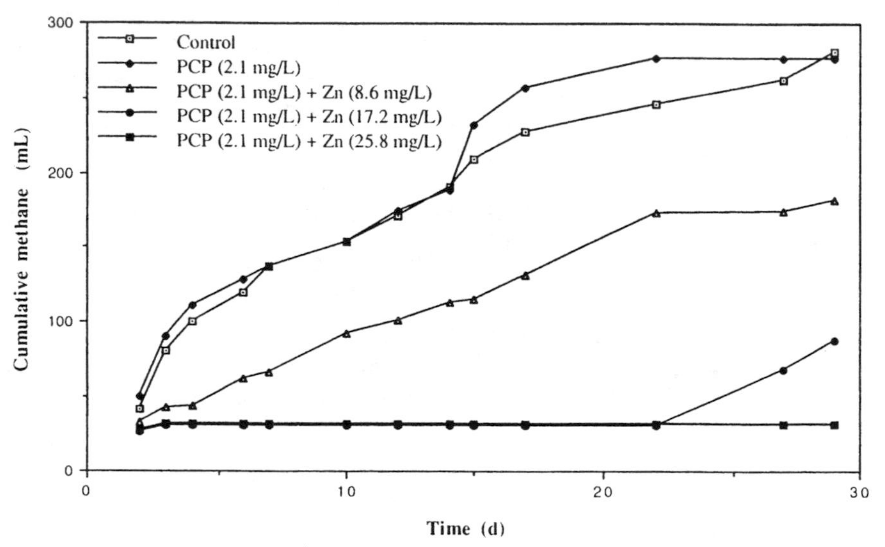

FIGURE 6. Effect of zinc on degradation of 25.8 mg/L of pentachlorophenol.

Figures 4 through 6 show the anaerobic degradation of 1.1, 1.4, and 2.1 mg/L of pentachlorophenol (PCP) in presence of 8.6, 17.2, and 25.8 mg/L of zinc, respectively. PCP alone was not toxic in the systems; HPLC data showed complete removal of PCP. With zinc concentration of 8.6 mg/L, the degradation of PCP up to 1.4 mg/L was not inhibited (Figures 4 and 5). The same zinc concentration showed partial effects on systems with 2.1 mg/L of PCP (Figure 6). Zinc at 17.2 and 25.8 mg/L showed either partial or complete inhibition of PCP degradation. Table 1 summarizes all the results. Measured total soluble zinc concentrations showed no correlation with toxicity (Table 1). This was expected based on recent free metal studies (Bhattacharya et al. 1988, Haghighi-Podeh 1991). Currently, the free zinc concentrations are being measured in all systems using a method combining dialysis and ion exchange following Haghighi-Podeh (1991).

REFERENCES

Bhattacharya, S. K., R. G. Janga, G. F. Parkin, and J. M. Morand. 1988. "Toxic Effects of Nickel on Anaerobic Treatment." In M. Astruc and J. N. Lester (Eds.), *Heavy Metals in the Hydrological Cycle*, pp. 577-582. Selper Ltd., London, UK.

Haghighi-Podeh, M. R. 1991. "Fate and Toxic Effects of Cobalt, Cadmium, and Nitrophenols on Anaerobic Treatment Systems." Ph.D. Thesis, Civil and Environmental Engineering Department, Tulane University, New Orleans, LA.

Yuan, Q. 1991. "Fate and Effects of Pentachlorophenol on Combined Anaerobic-Aerobic Treatment Systems." M.S. Thesis, Civil and Environmental Engineering Department, Tulane University, New Orleans, LA.

BIODEGRADATION OF HAZARDOUS ORGANIC COMPOUNDS BY SULFATE-REDUCING BACTERIA

M. M. Dronamraju and S. K. Bhattacharya

INTRODUCTION

Methanogens use carbon dioxide, acetate, or other organics as an electron acceptor, whereas sulfate-reducing bacteria (SRB) use sulfate as the electron acceptor. Both can compete for the same organic compound (substrate and electron donor). Although it is well established that SRB have an edge over methanogens, both kinetically and thermodynamically, comparatively much less research has been conducted on the ability of SRB to degrade hazardous organic compounds. It has been reported that much higher concentrations of CCl_4 can be degraded by SRB as compared to methanogens (Bhattacharya 1989). Other researchers have reported the effects of 2,4-dichlorophenol; 2,4-dinitrophenol; and 2-nitrophenol on methanogens (Haghighi-Podeh 1991, Jain 1992). Parkin et al. (1990) studied the effect of feed chemical oxygen demand (COD)/S ratio on the interaction between SRB and methanogens. The objective of this study was to determine the optimum COD/S ratio for growing SRB and to determine the percent degradation of selected concentrations of 2,4-dichlorophenol; 2,4-dinitrophenol; and 2-nitrophenol by the enrichment culture.

MATERIALS AND METHODS

Serum bottle studies were performed using a mixed culture of SRB and methanogens, being maintained in our laboratory for several years. The culture was obtained from the University of Iowa (Parkin 1991). Acetate was the feed carbon source. The selected range of feed COD/S ratios was varied between 1:1 and 80:1. The concentrations of 2,4-dichlorophenol; 2,4-dinitrophenol; and 2-nitrophenol used for spiking are 20, 40, 60 mg/L; 10, 20, 30 mg/L; and 10, 20, 30 mg/L, respectively. The gas volume (hydrogen sulfide + methane + carbon dioxide) was measured manometrically. Sulfate was analyzed by high-performance liquid chromatography (HPLC) equipped with a conductivity detector. The organic compounds were analyzed by HPLC equipped with a UV detector. The volatile acids were measured using gas chromatography (GC) or following a titration technique (Jenkins et al. 1983). Acetic acid was added daily to the serum bottles to maintain an acetic acid concentration of 1000 mg/L in the bottles.

RESULTS AND DISCUSSION

Results showed that sulfate utilization was maximum when the feed COD/S ratio was between 8:1 and 16:1 (Table 1). After additional experiments it was found that 10:1 ratio was optimum for maximum utilization of sulfate (Table 1). All subsequent experiments were performed with this feed COD/S ratio. Figure 1 shows the plot between cumulative gas production and time in days for SRB spiked with 2,4-dichlorophenol. The bottles were spiked on Day 13. There was no significant variation in gas production between the control bottles and spiked bottles even 27 days after spiking. This indicates that 2,4-dichlorophenol was not toxic to SRB up to a concentration of 60 mg/L. HPLC analyses for 2,4-dichlorophenol showed that for initial concentrations of 20, 30, and 40 mg/L, 80%, 93%, and 95%, respectively, of 2,4-dichlorophenol had been degraded 7 days after spiking.

Cumulative gas production vs. time plot for SRB spiked with 2,4-dinitrophenol is shown in Figure 2. The bottles were spiked on Day 13. From the plot, it is evident that 2,4-dinitrophenol was toxic to SRB even at a concentration of 10 mg/L. The bottles did not recover even after 27 days. The minor difference in gas volume in the 30 mg/L-bottle can be attributed to measuring cumulative gas production before all the bottles were fully stabilized. HPLC data indicated that no significant amount of 2,4-dinitrophenol was degraded.

TABLE 1. Feed COD/S ratios and corresponding amount of sulfate utilized.

Feed COD/S ratio	Sulfate utilized (mg/L) (over 8 days for Experiment 1 and over 5 days for Experiment 2)
Experiment 1	
1 : 1	180
2 : 1	60
4 : 1	180
8 : 1	420
16 : 1	420
40 : 1	350
80 : 1	245
Experiment 2	
8 : 1	50
10 : 1	290
12 : 1	110
14 : 1	230
16 : 1	230

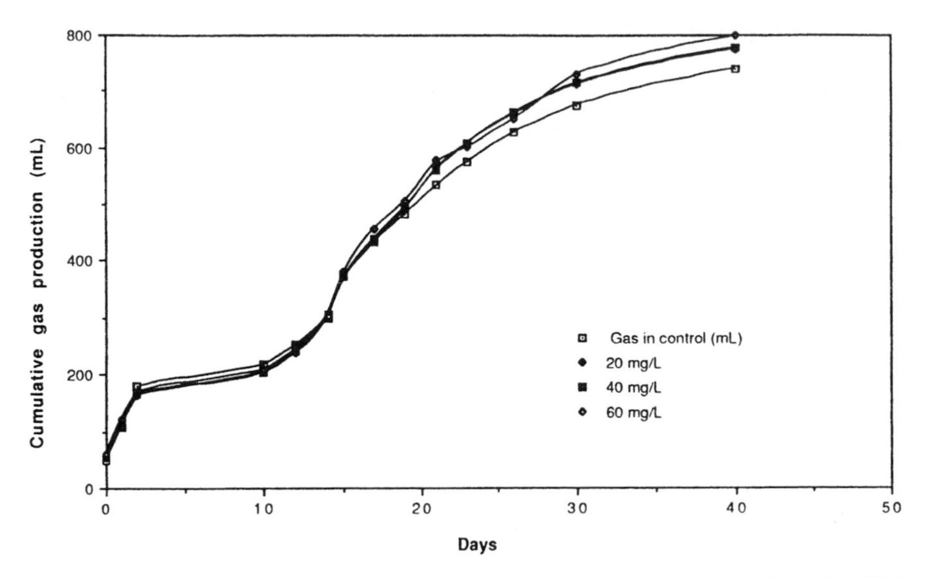

FIGURE 1. Plot of cumulative gas production against time in days for SRB spiked with 2,4-dichlorophenol.

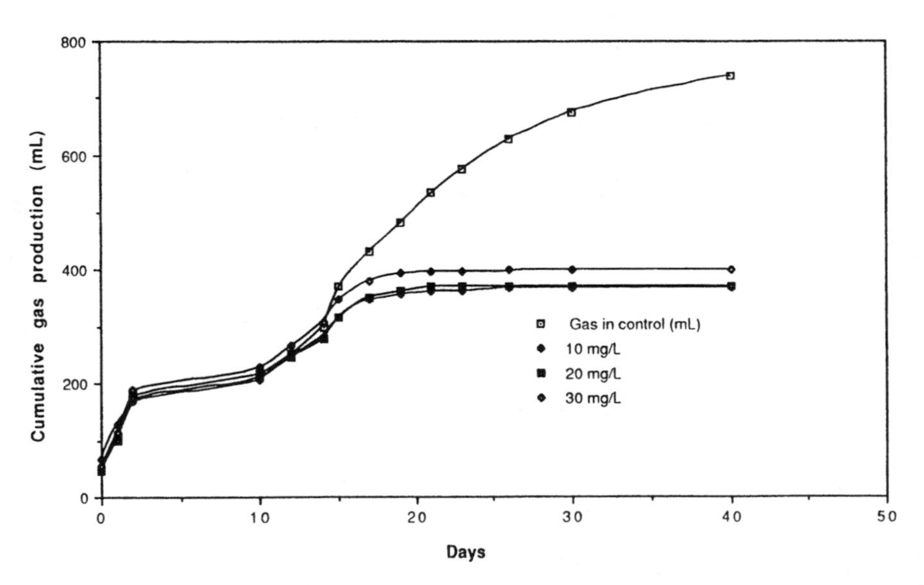

FIGURE 2. Plot of cumulative gas production against time in days for SRB spiked with 2,4-dinitrophenol.

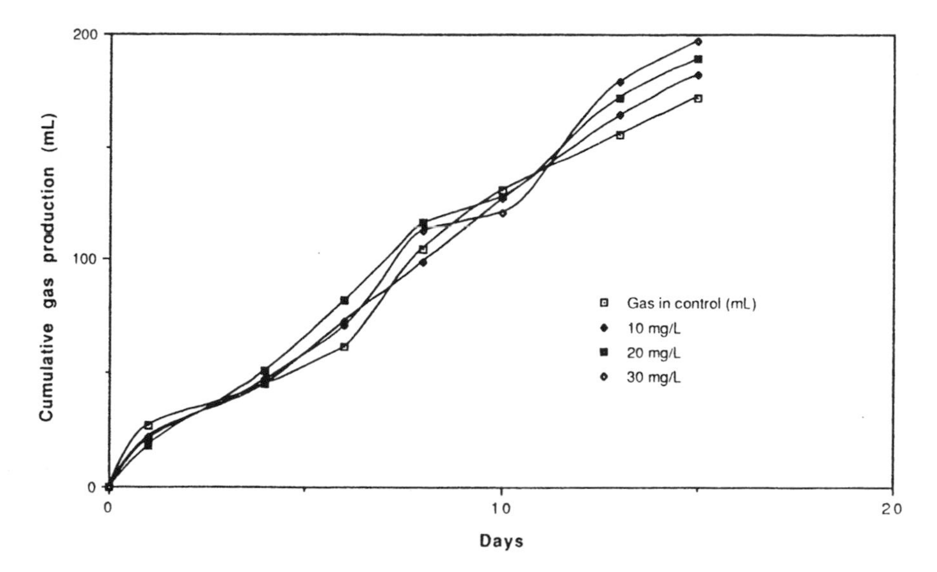

FIGURE 3. Plot of cumulative gas production against time in days for SRB spiked with 2-nitrophenol.

Figure 3 shows the plot of cumulative gas production against time in days for SRB spiked with 2-nitrophenol. The 2-nitrophenol was not toxic to SRB up to a concentration of 30 mg/L. HPLC data showed that when 10, 20, and 30 mg/L of 2-nitrophenol was spiked, 80%, 95%, and 90% was degraded over a period of 7 days. A gradual increase in total volume of gas was noticed in bottles spiked with 10, 20, and 30 mg/L over the control bottle, which must be due to the degradation of 2-nitrophenol.

CONCLUSIONS

The following conclusions were drawn from this study:

1. The optimum ratio of COD/S for maximum sulfate use was found to be 10:1.
2. At concentration levels of 60 mg/L or lower, approximately 80% to 95% of 2,4-dichlorophenol was found to be degraded by SRB.
3. 2,4-dinitrophenol was found to be toxic to SRB in as low a concentration as 10 mg/L.
4. At concentration levels of 30 mg/L or less, approximately 80% to 95% of 2-nitrophenol was found to be degraded by SRB.

ACKNOWLEDGMENT

This study was supported by the Board of Regents of Louisiana.

REFERENCES

Bhattacharya, S. K. 1989. "Fate and Effect of Carbon Tetrachloride and O-Xylene on Anaerobic Systems with Sulfate Reducing Bacteria." *Presented at 1989 National Conference on Environmental Engineering*, Sponsored by Environmental Engineering Division, ASCE, Austin, TX.

Haghighi-Podeh, M. R. 1991. "Fate and Toxic Effects of Cobalt, Cadmium, and Nitrophenols on Anaerobic Treatment Systems." Ph.D. Dissertation, Civil and Environmental Engineering Department, Tulane University, New Orleans, LA.

Jain, V. 1992. "Fate and Effect of 2,4-Dichlorophenol on Anaerobic Treatment Systems." M.S. Thesis, Tulane University, New Orleans, LA.

Jenkins, S. R., J. M. Morgan, and C. L. Sawyer. 1983. "Measuring Anaerobic Sludge Digestion and Growth by a Simple Alkalimetric Titration." *Journal of Water Pollution Control Federation 55* (2).

Parkin, G. F. 1990. Personal communication.

Parkin, G. F., N. A. Lynch, W. C. Kuo, V. Keuren, and S. K. Bhattacharya. 1990. "Interaction Between Sulfate Reducers and Methanogens Fed Acetate and Propionate." *Journal of Water Pollution Control Federation 62*(6): 780-788.

BIODEGRADATION OF TNT (2,4,6-TRINITROTOLUENE) IN CONTAMINATED SOIL SAMPLES BY WHITE-ROT FUNGI

A. Majcherczyk, A. Zeddel, and A. Hüttermann

INTRODUCTION

Until now, TNT (2,4,6-trinitrotoluene) has been the predominant conventional explosive used by military forces and for mining technologies. The disposal practices by manufacturing and loading facilities over the past 50 years have led to various levels of contamination of soils and groundwater (Pennington & Patrick 1990). In many countries, but especially in Germany, the destruction of TNT-producing plants during and after World War II left hundreds of highly contaminated sites and, consequently, highly contaminated groundwater reservoirs and sea and river sediments. The potential hazard of TNT and other nitroaromatics to human health and their toxicity for higher and lower animals and plants has been demonstrated in recent years (e.g., Rickert et al. 1984). In addition, the mutagenic activity of this compound has been demonstrated by experiments with bacteria (Hankenson & Schaeffer 1991).

The nitroaromatic compounds (e.g., TNT and dinitrotoluene) are recalcitrant to complete degradation in the environment. The abilities of biological systems to effect their degradation and other possible means of diminishing its concentration, e.g., photodegradation, were the subjects of many studies (Parrish 1977, Lipczynska-Kochany 1991, Kaake et al. 1992). The usual transformation pathway of these compounds by biological systems is the reduction of one or two nitrogroups to amino and hydroxyamino residues, but compounds are produced with at least the same toxicity as the original ones. The biological interactions influence the further transformation of these products to azoxycompounds and probably azoxypolymers, developing a source of long-term release of toxic compounds into the environment (McCormick et al. 1978). Even at low concentrations, TNT was reported to be inhibitory to many bacteria, actinomycetes, yeast, and fungi and until the last decade only a few organisms had been found that could degrade this compound (Parrish 1977, Naumova et al. 1982).

Positive results were reported on the application of enriched microbial cultures from sludge and the use of *Pseudomonas* species in degrading TNT. The mineralization of TNT was detected at only very low levels, and transformation of TNT to macromolecular structures of the polyamide type, formed by reaction of biotransformation products with lipids and protein constituents of the microflora, were published (Carpenter et al. 1978).

Studies during recent years have shown a high degree of degradation of TNT and dinitrotoluenes by the white-rot fungus *Phanerochaete chrysosporium* (Fernando et al. 1990, Valli et al. 1992). The present communication reports on studies of degradation of TNT by other white-rot fungi, *Pleurotus ostreatus* and *Trametes versicolor*.

MATERIALS AND METHODS

For a preliminary study in liquid cultures, 12-day-old stationary cultures of *Pleurotus ostreatus* on 50 mL of basal salts medium (BSM) under ambient air conditions and 25°C were used for experiments. Above culture conditions were also used for experiments. *Phanerochaete chrysosporium* was cultivated at 35°C and atmosphere of 70% oxygen in 50-mL nutrient nitrogen-limited medium. The 6-day-old stationary cultures were incubated with TNT. Trinitrotoluene was added to all cultures to give a final concentration of 100 mL/L. Cultures, in triplicates, were harvested after 0, 4, 16, 24, 45, and 120 hours, and the mycelia were separated from the culture medium. Both mycelia and medium filtrates were extracted three times with 20 mL of dichloromethane. Internal standards (dinitrobenzene and nitroaniline) were added before the extraction step. Samples were analyzed with gas-chromatography/mass spectrometric detection (GC-MSD), and the TNT concentration was calculated according to the recovery of the internal standard.

Contaminated soil sample was extracted with dichloromethane in Soxhlet and resulting extract evaporated. In parallel experiments, under analogue culture conditions, an extract was added from original TNT-contaminated soil in a concentration corresponding to 50 mg/L of pure TNT. Samples were processed as described above.

In the next step, the studies were extended to original soil samples contaminated with TNT collected from a manufacturing facility destroyed in 1945. Samples of soil containing 800 mg/kg TNT were mixed with water to obtain a slurry, supplemented with potato pulp, and solidified with wood chips to an absolutely homogeneous, crumbly structure. The material was inoculated with 3% millet culture of fungi and incubated at 25°C. *Pleurotus ostreatus* used for the preliminary studies in liquid cultures and additionally *Pleurotus* spec. and *Trametes versicolor* were used for these experiments. Soil samples prepared in this way had been fully penetrated by fungi after 2 weeks. Control samples were treated as above but not inoculated with fungi. Additionally, untreated soil samples were incubated in the same conditions. Samples of soil (inoculated, uninoculated and untreated) were extracted and analyzed after 2, 4, and 6 weeks. Concentration of TNT was calculated as percentage of untreated soil samples.

RESULTS AND DISCUSSION

Results of TNT degradation in liquid cultures are presented in Figures 1a, 1b, 2a, and 2b. *Pleurotus* and *Phanerochaete* were able to degrade TNT in only 120 hours but differed significantly in the kinetics of the process. *Pleurotus*

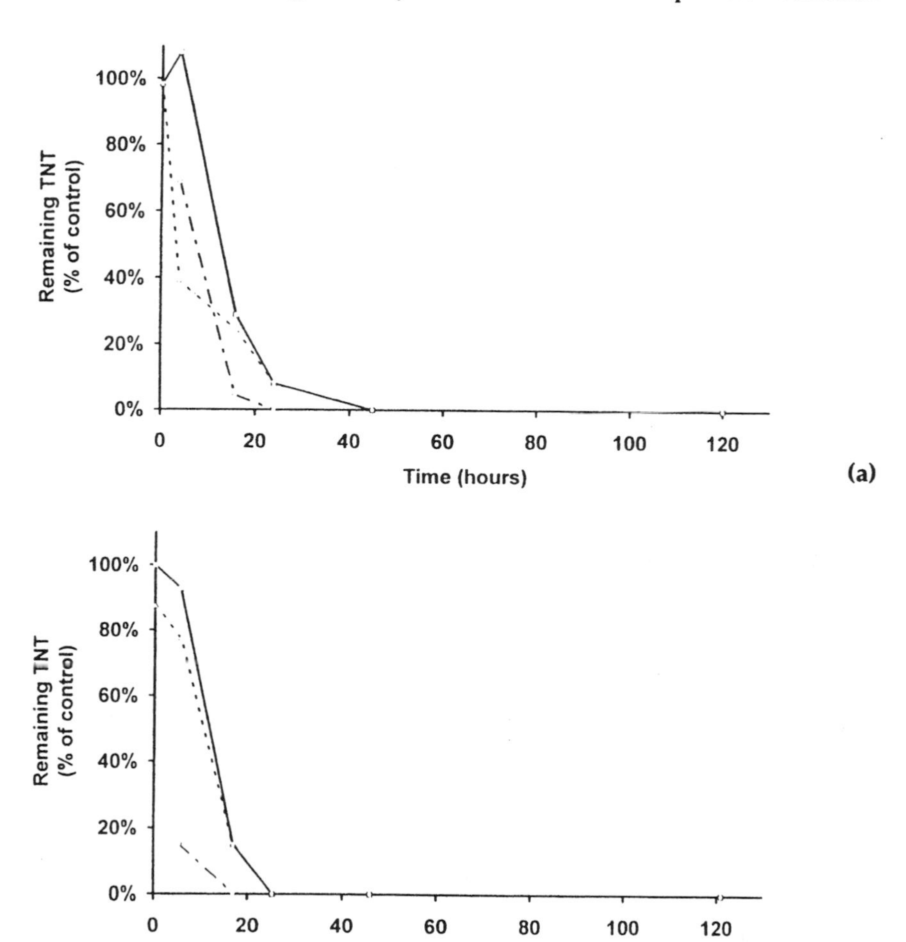

FIGURE 1. (a) Degradation of synthetic TNT in stationary liquid cultures of *Pleurotus ostreatus*. Starting concentration of TNT was 100 mg/L. Determined in triplicate as a disappearance of 2,4,6-trinitrotoluene in culture media and as bounded to mycelia; calculated as percentage of corresponding control flasks without fungi. Solid line: total TNT; short dashes: TNT in culture medium; short dashes and long dashes: TNT found in filtered and washed mycelia. (b) Degradation of TNT added as an extract from soil samples in stationary liquid cultures of *Pleurotus ostreatus*. Starting concentration of TNT was 50 mg/L; other parameters as in Figure 1a.

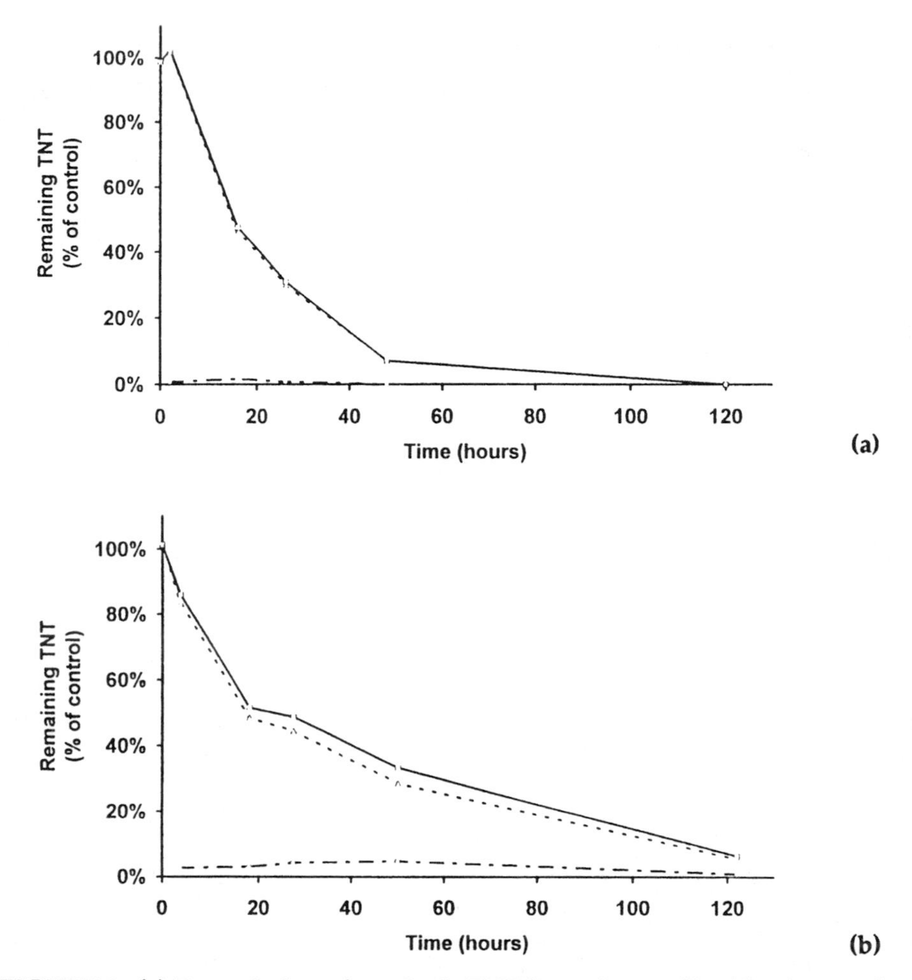

FIGURE 2. (a) Degradation of synthetic TNT in stationary liquid cultures of *Phanerochaete chrysosporium*. For parameters see Figure 1a. (b) Degradation of TNT added with an extract from original soil sample in stationary liquid cultures of *Phanerochaete chrysosporium*. Starting concentration of TNT was 50 mg/L; other parameters as in Figure 1a.

ostreatus degraded TNT almost completely after only 48 hours, in both the synthetic compound and soil extracts. *Phanerochaete chrysosporium* in this time degraded about 93% and 70% of TNT respectively. Mineralization of TNT was not studied, and the degradation was interpreted as disappearance of TNT altogether. The degradation pathways of these two fungi probably differ. A large amount of TNT was found to be adsorbed to the mycelium of *Pleurotus* after only 4 h. The mycelium of *Phanerochaete* did not adsorb TNT significantly. The degradation of TNT in the original soil extract was slower than in the synthetic compound.

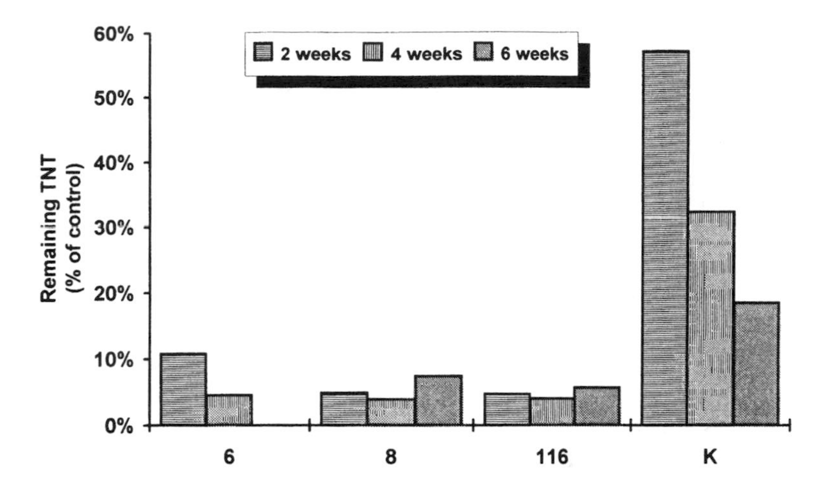

FIGURE 3. Degradation of TNT in original soil samples by white-rot fungi: 6 – *Trametes versicolor*, 8 – *Pleurotus ostreatus* Florida, 116 – *Pleurotus* spec., K – control with potato pulp but without fungi. Measured as disappearance of TNT in total sample and calculated as percentage of corresponding untreated soil samples; results of triplicate experiments.

The potential of *Phanerochaete* to degrade trinitrotoluene in a solid-state system (Fernando et al. 1990) was tested for *Pleurotus ostreatus* (two strains) and *Trametes versicolor* in a separate series of experiments. Results are presented in Figure 3. Primary degradation of more than 90% of the initial TNT concentration was already reached after 4 weeks by all fungi. *Trametes versicolor* removed TNT completely after 6 weeks of cultivation. Transformation products and mineralization of trinitrotoluene by these fungi are under investigation. Control samples used in these experiments were not sterile and showed, probably due to activated soil microorganisms, a significant potential of TNT degradation. Untreated soil samples revealed no loses of TNT.

The extraction of soil samples showed, together with a high TNT concentration (500 to 8,000 mg/kg), a large amount of brown-red, probably already polymerized degradation or even biodegradation products. Currently under study are the ability to degrade these compounds, the kinetics of TNT degradation, identification of intermediate degradation products and a resulting toxicity. Also under investigation is a comparison of the degradation ability and the degradation pathways of white-rot fungi with those of naturally occurring soil microflora.

REFERENCES

Carpenter, D. F., N. G. McCormick, J. H. Cornell, and A. M. Kaplan. 1978. "Microbial Transformation of 14C-Labeled 2,4,6-Trinitrotoluene in an Activated-Sludge System." *Applied and Environmental Microbiology* 35(5): 949-954.

Fernando, T., J. A. Bumpus, and S. D. Aust. 1990. "Biodegradation of TNT (2,4,6-Trinitrotoluene) by *Phanerochaete chrysosporium*." *Applied and Environmental Microbiology 56*(6): 1666-1671.

Hankenson, K., and D. J. Schaeffer. 1991. "Microtox Assay of Trinitrotoluene, Diaminonitrotoluene, and Dinitromethylaniline Mixtures." *Bull. Environ. Contam. Toxicol. 46*: 550-553.

Kaake, R. H., D. J. Roberts, T. O. Stevens, R. L. Crawford, and D. L. Crawford. 1992. "Bioremediation of Soils Contaminated with the Herbicide 2-*sec*-Butyl-4,6-Dinitrophenol (Dinoseb)." *Applied and Environmental Microbiology 58*(5): 1683-1689.

Lipczynska-Kochany, E. 1991. "Degradation of Aqueous Nitrophenols and Nitrobenzene by Means of the Fenton Reaction." *Chemosphere* 22(5-6): 529-536.

McCormick, N. G., J. H. Cornell, and A. M. Kaplan. 1978. "Identification of Biotransformation Products from 2,4-Dinitrotoluene." *Applied and Environmental Microbiology 35*(5): 945-948.

Naumova, R. P., T. O. Belousova, and R. M. Gilyazova. 1982. "Microbial Transformation of 2,4,6-Trinitrotoluene." *Applied Biochemistry and Microbiology 18*(1): 73-77.

Parrish, F. W. 1977. "Fungal Transformation of 2,4-Dinitrotoluene and 2,4,6-Trinitrotoluene." *Applied and Environmental Microbiology* 34(2): 232-233.

Pennington, J. C., and W. H. Patrick, Jr. 1990. "Adsorption and Desorption of 2,4,6-Trinitrotoluene by Soils." *J. of Environmental Quality 19*(3): 559-567.

Rickert, D. E., B. E. Butterworth, and J. A. Popp. 1984. "Dinitrotoluene: Acute Toxicity, Oncogenicity, Genotoxicity, and Metabolism." *CRC Critical Reviews in Toxicology 13*: 217-234.

Valli, K., B. J. Brock, D. K. Joshi, and M. H. Gold. 1992. "Degradation of 2,4-Dinitrotoluene by the Lignin-Degrading Fungus *Phanerochaete chrysosporium*." *Applied and Environmental Microbiology 58*(1): 221-228.

LANDFILL LEACHATE-POLLUTED GROUNDWATER EVALUATED AS SUBSTRATE FOR MICROBIAL DEGRADATION UNDER DIFFERENT REDOX CONDITIONS

H.-J. Albrechtsen, J. Lyngkilde, C. Grøn, and T. H. Christensen

INTRODUCTION

Organic matter in landfill leachate that migrates into an aquifer will be degraded by bacteria using the available electron acceptors in the aquifer. This process leads to the formation of a sequence of redox zones, with the most reduced (methanogenic or sulfate-reducing) zone close to the landfill and an aerobic zone in the front of the plume (if the aquifer is naturally oxidized) as the thermodynamically most favorable electron acceptor is used first. A sequence of redox zones containing a methanogenic/sulfate-reducing, an iron/manganese-reducing, a denitrifying and an aerobic zone has been identified and studied in detail at the Vejen landfill in Denmark (Lyngkilde & Christensen 1992a) and summarized for several cases in Christensen et al. (1993).

The availability of the electron acceptors may not be the only factor controlling the redox zones. Factors such as flowrate, microbial and physicochemical conditions (dilution, pH), and the value of the substrate as a C-source with the actual electron acceptors also may be controlling factors.

The purpose of this ongoing study is to simulate in the laboratory three redox level changes: leachate migrating into the methanogenic/sulfate-reducing zone in the aquifer, polluted groundwater from the methanogenic/sulfate-reducing zone migrating into the iron-reducing zone, and polluted groundwater from the iron-reducing zone migrating into the denitrifying zone.

MATERIALS AND METHODS

Sediment samples were collected from the methanogenic, iron-reducing, and denitrifying zones of the aquifer downstream from the Vejen landfill (Lyngkilde & Christensen 1992a). To obtain the main part of the bacteria from the sediment, the fine fraction (clay- and silt-sized) with the bacteria was extracted by landfill leachate or groundwater corresponding to the final setup (for further details see

Holm et al. 1992). This extraction was carried out in an anaerobic box, using aseptic techniques.

Leachate was collected from the landfill, and groundwater was collected from the methanogenic/sulfate-reducing and iron-reducing zones. The water samples were kept anaerobic, and before transferred to stirred glass reactors (10 L) with a headspace of N_2-CO_2 (80-20%), they were sterilized by filtering (0.45/0.2 µm Sartobran capsule, Sartorius) and inoculated with the bacteria in the fine fraction extracted from the sediment from the receiving redox environment. Sulfate (final concentration approx. 30 mg/L SO_4^{2-}-S), amorphous iron oxides (prepared as described by Lovley & Phillips (1986)) (final concentration approx. 220 mg/L Fe[III]) and nitrate (final concentration approx. 20 mg/L NO_3^--N) were added to the respective experiments to avoid limitations in available electron acceptors. For each set of redox conditions, an abiotic control (added 2 g/L of sodium azide) and a control without groundwater but with Millipore-filtered water were run in duplicate. The reactors were incubated at the actual groundwater temperature (10°C) and as of June 1992 the experiment had run for more than 450 days.

The concentration of specific organic compounds actually present in the leachate was followed by gas chromatographic/mass spectrometric analysis (Lyngkilde & Christensen 1992b) during the experimental period. The groundwater was low in content of specific organic compounds and were therefore spiked with the compounds at the start of the experiment. The bacteria were enumerated by direct microscopic counting after staining by acridine orange (AODC) as described in Albrechtsen and Winding (1992). The inorganic parameters (NO_2, NO_3, SO_4^{2-}), were determined by autoanalysis as described by Lyngkilde and Christensen (1992a). Fe[II] was determined by atomic absorption spectroscopy after filtration (0.45 µm, Minisart SRP 15, Sartorius). Dissolved organic carbon (DOC) was quantified as described in Lyngkilde and Christensen (1992a) and fractionated by the XAD-8 method (Leenheer 1981, modified by Grøn unpubl.) to yield fractions of hydrophilic compounds (capacity factor $k'<17$), hydrophobic acids at pH = 7.2 (carboxylic hydrophobic acids) and at pH = 11.0 (phenolic hydrophobic acids), followed by quantification by DOC measurements. The molecular weight distribution of DOC was characterized by size-exclusion chromatography on TSK G2000SW_{XL} and TSK G3000$_{XL}$ columns in tandem eluated with 0.2 N NaCl buffered to pH = 7.0 and with ultraviolet detection at 254 nm (Vartiainen et al. 1987, Grøn unpubl.). Molecular weight calibration was done with polystyrene-sulfonates.

RESULTS AND DISCUSSION

Compared to water samples collected in the field, the redox conditions (Eh, concentrations of inorganic electron acceptors) obtained in the laboratory experiments were very similar, and the redox conditions remained stable during the experimental period (see Table 1). The apparent presence of trace amounts of oxygen reflects only the detection level of the electrodes; the anaerobic conditions were verified by the low redox potential (Eh). The measured Eh values were

TABLE 1. Evaluation of the simulated redox conditions in the laboratory. Redox parameters are compared between field measurement of the redox parameters in water samples from the receiving redox environment and the average concentrations of these parameters during the initial period (approx. 100 days) of the laboratory experiment.

		Denitrifying conditions				Iron-reducing conditions				Methanogenic/sulfate-reducing conditions			
		Field	Laboratory experiment			Field	Laboratory experiment			Field	Laboratory experiment		
			GW[a]	GW + azide	C-free		GW	GW + azide	C-free		GW	GW + azide	C-free
Redox	mV	288	132	59	130	172	–16	23	82	144	–147	–132	21
O_2	mg/L	<1	<1	<1	<1	<1	<1	<1	<1	<1	<1	<1	<1
NO_3^--N	mg/L	4.8	17.8	12.4	18.1	0.05	<0.05	<0.05	<0.05	<0.05	NM[b]	NM	NM
Fe tot.	mg/L	NM	NM	NM	NM	NM	186	156	211	NM	83.4	73.3	0.3
Fe[II]	mg/L	1.3	10.7	6.7	6.0	9	NM	NM	NM	NM	NM	NM	NM
SO_4^{2-}-S	mg/L	14.6	NM	NM	NM	14.4	30.8	29.8	15.4	14.6	30.5	30.2	31.4
pH		5.7	5.9	6.0	5.8	6.5	6.6	6.5	6.5	6.3	6.8	6.8	7.0
DOC	mg/L	3.2	8.2	8.9	0.9	9	284	302	2	146	138	139	3.8
SS[c]	mg/L	NM	1.0	3.0	0.8	NM	3.4	5.3	1.9	NM	3.1	5.9	3.2

(a) GW = groundwater.
(b) NM = not measured.
(c) SS = suspended solids.

lower in the laboratory experiment (−16 to 82 mV) than in the field measurement (172 mV) under, e.g., iron-reducing conditions as a result of transferring groundwater from the methanogenic/sulfate-reducing zone (with a lower redox potential) to iron-reducing conditions by adding ferric oxides and sediment. The organic matter in the water samples was characterized initially before changing the redox conditions (Table 2). Under denitrifying conditions, the concentration of DOC decreased, verifying that the DOC could be used as a C source for the bacteria, but the degradation was very slow (less than 40% was degraded during 1 year). During the first year of the experiment, virtually no decrease of the concentration of DOC was observed under either iron-reducing or methanogenic/sulfate-reducing conditions.

The number of bacteria increased in the reactors (Figure 1), reflecting that at least part of the DOC or the carbon adsorbed to sediment particles could be used as substrate for growth. The increase in number of bacteria in the suspensions with millipore-filtered water indicated that part of the organic matter sorbed to the sediment particles was available for the bacteria. Under methanogenic/sulfate-reducing conditions, only minor increases in cell numbers were observed. The fact that an increase of the number of bacteria not always was reflected in a decrease in DOC may reflect that the DOC measurements were not sensitive enough to detect small changes relative to a large background. The number of bacteria increased the most under denitrifying conditions, which may reflect either that the DOC present was more degradable or that nitrate is the most favorable electron acceptor studied and therefore yields more energy and bacteria than the other electron acceptors.

To verify that the microbial processes taking place under the different redox conditions actually used the dominating electron acceptor under these conditions, inorganic compounds characterizing the processes were determined. Under denitrifying conditions the concentration of nitrate decreased and nitrous oxide was detected at the end of the experiment to verify nitrate reduction, and under iron-reducing conditions Fe[II] accumulated to verify reduction of Fe[III] (for further details see Albrechtsen [1994]). Methane was present in the reactors under methanogenic/sulfate-reducing conditions, but because the headspace increased due to sampling of water, it was not possible to detect a significant increase in methane. No depletion of sulfate, indicating sulfate-reduction, was observed, nor was the odor of sulfide or significant black precipitates detected.

To investigate if only the C source was limiting for the microbial activity, subsamples from the different redox levels were collected and enriched by acetate and benzoate. Under these conditions a potential for denitrification (by acetylene blockage), for iron reduction or methane production, but not for sulfate-reduction (after addition of $^{35}SO_4^{2-}$), was shown in the experiments under the respective conditions, through the entire experimental period. These observation verify the existence of a viable microbial population at the dominating redox conditions and that the degradability of the organic matter controlled the microbial activity.

Field investigations of the Vejen plume (Lyngkilde & Christensen 1992b) have shown that the concentration of organic matter and xenobiotic compounds in the leachate decreased more than could be explained by dilution and sorption.

TABLE 2. Characterization of the composition of the dissolved organic carbon (DOC) in the wells sampled for the laboratory experiment.

Samples	DOC	Polarity/pK_a			Molecular weight distribution		
					% UV absorbance		
	mg C/L	Hydrophilic compounds (%)	Carboxylic hydrophobic acids (%)	Phenolic hydrophobic acids (%)	> 100,000 D	5,000-100,000 D	< 5,000 D
Leachate (LG1)	140	47	46	11	<1	<1	100
Methanogenic/ sulfate-reducing	340	22	83	5.6	4.5	19	77
Iron-reducing	6.8	29	53	9.0	<1	<1	100

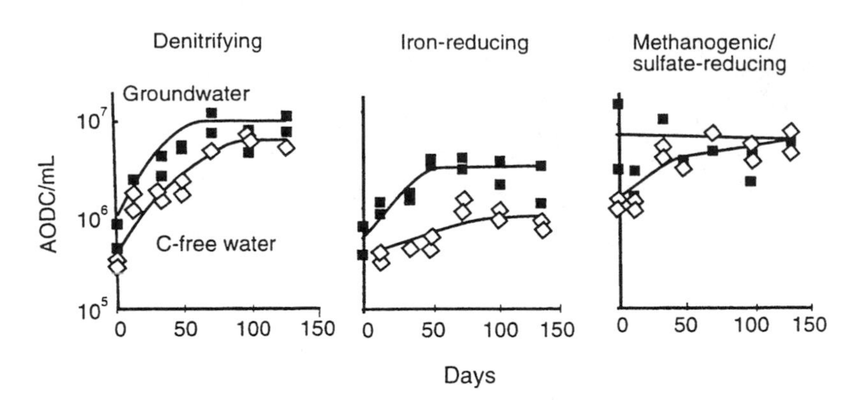

FIGURE 1. The number of bacteria (logarithmic scale) in reactors (duplicate) under three different redox conditions during the initial period of the experiment. The bacteria were enumerated by acridine orange direct counting (AODC), and each measurement represents one of the duplicate reactors. The results from a set of suspensions with groundwater and a set of suspensions with synthetic, C-free groundwater are presented.

To investigate if the organic matter sustained the growth of the bacteria degrading the xenobiotics, the concentrations of a range of compounds present in the leachate were followed at their naturally low concentrations (µg/L) during the experiment (Table 3). Under denitrifying conditions toluene and *p*/*m*-xylenes seemed to be degraded, and under iron-reducing conditions toluene was degraded. Under methanogenic/sulfate-reducing conditions, toluene, 1-propenylbenzene, and *o*-xylene seemed to be degraded in the mixture of leachate and sediment. In the mixture of C-free groundwater and sediment, the concentrations of benzene, ethylbenzene, *p*/*m*-xylene, naphthalene, and 2-methylnaphthalene all seemed to decrease.

CONCLUSION

Under these experimental anaerobic redox conditions, the organic matter in the leachate plume seems to be close to recalcitrant and only partially degradable under denitrifying conditions. Although degradation under methanogenic/sulfate-reducing and iron-reducing conditions was slower than expected from observations in the field, these microbial processes did take place as the number of bacteria increased and some of the specific contaminants from the polluted groundwater were degraded. The experiments stress not only the importance of redox environments but also the availability of organic matter as a microbial substrate in evaluating the degradation of specific organic compounds in polluted aquifers.

TABLE 3. Investigated organic components present in the leachate (µg/L). A mixture of the compounds was added to the bottles run under denitrifying conditions because their concentrations were too low to be studied with respect to degradation. Preliminary results from the degradation study are indicated: no degradation (–); significant degradation (+) or indication for degradation, not yet significant (?).

	Denitrifying				Iron-reducing				Methanogenic/ sulfate-reducing			
	Initial conc.	Ground-water	Groundwater + azide	C-free	Initial conc.	Ground-water	Groundwater + azide	C-free	Initial conc.	Ground-water	Groundwater + azide	C-free
benzene	4	–	–	–	43	–	–	–	50	–	–	?
ethylbenzene	43	–	–	–	430	–	–	–	380	–	–	?
1,3,5-trimethylbenzene	11	–	–	–	110	–	–	–	90	–	–	–
1-propenylbenzene	9	–	–	–	110	–	–	–	100	?	–	?
toluene	16	+	–	–	180	+	–	–	160	+	–	+
2-ethyltoluene	23	–	–	–	230	–	–	–	190	–	–	–
p/*m*-xylene	100	+	–	–	1100	–	–	–	920	–	–	?
o-xylene	21	–	–	–	220	–	–	–	210	?	–	–
naphthalene	22	–	–	–	260	–	–	–	240	–	–	?
2-methylnaphthalene	10	–	–	–	110	–	–	–	70	–	–	?
1-methylnaphthalene	11	–	–	–	120	–	–	–	80	–	–	–
camphor[a]	110	–	–	–	1500	–	–	–	1300	–	–	–

(a) 1,7,7-trimethylbicyclo-(2,2,1)-heptane-2-one.

ACKNOWLEDGMENTS

We thank Mona Refstrup, Lajla Olsen, and Gitte Brandt for their technical assistance. This work was part of a major research program focusing on the effects of waste disposal on groundwater. The program is funded by the Danish Technical Research Council, the Technical University of Denmark, and the Commission of the European Communities.

REFERENCES

Albrechtsen, H.-J. 1994. "Bacterial degradation under iron reducing conditions." In R. E. Hinchee, B. C. Alleman, R. E. Hoeppel, and R. N. Miller (Eds.), *Hydrocarbon Bioremediation.* Lewis Publishers, Ann Arbor, MI.

Albrechtsen, H.-J., and A. Winding. 1992. "Microbial biomass and activity in subsurface sediment from Vejen, Denmark." *Microbial Ecology 23(3)*: 303-317.

Christensen, T. H., P. Kjeldsen, H.-J. Albrechtsen, G. Heron, P. N. Nielsen, P. L. Bjerg, and P. E. Holm. 1993. "Attenuation of pollutants in landfill leachate pollution plumes." (A review). Submitted to *Critical Reviews in Environmental Control.*)

Holm, P., P. H. Nielsen, H.-J. Albrechtsen, and T. H. Christensen. 1992. "Importance of unattached bacteria and bacteria attached to sediment in determining potentials for degradation of xenobiotic organic contaminants in an aerobic aquifer." *Applied and Environmental Microbiology 58(9)*: 3020-3026.

Leenheer, J. A. 1981. "Comprehensive approach to preparative isolation and fractionation of dissolved organic carbon from natural waters and wastewaters." *Environmental Science & Technology 15(5)*: 578-587.

Lovley, D. R., and E. J. P. Phillips. 1986. "Availability of ferric iron for microbially reducible ferric iron in bottom sediments of the freshwater tidal Potomac River." *Applied Environmental Microbiology 52*: 751-757.

Lyngkilde, J., and T. H. Christensen. 1992a. "Redox zones of a landfill leachate pollution plume (Vejen, Denmark)." *Journal of Contaminant Hydrology 10*: 273-289.

Lyngkilde, J., and T. H. Christensen. 1992b. "Fate of organic contaminants in the redox zones of a landfill leachate pollution plume (Vejen, Denmark)." *Journal of Contaminant Hydrology 10*: 291-307.

Vartiainen, T., A. Liimatainen, and P. Kauranen. 1987. "The use of TSK size exclusion columns in determination of the quality and quantity of humus in raw waters and drinking waters." *The Science of the Total Environment 62*: 75-84.

USE OF PURE OXYGEN DISSOLUTION SYSTEM ENHANCES IN SITU SLURRY-PHASE BIOREMEDIATION

T. J. Bergman, Jr., J. M. Greene, and T. R. Davis

INTRODUCTION

The French Ltd. Task Group (FLTG) selected ENSR Consulting and Engineering (ENSR) to conduct a field-scale, in situ, slurry-phase bioremediation demonstration at the French, Ltd. Superfund site in Crosby, Texas. As part of this study, several aeration technologies were compared based on reliability, emissions to the atmosphere, and mixing and oxygenation characteristics. An economic and technical comparison was drawn between these technologies and several high-purity-oxygen-based technologies. Based on cost and air emissions, the Mixflo™ system was selected and a field-scale system was installed at the French Ltd. site. This system has successfully treated the first of two cells in less than 12 months, and is remediating the second cell.

SITE HISTORY

The French Ltd. Superfund Site is a former sand pit and chemical disposal lagoon located Crosby, Texas (30 mi [48 km] east of Houston). Sand mining activities from the late 1950s until about 1965 left a 7.3-acre (30-m^2) lagoon which is up to 25 ft (7.6 m) deep. From 1966 through 1971, French Limited, Inc. of Houston, Texas, operated the site as a licensed waste disposal facility, depositing approximately 70 million gal (265 million L) of industrial and commercial wastes from refineries and petrochemical plants located along the Houston Ship Channel.

In 1983, the Environmental Protection Agency (EPA) removed over 600 tons (536 tonnes) of floating sludge from the lagoon. Heavier constituents formed a 4- to 5-ft (1.2- to 1.5-m) deep, waste layer on the lagoon bottom. More than 20,000 tons (17,900 tonnes) of kiln dust are also present as a result of EPA's successful effort to control the lagoon's acidity.

EPA initially recommended incineration at an estimated cost between $75 and $125 million (excluding groundwater remediation costs). It was decided to explore the use of indigenous microorganisms to destroy the organic waste materials.

A demonstration program was performed in a test cell of the lagoon to show in situ bioremediation could effectively destroy organic waste and clean up contaminated soil. Five technical challenges for this program were identified: timely remediation, potential toxicity, suspension of tarry wastes, aeration of mixed liquor (8 to 12% solids), and air emissions management. The test was successful and EPA issued the ROD recommending bioremediation.

The bioremediation process consists of aeration, mixing, chemical addition, mixed liquor sampling, and water level control. Engineering designs were developed for each of these operations. Among these, aeration was considered particularly important because of concerns raised during the demonstration phase that significant hazardous emissions could be created by air stripping.

CHOOSING AN AERATION METHOD ALTERNATIVE

Aeration equipment was evaluated based on tarry waste mixing, aeration efficiency, and maintenance in a high-solids environment. Praxair's Mixflo™ system and merchant oxygen were selected and EPA approval was obtained. Praxair, Inc. completed process design and ENSR completed detailed mechanical design.

Mixflo™ Process Description. The Mixflo™ process dissolves oxygen in a two-stage process. In the first stage, slurry is pumped from the lagoon and pressurized to between 2 and 4 atm in a pipeline. Oxygen is then injected into the pipeline as finely dispersed bubbles. The resulting two-phase mixture turbulently flows through a pipeline designed to provide sufficient contact time to dissolve 60% of the injected oxygen. At the pipeline's elevated pressure, oxygen's solubility in the slurry increases substantially. In the second stage, the oxygen/slurry dispersion is reinjected into the lagoon using a liquid/liquid eductor which (1) dissipates the pumping energy into the oxygen/slurry mixture, forming a fine bubble dispersion; and (2) ingests unoxygenated slurry, mixes it with oxygenated slurry and then disperses the mixture throughout the holding cell. Typically, 75% of the oxygen not dissolved within the pipeline contactor dissolves in the lagoon using these eductors; therefore, the first and second stages typically combine to dissolve 90% of the injected oxygen.

Air/Oxygen Comparison. Air contains 79% nitrogen (essentially useless in bioremediation except in removing dissolved carbon dioxide to aid in pH control). Because pH can be controlled chemically, injecting large volumes of nitrogen into a waste treatment system is unnecessary. This excess gas volume also increases foaming and chemical emissions into the atmosphere.

Nitrogen also reduces the achievable dissolved oxygen level in a body of waste by a factor of five. This reduces the oxygen dissolution rate, as illustrated by:

$$\text{RATE} = K_L A\,(C^{*} - C)$$

where K_LA = constant
C^* = maximum achievable dissolved oxygen level
C = actual dissolved oxygen level present

This lower oxygen dissolution rate may limit the waste's biodegradation rate and make dissolved oxygen concentration control more difficult.

The use of oxygen in the Mixflo™ system reduced the off gas volume over other technologies using air by over 99%. In addition, many aqueous, oily organic wastes readily foam when bioremediated. A system which holds down off gas volumes minimizes this problem.

French Ltd. Design. In designing a Mixflo™ system for the French Ltd. Superfund site, the following "worst-case" design criteria were used:

Oxygen Requirement = 2,500 lb/hr (1,100 kgm/h)
Temperature = 104°F (40°C)
Slurry Depth = 12 ft (3.7 m)

$$\alpha = \frac{K_L a_{\text{Dirty Water}}}{K_L a_{\text{Clear Water}}} = 0.6$$

$$\beta = \frac{\text{Saturation dissolved oxygen level dirty water}}{\text{Saturation dissolved oxygen level clean water}} = 0.9$$

The pipeline contactor pressure was chosen to minimize the sum of the capital and operating costs. A minimum in total costs occurs for this site at a pressure of 30 psig (2.0 atm). At this pressure (given the specified temperature and β value), the saturation dissolved oxygen level is 82.4 ppm. This results in required slurry pumping rate of 60,000 gpm (3,800 L/sec) and a pumping power requirement of approximately 1,500 hp (1,100 kW).

The required pumping power decreases with decreasing temperature because of the increased dissolved oxygen capacity of the slurry. At the average temperature (approximately 75°F) (23°C), the pumping power required is approximately 1,000 hp (746 kW).

Based on available eductor, pump, and pipe sizes, the required slurry pumping rate of 60,000 gpm (3,800 L/sec) was divided into eight Mixflo™ units, each consisting of one pump and pipeline contactor and three eductors. The slurry flowrate to each Mixflo™ unit is 7,500 gpm (470 L/sec).

The pipeline contactor is constructed from 18-in. (46-cm) carbon steel pipe, giving a pipeline velocity of 11.5 ft/s (3.5 m/s). The contactor length was calculated by dividing the contact time necessary to oxygenate clean water by the α factor. Each of twenty-four 10-in. (25-cm) eductors processes 2,500 gpm (160 L/sec) slurry each with a minimum pressure loss of 25 psi (1.7 atm). Because

90% of the injected oxygen is dissolved at the design point, 70 scfm (31 L/sec) oxygen are injected in each Mixflo™ unit. Two flow-control skids are installed, each supplying four Mixflo™ units. Oxygen flow to each injection point is controlled manually.

To supply oxygen to the project, liquid oxygen is trucked from a Praxair facility in the Houston area to an 11,000-gal (42,000-L) cryogenic storage tank. This tank, when full, holds approximately 1½ days of the peak process requirement.

Safety. Because high-purity oxygen is in contact with organic chemical-containing slurry in the Mixflo™ units, the following safety devices were designed into the oxygen flow-control skid system:

- Oxygen is not delivered to the Mixflo™ unit until slurry is pumped.
- Oxygen supply is terminated to any Mixflo™ unit with a non-operating pump.
- Oxygen supply is terminated to any Mixflo™ unit in which the operating pressure is below a minimum allowable value.
- Oxygen supply is terminated if the pressure in the oxygen supply line is too low (indicating line breakage) or too high (indicating oxygen injector plugging).
- A novel vertical, bubble trap system is used for the oxygen supply piping to prevent slurry from entering solenoid valves and other moving parts.
- Oxygen piping at the injection point is made of a low flammability material.
- Lines are blown out with nitrogen before startup.
- The oxygen supply skid was designed, manufactured, and tested by Praxair.

REMEDIAL ACTION

ENSR carried out procurement and final design of the Mixflo™ system based on Praxair's specification of process parameters. Major procurements included mixed liquor pumps, pump pontoons, and eductors. Sala™ vertical cantilever pumps were chosen because of their reliability and low-maintenance requirements. The pumps are rated at 7,500 gpm (470 L/sec) at a 70-ft (21.3-m) total dynamic head. They operate at 900 rpm with 191 slurry horsepower (142 kW) and have a 77% pump efficiency at an impeller tip speed of 4,640 fpm (140 m/min); 300-hp (224-kW) motors were specified for the pumps to provide reserve power. All wetted parts are constructed of high chrome castings with a Brinell hardness of 550. The pumps are belt-driven with variable-speed pulleys, which will allow the pumps to be fine tuned to the required pressure and liquid flowrate if wear reduces the pressure losses in the Mixflo™ system.

The pumps are mounted on two mixed-liquor pump pontoons. A floating (as opposed to fixed) pump installation was specified because (1) the pumps were to be moved to Cell F when Cell E's remediation is complete; (2) the lagoon water level may change ±1 ft (0.3 m) during remedial operations; and (3) used pontoons have a higher resale value than metal supports for a fixed system.

The mixed liquor pump pontoons (designed and fabricated by Dredging Supply Corporation (DSC) of Harvey, Louisiana) are 40 ft (12.2 m) by 30 ft (9.1 m) with a molded depth of 7 ft (2.1 m). The pontoons are equipped with removable ¾-inch (1.9-cm) mesh screens, safety hand rails, man-overboard rails, lighting, a utility station, and power outlets.

The eductors experience higher abrasion than any other equipment in the system (due to the high velocity at the eductor nozzle). Schutte & Koerting was chosen as the eductor supplier based on cost and ability to meet specifications. The eductors are supported on sliding supports mounted on pipe piles for accessibility.

The pipeline contactor is constructed of abrasion-resistant, spiral-wound carbon steel pipe supported on sleepers with mechanical couplings. Blinded tees were used instead of elbows for improved abrasion resistance.

Construction proceeded without incident and the facility was mechanically complete on December 20, 1991 and started up on January 13, 1992. Cell E remediation was completed in November 1992. The equipment was moved to Cell F and restarted in January 1993. The estimated cost for remediation of approximately 300,000 tons of waste (including almost $20 million spent on the remedial investigation/feasibility study and pilot study) is $50 million.

CONCLUSION

The robust, mechanically simple Mixflo™ system enabled French Ltd. to achieve their remedial objectives while cost-effectively minimizing air emissions. Adequate dissolved oxygen levels were attained with no foaming and minimal emissions to the atmosphere.

BIODEGRADATION OF NOVEL AZO DYES

M. B. Pasti-Grigsby, A. Paszczynski, S. Goszczynski, D. L. Crawford, and R. L. Crawford

INTRODUCTION

Azo dyes are the most numerous of the manufactured synthetic dyes. Currently, at least 3,000 azo dyes are in use (Chung & Cerniglia 1992). They are extensively employed in the textile, food, printing, leather, mouthwash, cosmetic, pharmaceutical, and paper-making industries and are used in protective lacquers for the transparent coating of metal (aluminum) foils and other materials such as wood (greening lacquers), and as additives in petroleum products (Hunger et al. 1985). Commonly used waste-removal treatments do not adequately eliminate many azo dyes from the effluent waters of textile mills and dyestuff factories (Kimura 1980). Though by volume, synthetic organic colorants are not a major pollutant—0.06 million tons per year (1975) against 1.1 million tons of pesticides or 290 million tons of CO auto exhaust (Meyer 1981)—azo dyes can be carcinogenic and mutagenic to animals and humans (Combes & Haveland-Smith 1982), or they can become mutagenic when metabolized anaerobically (Chung & Cerniglia 1992). Anaerobic transformations lead to the formation and accumulation of colorless aromatic amines, which can be highly toxic and carcinogenic (Chung & Cerniglia 1992). Extensive reviews of the anaerobic metabolism of azo compounds have been published (Chung et al. 1992, Walker 1970).

Under aerobic conditions, azo dyes had been considered essentially nondegradable by bacteria (Michaels & Lewis 1986, Pagga & Brown 1986). Within the last decade, Kulla et al. (1983) described a degradative pathway for sulfonated azo dyes by *Pseudomonas* strains. However, despite the presence of oxygen, the first step was still to reduce the azo linkage to subsequently form sulfonated aromatic amines that would react in the cell with nonsulfonated intermediates of dye degradation, interrupting the degradation process.

The aerobic transformation of three azo dyes, Tropaeolin O, Congo Red, and Orange II, by the fungus *Phanerochaete chrysosporium* was first reported by Cripps et al. (1990). Although neither the metabolic pathways nor potential end products were discussed, the fungal lignin-degrading system was implicated in the biotransformation process.

Recently we described a new approach to increasing the biodegradation of azo dyes by *P. chrysosporium* and *Streptomyces* spp. (Paszczynski et al. 1991). The method is based on the introduction of a metabolizable substituent into the dye's chemical structure without affecting its property as a dye. In the same work we showed that while *P. chrysosporium* ligninase recognized Acid Yellow 9 as a substrate, Mn(II)-peroxidase oxidized the other azo dyes. In a subsequent paper, Paszczynski and Crawford (1991) showed the involvement of veratryl alcohol during the degradation of some azo compounds by *P. chrysosporium*. To study the influence of aromatic substitution patterns on the degradability of azo dyes by *P. chrysosporium* and *Streptomyces* spp., 22 azo dyes were investigated (Pasti-Grigsby et al. 1992). Five novel synthesized azo dyes which had the common structural pattern of a hydroxy group in the *para* position relative to the azo linkage, and at least one methoxy and/or one alkyl group in an *ortho* position relative to the hydroxy group, were identified as the most decolorized by the selected strains. The *Streptomyces* spp. as well the *P. chrysosporium* peroxidases were shown to be involved in the initial transformation process.

Because decolorization is not synonymous with complete degradation, the next step was to follow the mineralization of the five selected sulfonated azo dyes by *Streptomyces* spp. (Pasti et al. 1991, Paszczynski et al. 1992) and by *P. chrysosporium* (Paszczynski et al. 1992). Although *P. chrysosporium* mineralized all the sulfonated azo dyes, the ability of *S. chromofuscus* A11 to degrade selected ^{14}C-labeled azo dyes was significantly influenced by substitution pattern. Spadaro et al. (1992) reported on the degradation of nonsulfonated azo dyes by the lignin-degrading fungus *P. chrysosporium*. Compounds including aromatic rings with substituents such as hydroxyl, amino, acetamido, or nitro functions were mineralized to a greater extent than those with unsubstituted rings.

In the present preliminary study we addressed two major points needing clarification before newly designed dyes may be considered as an alternative for industrial application: (1) the biodegradability of the newly synthesized dyes by a broader range of soil microorganisms, and (2) the identification of biotransformation intermediates or end products compared to those expected under anaerobic conditions.

MATERIALS AND METHODS

Microorganisms and Culture Maintenance. The wild type actinomycete *S. chromofuscus* A11 (ATCC 55184) and the fungus *P. chrysosporium* Burds BKM-1667 (ATCC 24725) were used. Stock cultures of the actinomycete (Pasti et al. 1990) and the fungus (Paszczynski et al. 1991) were maintained at 4°C as previously described.

Substrates. Two radiolabeled compounds were used in the soil mineralization studies: the azo dye 4-(3-sulfo-4-amino-phenyl-azo)-[U-^{14}C] benzenesulfonic acid (commercial Acid Yellow 9 in Table 1) and the novel azo dye 4-(3-methoxy-4-hydroxy-phenyl-azo)-[U-^{14}C] benzenesulfonic acid (azo dye 3 in Table 1). The metabolic studies were limited to the unlabeled azo dye 1 (Figure 1). The syntheses of the radiolabeled azo dyes (Paszczynski et al. 1992) and of the novel azo dye 1 (Pasti-Grigsby et al. 1992) have been described previously.

TABLE 1. Percentage of radioactivity incorporated into CO_2 and recovered after washing the soil at the end of the 4-week incubation period.

	% Radioactivity in:			
	Sterile soil		Unsterilized soil	
Azo dye	Soil	CO_2	Soil	CO_2
Acid Yellow 9[a]	99.0	0.3	96.6	2.9
Azo dye 3 [b]	97.0	0.4	53.3	14.0

(a) HO_3S N N NH_2 SO_3H

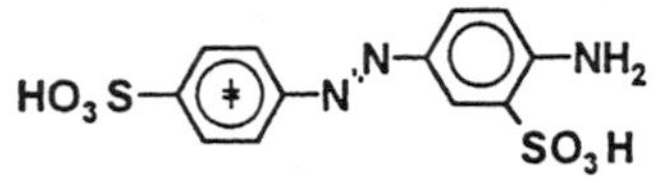

(b) HO_3S N N OH OCH_3

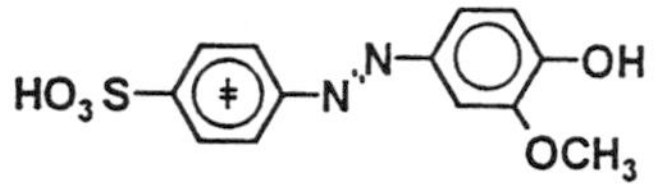

Mineralization Studies in Soil. Soils were collected from the University of Idaho Arboretum; its physicochemical properties have been described by Wang and Crawford (1989). It is a soil never previously contaminated by azo dyes; thus, there was no potential for the presence of an adapted microflora. Two ^{14}C-labeled azo dyes were used: Acid Yellow 9, as a model for common, commercially available dyes, and the newly modified azo dye 3. The dyes (0.25 mg/gr of soil) were mixed with 5 g of soil (dry weight) in a 250-mL flask. Each soil sample was brought to approximately 60% of water-holding capacity (WHC) by addition of sterile water, and a suitable carbon source was added. Each flask was fitted with a sterile gassing device that allowed for continuous passage of sterile, CO_2-free, humidified air into the flask, and all the exit gases from each flask were continuously passed through a CO_2-trapping solution (10 mL of 2N NaOH) (Wang & Crawford 1989). Experiments were incubated for 28 days and were run in triplicate. Trapped $^{14}CO_2$ was quantified as described by Pasti et al. (1990).

Pathway Studies. The oxidations of azo dye 1 (5 mg/10 mL enzyme reaction) by *S. chromofuscus* A11 and by *P. chrysosporium* crude enzyme preparations were carried out as described by Pasti-Grigsby et al. (1992). After a reaction was completed, 10 mg of tetrabutylammonium hydrogen sulfate (TBHS) was added to a 5-mL sample. The solution was then extracted five times with methylene chloride in an amount equal to the sample volume. The methylene chloride layer was dried with anhydrous sodium sulfate and concentrated to 500 μL under a gentle air stream. Derivatization of the same samples was performed by treating with methyl fluorosulfonic ester according to Sugiura and Whiting (1980). The methylene chloride layer was dried with anhydrous sodium sulfate and then under a gentle airstream. After dissolving in acetonitrile, the samples were analyzed by high-pressure liquid chromatography (HPLC) (Hewlett Packard Series 1050) or gas chromatography (GC) (Hewlett Packard Series II 5890) interfaced to a mass spectroscopy (MS) quadruple detector (Hewlett Packard Series II 5989A) (Paszczynski et al., manuscript in preparation).

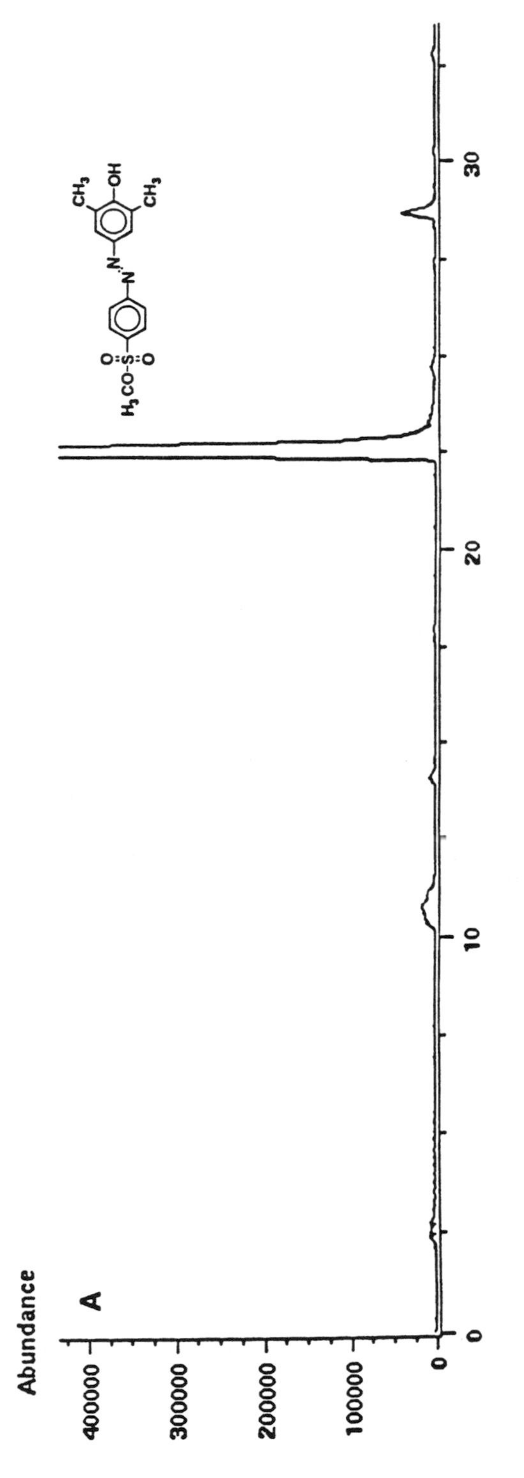

FIGURE 1A. Total ion chromatograms (TIC:150-450) HPLC/MS analysis of azo dye 1 incubated with an *S. chromofuscus* A11 extracellular enzyme preparation in the absence of H_2O_2.

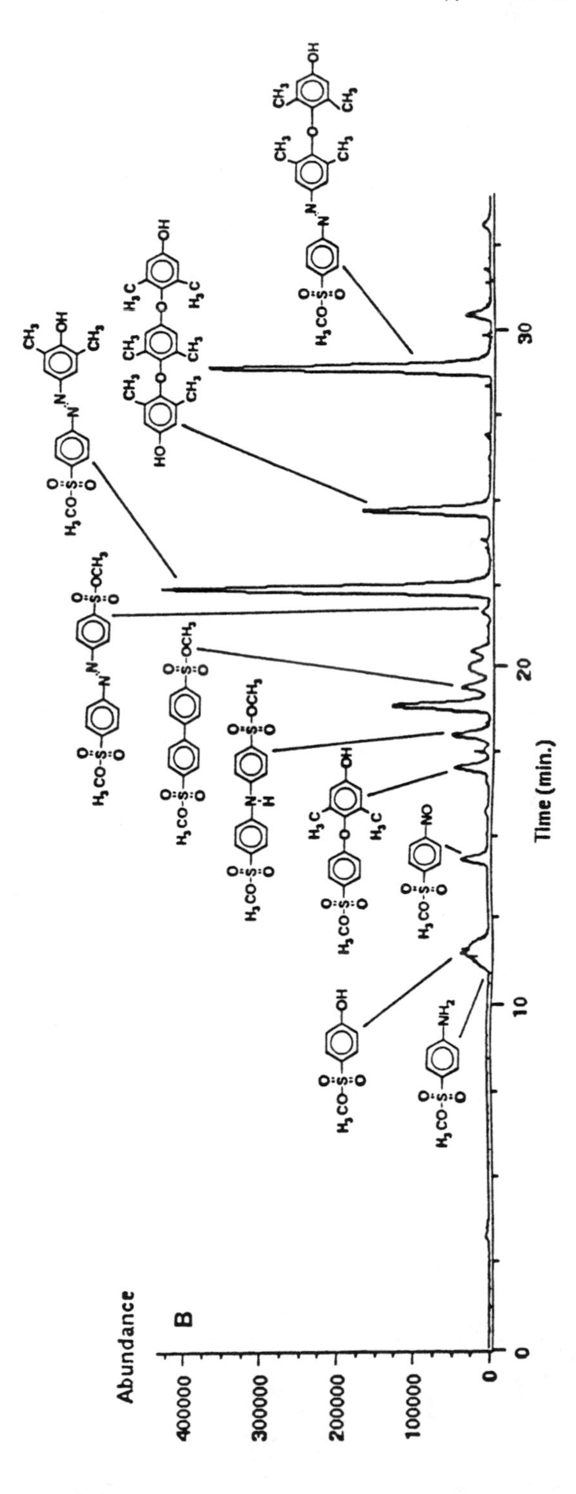

FIGURE 1B. Total ion chromatograms (TIC:150-450) HPLC/MS analysis of azo dye 1 incubated with an *S. chromofuscus* A11 extracellular enzyme preparation in the presence of H_2O_2.

RESULTS AND DISCUSSION

Table 1 shows the transformation and the mineralization of Acid Yellow 9 and azo dye 3 by an unacclimated soil microflora. The cumulative release of $^{14}CO_2$ from soil containing azo dye 3 was significantly greater than from soil mixed with Acid Yellow 9. Although the potential effects of varying supplemental carbon sources, pH, humidity, and temperature have not yet been examined, the increased biodegradability of the modified dye was confirmed using a natural soil community in place of pure cultures of *Streptomyces* spp. and the fungus *P. chrysosporium*. Further experiments with soil samples taken from different environments are planned.

Figure 1 shows the total ion chromatograms (TIC:150-450) by HPLC/MS analysis of azo dye 1 after incubation with an *S. chromofuscus* A11 extracellular enzyme preparation in the absence (Figure 1A) and presence (Figure 1B) of H_2O_2. Oxidation of the dye required peroxide and is therefore most likely catalyzed by peroxidase. The metabolites identified so far during the oxidation of azo dye 1 by the actinomycete enzyme preparation seem to be the same as isolated in the fungal oxidation reactions (Paszczynski et al., manuscript in preparation). One of the major products in both microorganisms was identified by GC analysis as 2,6-dimethyl benzoquinone (data not shown). Based on the recent study of Chung and Cerniglia (1992), the compounds identified so far should not be toxic or mutagenic. Further studies are required to define the complete aerobic metabolic pathway of the modified new azo dyes. However, with this work we have shown the applicability of our method for enhancing the biodegradability of azo dyes while minimizing potential negative environmental impacts.

REFERENCES

Chung, K.-T., and C. E. Cerniglia. 1992. "Mutagenicity of Azo Dyes: Structure-Activity Relationships." *Mutation Research 277*: 201-220.

Chung, K.-T., S. E. Stevens, Jr., and C. E. Cerniglia. 1992. "The Reduction of Azo Dyes by the Intestinal Microflora." *Crit. Rev. Microbiol. 18*(3): 175-190.

Combes, R. D., and R. B. Haveland-Smith. 1982. "A Review of the Genotoxicity of Food, Drug and Cosmetic Colours and Other Azo, Triphenylmethane and Xanthine Dyes." *Mutation Res. 98*: 79-84.

Cripps, C., J. A. Bumpus, and S. D. Aust. 1990. "Biodegradation of Azo and Heterocyclic Dyes by *Phanerochaete chrysosporium*." *Appl. Environ. Microbiol. 56*: 1114-1118.

Hunger K., P. Mischke, W. Rieper, and R. Raue. 1985. "Azo Dyes." In F. T. Campbell, R. Pfefferkorn, and J. F. Rousanville (Eds.), *Ulmann's Encyclopedia of Industrial Chemistry*, 5th ed. A3: pp. 246-323.

Kimura, M. 1980. "Prospects for the Treatment and Recycle of Dyeing Wastewaters." *J. Soc. Fiber Sci. Technol. Jpn. 36*: 69-73.

Kulla, H. G., F. Klausener, U. Meyer, B. Lüdeke, and T. Leisinger. 1983. "Interference of Aromatic Sulfo Groups in the Microbial Degradation of the Azo Dye Orange I and Orange II." *Arch. Microbiol. 135*: 1-7.

Meyer, U. 1981. "Biodegradation of Synthetic Organic Colorants." In T. Leisinger, A. M. Cook, R. Hütter, and R. Nüesch (Eds.), *Microbial Degradation of Xenobiotics and Recalcitrant Compounds*, pp. 371-385. Academic Press, Inc., Ltd., London.

Michaels, G. B., and D. L. Lewis. 1986. "Microbial Transformation Rates of Azo and Triphenylmethane Dyes." *Environ. Toxicol. Chem. 5*: 161-166.
Pagga, U., and D. Brown. 1986. "The Degradation of Dyestuffs: Part II. Behaviour of Dyestuffs in Aerobic Biodegradation Tests." *Chemosphere 15*(4): 479-491.
Pasti, M. B., A. L. Pometto III, M. P. Nuti, and D. L. Crawford. 1990. "Lignin Solubilizing Ability of Actinomycetes Isolated from Termite (Termitidae) Gut." *Appl. Environ. Microbiol. 56*: 2213-2218.
Pasti, M. B., S. Goszczynski, D. L. Crawford, and R. L. Crawford. 1991. "Mineralization of Azo Dyes in Liquid Cultures by *Streptomyces chromofuscus* A11." *Abstr. 8th Int. Symp. Biol. Actinomycetes.* Madison, Wis.
Pasti-Grigsby, M. B., A. Paszczynski, S. Goszczynski, D. L. Crawford, and R. L. Crawford. 1992. "Influence of Aromatic Substitution Patterns on Azo Dye Degradability by *Streptomyces* spp. and *Phanerochaete chrysosporium*." *Appl. Environ. Microbiol. 58*: 3605-3613.
Paszczynski, A., and R. L. Crawford. 1991. "Degradation of Azo Compounds by Ligninase from *Phanerochaete chrysosporium*: Involvement of Veratryl Alcohol." *Biochem. Biophys. Res. Commun. 178*: 1056-1063.
Paszczynski, A., M. B. Pasti, S. Goszscynski, D. L. Crawford, and R. L. Crawford. 1991. "New Approach to Improve Degradation of Recalcitrant Azo Dyes by *Streptomyces* spp. and *Phanerochaete chrysosporium*." *Enzyme Microb. Technol. 13*: 378-384.
Paszczynski, A., M. B. Pasti-Grigsby, S. Goszczynski, R. L. Crawford, and D. L. Crawford. 1992. "Mineralization of Sulfonated Azo Dyes and Sulfanilic Acid by *Phanerochaete chrysosporium* and *Streptomyces chromofuscus*." *Appl. Environ. Microbiol. 8:* 3598-3604.
Sugiura, T., and M. C. Whiting. 1980. "The Identification of Azo Dyes. Part 3. Methylation of Sulphonate Groups and Mass Spectrometry." *J. Chem. Research, Synop. 5*: 164-165.
Spadaro, J. T., M. H. Gold, and V. Renganathan. 1992. "Degradation of Azo Dyes by the Lignin-Degrading Fungus *Phanerochaete chrysosporium*." *Appl. Environ. Microbiol. 58*: 2397-2401.
Walker, R. 1970. "The Metabolism of Azo Compounds: A Review of the Literature." *Food Cosmet. Toxicol. 8:* 659-676.
Wang, Z., and D. L. Crawford. 1989. "Survival and Effects of Wild-Type, Mutant, and Recombinant *Streptomyces* in a Soil Ecosystem." *Can. J. Microbiol. 35*: 535-543.

BIOPUR®, AN INNOVATIVE BIOREACTOR FOR THE TREATMENT OF GROUNDWATER AND SOIL VAPOR CONTAMINATED WITH XENOBIOTICS

E. H. Marsman, J. M. M. Appelman, L. G. C. M. Urlings, and B. A. Bult

INTRODUCTION

Soil contamination by xenobiotics is widespread. In the Netherlands, contaminated soils have been registered at more than 100,000 sites (Committee BSB 1991). At about 80% of these sites, the soil is contaminated with hydrocarbons and sometimes with halogenated hydrocarbons (8,000 sites). Due to the physicochemical properties of many of these compounds, the contamination is likely to extend into soil vapor, groundwater, and soil.

At about 13,000 contaminated sites in the Netherlands, in situ soil remedial techniques can be used such as in situ biostimulation, soil venting, and bioventing (TAUW Infra Consult B.V. 1992). Both techniques involve injection of air into the aquifer to stimulate biological activity in the soil or to strip volatile compounds from the groundwater (Urlings et al. 1991). With this type of in situ remediation, groundwater often has to be withdrawn to control the contamination. At present, the treatment of soil vapor and groundwater is by far the most expensive part of in situ soil remediation. Consequently, it is a challenge for environmental engineers to develop simple and cost-effective systems to treat soil vapor and groundwater. Once such a system exists, the application of in situ soil remedial techniques will be much more attractive (Vree et al. 1992).

Under the proper environmental conditions (Bouwer 1992), many xenobiotics can be biodegraded partially or fully, which makes biological processes potentially suitable for groundwater and soil vapor treatment. Xenobiotics which can be biodegraded are halogenated and nonhalogenated volatiles and semivolatiles, polynuclear aromatics, and BTEX.

As many xenobiotics are present in the groundwater in low concentrations (range µg/L-mg/L), biological reactors based on the fixed-film concept are particularly suitable (Bouwer et al. 1988). Our first experiments with the full-scale biological groundwater treatment took place in 1986 when we first applied rotating biological contactors (RBC) to remove benzene, monochlorobenzene, and lindane from contaminated groundwater (van der Hoek et al. 1989).

We have developed a new biofilm reactor that is suitable for the simultaneous treatment of soil vapor and groundwater that are contaminated with volatile and nonvolatile organic compounds. In 1989, the first full-scale application of the biofilm reactor for the simultaneous treatment of groundwater and soil vapor contaminated with benzene, toluene, ethylbenzene, and xylene (BTEX) and total petroleum hydrocarbons took place at a petrol station (Coffa et al. 1991). In the United States, BIOPUR® is patented, and in Europe a patent is pending.

This technical note explains the biofilm reactor system and gives some removal efficiencies obtained in groundwater treatment and in the simultaneous treatment of groundwater and soil vapor.

BIOFILM REACTOR

Our reactor is an aerated, packed-bed biofilm reactor using reticulated polyurethane (PUR) as a carrier material for microorganisms (Oosting et al. 1992). PUR is a foam with a very open structure; consequently, the specific surface is large (500 m^2/m^3). High biomass concentrations can be obtained (5-15 g dm/L PUR).

Biomass grows as a thin biofilm on PUR. The retention time of biomass is longer in biofilm reactors than in suspended-sludge reactors. Consequently, even microorganisms that grow slowly are kept in the reactor. Using fixed-film processes, the hydraulic loading is independent of the growth rate of the biomass as it is in suspended-sludge systems. A variety of microorganisms can survive on the biofilm, and each population will mineralize contaminants or degrade products of contaminants.

The biofilm reactor consists of several compartments in series to model a plug flow pattern in the reactor (see Figure 1). The advantage of a plug flow reactor is that it can degrade xenobiotics to a great extent. Moreover, completely different populations of bacteria can reside in each compartment. Water and air flow concurrently upward through the compartments. The water phase flows through the compartments of the biofilm reactor by means of gravity, and air is forced from one compartment to the next via blowers. In this way, volatile components that are stripped in the first compartment are forced into the second compartment. Due to consumption of contaminants, the volatile compounds in the vapor phase will dissolve in the water phase. Whenever soil vapor is extracted in the course of an in situ soil remediation, the soil vapor can also be treated in the biofilm reactor.

The biofilm reactor can also achieve process conditions required to cultivate specific microorganisms which biodegrade recalcitrant compounds such as perchloroethylene (TeCE), trichloroethylene (TCE), and 1,1,1-trichloroethane (TCA). We are now investigating the biodegradation of TeCE, TCE, and TCA under anaerobic and methanotrophic process conditions on a pilot plant scale.

RESULTS

Groundwater Treatment. Pilot plant research was carried out at a former gasworks, at an asphalt factory (van der Hoek et al. 1989), and at a former pesticides

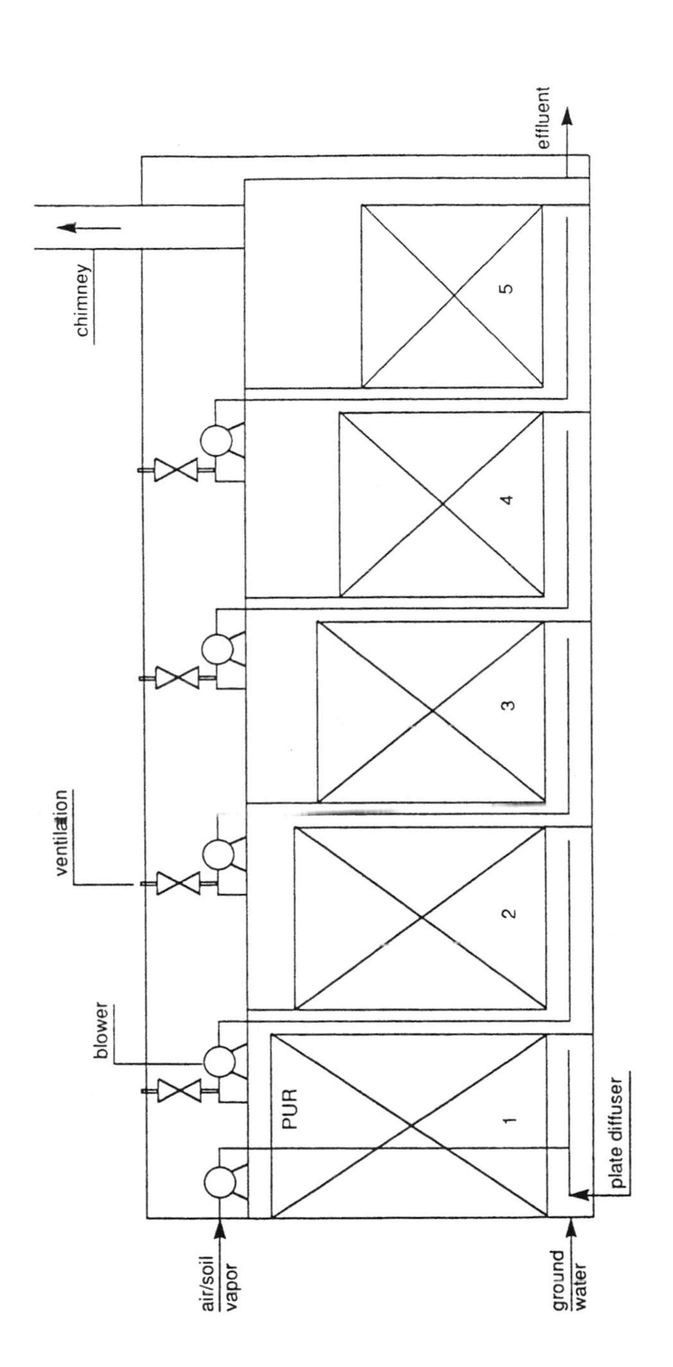

FIGURE 1. Diagram of the biofilm reactor.

factory (Veen et al. 1989). The groundwater at these sites was contaminated mainly with BTEX, total petroleum hydrocarbons, PAH, phenolic compounds, chlorobenzene, and lindane (HCH).

The pilot plant research showed that the fixed-film systems (RBC and the biofilm reactor) obtained excellent removal efficiencies. More than 99% of BTEX, naphthalene, and chlorobenzene were removed, up to 70% of alpha and gamma HCH were removed and more than 70% of phenolic compounds were removed. Moreover, stripping of volatile compounds hardly occurred (<1% to 8%).

On the full-scale, the biofilm reactor is used mainly at gasoline stations to treat groundwater contaminated with aromatics (BTEX) and volatile and nonvolatile hydrocarbons; gasoline, kerosene and diesel fuel. In Table 1 some results of full-scale biofilm reactor applications are presented.

The results show high removal efficiencies (55% >99.9%) at short hydraulic retention times (0.25-1.2 hour), which means that the hydraulic loading of the biofilm reactor can be high. The concentration of contaminants in the groundwater and soil vapor is often very low resulting in low organic loading. It should be noted that the temperature in the biofilm reactors varies between 9 and 12°C.

With a contractor, we have been undertaking full-scale experiments with the biofilm reactor for more than 3 years at 9 treatment plants. Throughout that period, attention has been paid to the removal efficiencies, to the stripping of volatile compounds, and to process technology aspects of the bioreactor. Exhaust gas monitoring showed that stripping occurred only in a range of fractions of a percent of the offered load. Maintenance of the biofilm reactor has been limited to reading water meters and filling the nutrient stock solution. No problems were encountered with the formation of iron oxides in the PUR. Groundwater containing iron in concentrations of up to 25 mg/L has been treated in the biofilm reactor without an iron removal step. Eventually, the PUR may be removed from the reactor to remove the excess sludge attached to it by means of a high-pressure spraying pistol. The cleaned PUR can be reused. The installations are cleaned twice a year by a qualified technician.

At certain sites, treated groundwater can be discharged into the surface water. At other sites, the effluent of the bioreactor must be treated in a sand filter and with activated carbon before it is discharged.

The results show that the biofilm reactor is an excellent technique for biological treatment of groundwater contaminated with different kinds of hydrocarbons. The fixed-film system is compact, and high bioactivities and removal efficiencies are achieved.

Simultaneous Treatment of Groundwater and Soil Vapor. The biofilm reactor can mineralize both the volatile compounds in extracted soil vapor and the dissolved contaminants in the withdrawn groundwater. The first full-scale treatment unit with an effective volume of 10 m^3 was tested at a petrol station in Raalte (see Figure 2). The groundwater and the soil vapor were contaminated with BTEX and with total petroleum hydrocarbons.

The HRT was approximately 15 minutes, and the gas retention time was less than 3 minutes. Ambient air was supplied to the different reactor compartments.

TABLE 1. Removal efficiencies of the laboratory and full-scale biofilm reactor process.

Site	Capacity (m^3/h)	Hydraulic Retention Time (hour)	Influent (μg/L)	Effluent (μg/L)	Efficiency (%)
Raalte 1	15	0.25			
BTEX			650	0.5	>99.9
Raalte 2	4	0.9-1.2			
BTEX			300-1,980	<1	99.6
TPH			40-330	20-110	>66
Overslag	6	0.87-1			
BTEX			135-460	<1	<79
TPH			50-2,300	<100	>84
Utrecht	13	0.5			
BTEX			390-1,300	3-5	99
TPH			323-1,000	>50	>80
Amersfoort	10	0.4			
BTEX			10,390	7	>99
TPH			1,300	<100	>92
Borculo	5	1			
BTEX			420-2,600	0.4-17	>95
TPH			6,000-18,000	40-2,800	>55
Olst	0.05	1			
BTEX			7,390	295	>96
naphthalene			4,180	41	>99
phenol			180	32	>82
o-cresol			700	210	>70
m-cresol			370	59	<84
p-cresol			85	28	>66
2,5-dimethylphenol			360	111	>69
2,4-dimethylphenol			100	20	>80
2,3-dimethylphenol			150	60	>60
3,4-dimethylphenol			120	12	>90
m-ethylphenol			73	12	>67
p-ethylphenol			4,000	1,320	>76

The treatment efficiency with regard to the total load (220 g TPH/hour) extended to more than 98%. All the requirements were met for groundwater to be discharged, e.g., 100 μg aromatics/L and 1 mg TPH/L.

Figure 3 shows a mass balance of the biofilm reactor in a steady state. The mass balance is representative of a significant period of remediation. In the exhaust gas, no volatile compounds were detected (<0.1 ppm). Figure 4 shows the chromatograms of the soil vapor and exhaust gas from the biofilm reactor.

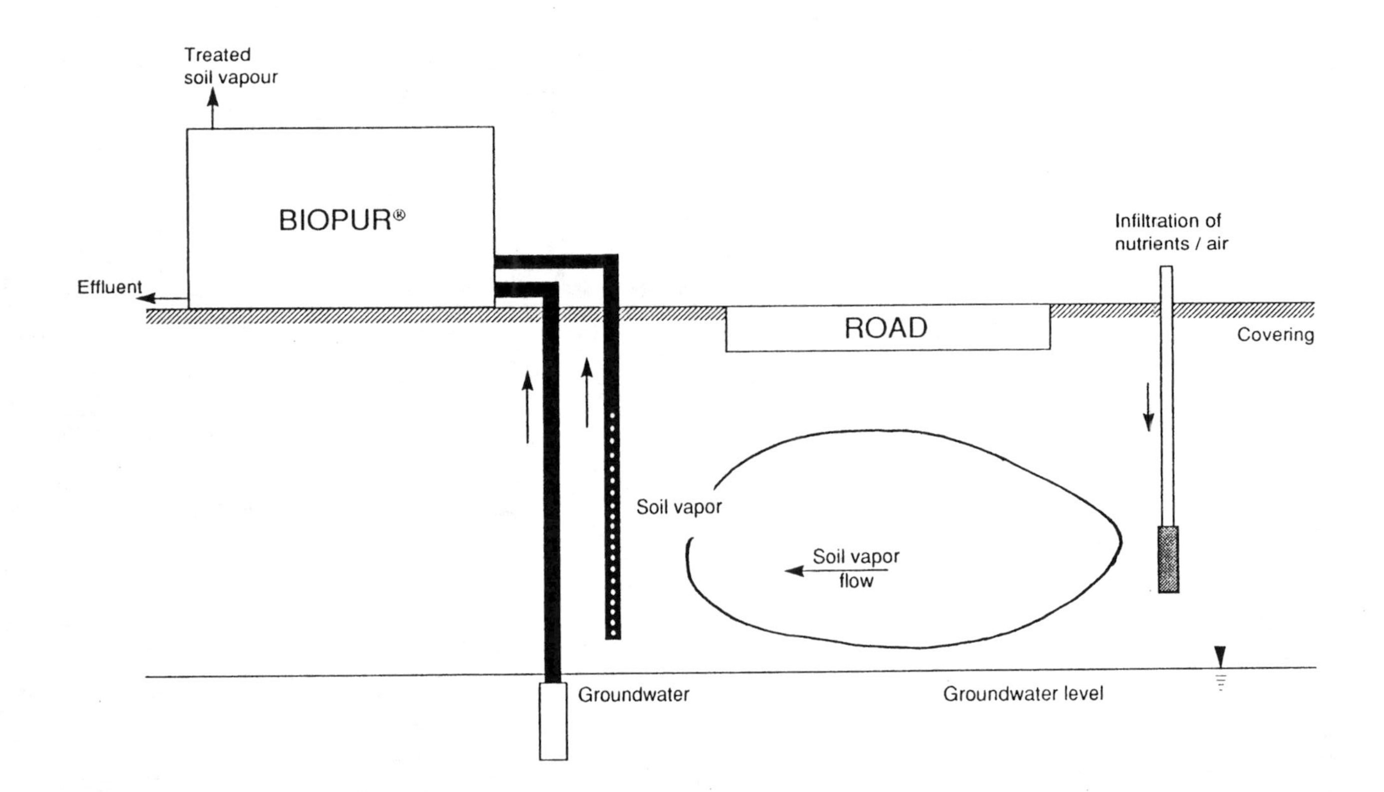

FIGURE 2. Cross-section of the site at Raalte.

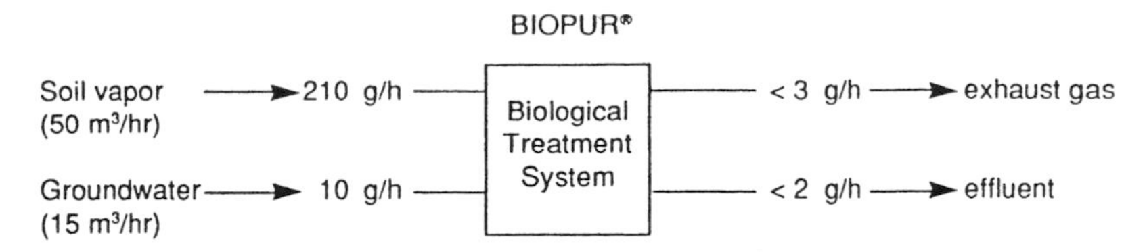

FIGURE 3. Mass balance of the biofilm reactor process.

In the effluent, the concentration of BTEX was usually <1 µg/L, which explains why the water board allows discharge into the surface water.

COSTS

To compare the costs of the biolfim reactor with those of alternative systems, the costs of four groundwater treatment techniques were estimated for a theoretical case. The case study involved groundwater contaminated with 5 mg BTEX/L and 10 mg total petroleum hydrocarbons/L to be withdrawn over a period of 2 years. Table 2 shows the results of the estimation. The estimation included installation, removal, and rent of the equipment; maintenance; and the possible addition of nutrients.

FURTHER RESEARCH

At gasoline stations, the biofilm reactor is inoculated with microorganisms present in the groundwater. However, it is also possible to inoculate with specific microorganisms requiring specific process conditions. By using specific biomass, recalcitrant halogenated hydrocarbons can be degraded. In our laboratory, pilot plant research is being carried out to investigate the use of fixed methanotrophic

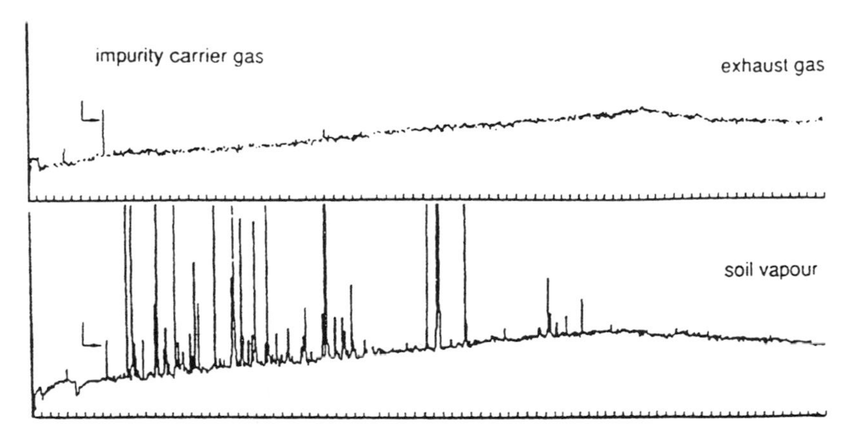

FIGURE 4. Chromatogram of soil vapor and exhaust gas.

TABLE 2. Estimation of the costs of several groundwater treatment systems in Dutch guilders ($ = 1.89 Dfl, March 1993).

Capacity (m^3/hour)	10	20	40
Stripper + activated carbon (vapor phase)	1.16	0.87	0.72
Stripper + compost filter (vapor phase)	0.85	0.54	0.36
Rotating biological contactor	0.78	0.65	0.59
BIOPUR® biofilm reactor	0.60	0.50	0.35

bacteria for the biodegradation. Preliminary results show removal efficiencies in excess of 90% for all compounds concerned (trichloroethylene, perchloroethylene, and 1,1,1-trichloroethane). In the near future, we will focus on full-scale research on the biodegradation of VOCs.

CONCLUSION

The patented biofilm reactor is available on the market and is suitable for the biological degradation of xenobiotics present in groundwater and soil vapor. High biomass concentrations are achieved so that the reactor is compact and short hydraulic retention times can be obtained. The cost estimation showed the biofilm reactor to be a cost-effective system, involving low maintenance costs. Moreover, the reactor can be used for the biological degradation of VOCs, which requires specific process conditions.

REFERENCES

Bouwer, E. J. 1992. "Microbial Remediation Strategies, Potentials and Limitations." Presented at Eurosol, European Conference on Integrated Research for Soil and Sediment Protection and Remediation, Maastricht.

Bouwer, E., J. Mercer, M. Kavanaugh, and F. Digiana. 1988. "Coping with Groundwater Contamination." *J. Wat. Poll. Contr. Fed. 60*: 1415-1427.

Coffa, S., L. G. C. M. Urlings and J. M. H. Vijgen. 1991. "Soil Vapour Extraction of Hydrocarbons In Situ and On Site Biological Treatment." NATO/CCMS International Meeting (Soil Remediation), Washington DC, November 18-22.

Committee BSB. 1991. "Bodemsanering in gebruik zijnde bedrijfsterreinen."

Hoek, J. P. van der, L. G. C. M. Urlings, and C. M. Grobben. 1989. "Biological Removal of Polycyclic Hydrocarbons, Benzene, Toluene, Ethylbenzene, Xylene and Phenolic Compounds from Heavily Contaminated Groundwater and Soil." *Environmental Technology Letters 10*: 105-194.

Oosting, R., L. G. C. M. Urlings, P. H. van Riel and C. van Driel. 1992. "BIOPUR®: Alternative Packaging for Biological Systems." In A. J. Draft and J. van Ham (Eds.), *Biotechniques for Air Pollution and Odour Control Policies*, pp. 63-70. Elsevier Science Publishers B.V.

Rittman, B. E., D. Jackson, and S. L. Storck. 1988. "Potential for Treatment of Hazardous Organic Chemicals with Biological Processes." In D. L. Wise (Ed.) *Biotreatment Systems*, Vol. III, pp. 15-64. CRC Press Inc. Boca Raton, FL.

TAUW Infra Consult B. V. 1992. *Possibilities of In Situ Soil Remediation in the Netherlands*. Report number R3206270.

Urlings, L. G. C. M., F. Spuij, S. Coffa, and H. B. R. J. van Vree. 1991. "Soil Vapour Extraction of Hydrocarbons: *In Situ* and On-Site Biological Treatment." In R. E. Hinchee and R. F. Olfenbuttel (Eds.), *In Situ Bioreclamation*, pp. 321-336, Butterworth-Heinemann, Stoneham, MA.

Veen, F. van, L. G. C. M. Urlings, and J. P. van der Hoek. 1989. "A Rotating Disc Biological Contactor Used on Pesticide Contaminated Groundwater Containing Chlorinated Organics." Presented at Hastech International Conference, San Francisco, CA. September 27-29.

Vree, H. B. J. R. van, L. G. C. M. Urlings, J. G. Cuperus, and P. Geldner. 1992. "In Situ Bioremediation of PAH, Applying Nitrate as an Alternative Oxygen Source at Laboratory and Pilot Plant Scale." Presented at Eurosol: European Conference on Integrated Research for Soil and Sediment Protection and Remediation, Maastricht, September 6-12.

A CRITICAL EVALUATION OF THE FUME PLATE METHOD FOR THE ENUMERATION OF BACTERIA CAPABLE OF GROWTH ON VOLATILE HYDROCARBONS

J. D. Randall and B. B. Hemmingsen

INTRODUCTION

Available methods for the enumeration of volatile hydrocarbon-utilizing bacteria are time consuming or may be inaccurate. For example, adding diluted sample to tubes of liquid mineral medium together with the desired hydrocarbon (with or without radiolabel) yields a pattern of growth measured as turbidity or the release of $^{14}CO_2$ from which can be calculated a most probable number (MPN) of hydrocarbon-degrading bacteria after weeks or months of incubation (Mills et al. 1977). Attempts to speed up the enumeration process include spread plating diluted sample onto agar-solidified mineral medium and incubation in fumes of the desired hydrocarbon (e.g., Horowitz & Atlas 1977a; Vecchioli et al. 1990). The colonies that appear are assumed to contain hydrocarbon-degrading bacteria, an assumption rarely tested despite the recognition some years ago that agar contains organic impurities (Leadbetter & Foster 1958) or can absorb volatile nutrients from the air (Hill & Postgate 1969) in amounts sufficient to support the growth of many environmental bacteria (Walker & Colwell 1976). Thus, agar plates incubated in fumes may have many colonies of non-hydrocarbon-degrading bacteria and, as a result, the population of hydrocarbon degraders may be overestimated. Sayler et al. (1985) provided evidence that such overestimation occurred when oiled laboratory microcosms were plated.

We report here a critical evaluation of the fume-plate method using water from the San Diego River and from aquifers contaminated with diesel fuel and gasoline. A more extensive account will be published elsewhere (Randall & Hemmingsen, manuscript submitted).

MATERIALS AND METHODS

Samples were collected from monitoring wells at each polluted site after about 20 L had been removed by hand-baling. The San Diego River was sampled about 9.1 km from the Pacific Ocean. The water was stored in sterile bottles in the dark between 4 and 8°C before analysis, which usually took place within 24 hours.

A sterile hand-held tissue homogenizer was used to break up visible clumps of particulate matter.

Serial 10-fold dilutions were made in filtered (0.22 µm pores), reverse osmosis water. Colony-forming units (CFUs) per mL were found by spreading aliquots of each dilution in duplicate onto the surface of one-tenth strength trypticase soy broth (Difco Laboratories, Inc., Detroit, Michigan) solidified with 2% (w/v) Baccto grade agar (1/10th TSB+A). After 14 days of incubation at 25°C, colonies were counted. Aliquots also were spread in sextuplicate onto mineral plates (Bogardt & Hemmingsen 1992) containing purified agar (Baltimore Biological Laboratories, Baltimore, Maryland) and no vitamins or cycloheximide. All mineral plates were incubated in 5-L Rubbermaid™ containers with lids at room temperature (RT, 22 to 24°C) for 26 days. Two sets of duplicate plates were incubated in the laboratory with small vials; one vial was empty and the other contained 50 µL of a 1:1 mixture of toluene and xylene (TX) added approximately biweekly. A third set of duplicates was incubated with an empty vial in a nearby office.

The MPN of TX-degrading bacteria was found by inoculating sets of 5 tubes of 10 mL each of the mineral medium (without agar and modified by substituting 0.5 g KNO_3 for NH_4Cl) from each dilution. Each tube initially received 10 µL of the TX mixture and subsequently as needed, usually biweekly; the tightly capped tubes were incubated at either RT or 25°C for up to 4 months. The appearance of turbidity in the liquid was scored as growth of TX-degrading bacteria. In the absence of TX, only those tubes inoculated with aliquots of the undiluted sample showed turbidity.

The MPN of bacteria containing TOL plasmids was found by first enriching this population (Assinder & Williams 1990) in 20 Mm *m*-toluate (Eastman Kodak, Rochester, New York) mineral medium. The sample was diluted in this medium, and 0.1 mL of each dilution added to each of 5 microtiter wells precharged with 0.1 mL of the same medium. After 10 to 14 days of incubation at RT, each well was probed for the presence of the TOL plasmid using the entire TOL plasmid isolated from *Pseudomonas putida* ATCC 33015, and the Genius System Nonradioactive DNA Labeling and Detection Kit (Boehringer Mannheim, Indianapolis, Indiana) (Korth 1992).

Two characteristics of the colonies growing on each set of plates were found by picking colonies at random and suspending each in 40 to 75 µL of 0.85% sterile saline; 7 to 10 µL each were then applied to two mineral agar plates and a 1/10th TSB+A plate. One mineral agar plate was incubated in TX fumes, the other in the no-fumes container. All plates were incubated at RT for 2 to 4 weeks. The patches on the 1/10th TSB+A plate were probed for TOL plasmids, as described earlier. Growth on all the mineral agar plates was compared visually; only those patches on the fume plates that were substantially heavier than those on the control plates were scored as positive for growth on TX fumes.

RESULTS AND DISCUSSION

For every water sample, the CFUs/mL on spread plates of mineral agar incubated in TX fumes were similar to those on the plates incubated without fumes

TABLE 1. Numbers of bacteria per milliliter of water from polluted and unpolluted sites using six different methods.

	Colony-forming units on				Most probable number of:	
		mineral agar plates incubated in:				
Site[a]	nonselective plates	TX fumes, in lab	no fumes, in lab	no fumes, office	TX-degrading bacteria[b]	bacteria with TOL plasmids[c]
MW1 (1/6/91)	7.1×10^6	5.8×10^6	4.3×10^6	5.6×10^6	8.0×10^5	1.3×10^6
TT1 (4/2/92)	2.6×10^4	6.6×10^4	5.8×10^4	5.9×10^4	8.0×10^2	1.1×10^3
SDR (3/5/92)	6.6×10^5	2.7×10^5	2.9×10^5	3.3×10^5	5.0×10^2	9.0×10^1

(a) MW1 water came from a gasoline- and diesel fuel-contaminated aquifer underwent bioremediation from 7/89 to 5/92. Downgradient water was extracted, nutrients added and the enriched water infiltrated through the site of the fuel tanks. MW1, upgradient of the site, had about 180 mg/L BTX (benzene, toluene, and xylenes found by gas chromatography as described in Korth [1992]) at the start, and about 50 mg/L at the end of nutrient addition. TT1, a monitoring well at a similar contaminated aquifer that is being treated by pump and treat with no nutrient addition; BTX was 13 mg/L. SDR, the San Diego River, a year-round urban stream, had a BTX concentration of 0.03 mg/L. Dates of sample collection are indicated.

(b) TX degrading bacteria: bacteria able to form visible turbidity in liquid mineral medium in the presence of toluene and xylene.

(c) Bacteria with TOL plasmids: bacteria that contain nucleotide sequences that hybridize with a probe made from the entire TOL plasmid of *Pseudomonas putida* ATCC 33015 (Korth 1992).

(Table 1). We thus conclude that $CFU_{FUMES} \approx CFU_{CONTROL}$, whereas if the fume plates were only growing TX-degrading bacteria, $CFU_{FUMES} >> CFU_{CONTROL}$. CFUs/mL found using the fume plate method were much higher than the MPN/mL of TX-degrading bacteria or TOL plasmid-containing bacteria (Table 1). We thus conclude that $CFU_{FUME} >> MPN_{TX\ or\ TOL}$, whereas if the fume plates grew only TX-degrading bacteria, $CFU_{FUME} \approx MPN_{TX\ or\ TOL}$.

The fume plates, control plates, and nonselective plates inoculated with each water sample grew colonies with approximately the same characteristics. For example, 54% of the 113 colonies taken from plates inoculated with MW1 water and incubated in TX fumes were positive both for growth in TX fumes and for the presence of the TOL plasmid; 40% of the 124 colonies from no fume control plates and 57% of the 80 colonies tested from 1/10th TSB+A also were double positive. When San Diego River water was used, none of the 312 colonies tested, regardless of their origin, was double positive, and >92% were negative for both characters.

Clearly, the use of agar-based medium and incubation in TX-fumes does not select for TX-degrading bacteria. Potentially any agar-based medium (e.g., oil agar, Horowitz & Atlas 1977b, Venkateswaran el al. 1991) inoculated with environmental samples will yield similar erroneous results if colony formation is used as the sole criterion for hydrocarbon utilization. If agar-based medium is to be used to enumerate specific environmental bacteria, it appears essential to conduct appropriate controls and experiments to prove that the physiological group desired is indeed being accurately counted.

ACKNOWLEDGMENTS

This work was supported by the Applied Microbiology Laboratory and the Associated Students of San Diego State University. We thank Kevin G. Korth for his advice and assistance with the TOL plasmid probe.

REFERENCES

Assinder, S. J., and P. A. Williams. 1990. "The TOL plasmids: Determinants of the catabolism of toluene and the xylenes." *Advances in Microbial Physiology 13*: 1-69.

Bogardt, A. H., and B. B. Hemmingsen. 1992. "Enumeration of phenanthrene-degrading bacteria by an overlayer technique and its use in evaluation of petroleum contaminated sites." *Applied and Environmental Microbiology 58*: 2579-2582.

Hill, S., and J. R. Postgate. 1969. "Failure of putative nitrogen-fixing bacteria to fix nitrogen." *Journal of General Microbiology 58*: 277-286.

Horowitz, A., and R. M. Atlas. 1977a. "Response of microorganisms to an accidental gasoline spillage in an Arctic freshwater ecosystem." *Applied and Environmental Microbiology 33:* 1252-1258.

Horowitz, A., and R. M. Atlas. 1977b. "Continuous open flow-through system as a model for oil degradation in the Arctic Ocean." *Applied and Environmental Microbiology 33:* 647-653.

Korth, K. G. 1992. "Microbial population changes during bioremediation of gasoline contaminated aquifer water: Use of a gene probe to enumerate toluene- and xylene-degrading bacteria containing TOL plasmids." M.S. thesis, San Diego State University, San Diego, CA.

Leadbetter, E. R., and J. W. Foster. 1958. "Studies on some methane-utilizing bacteria." *Archiv für Mikrobiologie 30:* 91-118.

Mills, A. L., C. Breuil, and R. R. Colwell. 1978. "Enumeration of petroleum-degrading marine and estuarine microorganisms by the most probable number method." *Canadian Journal of Microbiology 24:* 552-557.

Sayler, G. S., M. S. Shields, E. T. Tedford, A. Breen, S. W. Hooper, K. M. Sirotkin, and J. W. Davis. 1985. "Application of DNA-DNA colony hybridization to the detection of catabolic genotypes in environmental samples." *Applied and Environmental Microbiology 49:* 1295-1303.

Vecchioli, G. I., M. T. Del Panno, and M. T. Painceira. 1990. "Use of selected autochthonous soil bacteria to enhance degradation of hydrocarbons in soil." *Environmental Pollution 67*: 249-258.

Venkateswaran, K., T. Iwabuchi, Y. Matsui, H. Toki, E. Hamada, and H. Tanaka. 1991. "Distribution and biodegradation potential of oil-degrading bacteria in north eastern Japanese coastal waters." *FEMS Microbiology Ecology 86:* 113-122.

Walker, J. D., and R. R. Colwell. 1976. "Enumeration of petroleum-degrading microorganisms." *Applied and Environmental Microbiology 31*: 198-207.

MONITORING THE FATE OF BACTERIA RELEASED INTO THE ENVIRONMENT USING CHROMOSOMALLY INTEGRATED REPORTER GENES

C. W. Greer, L. Masson, Y. Comeau, R. Brousseau, and R. Samson

INTRODUCTION

Sensitive, reliable detection and monitoring of bacteria with the potential for environmental release is an essential element in the assessment of their ecological impact (Sussman et al. 1988). A direct monitoring system has been developed where the genes for lactose utilization (*lacYZ*) and for bioluminescence (*luxAB*) are integrated into a single site in the chromosome of the target organism providing a genetically stable, nontransferable marker system (Masson et al. 1993). Marked bacteria are virtually isogenic to the parent strain, and can be used as a model system to predict the environmental fate of the parent strain. The marked strains can be differentiated from indigenous bacteria and detected visually on solid media, as light-emitting blue colonies, at a level of sensitivity below 10 viable cells per gram of soil.

This study monitored the fate of a 2,4-dichlorophenoxyacetic acid (2,4-D) degrading *Pseudomonas cepacia* (BRI6001L), chromosomally marked with the *lac-lux* reporter genes, in an agricultural soil and in a soil from a wood treatment facility contaminated with pentachlorophenol (PCP), creosote, and copper/chrome/arsenic (CCA). The effects of temperature and of freezing and thawing also were examined.

EXPERIMENTAL

Pseudomonas cepacia BRI6001 was isolated from peat using 2,4-D as the sole carbon source (Greer et al. 1990). Construction of the *lac-lux* containing delivery vector, transformation and subsequent isolation of the chromosomally marked recipient, *P. cepacia* BRI6001L, has been described elsewhere (Masson et al. 1993).

Soil (50 or 100 g wet weight) microcosms were prepared in Erlenmeyer flasks for bacterial enumeration, or in serum bottles (20 g soil) for monitoring the consumption of [^{14}C]2,4-D. Serum bottles received a small tube containing 1 mL of 0.5 M KOH as a trap for evolved CO_2 (Comeau et al. 1993). Results are expressed as the percentage of introduced [^{14}C]2,4-D recovered as [^{14}C]CO_2. Sterile soils for microcosm experiments were prepared by g-irradiation as previously described (Comeau et al. 1993). Soil microcosms were incubated at 23°C or 10°C. In some cases (as indicated), the microcosms were frozen at −80°C on day 59, and were kept frozen for 3 weeks. All frozen microcosms were thawed at 23°C and sampled before being reincubated at the appropriate temperature (10 or 23°C).

Viable bacteria in soil, measured in colony-forming units (CFU) per gram, were determined by the spread-plate technique, using 1 to 2 g of soil (Greer et al. 1993). Aliquots (0.1 to 1.0 mL) of the dilution series in mineral salts medium (MSM) were plated on a rich nutrient medium (antibiotic medium no. 2, Oxoid) containing the chromogenic b-galactosidase substrate XGal (5-bromo-4-chloro-indolyl-b-D-galactopyranoside), or on a selective 2,4-D (3 mM) medium also containing XGal. Plates were incubated at 23°C for at least 1 week before counting. Colonies of BRI6001L (*lac-lux*) were differentiated from indigenous bacteria as blue colonies that emitted light when lifted onto membranes and exposed to *n*-decanal as previously described (Masson et al. 1993).

RESULTS AND DISCUSSION

The survival of BRI6001L in agricultural soil was monitored under various conditions (Figure 1). All soils were inoculated with 2.6×10^6 BRI6001L/g. In sterile soil containing 2,4-D (220 mg/kg) the population of BRI6001L increased to 10^8 CFU/g, a level expected, and previously observed (Masson et al. 1993) based on the amount of 2,4-D present. Mineralization of 2,4-D in this soil followed the growth curve of BRI6001L. In sterile soil in the absence of 2,4-D, BRI6001L was able to grow using the nutrients available in the soil, possibly derived from the indigenous bacteria killed during sterilization. In nonsterile soil, the population of BRI6001L decreased during the first 2 weeks, but did increase somewhat after this to achieve a stable population comparable to the starting level of 10^6 CFU/g. At 10°C in nonsterile soil, the population of BRI6001L followed the pattern observed at 23°C, but only slowly mineralized 2,4-D. Freezing the microcosms for approximately 3 weeks during the experiment and examining them after thawing and reincubation indicated that freezing had a minor effect on the total viable bacterial population (not shown) and the population of BRI6001L. The most notable change occurred in the sterile soils with and without 2,4-D where the population of BRI6001L decreased by less than 1 order of magnitude. BRI6001L also decreased in the nonsterile soil incubated at 23°C. The level of BRI6001L in nonsterile soil incubated at 10°C was unaffected by freezing and thawing.

These results indicate that BRI6001L survived and colonized sterile and nonsterile agricultural soil incubated at 23 or 10°C. An indigenous bacterial population capable of degrading 2,4-D (not shown) was detected in this soil, leading to an observed competition effect for nutrients. This was particularly evident when comparing the population levels of BRI6001L in sterile and nonsterile soil incubated at 23°C in the absence of 2,4-D.

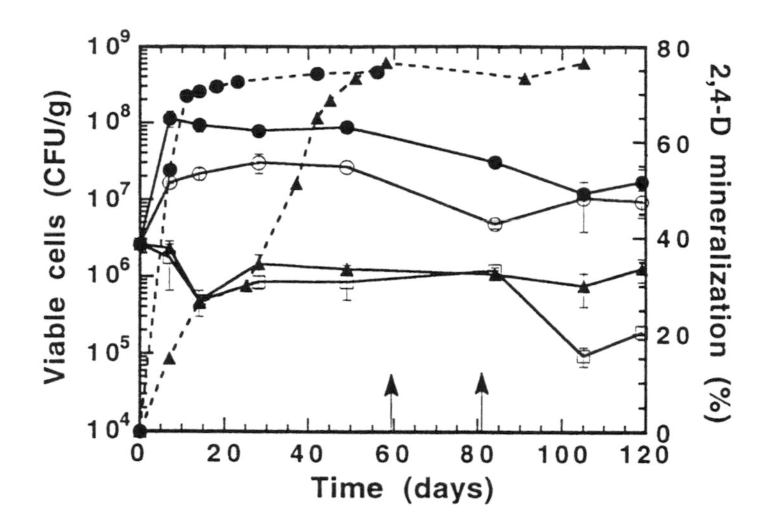

FIGURE 1. Survival and activity of BRI6001L in agricultural soil. Viable cells (solid lines) from microcosms incubated at 23°C were monitored in sterile soil with (J) and without 2,4-D (E), and in nonsterile soil without 2,4-D (G). Viable cells were monitored in nonsterile soil microcosms containing 2,4-D and incubated at 10°C (H). Mineralization of [^{14}C]2,4-D (dashed lines) was followed in sterile soil incubated at 23°C (J), and in nonsterile soil incubated at 10°C (H). Arrows delineate the time during which microcosms were kept frozen at −80°C.

The survival of BRI6001L also was monitored in a soil (YB) from a wood treatment facility contaminated with pentachlorophenol (PCP, 120 mg/kg), creosote (polycyclic aromatic hydrocarbons, or PAHs, 15 mg/kg), and copper/chrome/arsenic (CCA, 65.9/40.4/5.2), and in a soil (YC) collected from the forest adjacent to the treatment facility. BRI6001L increased in the presence of 2,4-D in YC soil, but mineralized 2,4-D at a slower rate than expected (Figure 2). This lower rate of substrate mineralization appears to be due to the acidity of the forest soil (pH 4.0), which is known to reduce the growth rate of BRI6001 (Greer et al. 1990). In YC soil without added substrate, the population declined by 1 order of magnitude after 30 days of incubation. In YB soil without 2,4-D, BRI6001L disappeared in 3 days. The inclusion of 2,4-D in this soil caused a substantial delay in the disappearance of BRI6001L, although the actual rate of substrate mineralization was extremely low. The role of 2,4-D in delaying the disappearance of BRI6001L is being examined. Additional data suggest that the toxic nature of the contaminated soil is not entirely due to the presence of wood-preserving compounds. Analysis of the granulometry of the soil indicated that it has a high concentration of fines (25%) relative to the forest soil (8.7%) suggesting that the physiochemical properties also are important.

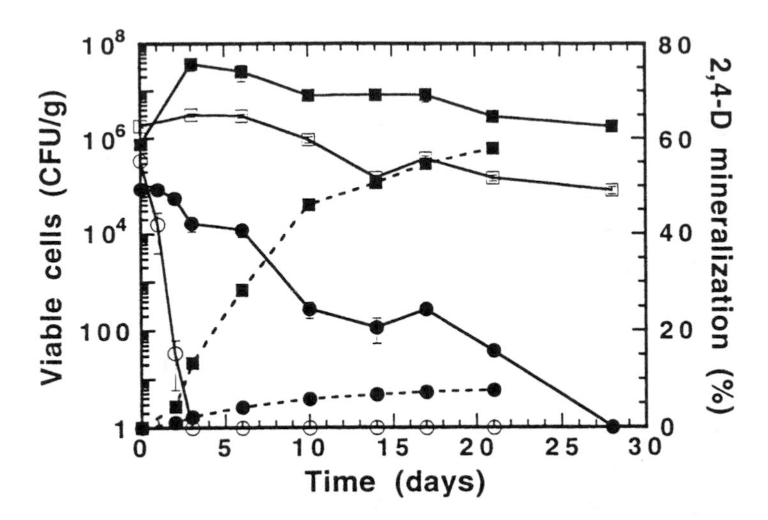

FIGURE 2. Survival and activity of BRI6001L in contaminated soil from a wood treatment facility. Viable cells (solid lines) were monitored in contaminated soil (YB) with (J) or without 2,4-D (E), and in pristine soil from an adjacent forest (YC) with (B) and without 2,4-D (G) incubated at 23°C. Mineralization of [^{14}C]2,4-D (dashed lines) was followed in YB soil (J) and YC soil (B).

CONCLUSIONS

A 2,4-D degrading bacterium, chromosomally marked with *lac-lux* reporter genes, provided an effective model system to monitor the survival and fate of this bacterium following its introduction into different soils. Although this bacterium survived in soil for extended periods (more than 4 months), the population levels that could be established were substrate (2,4-D) and contaminant-dependent. In addition, incubation temperature and competition by indigenous bacteria for available nutrients also affected the population level that could be established and maintained. A freeze/thaw cycle had a very minor impact on both total bacteria and BRI6001L.

ACKNOWLEDGMENT

The authors thank Anca Mihoc for her technical assistance.

REFERENCES

Comeau, Y., C. W. Greer, and R. Samson. 1993. "Role of inoculum preparation and density on the bioremediation of 2,4-D contaminated soil by bioaugmentation." *Appl. Microbiol. Biotechnol.* 5: 681-687.

Greer, C. W., J. Hawari, and R. Samson. 1990. "Influence of environmental factors on 2,4-dichlorophenoxyacetic acid degradation by *Pseudomonas cepacia* isolated from peat." *Arch. Microbiol. 154*: 317-322.

Greer, C. W., L. Masson, Y. Comeau, R. Brousseau, and R. Samson. 1993. "Application of molecular biology techniques for isolating and monitoring pollutant-degrading bacteria." *Water Poll. Res. J. Can.* (in press).

Masson, L., Y. Comeau, R. Brousseau, R. Samson, and C. W. Greer. 1993. "Construction and application of chromosomally integrated *lac-lux* gene markers to monitor the fate of a 2,4-dichlorophenoxyacetic acid-degrading bacterium in contaminated soils." *Microbial Releases* (in press).

Sussman, M., C. H. Collins, F. A. Skinner, and D. E. Stewart-Tull (Eds.). 1988. *The Release of Genetically-Engineered Micro-Organisms.* Academic Press, London, UK.

ENZYME-ENHANCED BIOREMEDIATION

R. H. Meaders

INTRODUCTION

FyreZyme™ is a proprietary bioremediation enhancing agent that is stated to contain extracellular enzymes, microbial nutrients, and bioemulsifiers. The product is diluted from concentrate to a 4% aqueous solution. The product has been shown to enhance bioremediation under aerobic conditions. Full-scale field trials have demonstrated rapid reduction of petroleum product contamination in soil. Bench and full-scale results are presented, as reported by end-users.

MATERIALS AND METHODS

In all test circumstances, the product is utilized in a 4% solution, calculated to deliver the equivalent of 1 gallon of concentrate per 8 yd^3 of contaminated soil. The total volume of the 4% solution is divided into equal aliquots, with an equal amount applied to the contaminated soil weekly. Some variation in volume is allowed, depending on test conditions. Treated soil is maintained at approximately 80% of soil water-holding capacity, and aerobic conditions are maintained. Six cases are presented in the following sections.

CASE 1: BENCH SCALE

A three-cell test was performed comparing the product with fertilizer and an untreated control in the remediation of fresh gasoline. Three test cells were constructed, each containing approximately 1 yd^3 of uncontaminated topsoil, and 1-1/2 gallon of unleaded gasoline was added to each cell. Then 2 pounds of soluble 20% nitrogen fertilizer and 5 gallons of water were added to cell no. 1. Only 5 gallons of water was added to cell no. 2, and 5 gallons of 4% aqueous solution of the product was added to cell no. 3. The addition of nitrogen was calculated to be approximately equal in cell nos. 1 and 3. Test materials were added at days 1, 7, and 14. All cells were maintained at approximately 80% of field holding capacity (no leachate formation). Grab samples were obtained from corresponding areas of each cell on days 1, 7, 14, and 21 for testing of TPH and individual contaminants (benzene, toluene, ethyl benzene, and total xylene). At the termination of the test (21 days), the treated cells were each flooded and leachate collected to analyze the contaminants as well as nitrogen compounds.

Test Results. Table 1 demonstrates that cell no. 3 proceeds to more complete remediation of the petroleum products than the control or fertilizer-treated cells. On testing of forced leachate on day 21, cell no. 1 demonstrated the highest residual nitrogen compound levels, whereas cell nos. 2 and 3 demonstrated significantly lower levels. The tests demonstrate "cycling" of levels of contaminants common to bioremediation, with spikes in value occurring as soil-bound contaminant is released. Not demonstrated are the bacterial total plate counts (TPC) which demonstrated a significant decrease in bacterial colonies in cell no. 1 and an increase in cells nos. 2 and 3.

CASE 2: BENCH SCALE

Soil contaminated with diesel fuel and chemicals from an oil company flare pit was subjected to a treatability study. A lined berm of containing approximately

TABLE 1. Three cell soil and leachate test.

	Start	Day 7	Day 14	Day 21
Total Petroleum Hydrocarbons (TPH) (mg/kg)				
Cell 1	53,000	230	2,000	2,400
Cell 2	60,000	220	17	870
Cell 3	72,000	1,300	88	56
Benzene (μg/kg)				
Cell 1	1,900,000	10	200	400
Cell 2	2,000,000	10	10	40
Cell 3	2,000,000	100	10	10
Ethylbenzene (μg/kg)				
Cell 1	1,500,000	2,000	49,000	50,000
Cell 2	1,700,000	1,700	120	9,900
Cell 3	1,800,000	15,000	400	10
Total Xylene (μg/kg)				
Cell 1	7,800,000	14,000	320,000	340,000
Cell 2	7,100,000	15,000	14,000	65,000
Cell 3	7,300,000	110,000	6,000	370
Leachate Testing		Cell 1	Cell 2	Cell 3
Nitrates		50	5.1	4.3
Nitrites		1.9	0.33	0.6
Total Nitrogen		1,500	27	15

Note: Cell 1 = Fertilizer, Cell 2 = Control, Cell 3 = Product.

TABLE 2. Total petroleum hydrocarbons (TPH) in mg/kg and polycylic aromatic hydrocarbons (PAH) in µg/kg.

	Pre-Treatment	Day 21
TPH	60,000	440
Acenaphthene	5,200	ND
Benzo-a-anthracene	1,100	ND
Benzo-b-fluoranthene	620	ND
Benzo-k-fluoranthene	750	ND
Benzo-g,h,i-perylene	2,100	ND
Dibenzo-a,h-anthracene	520	ND
Fluoranthene	15,000	ND
Naphthalene	700	ND
Phenanthrene	2,000	ND
Pyrene	5,200	ND

1 yd^3 was constructed. A 4% solution of the product was applied on days 1, 7, and 14. Between applications, the soil was maintained at approximately 20% of field holding capacity. No tilling of the soil was done. Samples were taken from mid-depth (7 inches) and at the bottom of the berm (14 inches) at the onset and on day 21.

Test Results. Table 2 demonstrates the reduction in total petroleum hydrocarbon (TPH) and in polynuclear aromatic hydrocarbons (PAH) in the 21 days of the test. Full-scale project awaits favorable climatic conditions.

CASE 3: FULL-SCALE

The oil disposal pit of a closed service station was excavated, and 30 yards of contaminated soil was placed inside a lined berm 14 inches deep. The product was applied on days 1 and 7. The soil was turned once on day 7 to break up a crust caused by rapid evaporation under extreme high temperature. Moisturization was maintained at 15-20% by weight. Samples were drawn from mid-depth and from the bottom of the berm at the side, middle, and ends on days 1, 7, and 14. At the client's request, no control berm was constructed.

Test Results. Figure 1 demonstrates the drop in TPH. At the end of the 14th day, TPH levels had fallen from the original 700 mg/kg to <10 mg/kg. The site has been closed and sold as commercial property.

CASE 4: FULL-SCALE

A leaking underground storage tank site was excavated. Approximately 400 yd^3 of soil was placed in a 14 inch-deep lined berm. The product was applied

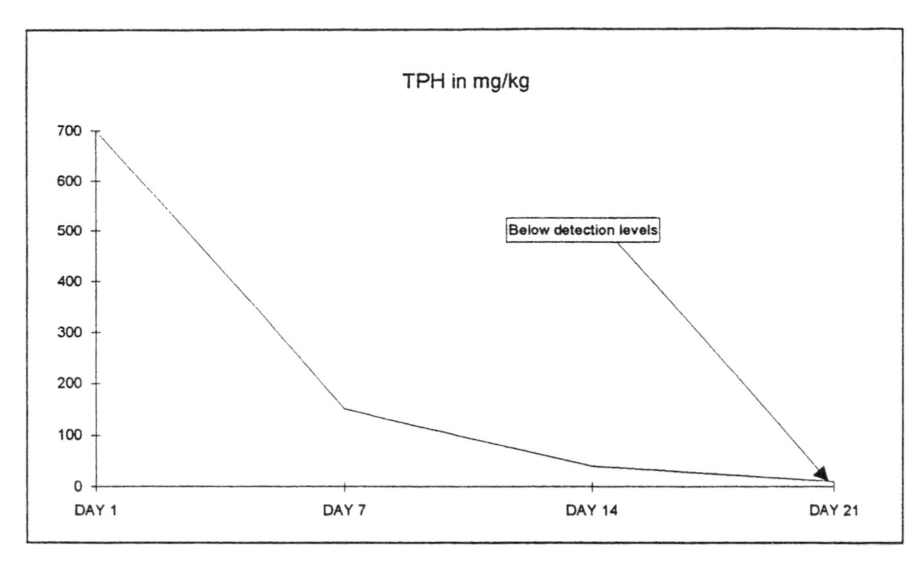

FIGURE 1. Service station disposal pit: motor oil, hydraulic fluids, and brake fluid.

on days 1, 14, and 21. The bottom of the pit, approximately 14 inches deep with loose soil, was used as an untreated control. Moisturization was maintained at approximately 15 to 20% by weight.

Test Results. Figure 2 demonstrates the initial and final TPH, benzene, and xylene determinations. The control area was essentially unchanged, while the treated berm was below regulatory limits. The testing demonstrated a need for further excavation of the pit bottom, which was accomplished as the second phase, now awaiting favorable climatic conditions for bioremediation.

CASE 5: FULL-SCALE

Soil from a chemical company blending site was contaminated with "mineral spirits," diesel fuel, TCE (trichloroethylene), and DCE (dichloroethylene). The soil was excavated and placed on two 6-ft-high lined soil piles, one of 2,000 yd^3 and one of 4,500 yd^3. Slotted polyvinyl chloride (PVC) pipe was placed in a manifolded grid at the bottom, at the 2-ft level, at the 4-ft level, and on the top. The product was introduced through the PVC in the top three manifolds on days 1 and 7. These same pipes were used to introduce tap water as needed to maintain an estimated moisture level of 15 to 20% by volume. Atmospheric air was pumped from a compressor through the bottom three manifolded layers daily for 8 hours. A low air volume was utilized to minimize volatilization. Test samples were drawn at the ratio of one test per 100 yd^3 from flagged sites at varying depths and averaged for reporting purposes.

Test Results. Figures 3 and 4 demonstrate the decrease in TPH as well as TCE and DCE over the 14-day course of the remediation. All tests were below detection at close of remediation.

CASE 6: PILOT SCALE

Approximately 250,000 individual barrels of a sludge, soil, and water mix from a refinery site were evaluated for reduction in TPH below 4,000. A single 200-L barrel was opened, product was applied to equal 4% by volume of the barrel contents, and the barrel was recapped loosely. No stirring of the mixture was done, and no aeration was provided.

Test Results. At 60 days, the barrel top was removed and TPH testing performed. An untreated, opened, and then closed barrel was used as a control. Table 3 demonstrates the TPH reduction from 250,000 to 3,500. Not demonstrated is a reduction in percent solids from 65% in the control barrel to less than 15% in the product barrel.

DISCUSSION

It is thought that the enzyme content and microbial nutrients of the product initiate abiotic transformation of petroleum product while stimulating exponential microbial growth. This combination is believed to favor petroleum degraders in the overall soil or water microbial community. Bioemulsifiers contained in the product are intended to increase the bioavailability of previously soil-bound

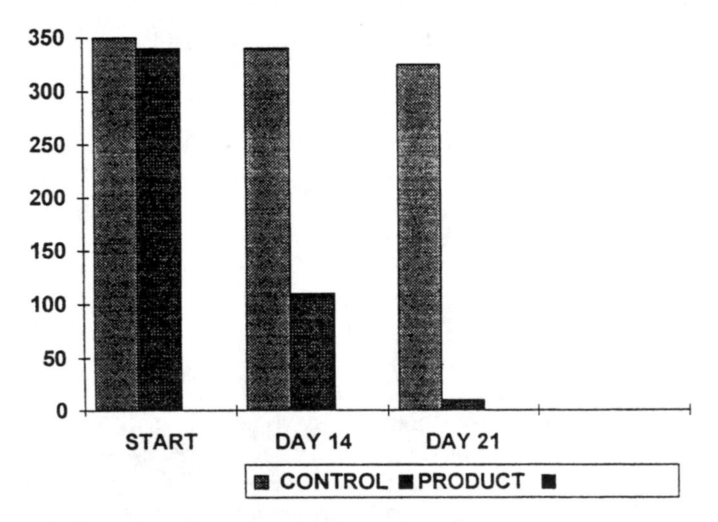

FIGURE 2. Underground storage tank.

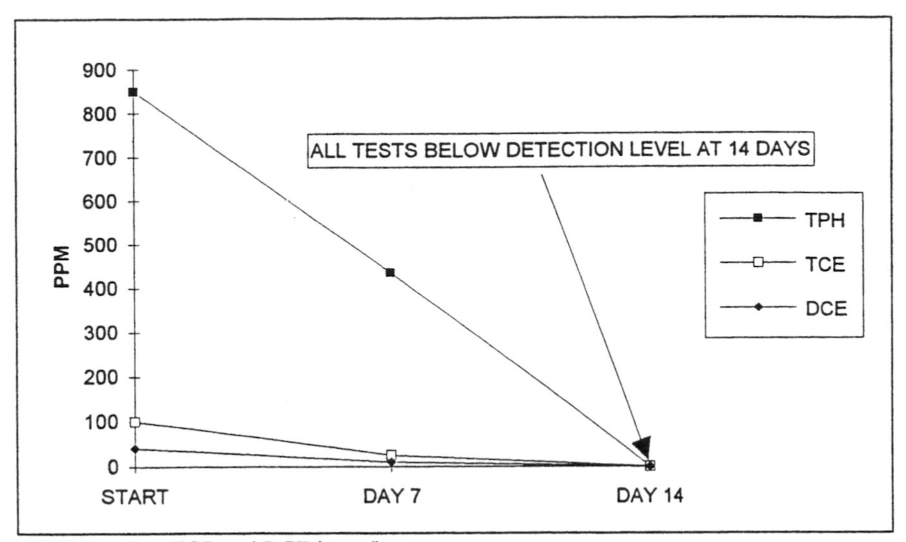

TPH in mg/kg, TCE and DCE in µg/kg

FIGURE 3. Chemical company blending area, soil pile no. 1.

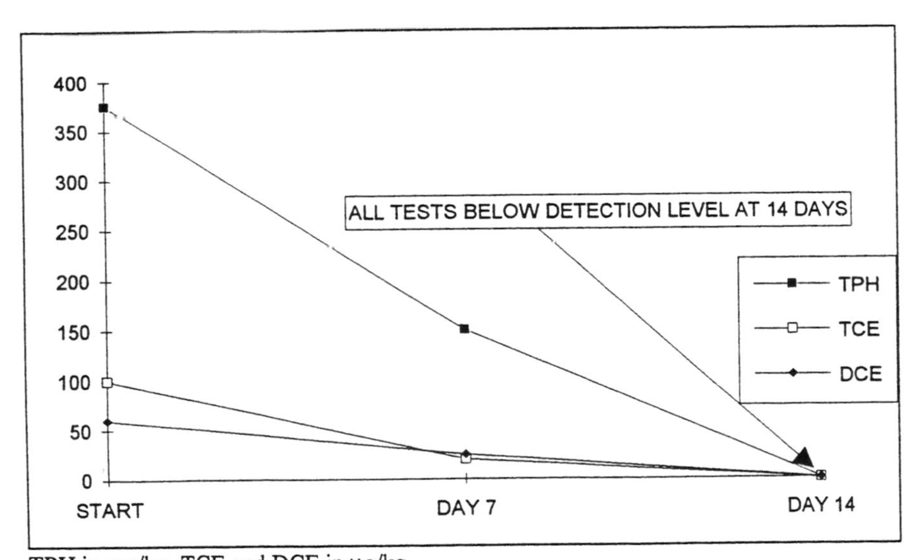

TPH in mg/kg, TCE and DCE in µg/kg

FIGURE 4. Chemical company blending area, soil pile no. 2.

TABLE 3. Refinery sludge, soil, water TPH reduction, mg/kg.

Start	60 Days
250,000	3,500

petroleum product and further enhance bioactivity. Careful environmental control is necessary to maintain aerobic conditions and adequate moisturization for best performance.

The product appears to promote rapid bioremediation of petroleum products, including some of the more recalcitrant polynuclear aromatic hydrocarbons and halogenated organic compounds. Carefully controlled studies are being undertaken in conjunction the Bioremediation Product Evaluation Center (BPEC) of the National Environmental Technology Applications Corporation (NETAC). These studies are utilizing appropriate controls and standards to compare the product's ability to enhance bioremediation of No. 2 fuel oil in a standard soil matrix compared to inorganic nutrient solutions against background biodegradation and abiotic loss.

MICROORGANISMS ADAPTATION FOR OIL-ORIGINATED SUBSTANCES TREATMENT

K. Miksch and J. Surmacz-Gorska

INTRODUCTION

When using biological methods to remove oil-originated pollutants, it is important to choose the appropriate inoculum and to carefully monitor the adaptation of the microorganisms. It is essential to control the degree of acclimatization of activated sludge or biofilm respectively to the specific contents of wastewater in the influent, and to precisely determine the time and concentration for possible further use in adaptation.

To determine the degree of acclimatization in the continuous activated sludge process, and to assess to what degree the contents of the refinery wastewater supplied in the influent could be increased, the dynamics of the batch process were studied. For this purpose, activated sludge was taken from the continuously activated models and placed in chambers for periodic testing. Next, the influent of the same composition as used in the given period, in continuous culture, was introduced to one of the chambers. At the same time, a nutrient was introduced to other chambers in which the refinery wastewater contents was higher, i.e., such as was to be used next in continuous testings. The process dynamics in the chambers containing increased amounts of oil pollutants were compared with the process dynamics taking place in the chamber of the composition used in continuous studies to estimate the probable effect of the increased contents of refinery on the wastewater biocenosis of activated sludge.

EXPERIMENTAL METHODS

The models of biological treatment of wastewater were studied in one- and two-stage activated sludge systems. The inoculating material was the activated sludge purifying the municipal sewage. The adaptation of the microorganisms to oil pollutants was started by introducing 0.5% of refinery wastewater to the municipal sewage feed. In the successive stages of adaptation, based on studies of the process dynamics during periodic introduction of the substrate, the amount of refinery wastewater supplied in the nutrient was increased.

If the experiments demonstrated that the activated sludge had not yet adapted completely to the wastes used, and the increased percentage of oil pollutants in the supply was not favorable to its living processes, a nutrient of the same composition as during the testings of the batch process was fed to the models for continuous testings. The experiments using the same substrates were repeated until the activated sludge became sufficiently acclimatized and the increased percentage of refinery wastewaters did not inhibit the living activities of microorganisms.

RESULTS

The studies of the dynamics of biodegradation had the specific aim of determining the ability of activated sludge for use as a substrate for the components of refinery wastewater. These studies primarily sought to control the degree of acclimatization of the microorganisms to these components and to ascertain the possibility of increasing their amount in the nutrient supplied. For this reason, not all the results of the investigation series are discussed here, but only the typical ones that illustrate achievement of the experimental assumptions.

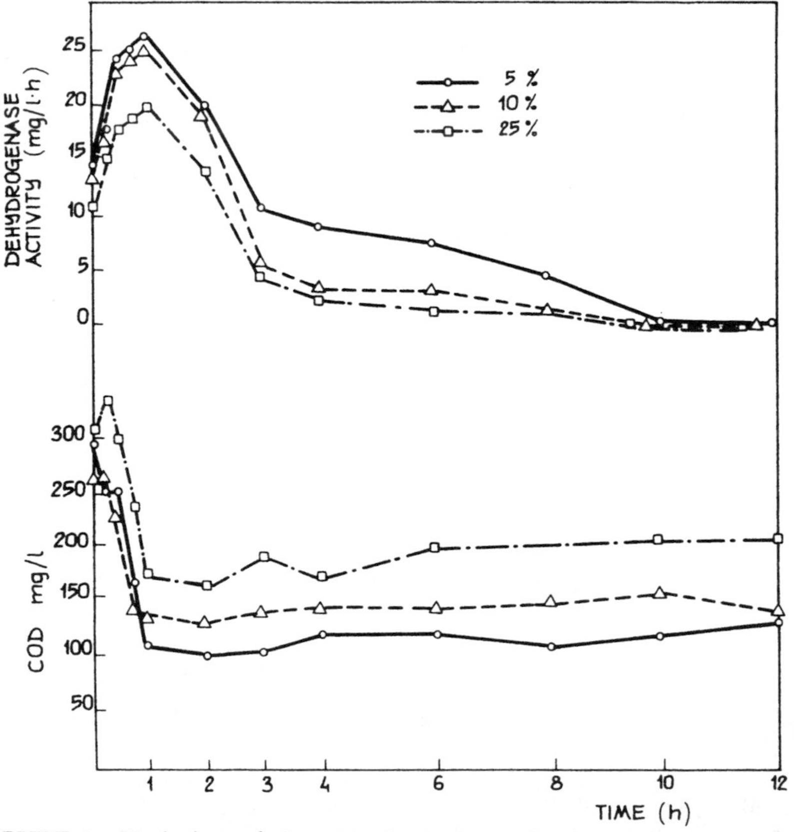

FIGURE 1. Variation of the organic compounds concentration and of the enzymatic activity in the batch process (well-adapted activated sludge, T = 35°C, MLSS = 2.3 g/L).

One example of successful adaptation of microorganisms to the components of refinery wastes occurred in the series using activated sludge taken from a one-stage system (Figure 1). After 1 hour of aeration, the total quantity of organic compounds that could possibly be used by the microorganisms was removed. The lack of available organic compounds in the nutrient set off nitrification processes in the chambers.

This series of tests was performed for the activated sludge after 4 days of adaptation to the nutrient containing 5% refinery wastes. The test results show that 4 days was sufficient for complete acclimatization of the sludge. The fact that a shorter period might prove insufficient is evidenced by the results of a series for the sludge taken from the aeration chamber of the first stage of a two-stage system 2 days after batching of the nutrient began, containing, as in the previous series, 5% of refinery wastewater. A decrease in chemical oxygen demand (COD) took place here much more slowly (Figure 2).

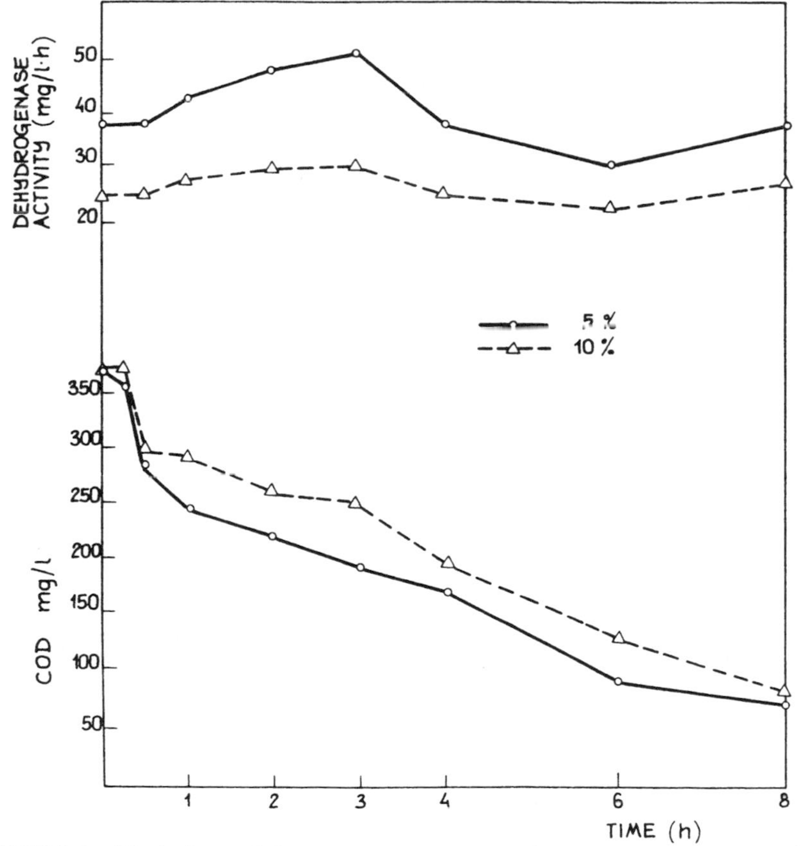

FIGURE 2. Variation of the organic compounds concentration and of the enzymatic activity in the batch process (not-too-well-adapted activated sludge, T = 35°C, MLSS = 3.2 g/L).

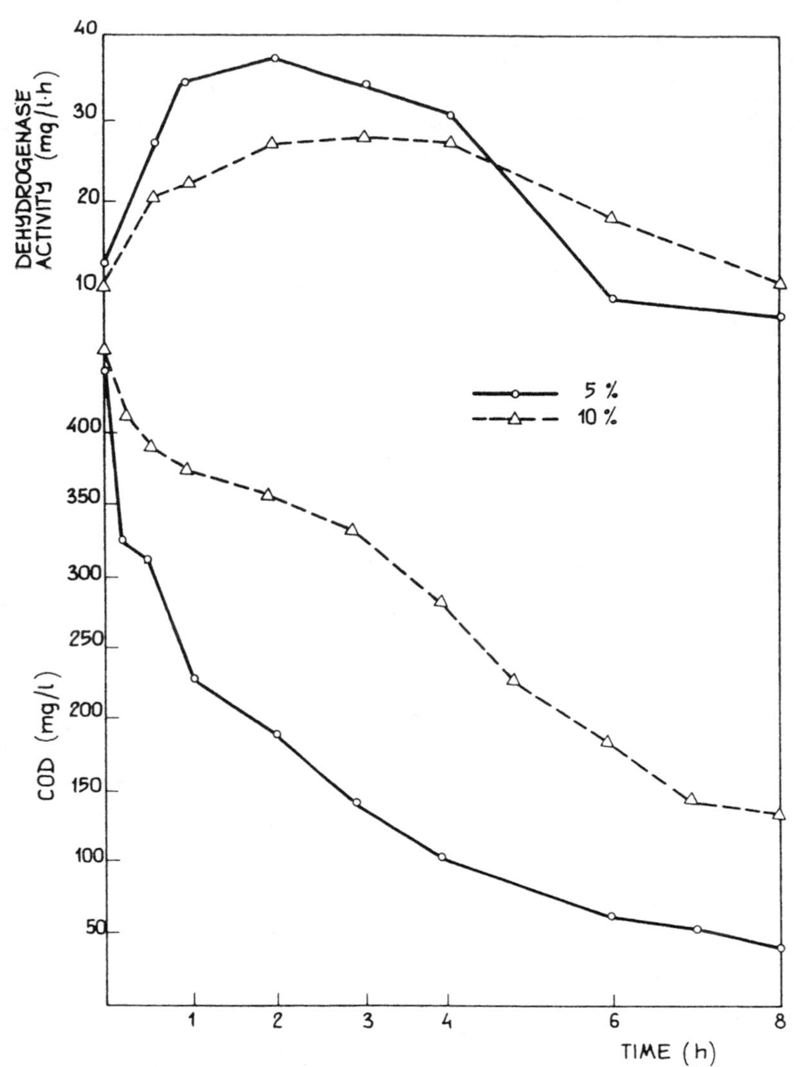

FIGURE 3. Variation of the organic compounds concentration and of the enzymatic activity in the batch process (activated sludge well-adapted to 5% contents of refinery wastewater in the influent, T = 35°C, MLSS = 3.1 g/L).

These tests show that an increased quantity of oil pollutants in the feed has no toxic effect on activated sludge. However, it also has been observed that the microorganisms still are not well adapted. This conclusion was confirmed a few days later (Figure 3). The rate of substrate removal greatly increased in the chamber with 5% contents of refinery wastewater, but in the chamber with 10% of these wastes, this rate was almost the same as in the previous series. For this reason, during the following days the amount of refinery wastewater in the influent was not further increased.

In the last testing, only industrial wastes were batched, with no admixture of municipal wastes. The average acclimatization time of the microorganisms lasted about 10 days, while the contents of industrial wastewater in the particular stages of acclimatization was 0.5, 1, 2, 3, 5, 10, 25, 50, 75, and 100%.

CONCLUSION

The investigations confirmed the usability of the proposed method of controlled adaptation of activated sludge for substances, that are difficult to assimilate in a continuous culture. The method is simple, does not require complex apparatus, and provides results that are easy to interpret. The method consist of conducting, during a continuous process, periodic tests of the process dynamics using various concentrations of assimilable substances to answer the following questions:

1. To what degree are the microorganisms adapted to the nutrient used in the particular stage of adaptation?
2. At which moment is it possible to begin the next stage of continuous adaptation while increasing the concentration of the substance to which the microorganisms are adapted?
3. What is the permissible increase of concentration of the substance in the nutrient supplied?

PENTACHLOROPHENOL DEGRADATION BY MICROENCAPSULATED FLAVOBACTERIA AND THEIR ENHANCED SURVIVAL FOR IN SITU AQUIFER BIOREMEDIATION

K. E. Stormo and R. L. Crawford

INTRODUCTION

Bacteria introduced to a foreign environment often fail to adjust to the new environmental conditions. This is especially true when the cells are grown under conditions that allow rapid growth and a high cell yield, and specific compounds are added to induce specific enzymatic pathways. When such cells are added to aquifers, they may undergo extreme physiological stress because the groundwater contains low concentrations of nearly all materials needed by them (Acea et al. 1988). Organisms introduced into the environment have a distinct advantage if they have been combined with some type of carrier material (Aarons & Ahmad 1986, Bashan & Levanony 1988, Berg et al. 1988, Fravel & Marois 1986, Jawson et al. 1989), which may be agar (Jawson et al. 1989), peat (Aarons & Ahmad 1986, Berg et al. 1988), alginate (Bashan 1986), alginate-clay (Fravel et al. 1985), or fluid gels (Jawson et al. 1989). These carriers allow a high concentration of organisms to be incorporated into the soil, and they often contain nutrients and serve as a moisture reserve for cells. They also help isolate the cells from predators. A number of researchers have shown that control of predation is an important biological factor in the survival of introduced organisms (Acea et al. 1988, Compeau et al. 1988, Corman et al. 1987, van Elsas et al. 1986, Vargas & Hattori 1986).

METHODOLOGY

We have developed methods to allow the microencapsulation of bacteria and nutrients in beads on a large scale with small enough bead diameters to travel through aquifers. These microbeads were introduced into aquifer microcosms that had pentachlorophenol (PCP)-contaminated groundwater flowing through at in situ flowrates.

For this work we employed *Flavobacterium* ATCC 39723, a Gram-negative aerobe that degrades a variety of chlorinated phenols such as PCP. The bacterium was grown in a defined mineral salts medium using Na-glutamate as its carbon and energy source (O'Reilly & Crawford 1989). When cells reached mid-logarithmic phase, PCP (50 mg L^{-1}) was added to induce the catabolic enzymes responsible for degradation of chlorinated phenols. When the PCP had degraded, the cells were harvested by centrifugation (O'Reilly & Crawford 1989).

The cells were immobilized following the procedure developed by Stormo and Crawford (1992) that allows large quantities of very small beads (2 to 50 µm) to be produced. An agarose-cell suspension at 45°C was sprayed through a low-pressure nozzle into cold buffer to solidify the beads. They were collected at 4°C by gravity settling and low-speed centrifugation, washed in buffer, and stored at 4°C.

Aquifer samples collected with an auger from three wells at the University of Idaho Ground Water Research Site at depths of up to 5 m (saturated zone in the aquifer begins at about 2.5 m) were stored at 4°C. Groundwater samples were pumped as needed from nearby wells with ample purging time to replace the water that had been in the well casing with fresh pore water. Samples were collected by completely filling containers that were stored at 4°C until needed. The water was spiked with various concentrations of PCP and introduced into the aquifer columns.

The previously collected aquifer material was packed into 24 columns (2.2 cm by 25 cm) with rubber stoppers and two pieces of Whatman 3MM filter paper at the outlet end. The columns were packed in three lifts, and each lift was compacted by sharply striking the outlet stopper on a table with the column in a vertical position. Normally only two impacts were needed before adding the next lift. Care was taken to keep void spaces from forming and to completely fill the columns. The top rubber stopper was placed on the column and the trapped air was vented from under the stopper as it was pushed down onto the saturated aquifer material. Then 22-gauge Vacutainer needles were inserted into tightly fitting Teflon™ tubing and into the stopper so that the outlet needle just protruded from the inside of the bottom stopper but did not pierce the filter papers. The inlet needle went into the side of the top stopper and just protruded into the top of the column. The inlet tubes from all columns were placed in a common container that contained groundwater spiked with PCP.

Some sterile columns were prepared by irradiation at the Washington State University Radiation Center. Columns placed near a ^{60}Co source received 5×10^6 rads of radiation. In some of the columns the free *Flavobacterium* cells were mixed with the aquifer material before packing the columns at a density of 5×10^7 cells per gram of aquifer material. Others had agarose microimmobilized *Flavobacterium* mixed into the aquifer material at the same cell density as above. The majority of columns received identical numbers of free cells or the agarose micro-immobilized *Flavobacterium* injected at a flowrate of 10 mL day^{-1}.

The outlet tubes were connected by 22-gauge needles into individual syringes mounted on an apparatus that allowed the syringe plunger to be withdrawn at a constant slow rate; 18 columns had 10-mL syringes with a flowrate of 5.0 mL day^{-1}, and additional columns had 3-mL syringes at 2.0 mL day^{-1} or 30-mL syringes at 12 mL day^{-1}. The effluent was collected daily for 170 days and analyzed by ultraviolet (UV) spectroscopy or by high-performance liquid chromatography (HPLC) for remaining PCP. Selected samples also were analyzed by a chloride electrode and ion chromatography for chloride concentration to confirm PCP degradation (Levinson et al. 1993).

RESULTS

No statistically significant differences were observed in the degradation rates between free and encapsulated *Flavobacterium* in sterile or native aquifer material as tested in these experiments. Two representative figures show the degradation of PCP is by free *Flavobacterium* in sterile microcosms (Figure 1) and by *Flavobacterium* in beads injected into native aquifer microcosms (Figure 2). The columns in Figure 2 had a number of interesting effects due to the flow stopping, starting again, and then reducing to 2 mL day^{-1} for the remaining time. This set of columns was able to maintain an outlet concentration of 50 mg L^{-1} throughout the final 20 days, despite an inlet concentration of 275 mg L^{-1}. The introduced bacteria were capable of reducing the PCP levels to near-detection limits until the inlet concentration exceeded 150 mg L^{-1} of PCP in groundwater.

After completion of the degradation studies in the columns, the PCP-contaminated groundwater flow into the columns was stopped and the columns

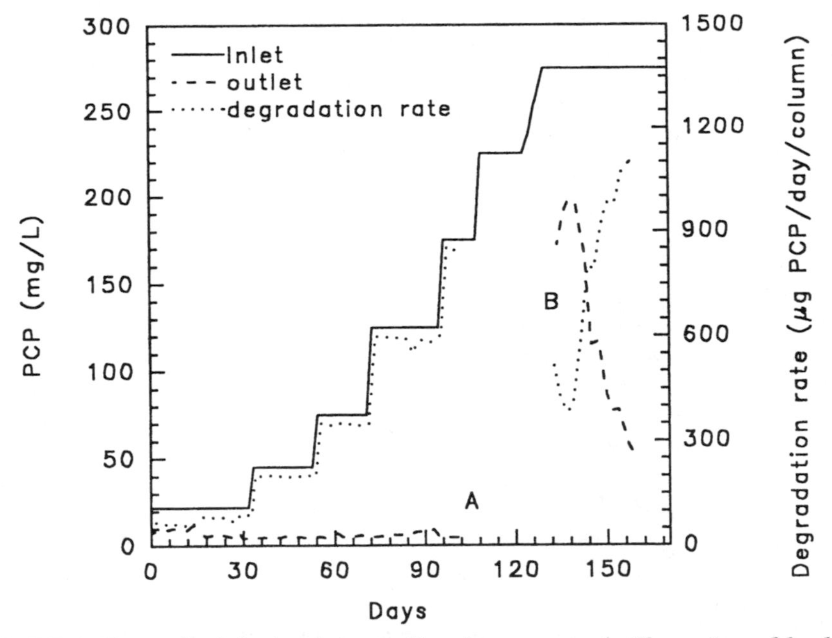

FIGURE 1. Free cells injected into sterile microcosms. A. Flowrate suddenly stopped. B. Flowrate started again at 5 mL day^{-1}.

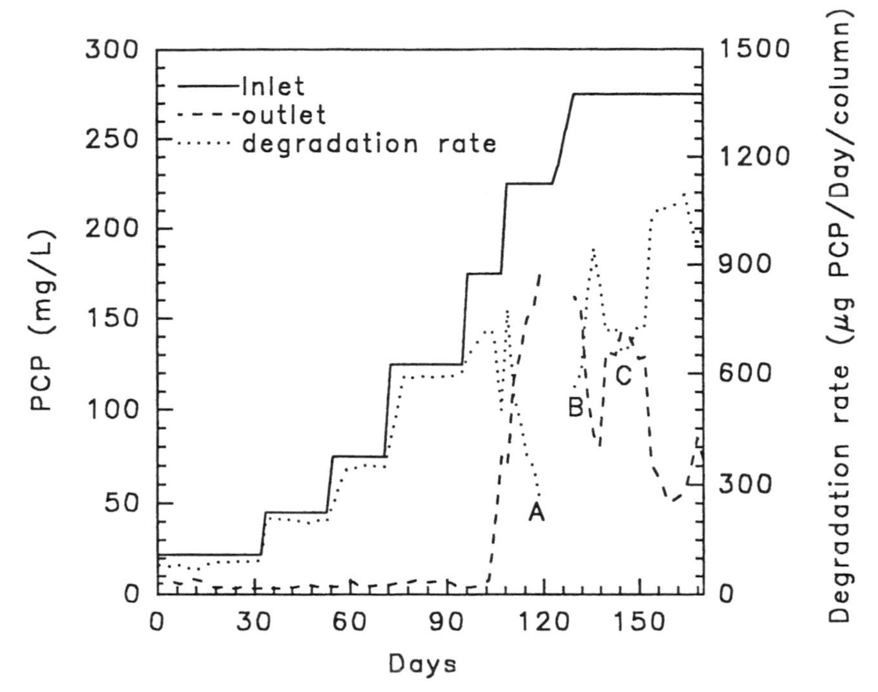

FIGURE 2. Cells in beads injected into native aquifer microcosms. A. Flowrate suddenly stopped. B. Flowrate started again at 5 mL day^{-1}. C. Flow slowed to 2 mL day^{-1}.

were stored at 20°C. Selected columns were sacrificed at 13 and 24 months after storage to determine the survivability of the introduced *Flavobacterium* under conditions of starvation, predation, and no water flow. After 13 months, replicate enrichment flasks were made with 30 g of aquifer material from selected columns removed from storage and placed in 50-mM HEPES at 7.2 pH and supplemented with 0.5 g L^{-1} of glutamate and 50 mg L^{-1} of PCP. The only enrichments from columns after 13 months of storage that showed PCP degradation were columns inoculated with agarose-immobilized *Flavobacterium* or injected with agarose-immobilized *Flavobacterium* (Figure 3). Similar enrichments were made at 24 months, except that 30-, 3-, 0.3-, and 0.03-g enrichments were made from a larger selection of treatments and the PCP was 75 mg L^{-1}. The only additional treatment that showed enrichment for PCP degradation after 24 months was from the free *Flavobacterium* injected into sterile columns (Figure 3). The agarose-immobilized *Flavobacterium* columns that could be enriched for PCP degradation at 13 months also could be enriched at 24 months using 30-g inoculations. The other inoculation amounts and treatments did not yield positive enrichments for PCP degradation at 24 months.

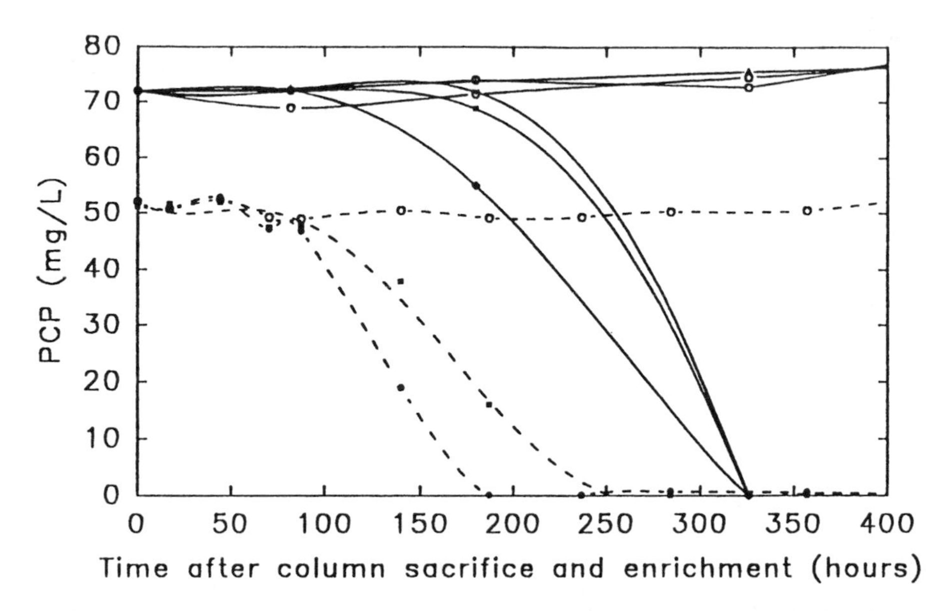

FIGURE 3. Enrichment flask PCP concentrations with 30-g inoculation from selected columns. Dashed lines are 13 months storage and solid lines are 24 months storage.

DISCUSSION

This research confirmed the enhanced survivability of agarose-immobilized *Flavobacterium* when compared to free cells in a simulated in situ microcosm study. A significant improvement in the long-term survival for introduced microorganisms by microencapsulation should be beneficial if no native organisms exist that can degrade the pollutant. We are currently developing a polymerase chain reaction (PCR) method for accurate detection and quantitation of this *Flavobacterium* to be used to further characterize the enhanced survivability and to track this organism in field transport and survival studies.

REFERENCES

Aarons, S., and M. H. Ahmad. 1986. "Examining Growth and Survival of Cowpea Rhizobia in Jamaican Peat." *Letters in Applied Microbiology* 2: 115-118.

Acea, M. J., C. R. Moore, and M. Alexander. 1988. "Survival and Growth of Bacteria Introduced into Soil." *Soil Biology and Biochemistry 20*: 509-515.

Bashan, Y. 1986. "Alginate Beads as Synthetic Inoculant Carriers for Slow Release of Bacteria that Affect Plant Growth." *Appl. Environ. Microbiol. 51*: 1089-1098.

Bashan, Y., and H. Levanony. 1988. "Adsorption of the Rhizosphere Bacterium *Azospirillum brasilense* Cd to Soil, Sand and Peat Particles." *J. Gen. Microbiology 134*: 1811-1820.

Berg, R. K., T. E. Loynachan, R. M. Zablotowicz, and M. T. Lieberman. 1988. "Nodule Occupancy by Introduced *Bradyrhizobium japonicum* in Iowa Soils." *Agronomy Journal 80*: 876-881.

Compeau, G., B. J. Al-Achi, E. Platsouka, and S. B. Levy. 1988. "Survival of Rifampin-Resistant Mutants of *Pseudomonas fluorescens putida* in Soil Systems." *Appl. Environ. Microbiol. 54*: 2432-2438.

Corman, A., Y. Crozat, and J. C. Cleyet-Marel. 1987. "Modeling of Survival Kinetics of Some *Bradyrhizobium japonicum* Strains in Soils." *Biology and Fertility of Soils 4*: 79-84.

Fravel, D. R., and J. J. Marois. 1986. "Edaphic Parameters Associated with Establishment of the Biocontrol Agent *Talaromyces flavus*." *Phytopathology 76*: 643-646.

Fravel, D. R., J. J. Marois, R. D. Lumsden, and W. J. Connick, Jr. 1985. "Encapsulation of Potential Biocontrol Agents in Alginate-Clay Matrix." *Phytopathology 75*: 774-777.

Jawson, M. D., A. J. Franzluebbers, and R. K. Berg. 1989. "*Bradyrhizobium japonicum* Survival in Soybean Inoculation with Fluid Gels." *Appl. Environ. Microbiol. 55*: 617-622.

Levinson, W. E., K. E. Stormo, H. L. Tao, and R. L. Crawford. 1993. "Hazardous Waste Treatment using Encapsulated or Entrapped Microorganisms." In G. R. Chaudhry (Ed.), *Biological Degradation and Bioremediation of Toxic Chemicals*. Timber Press, Portland, OR. In press.

O'Reilly, K. T., and R. A. Crawford. 1989. "Degradation of Pentachlorophenol by Polyurethane-Immobilized *Flavobacterium* Cells." *Appl. Environ. Microbiol. 55*: 2113-2118.

Stormo, K. E., and R. L. Crawford. 1992. "Preparation of Encapsulated Microbial Cells for Environmental Applications." *Appl. Environ. Microbiol. 58*: 727-730.

van Elsas, J. D., A. F. Dijkstra, J. M. Govaert, and J. A. van Veen. 1986. "Survival of *Pseudomonas fluorescens* and *Bacillus subtilis* Introduced into Two Soils of Different Texture in Field Microplots." *FEMS Microbiology Ecology 38*: 151-160.

Vargas, R., and T. Hattori. 1986. "Protozoan Predation of Bacterial Cells in Soil Aggregates." *FEMS Microbiology Ecology 38*: 233-242.

A BENCH-SCALE ASSESSMENT OF NUTRIENT CONCENTRATION REQUIRED TO OPTIMIZE HYDROCARBON DEGRADATION AND PREVENT CLOGGING IN FULLY SATURATED AEROBIC SANDS

L. P. Olmsted

INTRODUCTION

In situ bioremediation schemes can involve the addition of nutrient supplements to enhance the microbial population with the intention of increasing the degradation rate of the pollutant. Clogging in the near-well zone is often experienced because nutrient solutions are injected into the aquifer without quantifying the optimum concentration before field applications commence.

This study was designed to show the change in aquifer permeability resulting from the application of three concentrations of nutrients to aquifer material contaminated with two concentrations of oil. Fully saturated, aerobic conditions in an oil-contaminated aquifer were simulated in glass columns in a laboratory. The sand was inoculated with a natural bacteria population typically found in soil. Numerous tests were conducted on the system to verify the induced constant-head conditions and to investigate the primary relationships established in the columns (Columns T-1 to T-7). Flowrates were measured under an induced hydraulic gradient between the influent and effluent tubes. This was then converted to a change in permeability by applying Darcy's law (Figure 1).

Control columns were run to measure the permeability changes in a simulated "clean aquifer." Any effects of the oil on the dissolved oxygen levels were checked. The packing was refined to ensure consistency between columns. Once the system was operating consistently and the methods of measurement were optimized, four columns, Columns R-1 to R-4, were tested to measure permeability changes. Two concentrations of nutrients were fed into columns containing two concentrations of oil-contaminated sand. Column characteristics are summarized in Table 1. The nutrient used is a commercially available solution, Alpha Nutrient Tea 2. Table 2 details relative amounts of the major components in the nutrient solution.

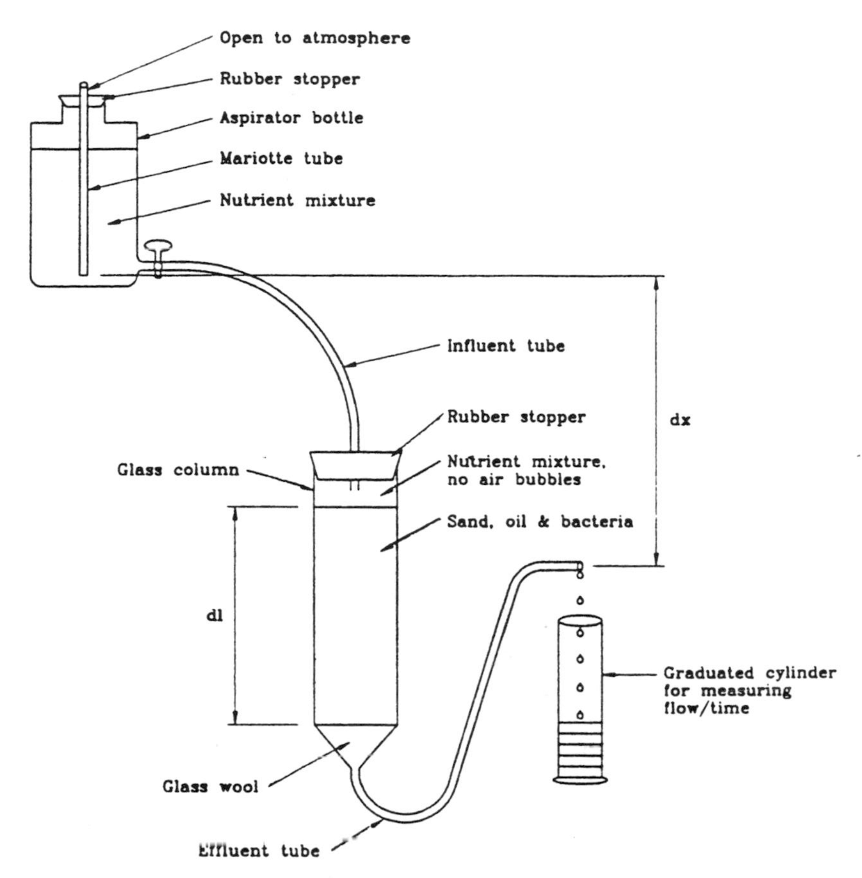

FIGURE 1. Schematic diagram of laboratory apparatus.

RESULTS

Permeability reduction is believed to be linked to an increase in biomass (Figures 2 and 3), based on the observation of a decrease in dissolved oxygen in the effluent when compared to the influent. In columns containing low nutrient concentrations, T-4 and T-6, a decrease in permeability was measured with time, although the reduction was very slight and extremely variable. However, Columns R-1, R-2, and R-4 exhibited an exponential rate of reduction and a marked overall decrease in permeability. Column R-3 displayed a more linear reduction in permeability.

Both Columns R-4 and T-5 displayed a pattern also observed by Hijnen and Van Der Kooij (1992) where the permeability declines, then partially rebounds with an increase in permeability. It is thought that once the biomass within a column increases to the point of oxygen limitation, a subsequent depletion of that oxygen supply ultimately leads to an increase in permeability (Figure 4) as a result of flushing or some of the microbes from the aquifer system.

of

TABLE 1. Conditions present in all columns.

Column	Oil Content (%)	Volume of Inoculant (mL)	Nutrient Concentration	Operational Time (days)
T-1	0.1	17	Variable	3
T-2	1	17	Variable	3
T-3	None	None	Tap water then nutrients	10
T-4	0.1	34	Medium for 4 days, then low	18
T-5	1	34	Medium for 2 days, then low	17
T-6	1	34	Low	11
T-7	None	None	Deionized water only	10
R-1	0.1	30	High	21
R-2	1	30	High	21
R-3	0.1	30	Medium	20
R-4	1	30	Medium	20

When the concentrations of NO_3^-, SO_4^{-2}, and PO_4^{-2} were monitored in the influent and effluent, there did not appear to be any relationship between the permeability decline and the nutrients consumed. A possible explanation for this is that some of the necessary elements required for biomass growth were supplied by the oil as it was degraded. Perhaps if the test was extended in operational time, a clear relationship may be obvious.

Total organic carbon (TOC) analyses were conducted to roughly estimate the amount of oil degraded over the time of the study. One must recognize that while the carbon attributed to the oil declined with time, some of this carbon will have been used by the bacteria to increase its population. Hence, the net reduction in TOC is due to both degradation of oil and re-use of carbon by the biomass. The initial oil concentration was tested for consistency between the columns, and final TOC samples were taken at discrete depths through the columns when they were dismantled at the end of the test. When the total amount of carbon in the column was compared before and after the test, Columns R-2 and R-4, containing high oil concentrations, illustrated a relationship that varied with the concentration of the nutrient (Figure 5). High nutrient concentration

TABLE 2. Relative amounts of nutrient supplement components.

Level	NO_3(mg/L)	PO_4(mg/L)	SO_4(mg/L)
Low	100	60	10
Medium	350	220	45
High	725	460	95

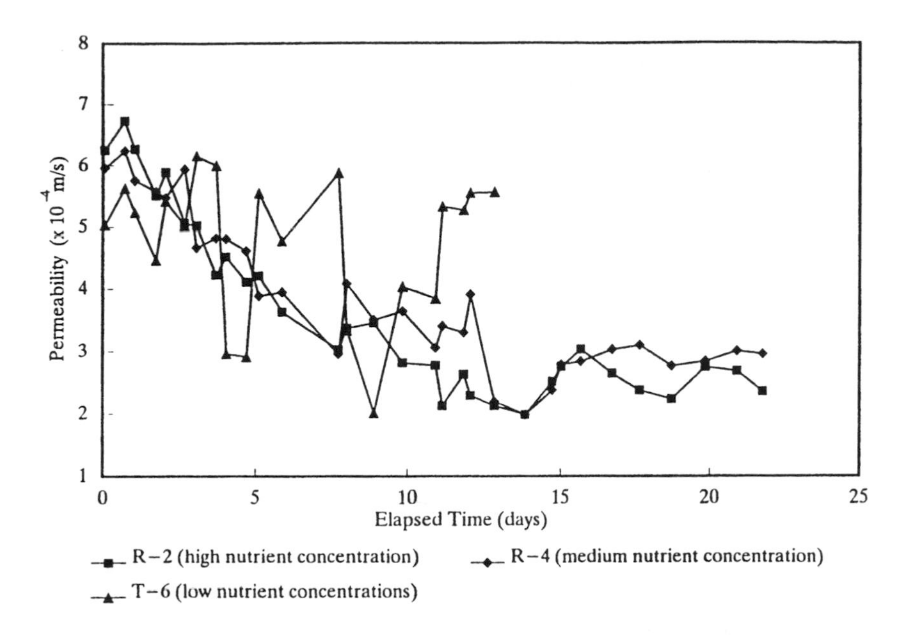

FIGURE 2. Permeability changes in columns with 1% oil concentrations.

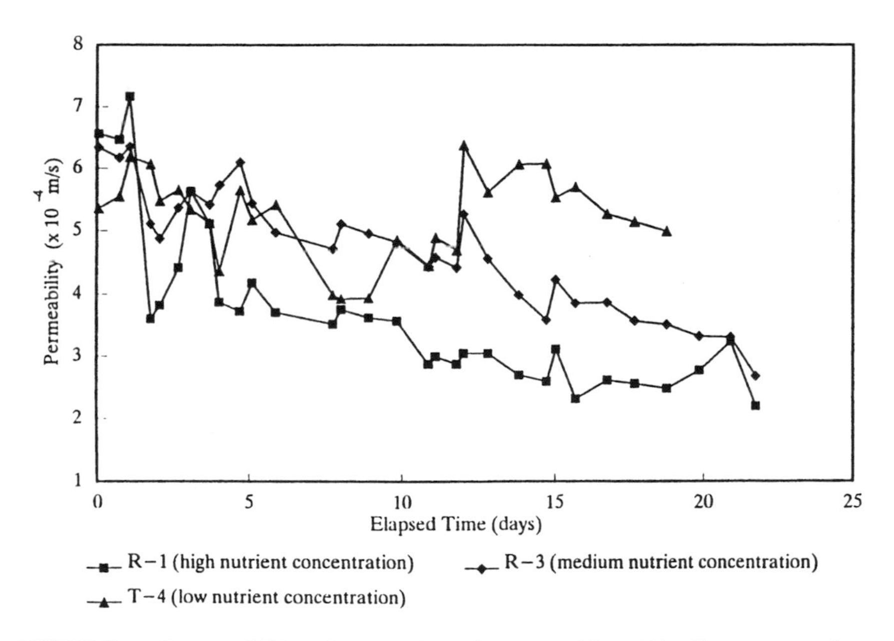

FIGURE 3. Permeability changes in columns with 0.1% oil concentrations.

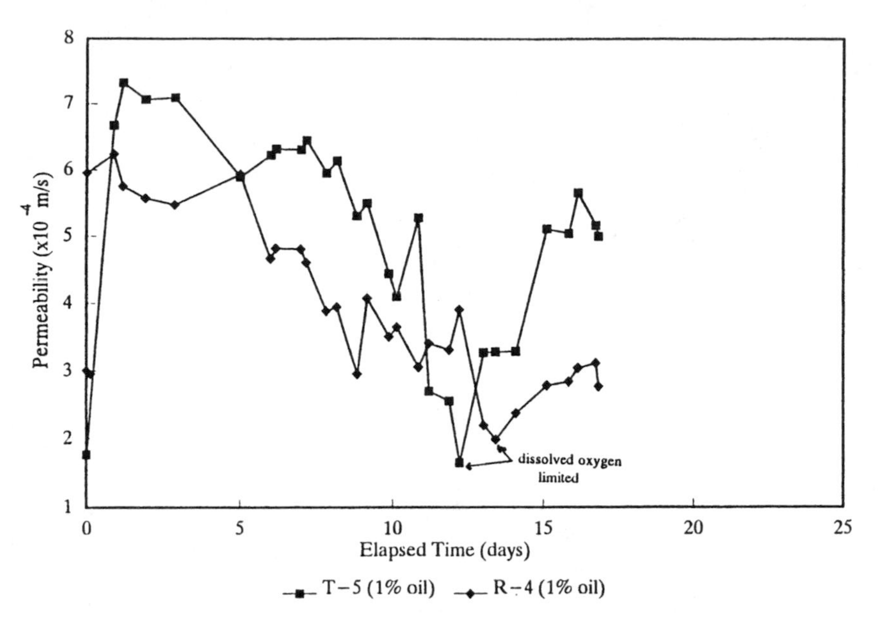

FIGURE 4. Permeability of Columns T-5 and R-4 showing an increase in permeability when dissolved oxygen is limited.

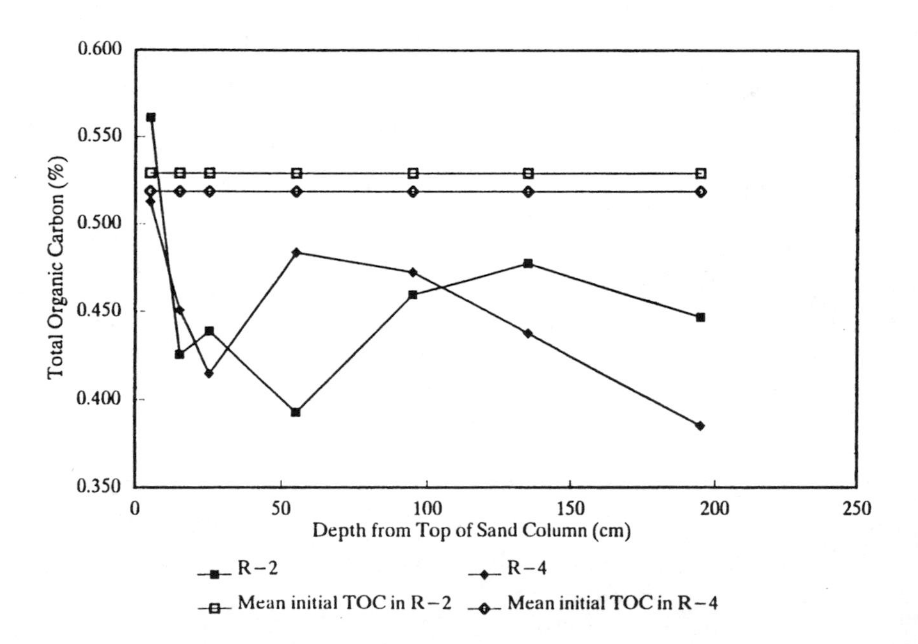

FIGURE 5. TOC for columns with 1% oil concentrations, at the end of the trial.

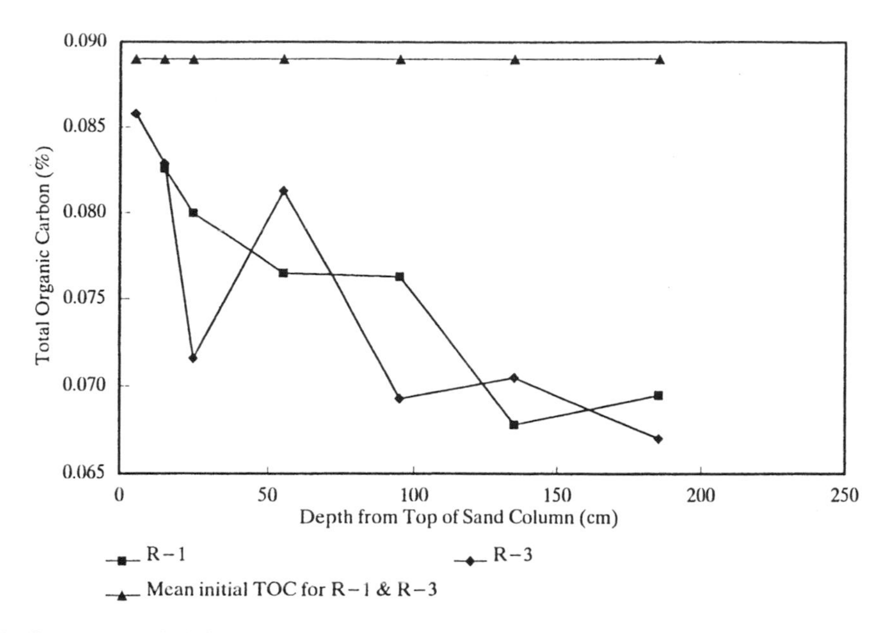

FIGURE 6. TOC for columns with 0.1% oil concentrations, at the end of the trial.

resulted in a reduction of 15.2% TOC, whereas the medium nutrient level only reduced the TOC by 9.3%. Conversely, the effect of nutrient concentration is insignificant at low oil concentrations, reducing the TOC by approximately the same amount in Columns R-1 and R-3, by 17.9 and 17.2%, respectively (Figure 6). On the graphs, it appears as though a low TOC zone is significantly more extensive in the columns with high oil concentrations, suggesting that the larger bacteria population supported by the higher oil and nutrient concentrations have led to a greater consumption of oil. This zone is in the upper half of the column where one would expect the population to have proliferated most. In field situations, this zone would be the equivalent of the near-well zone.

Fundamental relationships can be demonstrated in an idealized laboratory environment from which we may extrapolate important in situ interactions. If one is dealing with oil concentrations of approximately 1% oil by weight, the use of a medium concentration of nutrient will reduce the permeability at the same rate as a high concentration, however, the high nutrient will clean up the oil faster. At low oil concentrations, the rate of degradation is approximately the same, hence it would be cheaper and have less impact on permeability in the long term if the medium concentration of nutrient were selected. But, if reduction in permeability is of prime concern, the use of a low nutrient level will minimize the clogging effects. Also, if clogging becomes a problem during remediation, it would be advantageous to lower the nutrient concentration, or drastically reduce the dissolved oxygen content available to the system, with the intention of killing and flushing some of the bacteria population from the system. This starve-and-flush technique is probably more effective as an ongoing maintenance technique if clogging is a problem. Ideally, it would be preferable to maintain a sustained low level of nutrients, monitoring permeability and contaminant reduction on a regular basis in order to adjust any inputs to the system before clogging gets out of control.

REFERENCE

Hijnen, W. A. M., and D. Van Der Kooij. 1992. "The Effect of Low Concentrations of Assimilable Organic Carbon (AOC) in Water on Biological Clogging of Sand Beds." *Water Research* 26(7): 963-972.

BIOREMEDIATION: THE STATE OF USAGE

K. Devine

INTRODUCTION

The Data Identification and Collection Subcommittee, of the U.S. Environmental Protection Agency's Bioremediation Action Committee, is tasked with collecting bioremediation information that exhibits the technology's costs and benefits for dissemination to the public and private sectors. As a means of accomplishing this goal, a call for bioremediation case study information was issued in August 1991, to augment the Office of Research and Development's network retrieval system, the Alternative Treatment Technology Information Center (ATTIC). The ATTIC system provides technical information on innovative treatment methods for hazardous waste and other contaminants. Included in this study are 132 case studies from 10 U.S. remediation companies that had enough pertinent information for the ATTIC system.

RESULTS OF DATA COLLECTION

More than 80% of the 132 cases (106) were at field pilot or full-scale level. One-quarter of the cases (33) reported bioremediation activity commencing earlier than 1989, 12% (16 cases) commenced in 1989, 24% (32 cases) commenced in 1990, and 4.5% (6 cases) were started in 1991 (as is the case with other statistics reported, the remainder of the total did not report information on this parameter). About 68% of the cases (90) were completed as of the end of 1991, and about 24% (31 cases) were still in progress. Another 5% (7 cases) are ongoing and had set no completion date because they involve treatment of a continual flow of contaminants.

Almost 60% (79 cases) of the bioremediation activity was located in EPA Regions 1 (Connecticut, Massachusetts, Maine, New Hampshire, Rhode Island, Vermont); 2 (New Jersey, New York, Puerto Rico, and the Virgin Islands); 6 (Arkansas, Louisiana, New Mexico, Oklahoma, Texas); and 9 (Arizona, California, Guam, Hawaii, and Nevada).

More than 62% of the cases (82) reported treatment of a petroleum-related waste. Almost 10% (13 cases) involved treatment of solvents. Wood-preserving and agricultural (pesticides) chemicals constituted about 11% (15 cases) and 6% (8 cases), respectively, of the total 132 cases. Munitions and coal tar/tar each comprised 4% (5 cases). See Table 1.

TABLE 1. ATTIC Bioremediation Case Study Project: Cases by contaminant treated.

Contaminant/Contaminant Use	Number of Cases	Percent of Total
Petroleum-related[a]	75	56.8
Wood preservatives[b]	13	9.8
Solvents	10	7.6
Other	9	6.8
Agricultural chemicals	7	5.3
Coal tar/Tar	5	3.8
Munitions	4	3.0
Petroleum-related/Other	3	2.3
Petroleum-related/Wood preservatives	2	1.5
Petroleum/Solvents	2	1.5
Munitions/Agricultural chemicals	1	0.8
Solvents/Other	1	0.8
Total	132	100.0%[c]

(a) Twenty-five cases (19% of the total and 31% of the petroleum-related cases) were reported as UST sites.
(b) Creosote and pentachlorophenol.
(c) Does not add to 100% due to rounding.

In more than 74% of the cases (98), soil was at least one of the media that was bioremediated. Treatment of soil alone constituted 46% of the cases reported (61), whereas soil and groundwater bioremediation at a single site accounted for 23% of the total number of cases (30). Almost 5% of the cases (6) involved soil and sludge bioremediation at a single site. Cases involving bioremediation on only groundwater/water made up 16% of the sites reported (21 cases), whereas bioremediation of groundwater/water or other media comprised 40% (53 cases). Treatment of sludge constituted more than 14% of total cases (19 cases).

More than 60% of the cases (48) involving soil at field pilot or full scale involved solely the use of solid phase/land treatment, whereas in more than 25% (21 cases), soil was treated in situ. Also, 4% (3 cases) reported bioreactor/bioslurry usage for soil treatment and another 5% (4 cases) utilized both in situ and some type of aboveground treatment.

Of the 41 cases that indicated the treatment of groundwater/water at field pilot or full scale, 46% (19 cases) included the treatment of bringing the water to the surface and reinjecting without the use of a bioreactor; 42% (17 cases) indicated the use of bioreactor for treatment, and 12% (5 cases) involved treatment of groundwater totally below the surface.

Almost two-thirds of the 16 cases (10 cases) that dealt with sludge reported bioreactor/bioslurry-type treatment. The remainder (6 cases) were treated by either slurry followed by land treatment or land treatment alone. See Table 2.

TABLE 2. ATTIC Bioremediation Case Study Project: Cases by treatment method used for soil, groundwater, and sludge.

Treatment Method	Number of Cases	Percent of Total
Soil		
In Situ[a]	21	26.6
Aboveground		
Solid Phase/Land Treatment[b]	48	60.8
Bioreactor/Bioslurry[c]	3	3.8
Solid Phase/Land Treatment and Bioreactor/Bioslurry	3	3.8
In Situ and Aboveground	4	5.1
Total	79[d]	100.0%[e]
Groundwater[f]		
No Bioreactor/Recirculation[g]	19	46.3
Bioreactor/Bioslurry[c]	17	41.5
Other[h]	5	12.2
Total	41[i]	100.0%
Sludge		
Solid Phase/Land Treatment	3	18.8
Bioreactor/Bioslurry[c]	10	62.5
Bioreactor/Bioslurry and Solid Phase/Land Treatment	3	18.8
Total	16[k]	100.0%[e]

(a) Contaminated medium is treated in place, usually through the delivery of oxygen and other nutrients.
(b) Contaminated medium is spread over a prepared area and treated through optimization of microbial requirements for biodegradation. Includes composting.
(c) Contained systems, typically used for treatment of aqueous and slurry media. Water treated can be reinjected.
(d) Percentages based upon total number of full-scale and field pilot-scale cases involving soil treatment.
(e) Does not add to 100% due to rounding.
(f) Includes water.
(g) Treatment includes bringing groundwater to the surface and making amendments prior to reinjection.
(h) Water fully treated in place, below the surface.
(i) Percentages based upon total number of full-scale and field pilot-scale cases involving groundwater treatment.
(j) Contaminated medium is placed in a prepared area and treated through optimization of microbial requirements for biodegradation. Includes composting.
(k) Percentages based upon total number of full-scale and field pilot-scale cases involving sludge treatment.

LIMITATIONS OF STUDY

Terminology was not consistent among companies, and thus in some cases, assumptions were made concerning the reported parameters.

These cases do not necessarily reflect all cleanup efforts of any one company. The reported cases are only those for which information existed, that were the most relevant for ATTIC inclusion, and that were readily available.

ACKNOWLEDGMENT

This study was funded by the U.S. Environmental Protection Agency under contract to DEVO Enterprises, Inc., Washington, DC.

REFERENCES

Devine, Katherine. 1992. *Bioremediation Case Studies: An Analysis of Vendor Supplied Data.* EPA/600/R-92/043.

Devine, Katherine. 1992. *Bioremediation Case Studies: Abstracts.* EPA/600/R-92/044.

MONITORING CATABOLIC GENE EXPRESSION BY BIOLUMINESCENCE IN BIOREACTOR STUDIES

R. S. Burlage, D. Kuo, and A. V. Palumbo

INTRODUCTION

Bioremediation has emerged as an attractive method for the elimination of hazardous wastes, due to the potential cost savings and ability to mineralize many compounds. The success of bioremediation depends on many factors, most of which are poorly defined. Given the complexity of biological systems, it is often difficult to determine if a process is efficient, or to identify which characteristics of the system are limiting.

To study the expression of specific catabolic genes under a variety of conditions, and to determine whether certain conditions tend to increase or decrease efficiency of expression, a bioreporter gene can be introduced into the microorganism. Activity from the bioreporter gene would indicate successful bioremediation.

Our laboratory has produced several bioreporter strains using the bioluminescent *lux* genes of *Vibrio fischeri* (Meighen 1991). These bioreporter strains produce visible light when genetic expression is induced. The advantages of a bioluminescent system include sensitivity of detection, analysis of response in real time, and on-line capability for use in bioreactor systems. Bioluminescent reporter strains specific for naphthalene degradation have been successfully useful in examining this catabolic system (Burlage et al. 1990, King et al. 1990).

We have constructed a bioreporter strain for following the degradation of toluene and related compounds. This study illustrates the use of this strain to study expression of the catabolic genes with various substrates and to optimize activity in a bioreactor.

MATERIALS AND METHODS

The bioreporter strain used in this study, RB1401, was constructed by fusing the promoter of the upper pathway for toluene degradation from the TOL plasmid (Assinder and Williams 1990) to the *lux* gene cassette. This plasmid was introduced into the host strain, *Pseudomonas putida* mt-2, to form the bioreporter strain (Burlage et al., manuscript in preparation). Relevant details of the plasmids are presented in Figure 1.

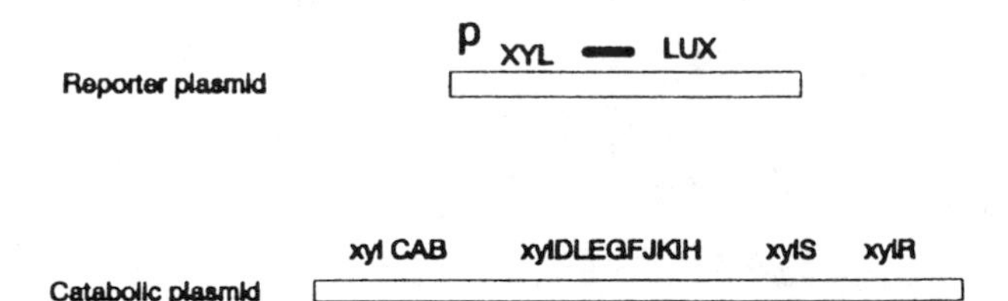

FIGURE 1. Plasmids of RB1401. The reporter plasmid contains the *lux* genes fused to the promoter (P) of the upper pathway from the TOL plasmid. The catabolic plasmid contains all the genes for the degradation of toluene and the xylenes, as well as the *xylS* and *xylR* regulatory genes.

RESULTS

To determine whether RB1401 could be induced by toluene and related compounds, and to produce a bioluminescent response as a result of this induction, batch cultures of the strain were grown in LB medium (Sambrook et al. 1989) amended with kanamycin (50 mg/L final concentration). Growth was measured as optical density in a Beckman spectrophotometer at 600 nm. Inducers (1 mM,

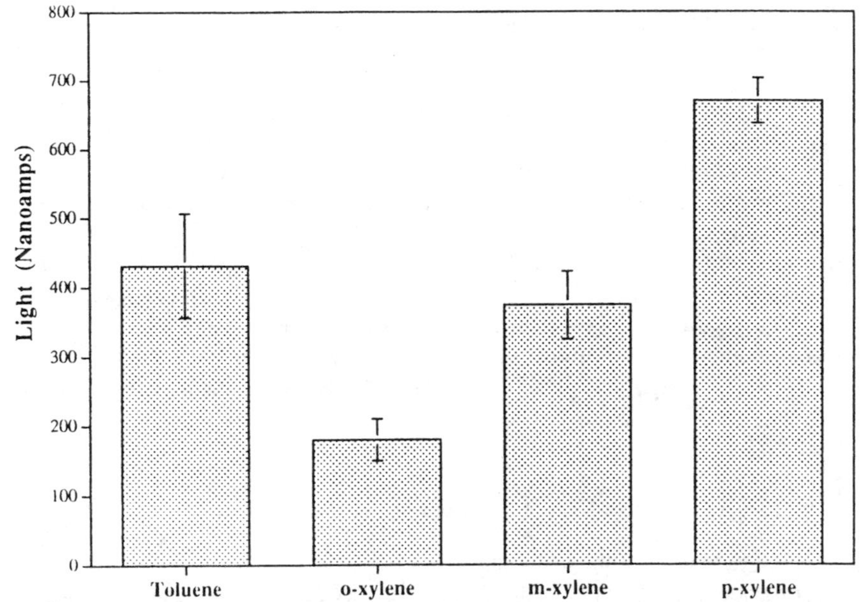

FIGURE 2. Bioluminescence from strain RB1401 when induced with various substrates. Cultures were grown to an O.D. (600 nm) = 1.5 and induced with the indicated substrates at a final concentration of 1 mM and aerated by vigorous shaking. Units are nanoamps of induced current (see text). Results are the mean of 5 samples; error bars indicate standard deviation.

final concentration) were added during stationary phase. Bioluminescence was measured using an Oriel photomultiplier model 7070 with a flexible liquid-light pipe. Bioluminescence is reported as nanoamps of induced current in the photomultiplier. All four compounds induce the strain to high levels of light production (Figure 2). The *o*-xylene is not a growth substrate for the parent strain, but is known to be a gratuitous inducer (Assinder & Williams 1990).

To study toluene induction, a bioreactor was used to cultivate RB1401 continuously, using NATE minimal medium (Little et al. 1988) amended with kanamycin. The design allowed constant monitoring of bioluminescence and a variable addition of selected nutrients (Kuo et al., manuscript in preparation). The growth rate of the culture was 0.083/hr, with a retention time of 12 hr. Toluene feed concentration was 1 mM. Only a single retention time was used between experiments to allow new conditions to establish, because the growth rate was very slow. A pronounced effect on light production was seen when the concentration of iron in the bioreactor feed was varied (Figure 3). Despite higher cell concentrations at high iron concentrations (data not shown), peak bioluminescent activity is seen at 10 micromolar.

Varying the concentration of phosphate in the bioreactor feed also resulted in clear differences (Figure 4). Here also, high concentrations led to high cell concentrations, but peak activity was seen at 0.025%. In fact, removal of phosphate

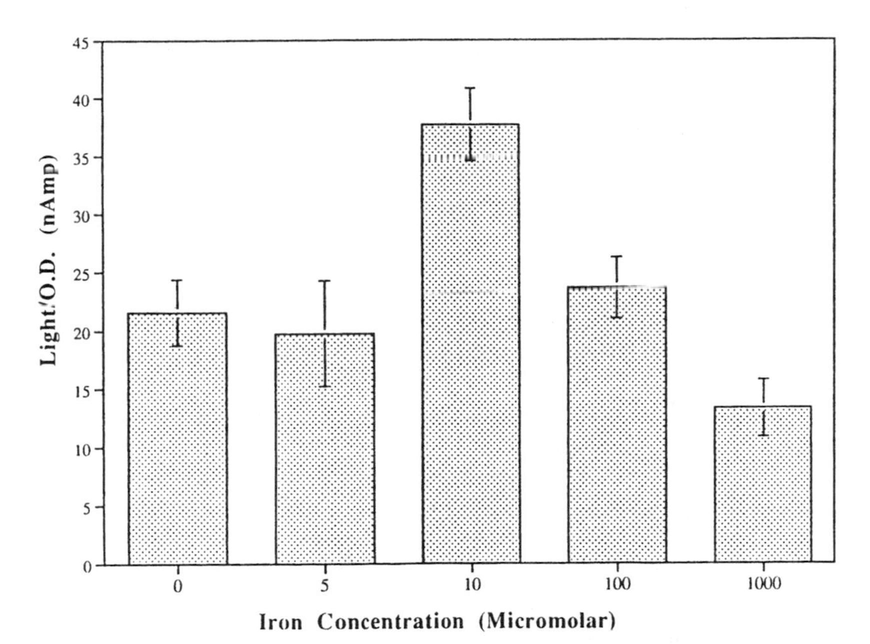

FIGURE 3. Effect of iron concentration on bioluminescence. A continuous culture of RB1401 was induced with toluene and bioluminescence was measured continuously as the iron concentration was varied. Data are corrected for cell count. Error bars indicate standard deviation.

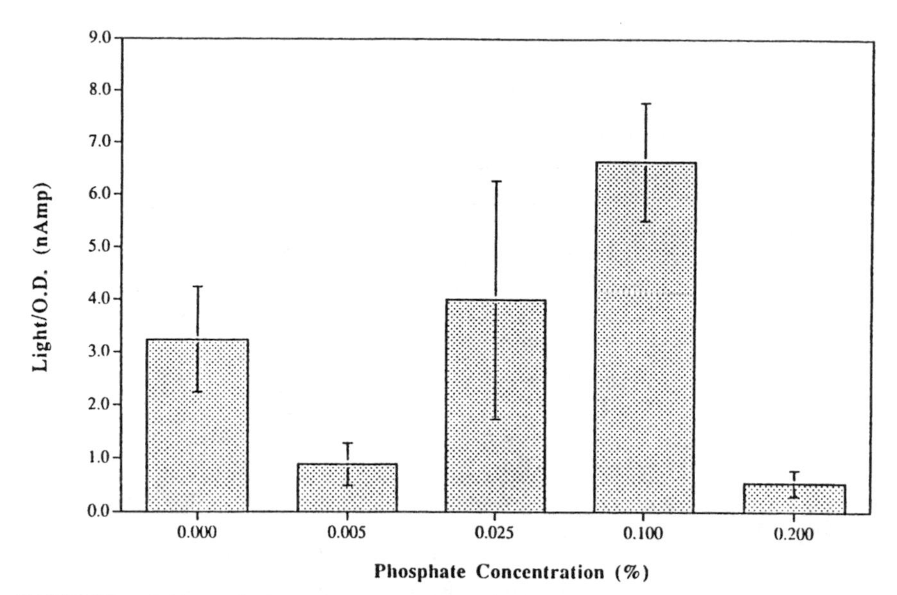

FIGURE 4. Effect of phosphate concentration on bioluminescence. A continuous culture of RB1401 was induced with toluene and bioluminescence was measured continuously as the phosphate concentration was varied. Data are corrected for cell count. Error bars indicate standard deviation.

from the feed medium, while reducing cell concentration greatly, still results in a strong bioluminescent signal.

SUMMARY

Using a bioluminescent reporter strain, we have been able to detect the induction of a specific operon in response to the addition of appropriate compounds. One of these compounds, *o*-xylene, is a gratuitous inducer of the catabolic genes, and produced a strong bioluminescent signal in these studies. We have varied the concentration of components of minimal medium for an induced bioreactor culture of RB1401, and our data suggest that conditions for optimal expression of the catabolic operon might not be identical with optimal growth conditions. This raises the question of appropriate operating conditions for efficient biodegradation to take place, which remains the long-term goal of this research.

ACKNOWLEDGMENT

Research sponsored by the Office of Energy Research, U.S. Department of Energy, under contract DE-AC05-84OR21400 with Martin Marietta Energy Systems, Inc.

REFERENCES

Assinder, S. J. and P. A. Williams. 1990. "The TOL plasmids: determinants of the catabolism of toluene and the xylenes." *Adv. Micobiol. Physiol.* 31: 1-69.

Burlage, R. S., G. S. Sayler, and F. Larimer. 1990. "Monitoring of naphthalene catabolism by bioluminescence with nah-lux transcriptional fusions." *J. Bacteriol.* 172: 4749-4757.

King, J. M. H., P. M. DiGrazia, B. Applegate, R. Burlage, J. Sanseverino, P. Dunbar, F. Larimer, and G. S. Sayler. 1990. "Rapid, sensitive bioluminescent reporter technology for naphthalene exposure and biodegradation." *Science* 249: 778-781.

Little, C. D., A. V. Palumbo, S. E. Herbes, M. E. Lindstrom, R. L. Tyndall, and P. J. Gilmer. 1988. "Trichloroethylene biodegradation by pure cultures of a methane-oxidizing bacterium." *Appl. Environ. Microbiol.* 54: 951-956.

Meighen, E. A. 1991. "Molecular biology of bacterial bioluminescence." *Microbiol. Rev.* 55: 123-142.

Sambrook, J., E. F. Fritsch, and T. Maniatis. 1989. *Molecular Cloning: A Laboratory Manual.* Cold Spring Harbor Laboratory, Cold Spring Harbor.

BIORECLAMATION AND BIOGAS RECOVERY IN DENMARK

R. B. Dean

INTRODUCTION

The oldest form of bioreclamation is the uncontrolled anaerobic decomposition of animal excreta and straw in a dunghill. Reclamation takes place by anaerobic decomposition of putrefactive organic matter with loss of biogas and some reduction of moisture. The final product has an improved odor and is thus less objectionable when spread on the fields.

Modern large dairies and cattle feed lots can no longer use such techniques because they use water instead of pitchforks to remove excreta and bedding in the form of a dilute slurry. The concentration of solids in such a slurry depends on the treatment used to produce a pumpable slurry and is usually below 100 g/L. The slurry contains potentially infectious microorganisms, weed seeds and organic matter that will rot and produce foul odors. It is usually unsuitable for spreading on fields until it has been stored for periods of up to a full year. Storage of large volumes of slurry is expensive especially when land spreading is limited to certain seasons of the year. Also, methane is liberated from storage tanks and becomes a serious component of the greenhouse gas emissions.

On the positive side, the use of slurry is a labor-saving method of handling animal wastes and, because the wastes are pumpable, they can be transported in tank trucks and treated by well-established methods developed for wastewater sludge. Bioreclamation of animal slurry and other bioorganic wastes with recovery of biogas, to be successful, should be done on a large scale. Early attempts in the late 1970s to manufacture biogas on individual farms in Denmark were almost uniformly unsuccessful if measured by cost and the value of the biogas produced. The present-day approach is to haul the slurry in tankers to a regional plant; digest it in a well-controlled, thermophilic anaerobic digester; and return an equal volume of treated slurry, now called manure or fertilizer, to the farmer along with an analysis of its nutrients and trace elements. The treated manure is returned in the emptied and cleaned tanker. It has been disinfected, the seeds have been killed, and its odor has been reduced so that it can be applied to fields whenever they can accept its water content. The treatment also produces useful quantities of biogas as a by-product.

OVERVIEW OF CURRENT PRACTICE

Gas production from the principal slurry-to-biogas plants in Denmark is shown in Figure 1 (Øllgaard 1992). Other sources of organic waste from abattoirs and fish-processing plants, and filter cake from edible oil pressing may be added to increase the volatile solids content. The inhibitory effect of oils in some of these wastes has been overcome by a proprietary process that includes the addition of adsorbent mineral particles such as bentonite (Ahring et al. 1992). Some plants accept limited quantities of sorted household waste that is prepared as if it were going to be composted (Andresen & Rasmussen 1992). One experimental plant uses a hydropulper to disperse the solid wastes. This makes it possible to separate small, heavy items by sedimentation, especially household batteries, and thus reduces the content of cadmium and mercury.

The preferred treatment is thermophilic digestion, usually between 55 and 60°C. Although the digester operates at temperatures considered adequate for pasteurizing compost and the digester has a sludge retention time of about 3 weeks, fresh slurry is added every few hours to the stirred digester and an equal volume of the digester contents is removed (Møller & Bertelsen 1992). Essentially the only reduction of heat-resistant microorganisms is by dilution (Srivastava & Lund 1980).

The operation of digesters with frequent additions of raw slurry, although it may permit the passage of infectious organisms, greatly helps to stabilize the performance of the bacterial consortium. The thermophilic anaerobic bacteria can become upset more easily than their mesophilic relatives by changes in temperature or concentration of raw materials. When pasteurization is required it may be done ahead of digestion so that the hot slurry goes almost directly into the digester as is done for sewage sludge in Switzerland (Clements 1982).

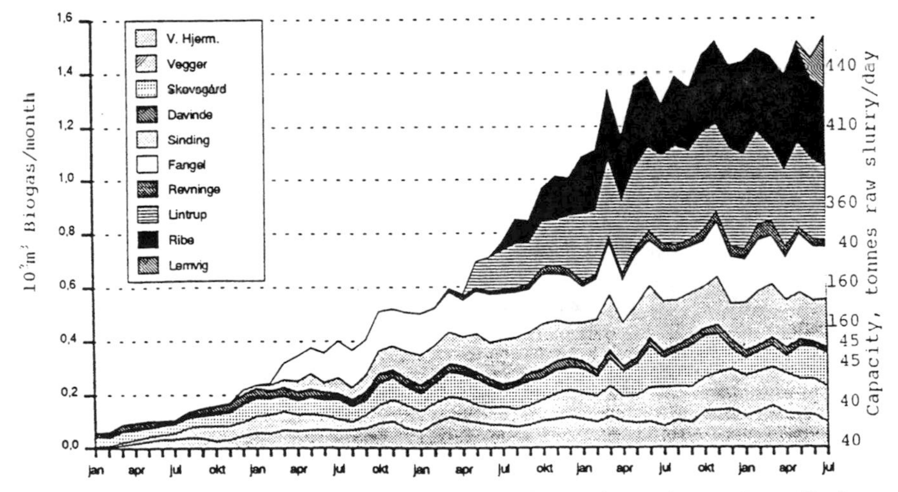

FIGURE 1. Gas production of Danish Collective Biogas Installations, January 1988-August 1992. Slurry capacity in tonnes per day.

PERFORMANCE

The three largest Danish plants, which are very similar in design, have capacities from 360 to 440 tonnes of wet slurry per day (Øllgaard 1991, 1992; Prisum 1992). These plants each achieve a gas production rate of about 1.5 digester volumes of gas per day when operating at a hydraulic retention time of around 20 days. Some plants burn the gas for heat, some of which may be used to produce hot water for district heating systems. Other plants purify the gas to remove objectionable trace compounds such as hydrogen sulfide before using the gas in internal combustion engines that generate electricity, and they use the engine cooling system to heat water for district heating. The economy of gas production is favored by the Danish tax on nonrenewable fuels (Øllgaard 1991, 1992).

Although they are supported by the Danish Energy Ministry as a source of renewable fuel, the biogas plants operate as highly efficient bioreclamation systems that convert an obnoxious and potentially infectious waste into useful by-products. A similar but somewhat smaller experimental plant has been installed near Elsinore to treat source-separated household waste in an area where there is negligible slurry from farm animals. This plant uses hydropulpers followed by alkaline hydrolysis with NaOH at 70°C to pasteurize and hydrolyze the waste for digestion (Øllgaard 1992, Røgen & Kræmer 1992).

REFERENCES

Ahring, B. K., I. Angelidaki, and K. Johansen. 1992. "Anaerobic Digestion of Combined Waste." In DAKOFA (Ed.), *Proceedings Biowaste '92*, 10 pp. ISWA, Copenhagen, DK.

Andresen, S., and K. Rasmussen. 1992. "Source Separation and Collection System for Biological Waste Collected from Households." In DAKOFA (Ed.), *Proceedings Biowaste '92*, 10 pp. ISWA Copenhagen DK.

Clements, R. P. L. 1982. "Sludge Hygenization by Means of Pasteurization Prior to Digestion." In A. M. Bruce, A. H. Havelar and P. L'Hermite (Eds.), *Disinfection of Sewage Sludge: Technical,Economic and Microbial Aspects*, pp. 37-52. Reidel Pub. Co., Boston MA.

Møller, H., and F. Bertelsen. 1992. "Biogas from Source Separated Household Waste." In DAKOFA (Ed.), *Proceedings Biowaste '92*, 7 pp. ISWA, Copenhagen, DK.

Øllgaard, K. 1991. (Chairman) *Biogasfælles anleg: Hovedreport fra Koordineringsudvalget* {Report of the Chairman of the Committee for Collective Biogas Installations of the Danish Energy Administration}. Copenhagen, DK.

Øllgaard, K. 1992. "Status for Biogasfælles anleg." Draft report.

Prisum, J. 1992. "State of the Art of Large Scale Biogas Plants." In *Proceedings CEC THERMIE Seminar Collective Biogas Plants*, Herning, DK, 22 October.

Røgen, K. L., and S. Kræmer. 1992. "Full Scale Biogas Plant for Source Separated Household Waste." In DAKOFA (Eds.), *Proceedings Biowaste '92*, 6 pp.

Srivastava, R. N. and E. Lund. 1980. "The Stability of Bovine Parvovirus and its Possible Use as an Indicator for the Persistance of Enteric Viruses." *Water Research 14*: 101-102.

THE PHYSICAL-CHEMICAL APPROACH TO ORGANIC POLLUTANT ATTENUATION IN SOIL

C. A. du Plessis, E. Senior, and J. C. Hughes

INTRODUCTION

This project forms part of a larger research program that aims to limit the environmental impacts of landfill leachate. Elucidating the attenuation of vertically migrating organic pollutants in landfill covering soils should provide valuable insight into similar problems associated with these molecules.

Many studies have been made on the effects of soil properties on metal attenuation (Fuller 1980). Organic pollutants, however, in addition to their interactions with soil components, are susceptible to catabolism by microorganisms. (For this discussion, "attenuation" will be the summation of adsorption and microbial catabolism.) Soil-microorganism-pollutant (SMP) interactions have been studied mainly in soil columns (Webb et al. 1991). One limitation of column and in situ studies is that the identity and quantity of specific components affecting or influencing attenuation are usually unknown and cannot readily be determined. Physical-chemical interactions also may be masked by purely soil physical factors. Attenuation effects of soil components and interactions with catabolic species are, therefore, difficult to interpret.

Importance of Adhesion. Although soil coverings (intermediate and final) constitute only a small fraction of the landfill volume, their total specific surface area greatly exceeds that of the refuse. Thus, soil must have a significant influence on both microorganisms and pollutants migrating through the landfill mass. Vertical leachate migration takes place mainly through capillary rise and via saturated landfill gas. The surface area (surrounding the pore)-to-pore volume ratio, in these micropores, is much higher than the corresponding pores involved in downward migration. Microbial adhesion to surfaces influences metabolic activity, particularly catabolic capabilities (biodegradation) (Van Loosdrecht et al. 1990). Adhesion may also afford greater resistance/protection to toxins by microorganisms (Kefford et al. 1982), and even enzymes are affected by adhesion (Gianfreda et al. 1992).

Influence of Soil Components in the SMP System. The effects of the silicate clay minerals and other inorganic components of the clay fraction on soil microorganisms and pollutant interactions are poorly understood. The permanent negative charge due to isomorphous substitution of the silicate clays results in a cation exchange capacity independent of pH. The oxide minerals, however, have a pH-dependent charge that commonly results in their having a positive charge. This anion exchange capacity is probably more important in interaction with biological entities. Catabolism is influenced by exchange phenomena and thus by the soil pH. Solution elemental ratios are controlled by the soil properties through selective adsorption and exchange of nutrients. These ratios also influence microbial growth and adhesion. Hydrophobic interactions are extremely important (Senesi & Chen 1989) in their effects on adsorption to soil components and thus to microbial catabolism. The influence of soil organic matter, particularly humus, also must be considered in relation to the reactions in the SMP system.

EXPERIMENTAL APPROACH

The principal objective of this study was to determine how various soils influence the pollutant depletion curves (attenuation) in a system that excludes the effects of in situ soil physical conditions. The effect of various leachate challenges such as high heavy metal concentrations and the interaction on the degradation processes was also investigated. Topsoils and subsoils from an Oxisol (Inanda subsoil), a Vertisol (Rensburg), and an Inceptisol (Inanda topsoil), chosen to represent a range of characteristics (e.g., clay mineralogy, particle size, surface area, cation exchange capacity) were used. All the samples were gamma irradiated to eliminate the indigenous microbial species; the inoculum used was taken from a non-irradiated soil. To minimize nutrient variations 50 mL of liquid medium were added to 2 g soil. The medium contained, in a liter of deionized water, 1.5 g K_2HPO_4, 0.5 g KH_2PO_4, 0.5 g $(NH_4)SO_4$, 0.2 g $MgSO_4.7H_2O$, 10 mL soil extract, and 30 mg naphthalene (chosen as the model hydrophobic landfill leachate molecule [Harmsen 1983]). Soil suspensions were sampled at various time intervals after incubation at 30°C and orbital shaking at 120 rpm. Samples from all the treatments were prepared for scanning electron microscopy (SEM) by critical point drying as well as cryo. Adsorption isotherms were also determined for the three soil types that were used.

DISCUSSION AND CONCLUSIONS

Adsorption isotherms revealed the expected increase of adsorption capacity with an increase in organic carbon content (Figure 1). Aerobic naphthalene degradation in the absence of soil typically revealed a curve such as given in Figure 2. This curve was significantly influenced by the different soil types (data not shown). Cadmium in the growth media seemed to become toxic to the naphthalene degrading organisms at different concentrations depending on the soil type (Figure 3). A unusual finding was that the Rensburg soil with a much higher CEC than the other two soils had a much higher naphthalene concentration (in solution) at the 100 ppm cadmium contamination level. This seems to indicate

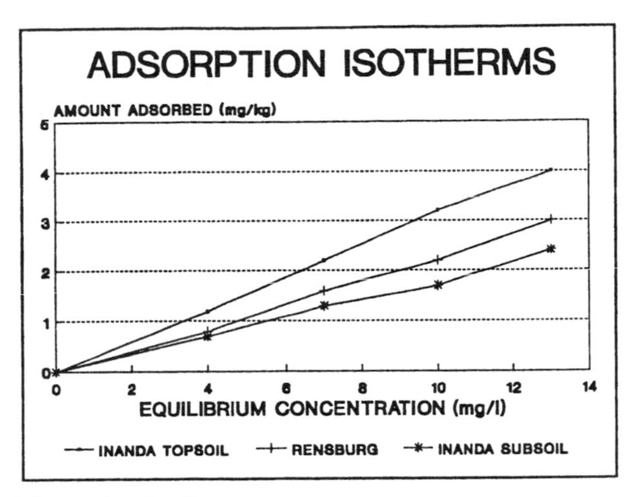

FIGURE 1. Adsorption isotherms.

that degradation is occurring on the surface of the soil particles. At the 400 ppm cadmium contamination level no growth occurred since these values reflect the adsorption isotherm experiment results. The different soil types also seemed to differ in the extent to which volatilization occurred from the soil suspensions (Figure 4). The retardation in volatilization was more pronounced in suspensions with a higher soil:solution ratio and in the higher organic content soil i.e., Inanda topsoil. The SEM revealed no microbial growth on any of the samples, despite the depletion of naphthalene. The SEM preparation procedure thus appears to be unsuitable and cryo-SEM was thus investigated as an alternative but revealed

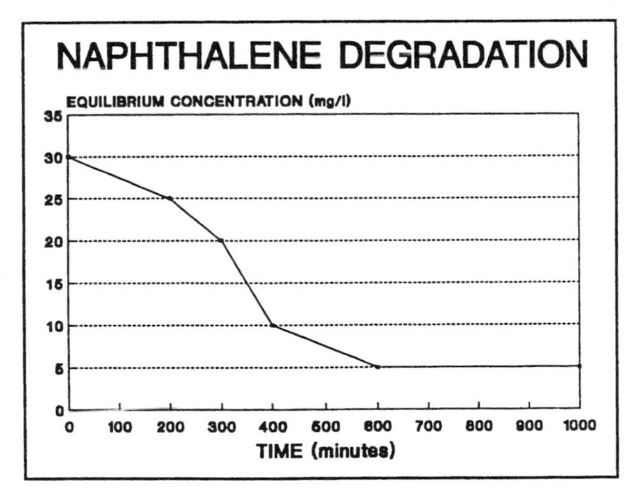

FIGURE 2. Aerobic degradation of naphthalene.

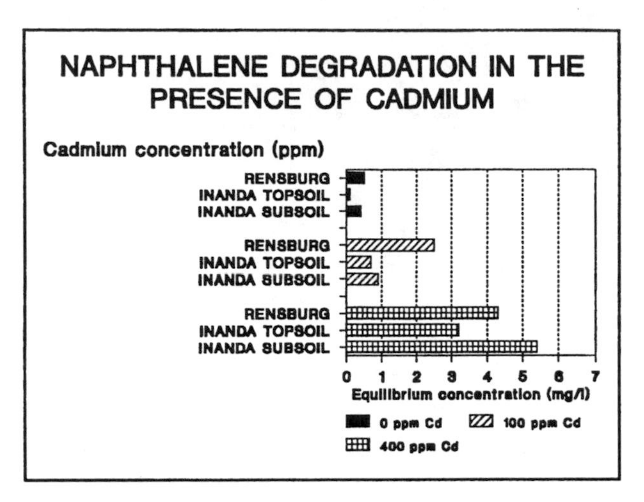

FIGURE 3. Naphthalene solution concentrations in the presence of cadmium after 24 hours aerobic growth.

similar results. This work represents ongoing research that could facilitate rapid evaluation of the physical-chemical interaction of various soil characteristics such as cation exchange capacity, total surface area, and surface structure, with microorganisms and pollutants. It is important that studies such as this should be pollutant- and soil-specific. Once a protocol has been developed, it is hoped that it can be used with various recalcitrant, hydrophobic, and charged organic compounds.

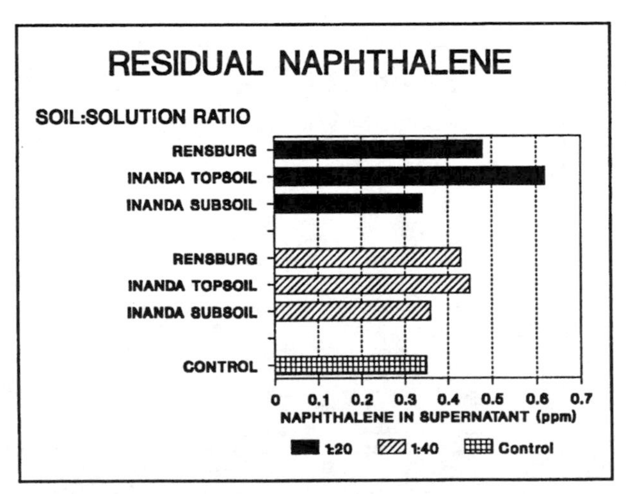

FIGURE 4. Naphthalene solution concentrations after 24 hours shaking (volatilization) in sterile conditions.

ACKNOWLEDGMENTS

This project is funded by the South African Foundation for Research Development and the University of Natal.

REFERENCES

Fuller, W. H. 1980. "Soil modification to minimize movements of pollutants from solid waste operations." *Critical Reviews in Environmental Control 9*: 213-270.

Gianfreda, L., M. A. Rao, and A. Violante. 1992. "Adsorption, activity and kinetic properties of urease on montmorillonite, aluminium hydroxide and $Al(OH)_x$-montmorillonite complexes." *Soil Biology and Biochemistry 24*: 51-58.

Harmsen, J. 1983. "Identification of organic compounds in leachate from a waste tip." *Water Resources 16*: 699-705.

Kefford, B., S. Kjelleberg, and K. C. Marshall. 1982. "Bacterial scavenging: Utilization of fatty acids localized at a solid-liquid interface." *Archives of Microbiology 133*: 257-260.

Senesi, N, and Y. Chen. 1989. "Interactions of toxic chemicals with humic substances." In Z. Gerstl, Y. Chen, U. Mingelgrin, and B. Yaron (Eds.), *Toxic Organic Chemicals in Porous Media*, pp. 37-90. Springer Verlag, Berlin.

Van Loosdrecht, M. C. M., J. Lyklema, W. Norde, and A. J. B. Zehnder. 1990. "Influence of interfaces on microbial activity." *Microbiological Reviews 54*: 75-87.

Webb, O. F., T. J. Phelps, P. R. Bienkowski, P. M. Digrazia, G. D. Reed, B. Applegate, D. C. White, and G. S. Sayler. 1991. "Development of a differential volume reactor system for soil biodegradation studies." *Applied Biochemistry and Biotechnology 28/29*: 5-19.

BIOREMEDIATION BENCH-SCALE TREATABILITY STUDY OF A SUPERFUND SITE CONTAINING OILY FILTER CAKE WASTE

P. C. Faessler, D. L. Crawford,
D. D. Emery, and R. D. Sproull

INTRODUCTION

A 13-week treatability study was performed to determine the feasibility of bioremediation of filter cake material contaminated with petroleum hydrocarbons. These contaminants include volatile organics, phenolic compounds, phthalate esters, polycylic aromatic hydrocarbons (PAHs), polychlorinated biphenyls (PCBs), and various inorganic constituents, especially lead. Bioremediation is being considered as part of the overall treatment train.

Site History. A Superfund site covering approximately 50 acres (20.2 ha) is the source of our study. Previous activities on this site began in 1941 with the establishment of a lubrication oil-recycling facility. Site operations included some recycling of waste solvents during the 1950s and 1960s. In 1979, the recycling operation turned to refining waste oil for use as fuel in industrial boilers. The refining operation continued until 1986 when recycling activities were discontinued.

Purpose. The purpose of this study was to optimize the biodegradation process that would reduce total petroleum hydrocarbons, including heavy oils and other hazardous constituents including PCBs, lead, and volatile and semi-volatile organic compounds, to acceptable levels for soil stabilization consistent with the requirements of the U. S. Environmental Protection Agency (EPA). These constituents were defined as those contaminants present in sufficient abundance and/or of sufficient toxicity to pose unacceptable risks to human health and/or which might prevent solidification/stabilization of the waste.

EXPERIMENT PREPARATION

Approximately 620 kg of soil was collected at four locations and shipped in four separate 55-gal (208.2 L) drums to our EPA-registered treatability laboratory, ORD987173440. The drums contained representative samples from four different regions at the site. The waste was homogenized by adding equal volumes from each drum to a Gibson 509015B mixer and mixing for 10 minutes. Twelve stainless steel trays (30.5 × 50.8 × 15.2 cm) were each filled with 15 kg of the homogeneous waste and placed three trays to a shelf in a plexiglass hood (120.7 × 137.2 × 182.9 cm). The trays were separated into four categories: A, B, AC, and BC. The trays in each category were then labeled 1 through 3, respectively. Trays A and AC would later be treated with surfactant, trays B and BC were not. Trays AC and BC served as controls; sodium azide was added to inhibit microbial growth. Composite samples were then collected and shipped to an independent laboratory (ToxScan Inc., Watsonville, California) for initial baseline characterization.

Nine weeks into the study, fluctuating analytical data indicated that even after extensive mixing, the soil was not homogeneous. It was also observed that the soil tended to agglomerate and reform into clumps. Since increased surface area and decreased particle size are very important for increasing the accessibility of hydrocarbons to the microbes, a decision was made to remix all the waste. A mixing canister with a high-speed blade was used to finely chop the contents of each bioreactor.

SOIL CONDITIONING

Soil conditioning began with pH adjustment. The pH was raised to 7.3 with hydrated lime (0.6%, or 90 g per tray). In addition, 1.0% chopped straw, 150 g, was added for stabilization and microbial colonization purposes. Trays were mixed 10 minutes and allowed to stand 12 to 16 hours.

A carbon:nitrogen:phosphorus ratio of 100:28:2 was established, using urea as the nitrogen source, and diammonium phosphate (DAP) as the phosphorus source. H_2O_2 (100 ppm) was initially added as an additional oxygen source. In addition, a nonionic surfactant, free of fatty acids, was added at a 1% concentration to the A and AC trays. In solution, 236.5 g urea, 86.7 g DAP, 10 mL 5% H_2O_2, and in trays A and AC 5 mL surfactant, were brought to a final volume of 500 mL and mixed into each tray. The trays were again allowed to stand 12 to 16 hours.

A microbial inoculum of the indigenous organisms was prepared by placing 10 g of waste in 1 L of oxygenated, sterile, enrichment broth. This inoculum was allowed to incubate at 80°F (26.7°C) for 12 to 16 hours. A 1% inoculum, 150 mL, was then mixed into the waste matrix of the A and B trays, while 15 g of sodium azide was brought to a volume of 150 mL with distilled water and added to the control, AC and BC, trays. Representative time zero samples were taken from each tray. Additional samples were taken every 7 days thereafter, and were shipped to ToxScan Inc. for analysis.

WASTE CHARACTERIZATION

Time Zero (with amendments) samples indicated total oil and grease (TOG) (ASTM 5520C) at 289,000 mg/kg and TPH (ASTM 5520F) at 266,000 mg/kg for sample A. Sample B results were 242,000 mg/kg and 199,000 mg/kg, respectively. An EPA 8015M test revealed that 47,000 mg/kg in A and 54,000 mg/kg in B were hydrocarbons shorter than 30 carbons. EPA Method 8240, volatile organics, found 13 of 34 compounds tested to be present. EPA Method 8080 found an average of 34,000 µg/kg of Aroclor™ 1260. Initial Baseline characterization by EPA Method 6010 detected varying amounts of 18 of 22 metals tested. EPA Method 350.1 indicated the samples contained 26 mg/kg of ammonia nitrogen, whereas EPA Method 300 indicated 33 mg/kg orthophosphate. EPA Method 150.1 found the pH to be 3.0. The Microtox bioassay, ASTM STP667, yielded an EC_{50} reading of 7.82. USFDA Method 9215A produced plate counts of 1.06×10^5 colony-forming units (CFUs) per gram. ASTM Method E 123 determined moisture to be 10%.

RESULTS

In a 13-week period, Bioreactor A (Figure 1) demonstrated significant degradation. TOG decreased from 289,000 mg/kg to 215,000 mg/kg, a 25.6% reduction. TPH decreased from 266,000 mg/kg to 164,000 mg/kg, a 38.4% reduction. Bioreactor B (Figure 2) demonstrated slower degradation trends. TOG decreased from 242,000 mg/kg to 164,000 mg/kg, a 14.9% reduction. TPH decreased from 199,000 mg/kg to 164,000 mg/kg, a 17.6% reduction. The control trays actually increased slightly (AC ≈2.0%) in TOG concentration. This apparent increase was created by the remixing at week 9 which decreased particle size thus increasing the efficiency of the analytical extraction procedure. This same effect was also seen in A and B trays after remixing. However, degradation proceeded at the same or a faster rate as compared to before remixing.

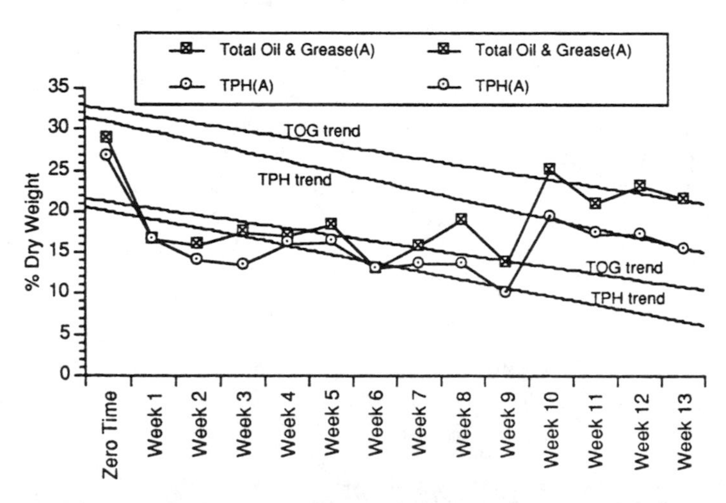

FIGURE 1. The trend developing in total oil and grease degradation (5520C) between composite samples from experiments A and B. The values are expressed as percent.

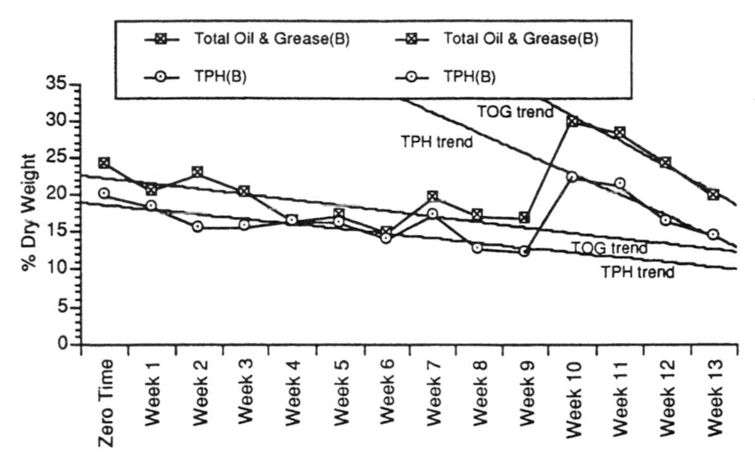

FIGURE 2. The trend developing in total petroleum hydrocarbon degradation (5520F) between composite samples from experiments A and B. The values are expressed as percent.

Figure 3 demonstrates the decrease in carbon content between bioreactor A and its corresponding control (AC). Hydrocarbon degradation was most effective on compounds with a carbon length ≤ C25.

Volatile compounds, including ethylbenzene, toluene, and total xylenes, decreased on average 74% in the A trays, 58% in the B trays, 44% in the AC, trays and 35% in the BC trays. Semivolatiles, including 2-methylnaphthalene, naphthalene and phenanthrene, decreased by approximately 50% in both A and B bioreactors.

PCBs were not anticipated to be significantly degraded during the course of the incubation. However a reduction of ≥ 50% was observed in Aroclor™ 1260 concentrations in 10 of the 12 bioreactors.

Lead concentrations, although unaffected by microbes, increased 10 to 20% in concentration in both A and B bioreactors. All results were subjected to statistical analysis by ANOVA for validation of the effects of time and treatment.

DISCUSSION

The decreases in TOG and TPH in both samples A and B indicate that hydrocarbons are being degraded by the indigenous microbe cultures added to the soil. The difference in the rate of degradation between A and B can be attributed to the surfactant addition in A. The surfactant reduces the hydrophobic condition of the hydrocarbons, making the hydrocarbons more miscible in water and, consequently, more accessible to the bacteria augmenting degradation. In the control trays azide addition reduced microbial numbers and simultaneously inhibited hydrocarbon degradation.

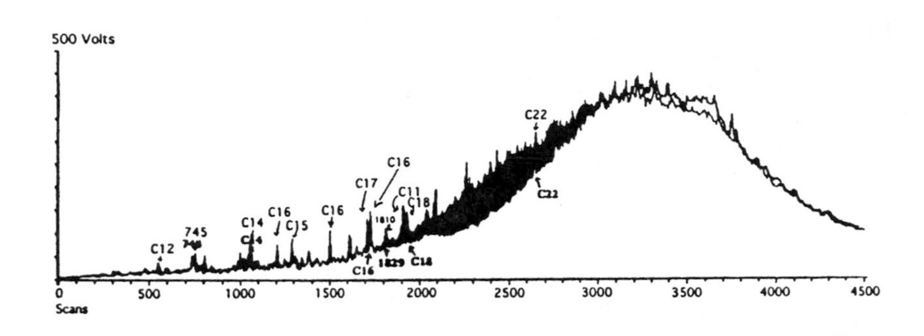

FIGURE 3. Gas chromatography/mass spectroscopy (GC/MS) chromatograms showing the makeup of the hydrocarbons present in the waste material in terms of relative abundance of different chain length hydrocarbons in bioreactor A and its subsequent control (AC).

The addition of the surfactant has acted as a catalyst to bioremediation in this experiment. The greater rates of degradation in TPH, TOG, and volatiles confirm the value of surfactant to this experiment.

It is believed that the PCBs were chemically dechlorinated by a chemical mechanism between hydrogen peroxide and iron (Fenton's reagent reaction).

CONDUCTING BIOREMEDIATION IN THE REGULATORY CLIMATE OF THE 1990s

K. Devine and D. E. Jerger

INTRODUCTION

Government regulations have significant effects on the marketplace. This paper presents several instances of the effects of some major environmental regulations promulgated by the U.S. Environmental Protection Agency (EPA) on bioremediation commercial operations that exemplify this fact. These examples are based on the experiences of a leading on-site remediation company who has conducted over 50 full-scale bioremediation projects.

LAND DISPOSAL RESTRICTIONS/CONTAMINATED SOIL PROPOSED RULEMAKING

A series of rules promulgated under the Resource Conservation and Recovery Act (RCRA), collectively known as the Land Disposal Restrictions (LDRs), stipulate that hazardous wastes must be treated to specified treatment standards, whether by a specific technology or to a specified concentration level (U.S. EPA 1990a). Treatment standards are based on the concept of best (statistically determined), demonstrated (currently at commercial level), available (can be purchased or leased) technology (BDAT). These rules also stipulate a fixed set of treatment standards for all multisource leachate and residues derived from the treatment of multisource leachate (U.S. EPA 1990a). BDAT levels established for organic wastes at RCRA treatment, storage and disposal facilities (TSDFs) Part B Permit and "generator-only" sites are based on data from treatment of industrial process wastes, typically by incineration. EPA has determined that contaminated soil and debris are generally more difficult to treat uniformly than RCRA wastes. In addition, LDRs for listed wastes are not relevant and appropriate for soil and debris contaminated with non-RCRA hazardous wastes. EPA considered the use of specific technologies as the standards for contaminated soil and debris, however, the data were inadequate to develop LDR treatment standards.

If the LDRs apply, compliance will involve either meeting BDAT treatment standards or an LDR compliance option, such as a treatability variance. Under a treatability variance, alternate treatment levels are established based on data from actual treatment of contaminated soil (U.S. EPA 1992).

Effect. Currently, a company must petition the EPA for a treatability variance if the waste is categorized as a Comprehensive Environmental Response, Compensation and Liability Act (CERCLA) (see below) or RCRA corrective action (a counterpart program to Superfund) cleanup. If the variance is granted, an exemption from the BDAT levels is achieved (U.S. EPA 1992). The estimated time to receive a variance is 1 year.

Current EPA LDR guidance 6A (U.S. EPA 1989b), which sets forth the procedures for granting site-specific treatability variances with options for concentration and percent reduction standards, is a viable BDAT scenario that will enable bioremediation to continue in the field. If inflexible, strict BDATs are enacted, the commercial use of innovative technologies, such as bioremediation, could be severely curtailed.

The EPA is promulgating treatment standards for multisource leachate, leachate that is derived from the treatment, storage or disposal of more than one listed hazardous waste and designated by its own hazardous waste code, "F039," that include concentration-based standards for the entire list of BDAT constituents (U.S. EPA 1989). Most of the multisource leachate nonwastewater treatment standards are based on direct transfer of RCRA-listed nonwastewater treatment standards, which were confirmed as being achievable by actual treatment performance data, most of which was incineration-based. Technology transfer of these multisource leachate standards to generic soil BDATs are under consideration. Bioremediation may not consistently meet multisource standards on difficult matrices and, therefore, if incineration is precluded, the only acceptable options may be stabilization or fixation followed by capping. If the F039 multisource standards become BDAT for soil, bioremediation technologies may be limited to gasoline- and diesel-contaminated sites and selected sandy soil matrices.

RD&D PERMITTING

In an attempt to alleviate some of the regulatory burden imposed by RCRA, the EPA instituted the RD&D permitting process. Companies testing experimental treatment technologies for hazardous waste require a Research Demonstration and Development (RD&D) permit, which provides for the construction of a facility and its operation for no longer than a year. Permit renewal is allowed no more than 3 times. RD&D facilities can receive only those wastes that are necessary to determine the efficacy of the treatment technology (U.S. EPA 1990c).

Effect. RD&D permit applications must include information such as a research plan, waste analysis plan, personnel training, emergency response health and safety plan, and a closure plan. Although RD&D permits were created to facilitate the development and demonstration of treatment technologies, requirements and regulations pertaining to procedures may be costly and time consuming. In the case of the RD&D permitting process for biological technology development and demonstration for wood-treating and wood-preserving contaminants, considered to be exemplary of such permitting, the process took approximately 1 year. The length of time necessary to obtain permit impacted the project schedule and budget.

SUPERFUND

CERCLA, which established the federal program for addressing the problem of hazardous substances that have been released into the environment, is funded by a significant amount of money called the "Superfund." The Superfund Amendments and Reauthorization Act (SARA) of 1986 established a preference for on-site treatment and significantly expanded research capabilities to develop cost-effective innovative treatment technologies. The process of determining the technology to be used for cleanup includes a remedial investigation (RI, a detailed investigation/site characterization)/feasibility study (FS, an evaluation, based on the data generated from the RI, of the alternative treatment options), thus RI/FS. A record of decision (ROD) is then prepared that lists the treatment(s) chosen for cleanup (Cooke 1990).

Effect. In one instance, ongoing full-scale soil washing and slurry-phase biological treatment of wood-preserving waste at a Superfund site by a fixed-price, performance-based contract, which was solicited by EPA, that allowed the contractor to propose the technology, thus encouraging technological development and innovation. However, the contractor was required to meet BDAT for "K001" wood-treating wastes without the benefits of a treatability study or treatability variance. With this type of contract, the contractor must be willing to assume the risk of achieving incineration-based BDAT standards by biological treatment. Failure in the field would be at the expense of the contractor and would be a setback to advancement of the technology even though a site-specific cleanup could be achieved.

PCB TREATMENT

The Toxic Substances Control Act (TSCA) requires that polychlorinated biphenyls (PCBs) disposal be approved by the EPA. A company wishing to treat PCBs by bioremediation may apply for a permit for PCB treatment as an alternative disposal method to incineration (U.S. EPA 1991).

Effect. As with RCRA permitting, a significant amount of permit-related time is necessary in order to perform commercial biological treatment of PCBs. An application for a TSCA R&D permit, containing information on experimental plans, sample storage and analytical and disposal requirements, is necessary to perform small-scale treatability studies to biologically treat PCBs in a variety of substrates. In addition, a final permit is necessary to operate any commercial process developed under this treatability permit. These upfront efforts are important in identifying treatment rates and efficiencies prior to technology scale.

CONCLUSIONS

Biological treatment technologies are increasing in demand to remediate RCRA facilities, underground storage tanks sites, Superfund sites and other hazardous contamination. Bioremediation can be successfully applied to achieve site-specific cleanup within the existing or proposed regulatory framework. However, important information must be obtained documenting the potential for success prior to full-scale field implementation. This information consists of CERCLA or RCRA treatability variances by RD&D permits, laboratory/treatability studies, or TSCA PCB treatment permits. Broad concentration-based standards, such as F039 multisource leachate standards, will be detrimental as acceptable technology-based BDAT for bioremediation.

REFERENCES

Cooke, Susan M. (Ed.). 1990. *The Law of Hazardous Waste*. Matthew Bender and Co., Inc., New York, NY.

U.S. Environmental Protection Agency. 1989a. "Land Disposal Restrictions; Final Rule." 40 CFR Part 268.

U.S. Environmental Protection Agency. 1989b. *Superfund LDR Guide 6A: Obtaining a Soil and Debris Treatability Variance for Remedial Action*. 93473.06-FS.

U.S. Environmental Protection Agency. 1990a. "Land Disposal Restrictions for Third Scheduled Wastes; Rule." 40 CFR Part 148 et al.

U.S. Environmental Protection Agency. 1990b. *Quality Assurance Project Plan for Characterization Sampling and Treatment Tests Conducted for the Contaminated Soil and Debris (CS&D) Program*. Office of Solid Waste, Washington, DC.

U.S. Environmental Protection Agency. 1990c. *RCRA Orientation Manual*. EPA/530-SW-90-038. Office of Solid Waste, Washington, DC.

U.S. Environmental Protection Agency. 1991. "Research and Development Permitting for Bioremediation under TSCA." *1*(3). *Bioremediation in the Field*. EPA/540/2-91/018.

U.S. Environmental Protection Agency. 1992. *Regional Guide: Issuing Site-Specific Treatability Variances for Contaminated Soils and Debris from Land Disposal Restrictions*. No. 9380.3-OBFS. Office of Solid Waste, Washington, DC.

DYNAMIC OPTIMAL CONTROL OF IN SITU BIOREMEDIATION

B. E. Minsker and C. S. Shoemaker

INTRODUCTION

Design of an in situ biorestoration program usually involves determining the location and pumping rates for injection and extraction wells. The injection wells are used to increase the supply of electron acceptors such as oxygen or nitrate, nutrients such as nitrogen or phosphorus, or additional carbon sources required to stimulate growth of the microbial population and accelerate degradation of the pollutant. Extraction wells are used to contain the contaminant plume, thereby preventing further spread of the contamination.

To assist in locating the pumping wells and determining pumping rates, engineers often use simulation models to predict the cost and effectiveness of various design options. The most well-known of these biodegradation models is BIOPLUME (Borden & Bedient 1986), but many other models are also available. The drawback of using these models alone, however, is that only a limited number of design options can be examined. Given the countless number of well locations and pumping rates possible, it is likely that any strategy obtained from a trial-and-error simulation will not be the least cost option.

Previous researchers have demonstrated that optimization techniques combined with numerical simulations can efficiently examine all available options and select the most cost-effective pump-and-treat design (e.g., Andricevic & Kitanidis 1990, Chang et al. 1992, Gorelick et al. 1984). In this research, we apply an optimization approach called optimal control to the design of in situ biorestorations. To our knowledge, this is the first application of optimization techniques to in situ bioremediation.

The optimal control algorithm uses a finite-element biodegradation model to simulate the effects of various pumping strategies and then to select the most cost-effective approach. This technical note presents the governing equations for the optimization and finite-element models. The computer code to couple the optimization and simulation is currently under development; preliminary results will be presented at the symposium.

FORMULATION OF THE OPTIMIZATION

For this analysis, we have used a type of optimal control called quasi-Newton differential dynamic programming (QNDDP) with management periods to compute the optimal bioremediation pumping strategy. As presented in Culver and Shoemaker (1992), the problem formulation is as follows:

$$\underset{U_1 \ldots U_k}{\text{Min}} \; J(U) + \sum_{k=1}^{K} G_k(X_k, U_k, k) \tag{1}$$

subject to:

$$X_{k+1} = Y(X_k, U_k, k), \qquad k = 1, \ldots, K \tag{2}$$

$$L(X_k, U_k, k) \leq 0, \qquad k = 1, \ldots, K \tag{3}$$

where J = total cost of a pumping strategy (the objective function);
U_k = control vector during management period k (dimension m);
X_k = state vector at the beginning of management period k (dimension n);
L = the set of r constraints on the control and state vectors;
G_k = cost of a strategy during management period k;
Y = the transition equation describing the change in X from one management period to the next, given U; and
K = the number of management periods.

In this analysis, the control vector represents rates of extraction or injection to be selected in each management period at all eligible pumping sites specified. The state vector consists of the hydraulic heads and concentrations of the contaminant (substrate), biomass, and electron acceptor (oxygen) at each node in the finite-element mesh. The transition equation is a finite-element model that simulates the changes in the state vector from the current management period to the next, given the pumping rates chosen for the current period.

The management period formulation allows a finer time discretization to be used in the simulation model than in the optimal control algorithm, which is particularly important for the bioremediation problem. Given that biomass growth and decay can occur quite rapidly relative to the time scale of the entire cleanup, maintaining numerical accuracy in the simulation model may require a far shorter time step than would be practical or feasible for changing the pumping rates.

The QNDDP algorithm consists of a forward sweep and a backward sweep. In the forward sweep, the simulation model (transition equation) determines the effects of the current pumping strategy on concentrations and hydraulic heads in the aquifer. Using these results, the backward sweep computes the derivatives of the objective and transition equations and selects a new strategy. The forward and backward sweeps iterate until the algorithm converges to the optimal bioremediation strategy. For more details on the algorithm, see Culver and Shoemaker (1992).

FORMULATION OF THE TRANSITION EQUATION

The transition equation is a simplified version of the two-dimensional finite-element biodegradation model described in Taylor and Jaffé (1991). The model, called BIO2D, predicts the hydraulic heads and substrate (contaminant), oxygen, and biomass concentrations that would result from the pumping rates selected by the optimization model. The two-dimensional, depth-averaged flow and transport equations for a confined aquifer are as follows:

$$\nabla \cdot (T \nabla h) + \sum_{i \in U} u_i \, \delta(x_i, y_i) = 0 \tag{4}$$

$$R_s b\theta \frac{\partial c_s}{\partial t} - \nabla \cdot (bD \cdot \nabla c_s) + bv \cdot \nabla c_s + b\theta\lambda(c_s, c_o, c_b) c_s + \sum_{i \in U} u_i (c_s - c_s') \, \delta(x_i, y_i) = 0 \tag{5}$$

$$b\theta \frac{\partial c_o}{\partial t} - \nabla \cdot (bD \cdot \nabla c_o) + bv \cdot \nabla c_o + b\theta\mu(c_s, c_o, c_b) c_o + \sum_{i \in U} u_i (c_o - c_o') \, \delta(x_i, y_i) = 0 \tag{6}$$

$$R_b b\theta \frac{\partial c_b}{\partial t} - \nabla \cdot (bD \cdot \nabla c_b) + bv \cdot \nabla c_b - b\theta\xi(c_s, c_o, c_b) c_b + \sum_{i \in U} u_i (c_b - c_b') \, \delta(x_i, y_i) = 0 \tag{7}$$

where c_s, c_o, and c_b are the concentrations of substrate (contaminant), oxygen, and biomass, respectively; h is the vertically averaged hydraulic head; U is the set of pumping sites; u_i is the pumping rate at (x_i,y_i); $\delta(x_i,y_i)$ is the Dirac delta function evaluated at (x_i,y_i); T is the transmissivity tensor; c_s', c_o', and c_b' are the concentrations of the contaminant, oxygen, and biomass, respectively, in the injection fluid; b is the saturated aquifer thickness; θ is the aquifer porosity; D is the dispersivity tensor; v is Darcy's velocity; and R_s and R_b are substrate and biomass retardation factors.

The terms $\lambda(c_s,c_o,c_b)$, $\mu(c_s,c_o,c_b)$, and $\xi(c_s,c_o,c_b)$ are nonlinear biodegradation terms. These terms include the rate of substrate use, r_s, which is modeled using the Haldane variant of the Monod equation (Taylor & Jaffé 1991):

$$r_s = \frac{\mu_m}{Y} \left(\frac{c_s}{K_s + c_s + \frac{(c_s)^2}{K_i}} \right) \left(\frac{c_o}{K_o + c_o} \right) \tag{8}$$

where μ_m is the maximum specific growth rate; K_s and K_o are Monod half-velocity coefficients for substrate and oxygen, respectively; K_i is an inhibition coefficient;

and Y is the yield coefficient. The inhibition coefficient decreases the microbial growth rate at high substrate concentrations, allowing toxicity effects to be included when appropriate.

A Galerkin finite-element formulation is used to solve these equations. Note that several assumptions are implicit in the formulation given above. First, note that equation (4) is a steady-state hydraulic head equation. This formulation assumes that the heads respond rapidly enough to the pumping rates selected in each management period to allow a steady-state approximation without significant loss of accuracy. The model also assumes that oxygen and substrate availability is the most important factor limiting microbial growth. Nitrogen and phosphorus are assumed to be present in sufficient quantities for the available oxygen and substrate.

As in Borden and Bedient (1986), the movement of microbes in the aquifer is assumed to be controlled by their tendency to adhere to the solid particles. This effect is modeled using a simple retardation factor, R_b, derived from a Freundlich adsorption isotherm. Substrate adsorption is also modeled using a simple retardation factor, R_s.

ACKNOWLEDGMENTS

This material is based upon work supported under a National Science Foundation Graduate Research Fellowship. Any opinions, findings, or recommendations expressed in this publication are those of the authors and do not necessarily reflect the views of the National Science Foundation.

REFERENCES

Andricevic, R., and P. K. Kitanidis. 1990. "Optimization of the Pumping Schedule in Aquifer Remediation Under Uncertainty." *Water Resour. Res.* 26(5): 875-885.

Borden, R. C., and P. B. Bedient. 1986. "Transport of Dissolved Hydrocarbons Influenced by Oxygen-Limited Biodegradation: 1. Theoretical Development." *Water Resour. Res.* 22(13): 1973-1982.

Chang, L.-C., C. A. Shoemaker, and P. L.-F. Liu. 1992. "Optimal Time-Varying Pumping Rates for Groundwater Remediation: Application of a Constrained Optimal Control Algorithm." *Water Resour. Res.* 28(12): 3157-3173.

Culver, T., and C. A. Shoemaker. 1992. "Dynamic Optimal Control for Groundwater Remediation with Flexible Management Periods." *Water Resour. Res.* 28(3): 629-641.

Gorelick, S. M., C. I. Voss, P. E. Gill, W. Murray, M. A. Saunders, and M. H. Wright. 1984. "Aquifer Reclamation Design: The Use of Contaminant Transport Simulation Coupled with Nonlinear Programming." *Water Resour. Res.* 20(4): 415-427.

Taylor, S. W., and P. R. Jaffé. 1991. "Enhanced In Situ Biodegradation and Aquifer Permeability Reduction." *Journal of Environmental Engineering* 117(1): 25-46.

AUTHOR LIST

M. D. Aitken
University of North Carolina at
Chapel Hill
Environmental Sciences & Engineering
Chapel Hill, NC 27599-7400 USA

H.-J. Albrechtsen
Technical University of Denmark
Department of Environmental
Engineering, Bldg. 115
DK-2800 Lyngby DENMARK

J. M. M. Appelman
TAUW Infra Consult B.V.
Post Office Box 479
7400 AL Deventer
THE NETHERLANDS

R. L. Autenrieth
Department of Civil Engineering
Texas A&M University
College Station, TX 77843-3156 USA

B.-H. Bae
Department of Civil Engineering
Texas A&M University
College Station, TX 77843-3156 USA

S. Bandyopadhyay
Civil and Environmental Engineering
Tulane University
New Orleans, LA 70118 USA

J. Bender
Research Center for Science and
Technology
Clark Atlanta University, Box 296
James P. Brawley Dr. at Fair St., SW
Atlanta, GA 30314 USA

I.P.A.M. Bennehey
DLO Winand Staring Centre for
Integrated Land, Soil and Water
Research (SC-DLO)
Post Office Box 125
6700 AC Wageningen
THE NETHERLANDS

T. J. Bergman, Jr.
Praxair, Inc.
777 Old Saw Mill River Road
Tarrytown, NY 10591 USA

S. K. Bhattacharya
Civil and Environmental Engineering
Tulane University
New Orleans, LA 70118 USA

G. J. Boettcher
Geraghty & Miller, Inc.
14497 N. Dale Mabry Hwy., Suite 115
Tampa, FL 33618 USA

J. S. Bonner
Department of Civil Engineering
Texas A&M University
College Station, TX 77843-3156 USA

G. D. Breedveld
Norwegian Geotechnical Institute
Post Office Box 40
Taasen N-0801 Oslo
NORWAY

T. Briseid
Center for Industrial Research
Post Office Box 124 Blindern
N-0314 Oslo
NORWAY

R. Brousseau
Department of Civil Engineering
Biotechnology Research Institute
National Research Council of Canada
6100 Royalmount Avenue
Montreal, Quebec
CANADA H4P 2R2

G. R. Brubaker
Remediation Technologies, Inc.
127 Kingston Drive
Chapel Hill, NC 27514 USA

B. A. Bult
TAUW Infra Consult B.V.
Post Office Box 479
7400 AL Deventer
THE NETHERLANDS

R. S. Burlage
Oak Ridge National Laboratory
Environmental Sciences Division
P.O. Box 2008, Mail Stop 6036
Oak Ridge, TN 3783111 USA

F. J. Castaldi
Radian Corporation
P.O. Box 201088
Austin, TX 78720-1088 USA

M. J. Chen
Department of Chemistry
University of Detroit Mercy
P.O. Box 19900
Detroit, MI 48219-0900 USA

T. H. Christensen
Technical University of Denmark
Department of Environmental
Engineering, Bldg. 115
DK-2800 Lyngby DENMARK

Y. Comeau
Department of Civil Engineering
École Polytechnique de Montréal
Post Office Box 6079, Station A
Montreal, Quebec
CANADA H3C 3A7

D. Cosgriff
Champion International, Inc.
952 East Spruce Street
Libby, MT 59923-2399 USA

D. L. Crawford
University of Idaho
Center for Hazardous Waste
Remediation Research
Food Research Center, Room 202
Moscow, ID 83843 USA

R. L. Crawford
University of Idaho
Center for Hazardous Waste
Remediation Research
Food Research Center, Room 202
Moscow, ID 83843 USA

J. Croonenberghs
Coors Brewing Company
Mail Stop CC290
Golden, CO 80401 USA

A. B. Cunningham
Montana State University
Center for Interfacial Microbial Process
Engineering
409 Cobleigh Hall
Bozeman, MT 59717-0007 USA

T. R. Davis
French Limited Project
1024 Gulf Pump Road
Crosby, TX 77532 USA

W. J. Davis-Hoover
U.S. Environmental Protection Agency
Risk Reduction Engineering Laboratory
26 West Martin Luther King Drive
Cincinnati, OH 45268 USA

R. B. Dean
Waste Management and Research
Bremerholm 1
DK-1069 Copenhagen K
DENMARK

J. B. DeBar
Montana State University
Center for Interfacial Microbial
Process Engineering
409 Cobleigh Hall
Bozeman, MT 59717-0007 USA

J.A.M. de Bont
Agricultural University of Wageningen
Division of Industrial Microbiology
Department of Food Science
P.O. Box 8129
6700 EV Wageningen
THE NETHERLANDS

K. A. DeFelice
Harding Lawson Associates
2400 ARCO Tower
707 Seventeenth Street
Denver, CO 80202 USA

E. de Jong
Agricultural University of Wageningen
Division of Idustrial Microbiology
Department of Food Science
P.O. Box 8129
6700 EV Wageningen
THE NETHERLANDS

B. Denovan
Battelle Marine Sciences Laboratory
1529 West Sequim Bay Road
Sequim, WA 98382-9099 USA

K. Devine
DEVO Enterprises, Inc.
704 9th Street, SE
Washington, DC 20003-2804 USA

J. M. Dougherty
Westinghouse Savannah River Co.
Building 773-42A
Savannah River Technology Center
Aiken, SC 29808 USA

J. R. Doyle
Woodward-Clyde Consultants
4582 S. Ulster Street
Denver, CO 80237-2637 USA

M. M. Dronamraju
Civil and Environmental
Engineering
Tulane University
New Orleans, LA 70118 USA

C. A. du Plessis
International Centre for Waste
Technology
University of Natal
P.O. Box 375
Pietermaritzburg 3200
SOUTH AFRICA

D. D. Emery
Bioremediation Service, Inc.
P.O. Box 2010
Lake Oswego, OR 97035 USA

D. A. English
B&V Waste Science and Technology
Corporation
4717 Grand Avenue, Suite 500
Kansas City, MO 64112 USA

D. C. Erickson
Harding Lawson Associates
2400 ARCO Tower
707 Seventeenth Street
Denver, CO 80202 USA

P. C. Faessler
Bioremediation Service, Inc.
P.O. Box 2010
Lake Oswego, OR 97035 USA

J. A. Field
Division of Industrial Microbiology
Agricultural University
Post Office Box 8129
6700 EV Wageningen
THE NETHERLANDS

C. B. Fliermans
Westinghouse Savannah River Co.
Building 773-42A
Savannah River Technology Center
Aiken, SC 29808 USA

M. M. Franck
Westinghouse Savannah River Co.
Building 773-42A
Savannah River Technology Center
Aiken, SC 29808 USA

R. R. Fulthorpe
Center for Microbial Ecology
Plant and Soil Sciences Building
Michigan State University
East Lansing, MI 48823 USA

C. Gao
University of Cincinnati
Department of Chemical Engineering
Cincinnati, OH 45221 USA

W. George
Department of Toxicology
Tulane School of Medicine
1430 Tulane Avenue
New Orlearns, LA 70112 USA

R. Gersonde
Universität Göttingen
Forstbotanisches Institut
Büsgenweg 2 D-3400 Göttingen
GERMANY

D. J. Glass
D. Glass Associates, Inc.
124 Bird Street
Needham, MA 02192 USA

R. Govind
University of Cincinnati
Department of Chemical Engineering
Cincinnati, OH 45221 USA

S. Goszczynski
University of Idaho
Center for Hazardous Waste
Remediation Research
Food Research Center, Room 202
Moscow, ID 83843 USA

D. A. Graves
IT Corporation
312 Directors Drive
Knoxville, TN 37923 USA

J. M. Greene
ENSR Consulting and Engineering
3000 Richmond Avenue
Houston, TX 77098 USA

Charles W. Greer
Biotechnology Research Institute
National Research Council of Canada
6100 Royalmount Avenue
Montreal, Quebec
CANADA H4P 2R2

C. Grøn
Technical University of Denmark
Department of Geology and
Geotechnical Engineering
Building 115
DK-2800 Lyngby
DENMARK

J. R. Haines
U.S. Environmental Protection Agency
Risk Reduction Engineering Laboratory
26 West Martin Luther King Drive
Cincinnati, OH 45268 USA

C. J. Hapeman
Pesticide Degradation Laboratory
Agricultural Research Service
Beltsville, MD 20705 USA

D. Hargens
IES Industries, Inc.
200 First Street
P.O. Box 351
Cedar Rapids, IO 52406 USA

J. Harmsen
DLO Winand Staring Centre for
Integrated Land, Soil and Water
Research (SC-DLO)
Post Office Box 125
6700 AC Wageningen
THE NETHERLANDS

T. D. Hayes
Gas Research Institute
8600 West Bryn Mawr Avenue
Chicago, IL 60631 USA

T. C. Hazen
Westinghouse Savannah River Co.
Building 773-42A
Savannah River Technology Center
Aiken, SC 29808 USA

E. Heessels
Division of Industrial Microbiology
Agricultural University
Post Office Box 8129
6700 EV Wageningen
THE NETHERLANDS

B. B. Hemmingsen
Department of Biology
San Diego State University
San Diego, CA 92182-0057 USA

S. E. Herbes
Environmental Sciences Division
Oak Ridge National Laboratory
Oak Ridge, TN 37831 USA

J. C. Hughes
International Centre for Waste Technology
University of Natal, P.O. Box 375
Pietermaritzburg 3200
SOUTH AFRICA

P. J. Hutchinson
The Hutchinson Group, Ltd.
5124 Scenic Drive
Murrysville, PA 15668 USA

A. Hüttermann
Forstbotanisches Institut der Universität Göttingen
Büsgenweg 2, D-3400 Göttingen
GERMANY

M. Islam
University of Cincinnati
Civil and Environmental Engineering
Cincinnati, OH 45221 USA

D. R. Jackson
Radian Corporation
P.O. Box 201008
Austin, TX 78720 USA

D. Jager
Bioclear Environmental Biotechnology
Zernikepark 2, Post Office Box 2262
9704 CG Groningen
THE NETHERLANDS

D. E. Jerger
OHM Remediation Services Corporation
16406 U.S. State Route 224 East
Findlay, OH 45840 USA

C. A. Jones
University of Cincinnati
Department of Civil and Environmental Engineering
Cincinnati, OH 45221 USA

R. H. Kaake
University of Idaho
Center for Hazardous Waste Remediation Research
Food Research Center, Room 202
Moscow, ID 83843 USA

M. Kadkhodayan
University of Cincinnati
Department of Civil and Environmental Engineering
Cincinnati, OH 45221 USA

S. Keuning
Bioclear Environmental Biotechnology
Zernikepark 2
Post Office Box 2262
9704 CG Groningen
THE NETHERLANDS

M. D. Klein
EG&G Rocky Flats, Inc.
P.O. Box 464
Building 051
Golden, CO 80402-0464 USA

M. Kotterman
Division of Industrial Microbiology
Agricultural University
Post Office Box 8129
6700 EV Wageningen
THE NETHERLANDS

D. Kuo
Oak Ridge National Laboratory
Environmental Sciences Division
P.O. Box 2008
Mail Stop 6036
Oak Ridge, TN 37831 USA

L. Lai
University of Cincinnati
Department of Chemical Engineering
Cincinnati, OH 45221 USA

W. F. Lane
Remediation Technologies, Inc.
127 Kingston Drive
Chapel Hill, NC 27514 USA

C. A. Lang
IT Corporation
312 Directors Drive
Knoxville, TN 37923 USA

R. J. Lang
California Polytechnic State University
Department of Civil and Environmental Engineering
San Luis Obispo, CA 93407 USA

K.-I. Lee
University of Southern California
Department of Civil and Environmental Engineering
224 A Kaprielian Hall
3620 South Vermont Avenue
Los Angeles, CA 90089-2531 USA

R. F. Lee
Skidaway Institute of Oceanography
P.O. Box 13687
Savannah, GA 31416 USA

A. Leeson
Battelle
505 King Avenue
Columbus, OH 43201-2693 USA

P. H. LeFevre
Coors Brewing Company
Mail Stop CC290
Golden, CO 80401 USA

Y. Li
University of Southern California
Civil and Environmental Engineering
224-A Kaprielian Hall
3629 South Vermont Avenue
Los Angeles, CA 90089-2531 USA

J. Lyngkilde
Technical University of Denmark
Department of Environmental Engineering, Bldg. 115
DK-2800 Lyngby
DENMARK

A. Majcherczyk
Forstbotanisches Institut der Universität Göttingen
Büsgenweg 2, D-3400 Göttingen
GERMANY

P. Majumdar
Civil and Environmental Engineering
Tulane University
New Orleans, LA 70118 USA

E. H. Marsman
TAUW Infra Consult B.V.
Post Office Box 479
7400 AL Deventer
THE NETHERLANDS

L. Masson
Biotechnology Research Institute
National Research Council of Canada
6100 Royalmount Avenue
Montreal, Quebec
CANADA H4P 2R2

I. Mawardi
Agency for the Assessment and Application of Technology (BPP Teknologi)
Jl. M.H. Thamrin 8, 14th Floor
Jakarta 10340 INDONESIA

P. C. McKinzey
Westinghouse Savannah River Co.
Building 773-42A
Savannah River Technology Center
Aiken, SC 29808 USA

R. H. Meaders
Ecology Technologies International, Inc.
4119 East Star Valley Street
Mesa, AZ 85205 USA

M. C. Medellin
California Polytechnic State University
Civil & Environmental Engineering
San Luis Obispo, CA 93407 USA

J. J. Meister
Department of Chemistry
University of Detroit Mercy
P.O. Box 19900
Detroit MI 48219-0900 USA

K. Miksch
Technical University of Silesia
Department of Environmental Biotechnology
PL-44 101 Gliwice POLAND

O. Milstein
Universität Göttingen
Forstbotanisches Institut
Büsgenweg 2 D-3400 Göttingen
GERMANY

B. E. Minsker
School of Civil and Environmental Engineering, Hollister Hall
Cornell University
Ithaca, NY 14853 USA

D. J. Mocsny
University of Cincinnati
Civil and Environmental Engineering
Cincinnati, OH 45221 USA

P. M. Molton
Battelle Pacific Northwest Laboratories
Battelle Boulevard
P.O. Box 999
Richland, WA 99352 USA

M. Mondecar
Delgado Community College
Science and Math Division
615 City Park Avenue
New Orleans, LA 70119 USA

R. E. Moon
Geraghty & Miller, Inc.
14497 North Dale Mabry Highway, Suite 115
Tampa, FL 33629 USA

C. M. Morrissey
Environmental Sciences Division
Oak Ridge National Laboratory
P.O. Box 2008
Oak Ridge, TN 37830-6035 USA

L. C. Murdoch
University of Cincinnati
Department of Civil and Environmental Engineering
Cincinnati, OH 45224 USA

N. J. Myers
Harding Lawson Associates
2400 ARCO Tower
707 Seventeenth Street
Denver, CO 80202 USA

M. Narayanaswamy
University of Cincinnati
Department of Civil and Environmental Engineering
Cincinnati, OH 45224 USA

D. Nigoyi
Battelle Marine Sciences Laboratory
1529 West Sequim Bay Road
Sequim, Washington 89382-0601 USA

E. A. Nowatzki
California Polytechnic State University
Civil & Environmental Engineering
San Luis Obispo, CA 93407 USA

E. K. Nyer
Geraghty & Miller, Inc.
P.O. Box 273630
Tampa, FL 33688 USA

L. P. Olmsted
University College London
Two-Ways Waverley Place
Leatherhead, Surrey
KT22 8AS England UK

A. V. Palumbo
Environmental Sciences Division
Oak Ridge National Laboratory
Oak Ridge, TN 37831 USA

J. K. Park
University of Southern California
Department of Civil & Environmental Engineering
224 A Kaprielian Hall
3620 S. Vermont Avenue
Los Angeles, CA 90089-2531 USA

M. C. Parsons
Woodward-Clyde Consultants
4582 S. Ulster Street
Denver, CO 80237-2637 USA

M. B. Pasti-Grigsby
University of Idaho
Center for Hazardous Waste Remediation Research
Food Research Center, Room 202
Moscow, ID 83843 USA

A. Paszczynski
University of Idaho
Center for Hazardous Waste Remediation Research
Food Research Center, Room 202
Moscow, ID 83843 USA

S. Pfanstiel
University of Cincinnati
Department of Chemical Engineering
Cincinnati, OH 45221 USA

T. J. Phelps
Environmental Sciences Division
Oak Ridge National Laboratory
Oak Ridge, TN 37831 USA

P. Phillips
Research Center for Science and Technology
Clark Atlanta University
James P. Brawley Drive at Fair Street, SW
Atlanta, GA 30314 USA

M. R. Piotrowski
Biotransformations, Inc.
1670 Newport Center Road, Suite 300
Colorado Springs, CO 80916 USA

A. Porta
Battelle Europe
7, route de Drize
1227 Carouge-Geneva
SWITZERLAND

J. Preslan
Tulane Medical School
1430 Tulane Avenue
New Orleans, LA 70112 USA

J. D. Randall
Department of Biology
San Diego State University
San Diego, CA 92182-0057 USA

G. O. Reid
Du Pont Environmental Remediation Services
7068 Koll Center Parkway, Suite 401
Pleasanton, CA 94566 USA

J. N. Rightmyer
IT Corporation
312 Directors Drive
Knoxville, TN 37923 USA

J. Ross
Biology Department
Xavier University of Lousiana
New Orleans, LA 70125 USA

R. Samson
Biotechnology Research Institute
National Research Council of Canada
6100 Royalmount Avenue
Montreal, Quebec
CANADA H4P 2R2

G. M. Seganti
Du Pont Environmental Remediation Services
12835 Main Street
Louviers, CO 80131-0067 USA

S. M. Sellers
California Polytechnic State University
Department of Civil & Environmental Engineering
San Luis Obispo, CA 93407 USA

E. Senior
International Centre for Waste Technology
University of Natal
Post Office Box 375
Pietermaritzburg, 3200
SOUTH AFRICA

R. R. Sharp
Montana State University
Center for Interfacial Microbial Process Engineering
409 Cobleigh Hall
Bozeman, MT 59717-0007 USA

D. R. Shelton
Pesticide Degradation Laboratory
Agricultural Research Service
U.S. Department of Agriculture
Beltsville, MD 20705 USA

C. S. Shoemaker
School of Civil and Environmental Engineering, Hollister Hall
Cornell University
Ithaca, NY 14853 USA

M. Silva
American Proteins, Inc.
P.O. Box 905
575 Colonial Park Drive
Roswell, GA 30075 USA

R. D. Sproull
Bioremediation Service, Inc.
P.O. Box 2010
Lake Oswego, OR 97035 USA

P. S. Stewart
Montana State University
Center for Interfacial Microbial Process Engineering
409 Cobleigh Hall
Bozeman, MT 59717-0007 USA

K. E. Stormo
University of Idaho
Center for Hazardous Waste Remediation Research
Food Research Center, Room 202
Moscow, ID 83843 USA

W. T. Stringfellow
University of North Carolina at Chapel Hill
Environmental Sciences & Engineering
Chapel Hill, NC 27599-7400 USA

P. J. Sturman
Montana State University
Center for Interfacial Microbial Process Engineering
409 Cobleigh Hall
Bozeman, MT 59717-0007 USA

J. Surmacz-Gorska
Technical University of Silesia
Department of Environmental Biotechnology
PL 44-101 Gliwice POLAND

C. M. Swindoll
Du Pont Environmental Remediation Services
Bellevue Park Corporate Center
300 Bellevue Parkway, Suite 390
Wilmington, DE 19809-3722 USA

H. H. Tabak
U.S. Environmental Protection Agency
Risk Reduction Engineering Laboratory
Cincinnati, OH 45268 USA

L.G.C.M. Urlings
Post Office Box 479
7400 Al Deventer
THE NETHERLANDS

H. J. Velthorst
DLO Winand Staring Centre for Integrated Land, Soil and Water Research (SC-DLO)
Post Office Box 125
6700 AC Wageningen
THE NETHERLANDS

A. D. Venosa
U.S. Environmental Protection Agency
Risk Reduction Engineering Laboratory
26 West Martin Luther King Drive
Cincinnati, OH 45268 USA

S. J. Vesper
University of Cincinnati
Department of Civil and Environmental Engineering
Cincinnati, OH 45221 USA

J. E. Wear
Westinghouse Savannah River Co.
Building 773-42A
Savannah River Technology Center
Aiken, SC 29808 USA

O. M. West
Oak Ridge National Laboratory
Environmental Sciences Division
P.O. Box 2008
Oak Ridge, TN 37831-6038 USA

R. Wijngaarde
Division of Industrial Microbiology
Agricultural University
Post Office Box 8129
6700 EV Wageningen
THE NETHERLANDS

W. S. Winanti
Agency for the Assessment and Application of Technology (BPP Teknologi)
Jl. M.H. Thamrin 8, 14th Floor
Jakarta 10340 INDONESIA

J. H. Wolfram
Montana State University
Center for Interfacial Microbial Process Engineering
409 Cobleigh Hall
Bozeman, MT 59717-0007 USA

J. Word
Battelle Marine Sciences Laboratory
1529 West Sequim Bay Road
Sequim, WA 98382-0601 USA

R. C. Wyndham
Center for Microbial Ecology
Michigan State University
East Lansing, MI 48824-1325 USA

X. Yan
University of Cincinnati
Department of Chemical Engineering
Cincinnati, OH 45221 USA

I. C.-Y. Yang
University of Southern California
Civil and Environmental Engineering
224 A Kaprielian Hall
3629 S. Vermont Avenue
Los Angeles, CA 90089-2531 USA

T. F. Yen
University of Southern California
Civil and Environmental Engineering
224 A Kaprielian Hall
3629 S. Vermont Avenue
Los Angeles, CA 90089-2531 USA

J. K. Young
Battelle Pacific Northwest Laboratories
Battelle Boulevard
P.O. Box 999
Richland, WA 99352 USA

A. Zeddel
Forstbotanisches Institut der Universität Göttingen
Büsgenweg 2, D-3400 Göttingen
GERMANY

INDEX